AF361910

Recent Developments on Protein–Ligand Interactions: From Structure, Function to Applications

Recent Developments on Protein–Ligand Interactions: From Structure, Function to Applications

Editor

Alexandre G. de Brevern

MDPI • Basel • Beijing • Wuhan • Barcelona • Belgrade • Manchester • Tokyo • Cluj • Tianjin

Editor
Alexandre G. de Brevern
DSIMB, INSERM UMR_S1134
Université de Paris
Université de la Réunion
Paris
France

Editorial Office
MDPI
St. Alban-Anlage 66
4052 Basel, Switzerland

This is a reprint of articles from the Special Issue published online in the open access journal *International Journal of Molecular Sciences* (ISSN 1422-0067) (available at: www.mdpi.com/journal/ijms/special_issues/protein_ligand).

For citation purposes, cite each article independently as indicated on the article page online and as indicated below:

LastName, A.A.; LastName, B.B.; LastName, C.C. Article Title. *Journal Name* **Year**, *Volume Number*, Page Range.

ISBN 978-3-0365-3394-0 (Hbk)
ISBN 978-3-0365-3393-3 (PDF)

Cover image courtesy of Alexandre G. de Brevern

Contents

About the Editor

Alexandre G. de Brevern

Alexandre G. de Brevern was trained as a Cell Biologist. He has been a Structural Bioinformatician for nearly 20 years. Senior Researcher at the French National Institute for Health and Medical Research (INSERM), he is the head of team 2 of INSERM UMR_S 1134 (Université de Paris and Université de la Réunion) located in Paris and Saint-Denis de la Réunion (Indian Ocean). He has two main axes of researches: (i) developing innovate methodologies useful for the scientific community and (ii) specific application to proteins implicated in diseases and pathologies. Concerning the first axis, he provided near 20 tools, webservers and databases mainly dealing with the analyses and prediction of protein behaviours. For instance, webservers were made to perform superimposition of protein 3D structures and prediction of protein flexibility. He recently proposed databases dedicated to essential blood group Rhesus and to Calreticuline variants associated to essential thrombocythemia. Concerning the second axis, he used mainly comparative modelling and molecular dynamics to analyse proteins for Red Blood Cells and platelets. We can notice a large number of publications dedicated on the modelling and dynamics of integrins associated to Glanzmann thrombocytopenia or Blood groups. He also extended his work to drug design with collaborations with companies and Next-Generation Sequencing. He had authored more than 150 papers, is editor in six peer-reviewed journals and is implicated in numerous scientific societies and institutes.

Preface to "Recent Developments on Protein–Ligand Interactions: From Structure, Function to Applications"

Protein–ligand interactions play a fundamental role in most major biological functions. The number and diversity of small molecules that interact with proteins, whether naturally or not, can quickly become overwhelming. They are as essential as amino acids, nucleic acids or membrane lipids, enabling a large number of essential functions. One need only think of carbohydrates or even just ATP to be certain. They are also essential in drug discovery. With the increasing structural information of proteins and protein–ligand complexes, molecular modelling, molecular dynamics, and chemoinformatics approaches are often required for the efficient analysis of a large number of such complexes and to provide insights. Similarly, numerous computational approaches have been developed to characterize and use the knowledge of such interactions, which can lead to drug candidates. For instance, one main application is to identify tractable chemical startpoints that non-covalently modulate the activity of a biological molecule. This new information brings questions that affect chemistry, biology, and even poses specific computer problems.

"Recent Developments on Protein–Ligand Interactions: From Structure, Function to Applications" was dedicated to the different aspect of protein–ligand analysis and/or prediction using computational approaches, as well as new developments dedicated to these tasks.

The 15 published papers clearly show the extent of such a focus, ranging from general analyses on a large dataset, to specific methodological developments, to applications of biomedical interest. It will interest both specialists and non-specialists, as the presented studies cover a very large spectrum in terms of methodologies and applications. It underlines the variety of scientific area linked to these questions, i.e., chemistry, biology, physics, informatics, bioinformatics, structural bioinformatics and chemoinformatics. I would like to use this editorial to thank all the researchers who submitted papers for this Special Issue and made it a success with work of great interest.

In the context of applications dedicated to a specific system, Aguero and Terreux were working on an explosive subject, 1,3,5,7-tetranitro-1,3,5,7-tetrazocane (HMX), an explosive that pollutes many sites. In order to decontaminate these sites, bioremediation was a promising approach. They therefore set out to improve a nitroreductase from Enterobacter Cloacaetowards HMX. With the Coupled Moves algorithm from Rosetta, they redesigned the active site around HMX, and analysed the results with Molecular Dynamics, showing encouraging results [1].

Bienboire-Frosini and co-workers were looking at cat allergies. The major cat allergen Fel d1 was a tetrameric glycoprotein of the secretoglobin superfamily, but its biological function was uncertain. They therefore used bioinformatics approaches to search for potential ligands and then experimentally tested them. The specific affinity of Fel d1 to semiochemicals supports a function of the protein in cat chemical communication, and pointed to a putative role for secretoglobins in protein semiochemistry [2].

Dharmatti and colleagues focussed on folate receptor (FR), a major target for cancer treatment and detection. They tried to enhance by click-reaction FR binding affinity by peptide conjugation. After multiple optimisations, the designed peptides resulted in an increase in the number of interaction sites, leading to potentially interesting drug developments [3].

Fan and co-workers worked on pimaricin, a polyene antibiotic of great pharmaceutical significance. Using Molecular Dynamics, they compared different stages of the molecules and showed how pimaricin thioesterase-catalyzed macrocyclization evolved, as the protein-polyketide recognition, and product release; they underlined potential residues for rational modification of pimaricin thioesterase [4].

Kim and colleagues have looked at monoclonal antibodies (mAbs) potentially interesting in cancer immunotherapy. mAb-based drugs have some drawbacks, e.g., poor tumour penetration. BMS-8, one of the potent small molecule drugs inhibits PD-1. Using in silico simulation, they optimized and successfully tested a derived compound five times more affine [5].

Nshogoza and co-workers analysed TDP-43, as an RNA-binding protein, implicated in neurodegenerative and cancer diseases. By combining Nuclear Magnetic Resonance and in silico approaches such as HADDOCK, they designed, tested and provided explanations of their binding to RNA Recognition Motifs of TDP-43 [6].

Pal and colleagues worked on galectins, a family of galactoside-recognizing proteins involved in different galectin-subtype-specific inflammatory and tumour-promoting processes. They synthetized and assessed interest of 3-C-methyl-gulopyranoside derivatives as galectin inhibitors with good affinity and selectivity [7].

Potthoff et al. applied molecular modelling approaches to build a structural model of full-length procollagen C-proteinase enhancer-1 (PCPE-1), which was not experimentally available. They characterized the interactions between the extracellular matrix PCPE-1 protein, and glycosaminoglycans (GAGs). They predicted GAG binding poses for various GAG lengths, types and sulfation pattern [8].

Reyes-Espinosa et al. focussed on an issue of economic importance, namely pesticide resistance. To do so, they modelled the complex-ligands of nine acetylcholinesterase structures of Lepidopteran organisms and 43 organophosphorus pesticides. The analysis of the complexes allowed a better understanding of the specificities of the variants of each species [9].

The enzyme phospholipase C gamma 1 (PLC1) was a potential drug target of interest for various pathological conditions such as immune disorders, systemic lupus erythematosus, and cancers. Tripathi and colleagues targeted its SH3 domain and various binding partners. They identified with molecular dynamic simulation the critical interacting essential residues leading to the possibility to also identify new inhibitors [10].

Caspases not only contributed to the neurodegeneration associated with Alzheimer's disease, but also played essential roles in promoting the underlying pathology of this disease. Xue et al. applied the Movable Type free energy method, a Monte Carlo sampling method extrapolating the binding free energy by simulating the partition functions for both free-state and bound-state protein and ligand configurations, to the caspase-inhibitor binding affinity study. They tested more than a hundred active inhibitors binding to caspase-3 on one side, and smaller well-characterized datasets on the other side. These studies revealed how small structural changes affected the caspase-inhibitor interaction energies [11].

Major difficulties for comparing docking predictions with experiments mostly came from the lack of transferability of experimental data and the lack of standardisation in molecule names. Gheyouche and colleagues have conceived the DockNmine platform to provide a service allowing an expert and authenticated annotation of ligands and targets. Researcher incorporated controlled information in the database using reference identifiers for the protein and the ligand, the data and the publication associated to it. It allowed the incorporation of docking experiments using forms that automatically parse useful parameters and results. Pre-computed outputs to assess the degree of correlations between docking experiments and experimental data were also provided [12].

Polishchuk and co-workers analysed the fact that pharmacophores derived from molecular dynamics simulations were more relevant than those just taken from rigid structures, but also generated a strong redundancy. They therefore proposed an approach to limit the number of pharmacophores, and showed its relevance [13].

Conserved three-dimensional (3D) patterns among protein structures provided valuable insights into protein classification, functional annotations or the rational design of multi-target drugs. Valdés-Jiménez and colleagues developed 3D-PP, a new free access web server for the discovery and recognition all similar 3D amino acid patterns among a set of protein structures independent of their sequence similarity. This new tool did not require any previous structural knowledge about ligands, and all data were organized in a high-performance graph database [14].

Finally, the number of available protein structures in the Protein Data Bank (PDB) had considerably increased in recent years. Here, with Nicolas Shinada and Peter Schmidtke, we presented a specific clustering of protein-ligand structures to avoid bias found in different studies. The methodology was based on binding site superposition, and a combination of weighted Root Mean Square Deviation assessment and hierarchical clustering. Defining these categories decreased by 3.84-fold the number of complexes, and offered more refined results compared to a protein sequence-based method [15].

References

[1] S. Aguero, R. Terreux, Degradation of High Energy Materials Using Biological Reduction: A Rational Way to Reach Bioremediation, International Journal of Molecular Sciences 20(22) (2019) 5556.

[2] C. Bienboire-Frosini, R. Durairaj, P. Pelosi, P. Pageat, The Major Cat Allergen Fel d 1 Binds Steroid and Fatty Acid Semiochemicals: A Combined In Silico and In Vitro Study, International Journal of Molecular Sciences 21(4) (2020) 1365.

[3] R. Dharmatti, H. Miyatake, A. Nandakumar, M. Ueda, K. Kobayashi, D. Kiga, M. Yamamura, Y. Ito, Enhancement of Binding Affinity of Folate to Its Receptor by Peptide Conjugation, International Journal of Molecular Sciences 20(9) (2019) 2152.

[4] S. Fan, R. Wang, C. Li, L. Bai, Y.-L. Zhao, T. Shi, Insight into Structural Characteristics of Protein-Substrate Interaction in Pimaricin Thioesterase, International Journal of Molecular Sciences 20(4) (2019) 877 .

[5] E.-H. Kim, M. Kawamoto, R. Dharmatti, E. Kobatake, Y. Ito, H. Miyatake, Preparation of Biphenyl-Conjugated Bromotyrosine for Inhibition of PD-1/PD-L1 Immune Checkpoint Interactions, International Journal of Molecular Sciences 21(10) (2020) 3639.

[6] G. Nshogoza, Y. Liu, J. Gao, M. Liu, S.A. Moududee, R. Ma, F. Li, J. Zhang, J. Wu, Y. Shi, K. Ruan, NMR Fragment-Based Screening against Tandem RNA Recognition Motifs of TDP-43, International Journal of Molecular Sciences 20(13) (2019) 3230.

[7] K.B. Pal, M. Mahanti, H. Leffler, U.J. Nilsson, A Galactoside-Binding Protein Tricked into Binding Unnatural Pyranose Derivatives: 3-Deoxy-3-Methyl-Gulosides Selectively Inhibit Galectin-1, International Journal of Molecular Sciences 20(15) (2019) 3786.

[8] J. Potthoff, K.K. Bojarski, G. Kohut, A.G. Lipska, A. Liwo, E. Kessler, S. Ricard-Blum, S.A. Samsonov, Analysis of Procollagen C-Proteinase Enhancer-1/Glycosaminoglycan Binding Sites and of the Potential Role of Calcium Ions in the Interaction, International Journal of Molecular Sciences 20(20) (2019) 5021.

[9] F. Reyes-Espinosa, D. Méndez-Álvarez, M.A. Pérez-Rodríguez, V. Herrera-Mayorga, A. Juárez-Saldivar, M.A. Cruz-Hernández, G. Rivera, In Silico Study of the Resistance to Organophosphorus Pesticides Associated with Point Mutations in Acetylcholinesterase of Lepidoptera: *B. mandarina, B. mori, C. auricilius, C. suppressalis, C. pomonella, H. armígera, P. xylostella, S. frugiperda*, and *S. litura*, International Journal of Molecular Sciences 20(10) (2019) 2404.

[10] N. Tripathi, I. Vetrivel, S. Téletchéa, M. Jean, P. Legembre, A.D. Laurent, Investigation of Phospholipase C1 Interaction with SLP76 Using Molecular Modeling Methods for Identifying Novel Inhibitors, International Journal of Molecular Sciences 20(19) (2019) 4721.

[11] S. Xue, H. Liu, Z. Zheng, Application of the Movable Type Free Energy Method to the Caspase-Inhibitor Binding Affinity Study, International Journal of Molecular Sciences 20(19) (2019) 4850.

[12] E. Gheyouche, R. Launay, J. Lethiec, A. Labeeuw, C. Roze, A. Amossé, S. Téletchéa, DockNmine, a Web Portal to Assemble and Analyse Virtual and Experimental Interaction Data, International Journal of Molecular Sciences 20(20) (2019) 5062.

[13] P. Polishchuk, A. Kutlushina, D. Bashirova, O. Mokshyna, T. Madzhidov, Virtual Screening Using Pharmacophore Models Retrieved from Molecular Dynamic Simulations, International Journal of Molecular Sciences 20(23) (2019) 5834.

[14] A. Valdés-Jiménez, J.-L. Larriba-Pey, G. Núñez-Vivanco, M. Reyes-Parada, 3D-PP: A Tool for Discovering Conserved Three-Dimensional Protein Patterns, International Journal of Molecular Sciences 20(13) (2019) 3174.

[15] N.K. Shinada, P. Schmidtke, A.G. de Brevern, Accurate Representation of Protein-Ligand Structural Diversity in the Protein Data Bank (PDB), International Journal of Molecular Sciences 21(6) (2020) 2243.

Alexandre G. de Brevern
Editor

International Journal of
Molecular Sciences

Article

Degradation of High Energy Materials Using Biological Reduction: A Rational Way to Reach Bioremediation

Stephanie Aguero [1,*] **and Raphaël Terreux** [1,2] 

1   PRABI-LG—Tissue Biology and Therapeutic Engineering Laboratory (LBTI) UMR UCBL CNRS 5305, University of Lyon. 7 Passage du Vercors, CEDEX 07, 69367 Lyon, France; raphael.terreux@ibcp.fr
2   Pharmaceutical and biological Research Institute (ISPB), CEDEX 07, 69367 Lyon, France
*   Correspondence: stephanie.aguero@ibcp.fr; Tel.: +33-(0)4-37-65-29-48

Received: 20 August 2019; Accepted: 4 November 2019; Published: 7 November 2019

Abstract: Explosives molecules have been widely used since World War II, leading to considerable contamination of soil and groundwater. Recently, bioremediation has emerged as an environmentally friendly approach to solve such contamination issues. However, the 1,3,5,7-tetranitro -1,3,5,7-tetrazocane (HMX) explosive, which has very low solubility in water, does not provide satisfying results with this approach. In this study, we used a rational design strategy for improving the specificity of the nitroreductase from *E. Cloacae* (PDB ID 5J8G) toward HMX. We used the Coupled Moves algorithm from Rosetta to redesign the active site around HMX. Molecular Dynamics (MD) simulations and affinity calculations allowed us to study the newly designed protein. Five mutations were performed. The designed nitroreductase has a better fit with HMX. We observed more H-bonds, which productively stabilized the HMX molecule for the mutant than for the wild type enzyme. Thus, HMX's nitro groups are close enough to the reductive cofactor to enable a hydride transfer. Also, the HMX affinity for the designed enzyme is better than for the wild type. These results are encouraging. However, the total reduction reaction implies numerous HMX derivatives, and each of them has to be tested to check how far the reaction can' go.

Keywords: bioremediation; High Energy Molecules; HMX; protein design; molecular dynamics; nitroreductase; flavoprotein; substrate specificity

---

1. Introduction

High Energy Molecules (HEMs) is a term that stands for the class of materials known as explosives, propellants, and pyrotechnics. HEMs are required for a wide range of purposes in the fields of construction, engineering, mining, quarrying, space sciences (propellants), pyrotechnics, and currency production. They are also known for their military purposes [1]. The large-scale manufacturing and extensive use of HEMs for military purposes since World War II (WWII) have contributed to a high level of environmental pollution [2]. Contaminated sites are not easy to identify because they are not only located in present and former war zones but are also present among the military firing ranges; manufacturing, handling, and storage sites; and areas where they are used for industrial purposes [3]. In Australia, Canada, and the US, most of these sites have been located, and a minimal clean-up is in process. In Germany, the situation is more confusing because many of the explosive manufacturing facilities were demolished at the end of WWII. In other countries worldwide, the extent of contamination by explosives is either undetermined or not available to the public [4]. Moreover, some wars are still ongoing at present. Therefore, the environmental issue due to explosives remains a hot topic [5].

Decontamination solutions exist. The first treatment process in use was the incineration of explosive-contaminated soils. This method had the advantage of offering a high level of process control and efficient destruction and removal. However, burning was relatively expensive and also polluting due to the production of ashes. Then came biochemical solutions such as composting, the first biological treatment process to be tested and approved for military sites [4]. Composting requires the addition of bulking agents, increasing the volume of material. Unlike the ash created by incineration, composted material can support vegetation and is less expensive. Bioslurry is another soil treatment. In this process, contaminated soil is mixed with water and nutrients to create a slurry that can be combined in a bioreactor and treated with various bio-organisms. However, these treatments are ex situ and require additional costs regarding the equipment and process controls. A solution to ex situ bioremediation is the phytoremediation, which explores the ability of plants to remove pollutants from contaminated soils. The main limitations of this method are the toxicity of the contaminants: treatments are possible only if toxicity is not a factor with the candidate species. Moreover, absorbed pollutants and their metabolites must move from the bulk soil to the zone of influence near the roots for phytoremediation to occur. This requires a good solubility of the toxic waste.

Among different forms of chemical explosives, 2,4,6-trinitrotoluene (TNT), hexahydro-1,3,5-trinitro-1,3,5-triazine (RDX), and 1,3,5,7-tetranitro-1,3,5,7-tetrazocane (HMX) are the most common (Figure 1). These explosives are highly stable compounds. They tend to blend with organic matter of soil and thereby to contaminate it [2]. Studies of toxicology on various organisms including bacteria, algae, plants, invertebrates, and mammals [6] have identified toxic and mutagenic effects of these common military explosives as well as their transformation products [4].

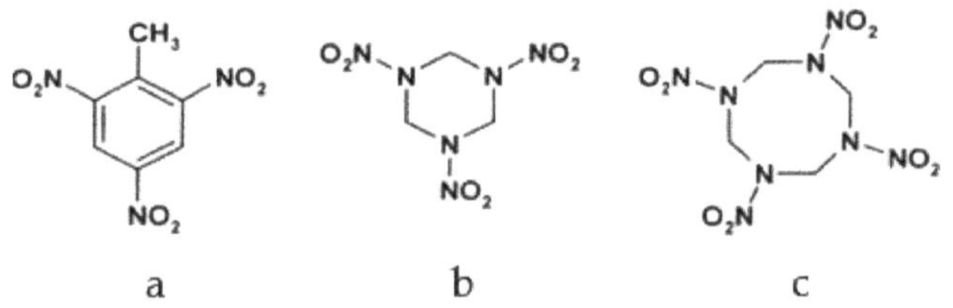

Figure 1. Two-dimensional chemical structures of (**a**) 2,4,6-trinitrotoluene (TNT), (**b**) hexahydro-1,3,5-trinitro-1,3,5-triazine (RDX), and (**c**) 1,3,5,7-tetranitro-1,3,5,7-tetrazocane (HMX).

Research on TNT and RDX detoxification is still going on, and various phytoremediation-based approaches look encouraging. Recently, green grass was created to degrade TNT efficiently [7,8], and a field-applicable grass species capable of both RDX degradation and TNT detoxification has also been engineered [9]. Phytoremediation of TNT is very well studied, and the results look promising. However, phytoremediation remains more complex for RDX and even more so for HMX. HMX and RDX are nitramine compounds. Nitramines are less stable compared to aromatic nitro compounds such as TNT. Recently, several studies involving different microbes have been carried out to determine RDX degradation potential. There is, however, a lesser number of studies on aerobic and anaerobic microbial degradation of HMX. Despite being a close homolog to RDX, HMX shows more resistance to chemical and biological degradation than RDX, due to its very low solubility [10]. Recently, a study investigating the HMX degradation potential of the native bacterial isolate *Planomicrobium flavidum* strain S5-TSA-19 was conducted under aerobic conditions [11]. This bacteria strain showed efficient degradation, although some secondary metabolites (like methylenedinitramine and N-methyl-N,N′-dinitromethanediamine) formed during biodegradation of HMX are toxic or have unknown toxicity.

Over the past few years, the NAD(P)H-dependent bacterial nitroreductases (NRs) have received particular attention for their potential use in biodegradation and bioremediation of nitroaromatics [12]. NAD(P)H-dependent bacterial NR, also named nfsB enzymes, are capable of using either NADH or NADPH as reducing equivalent, in opposition to nfsA, which only uses NADPH. The nfsB enzymes are dimeric proteins and encompass two flavin mononucleotide cofactor (FMN). Both types of NR reduce a broad range of nitroaromatic substrates [13]. Reaction kinetic studies of NR have shown a simple ping-pong mechanism (Figure 2) without gating steps able to enforce specificity [14]. Electrons are transferred pairwise from the NADH cofactor to the oxidized flavin, and from reduced anionic flavin to nitroaromatics substrates [14,15] (Figure 2a,c).The substrate reduction has been proposed to occur via hydride transfer where two electrons and one proton are transferred in a common way [16]. It has also been proposed, in the case of NRs, that protons were transferred as solvent-derived protons via electron-coupled proton transfer [17]. The occurrence of a hydride transfer is dependent on the distance between the N5 atom of the reduced flavin and the atom donating or receiving the proton (Figure 2b). A distance of 3.8 Å between the two entities is optimal [18]. It has also been shown that upon reduction, FMN adopts a butterfly-like bending of the isoalloxazine ring system [19].

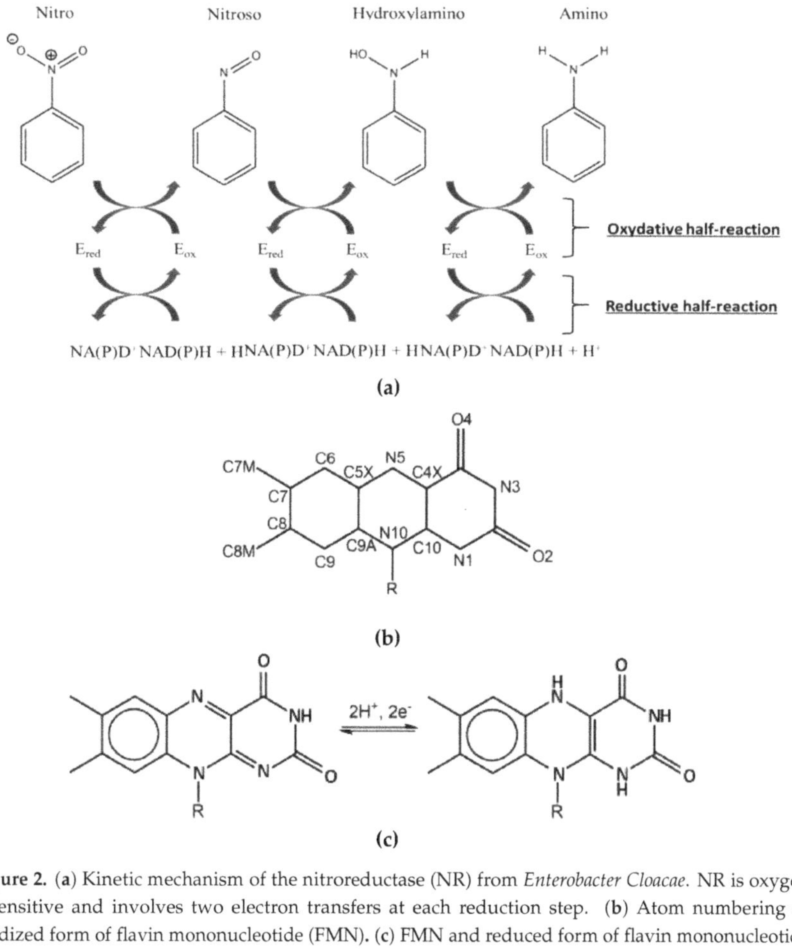

Figure 2. (**a**) Kinetic mechanism of the nitroreductase (NR) from *Enterobacter Cloacae*. NR is oxygen insensitive and involves two electron transfers at each reduction step. (**b**) Atom numbering of oxidized form of flavin mononucleotide (FMN). (**c**) FMN and reduced form of flavin mononucleotide (FMNH2) structures.

The reduction of nitro groups usually leads to hydroxylamines (Figure 2) and requires the transfer of four electrons. The two-electron nitroso reduction intermediate is not observed because the second two-electron reaction has a much faster rate than the first two-electron transfer [20,21]. It has been shown that the well-characterized NRs from *Escherichia Coli* and *Enterobacter Cloacae* (EcNfsB and

EntNfsB, respectively) were able to reduce nitroaromatics into the corresponding hydroxylamines, but not into the amines [14–21]. However, aromatic reduction to amine has been observed: two NRs from *B. subtilis* [22] and more recently one NR from *Gluconobacter oxydans* 621H reduce TNT into the corresponding amino products [23]. The advantage of using these enzymes, apart from their ability to reduce a broad range of nitro compounds including the explosive TNT, is that the reduction of the nitro group into amine buries any toxicity issues.

Miller et al. [24] have established a correlation between amine production and substrate properties, indicating that the best choices will have large pi systems and electron-withdrawing substituents. Likewise, they have shown that electron-withdrawing groups favor the reduction of nitro substrates. Thus, they suggest that smaller nitroaromatics will be less likely to undergo full reduction. This is why compounds as TNT undergo reductions to the corresponding amine. HMX has four nitro groups, which are electron withdrawing, but doesn't get a pi system. The pi system allows for pi stacking between the substrates and the flavin, stabilizing the complex. Our challenge was to create a stable complex without the solicitation of pi stacking.

We focused our research on the NAD(P)H-dependent bacterial NR from *Enterobacter Cloacae* [25]. The NO2 > NO reduction reaction has been observed in this NR on a similar substrate, the nitramine RDX [13]. However, HMX is relatively insoluble in water, and much more stable than RDX. Its metabolization by the same type of enzyme (for denitrification purposes) is therefore more complicated. Residues 1-nitroso, 2-nitroso, and 3-nitroso HMX have already been observed, but the rate of the reaction is lower than those observed for RDX. Thus, we computationally modified the structure of the protein to generate a substrate specificity toward HMX. We worked on the PDB structure 5J8G [26] and rationally redesigned the active site of the NR on its reduced form by using protein design methods (Figure 3). Then the mutated structure was pushed through molecular simulation to evaluate the stability of the newly designed active site. Docking of cofactor confirmed that the performed mutations did not alter the bonding of the FMN electron donor NAD(P)H. All the subtlety of this work was to improve the specificity of an unspecific enzyme and to allow for the long-term reduction of non-aromatic nitrated substrates into amine.

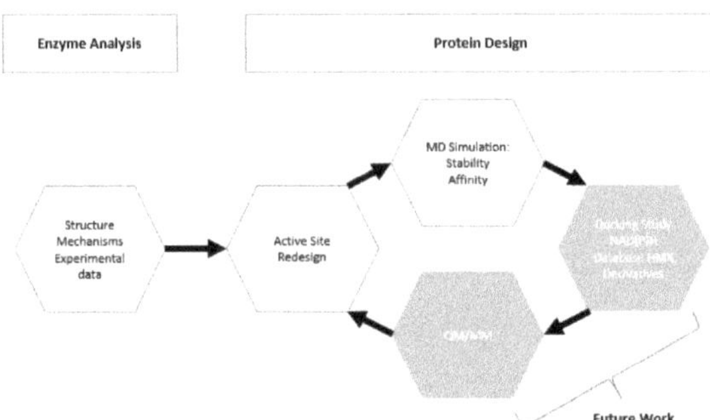

Figure 3. Pipeline/workflow of the study.

2. Results

2.1. Design of the Active Site

The native structure 5J8G shows the familiar symmetric dimeric fold of NRs [15]. The protein is composed of two intricated monomers. The H7 helix is central to the dimer interface. This configuration stabilizes the dimer even when bound to FMN or to various substrates [26]. The active site is a cleft at the protein surface. It is limited by helix 6 (amino acid 109–129) and 7 (134–148), strand 3, and helix 8 (156–175).

Multiple binding orientations have been observed in NfsB's active site, based on the crystal structure of the oxidized enzyme. However, there is a lack of information regarding the substrate binding to the reduced form of the NR. Drigger et al. [27] have shown that, for the SsuE FMN reductase, a protein related to the flavodoxin-like superfamily, as well as the NR, a stacking of the NADPH on top of the flavin was a non-productive mode of binding. Thus, a productive binding would be related to the possibility of a hydride transfer. This transfer depends on the distance between the N5 atom of the flavin and the atom donating or receiving the proton, with a distance of 3.8 Å being optimal. Thus, we decided to position one of the nitro groups of HMX close to the hydrogen bound to N5 at a distance inferior to 3.8 Å. Then, we selected, around the HMX molecule, every amino acid within a distance of 4.5 Å presenting a lateral chain facing the active site. From this ensemble of amino acids, we excluded Glu165 and Gly166, which support efficient hydride transfer from NADH to FMN. We also excluded Asn71, Thr41, and Phe124, which are involved in interactions between NADH and FMN [15–28]. Phe124 was also excluded, as the amino acid was identified as conferring NR activity [29]. Moreover, the side chain of Phe124 interacts with the nicotinamide ring of NADP [26]. These particularities are shared with several other NR homologs [17–30].

As a final verification, the remaining residues were cross-checked against a sequence alignment clustering the 50 most similar protein sequences in the UniProt Database filtered according to a UniRef90 BLAST Search. Highly conserved residues across the alignment were excluded. We finally allowed the modification of Tyr68, Phe 70, Glu120, Tyr123, and Ala125 into every natural amino acid (Figure 4). A restrained energy minimization was then performed to relax the structure in order to keep the HMX molecule in a relevant position.

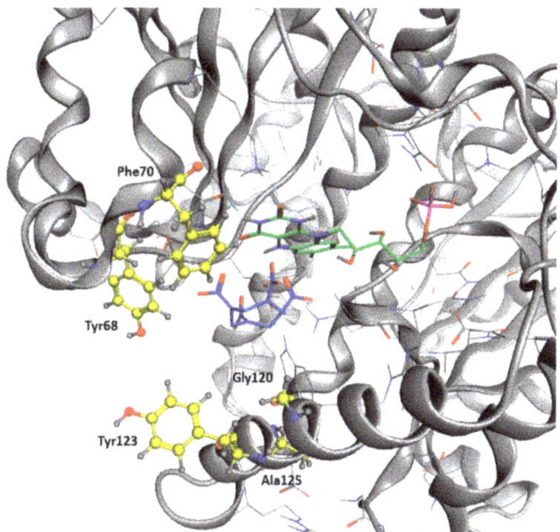

Figure 4. Target Residues for Design Trials. All the residues displayed in yellow are included in the resfile. FMNH2 cofactor is displayed in green, and HMX molecule in blue.

Then, we performed several trials using the Coupled Moves algorithm, using 30 identical starting structures (see Supplementary Materials for script and param files). A total of 10,000 steps per structure was conducted. Residues Tyr68, Phe70, Glu120, Tyr123, and Ala125 were allowed to be mutated into any natural amino acid, including their associated side chain rotamers. The selection of amino acid identity and rotamer at each step was based on the calculated ROSETTA energy scores and chosen according to Boltzmann weighted probability. The simulation produced 6093 low-energy sequences. The results of the trials are illustrated under a logo sequence (Figure 5). The logo depicts the amino acid conservation among the sequences previously generated.

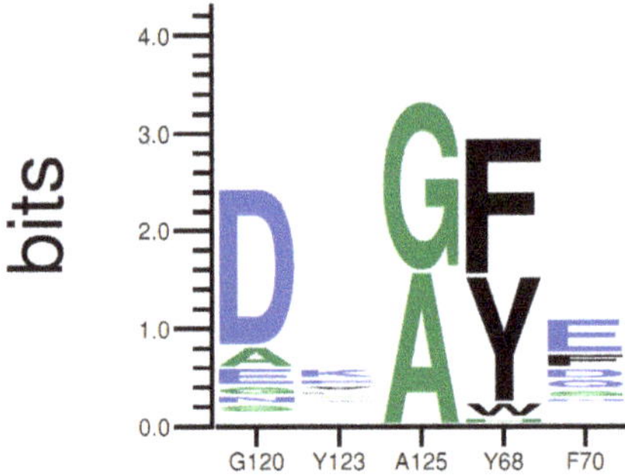

Figure 5. Relative Amino Acid Proportions at Positions 68, 70, 120, 123, and 125 in Low-Energy Structures. The relative size of each letter indicates their frequency in the sequences, and the total height of the letters shows the information content of the position, in bits.

As shown in Figure 5, the occurrence of each amino acid within the sequences is depicted by the total height of the letter. Biggest letters are identified as beneficial. Mutations G120D, A125G, Y68F, and F70E were accordingly proposed to improve the stability of HMX in the active site. Mutations of Y123 do not demonstrate clear results. The most frequently proposed mutation is the replacement of the tyrosine by an arginine. Mutations G120D, Y123K, and A125G are located within the H6 helix. It has been suggested that the substrate specificity was due to the plasticity of this helix. Indeed, H6 shows an elevated variability in amino acid and position for accommodating substrates of different sizes. Mutations Y68F and F70E are located on H4.

When compared to p-NBA (the bound substrate of the NR in 5J8G crystal structure used for mechanistic studies by Pitsawong et al. [26]), HMX shows a completely different structure: p-NBA is planar, allowing binding via H-bonds and pi staking above the re face of the FMN, as for the other substrates and analogs in the active sites of the NR and NfsB. However, the placement of p-NBA is not optimal for reactivity, as the nitro group to be reduced is too far from the reduction center (N5 of FMN). HMX does not have an aromatic structure. It consists of an eight-membered ring of alternated carbon and nitrogen atoms, with a nitro group attached to each nitrogen atom. This means that the molecule is not planar and can adopt four different crystalline conformations (alpha, beta, delta, and gamma). Stabilizations by pi stacking interaction are no longer possible. Transformation of one HMX conformation to another only occurs at high temperature. However, here, we consider the four conformations as only one possible substrate with several conformers for simplicity.

As shown in Figure 6b for the mutant NR, K14 forms H-bonds bridges between the protein, the FMNH2 cofactor, and HMX. This bridge maintains FMNH2 and HMX in close contact together. Additionally, there is an H-bond between non-cyclic part of FMNH2 and one nitro group of HMX. Finally, HMX is stabilized in the active site because of the electrostatic repulsion of the mutated G120D, which interacts with the hydrogens related to the carbon ring atoms. The rest of the molecule is exposed to the solvent. The residues Phe124, Thr41, Glu70, and Asn71 exhibit a side chain and a backbone segment, which are both very highly solvent exposed in the active site when HMX is not present. The presence of the ligand greatly reduces the solvent accessible surface area. HMX fits in a more homogeneous way in the active site of the mutant. Indeed, the WT NR stabilizes HMX with Asn71 and Thr 41, but also through Gly166 and Glu165 H-bonds, in a way similar to the binding mode of NADH to the NR (Figure 6a). The mutation G120D reduces the inner volume of the active site and allows for a better fit of HMX. In the mutant, one of the nitro groups of HMX is close enough to the N5 atom (4.35 Å) to allow hydride transfer. This first reduction step of HMX has been experimentally observed [31]. This situation is not surprising. However, the design of the active site aims to allow for both the experimental first step of the reaction and for the entire reduction of the nitro group.

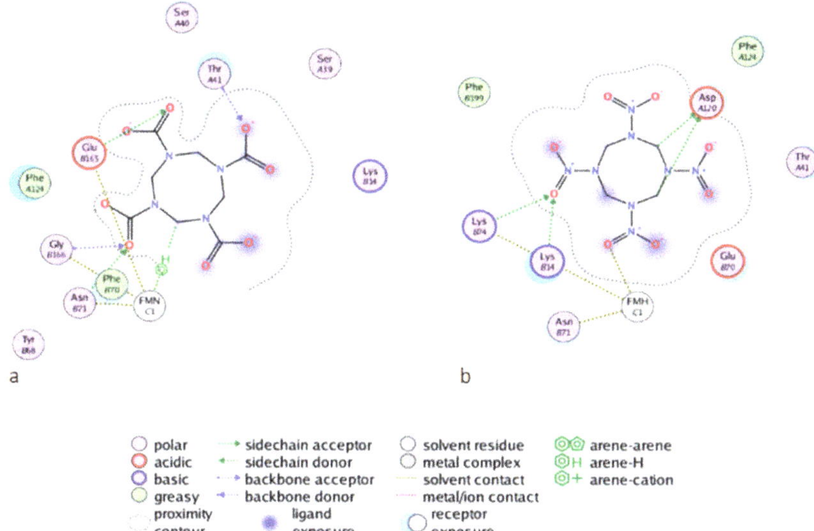

Figure 6. The Molecular Operating Environment software (MOE) Ligand Interactions application allows for the visualization of the protein active site in complex with HMX, in diagrammatic form. The diagram shows solvent interactions, H-bonds and surface of exposure. (**a**) Wild Type NR and HMX; (**b**) Mutant NR and HMX.

After the design phase of the NR being completed (Figure 7), we needed to explore the time dimension parameter of the 3D NR model. Thus, the behavior of HMX in the newly design active site was investigated. The objective was to see if the generated structure could reach an energetic balance and, if yes, in which way.

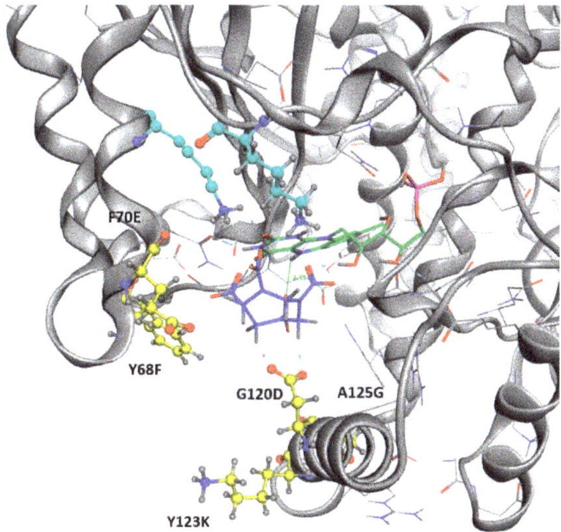

Figure 7. Mutated Residues after Design Trials. All the positions displayed in yellow were included in the resfile. FMNH2 cofactor is displayed in green and HMX molecule in blue. Non-mutated residues implied in H-bonds with the HMX are in cyan.

2.2. Molecular Dynamic Simulations

2.2.1. Global MD Analysis

A first MD simulation of 130 ns was performed on the NR in a ligand-bound complex with p-NBA (PDB ID: 5J8G). Pitsawong et al. [26] showed that the NR does not generate paminobenzoic acid and thus does not appear to reduce nitro groups into amines [14–20]. p-NBA has been co-crystallized with the NR from Enterobacter Cloacae. We used this crystal as a starting structure to understand how p-NBA was stabilized in the active site. The dynamics show that the p-NBA is not stable in the active site during the whole dynamic but moves in and out of the pocket. When in the pocket, the aromatic ring of the p-NBA is stabilized against the flavin ring, probably involved in pi stacking interactions. The nitro group is oriented in a non-productive way, meaning that the nitro group is too far from the N5 atom of the flavin to trigger a hydride transfer. These observations are consistent with the crystal description of Pitsawong et al. Affinity calculation of p-NBA/NR interaction were calculated (Table 1).

Table 1. Table of free-energy calculation of NR/ligand complex. MMGBSA is calculated as a sum of a conformational energy terms supplemented with a solvation free-energy term calculated using continuum electrostatics.

Complex NR/HMX	mmGBSA (kcal/mol)
Wild Type NR/p-NBA	−15.2117
Wild Type NR/HMX	−11.8261
Mutant NR/HMX	−29.0704

A similar molecular dynamic simulation of 130 ns was performed on the Wild Type NR in a ligand-bound complex with HMX. We used the 5J8G crystal structure as a starting point. The p-NBA was replaced with HMX positioned in a productive way, according to the original p-NBA orientation, with one of its nitro groups close to the N5 atom of the flavin at a distance of 3,8 Å.

We observe that HMX doesn't reach a stable position within the active site. It rolls in the space delimited by H6 and FMNH2 and struggles to find a stable position. As a consequence, none of the four groups is stabilized long enough within a radius of 3.8 Å around the reductive N5 of FMNH2 to consider a hydride transfer. Phe124 is positioned under one of the nitro groups and does not manage to stabilize it. Phe124 also cannot prevent HMX from leaving the active site, and HMX ends up escaping after 70.4 ns (frame 352) in one out of three simulations. During this particular dynamic, Tyr 123 failed to catch and get HMX into the site.

As a result, the WT NR does not successfully stabilize HMX because these weak interactions are not persistent in time and cause HMX to leave.

We did not use the docking method to place HMX in the NR active site because we had to dock the HMX between FMNH2 and H6. However, H6 is greatly variable in terms of position [15]. Such a variability, in terms of plasticity, may be directly related to the wide range of substrate accepted by the NR. Thus, while moving, the helix allows for the accommodation of more or less bulky substrates. The use of docking, even with induced-fit features, would not have given good results. By considering the helix as non-flexible, the pose wouldn't have reflected the reality of the NR flexibility. By using our method, thanks to the dynamics, the helix rearrangement can be considered as accommodating our large substrate, while respecting our positioning constraint.

Another 130 ns MD simulation was performed on the mutant NR previously designed, in a ligand-bound complex with HMX. The input .pdb structure was directly extracted from the coupled trials after the selection of the lower energy mutant. This mutant of Enterobacter Cloacae NR shows the following mutations: G120D, Y123K, A125G, Y68F, and F70E. Throughout the simulation, HMX remains trapped in the cleft formed by the active site. This cleft is delimited by Phe124 upstream. Laterally, Lys123 blocks HMX and prevents it from leaving due to H-bonds connections through the amine portion of its lateral chain and the hydrogens connected to the carbon atoms of the HMX cycle.

Alternatively, it also establishes H-bonds with the oxygen of the nitro groups. Downstream, Asn117 also makes H-bonds through its amino lateral chain. Phe68 occupies steric space above HMX, and places it against FMNH2. At no point during the dynamics does HMX go out of the molecule. It is stabilized in a relative way since it is able to roll in the previously described space. Thus, the nitro group presented at the reducing center N5 is never the same. This is partly due to conformational changes in the HMX.

2.2.2. Stability Studies

RMSD gives an overall picture of how much the protein structure has changed throughout the simulation. It provides an overview of protein stability over time. We captured RMSD of the protein, of the cofactor FMNH2 and each substrate (p-NBA and HMX). RMSD plots for all simulations are shown in Figure 8.

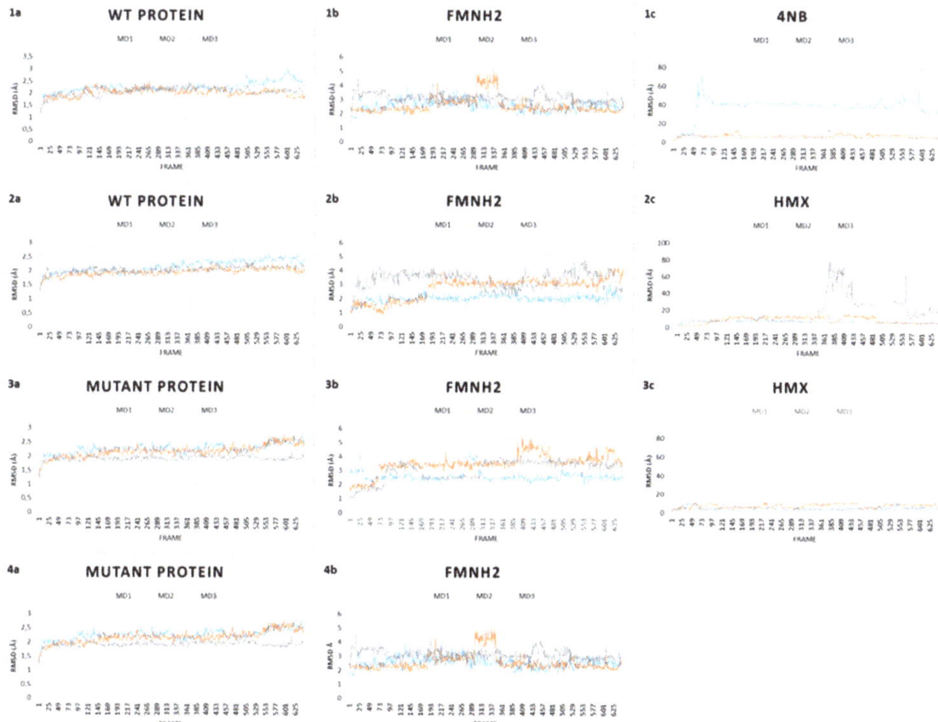

Figure 8. Root Mean Square Deviation (RMSD) plots over time of every simulation of both WT and mutant NR. (**1**) RMSD plots of WT protein in complex with 4NB. (**a**) Protein, (**b**) cofactor FMNH2, and (**c**) 4NB. (**2**) RMSD plots of WT protein in complex with HMX. (**a**) Protein, (**b**) cofactor FMNH2, and (**c**) HMX. (**3**) RMSD plots of mutant protein in complex with HMX. (**a**) Protein, (**b**) cofactor FMNH2, and (**c**) HMX. (**4**) RMSD plots of relaxed mutant protein without HMX (**a**) Protein, (**b**) cofactor FMNH2.

The first simulation captures the movements of the NR is complex with p-NBA, the initial structure of the NR co-crystallized acid para nitrobenzoic (Figure 8.1). All along the 130 ns of simulations, the WT NR protein is stable as shown by its low RMSD (2.2 Å). FMNH2 shows more movement relative to the protein (displacement of 3.5 ± 1.00 Å). This is because the cofactor is not covalently bound to the NR protein but through H bonds. This mode of binding allows for a certain flexibility of the structure, depending on the inner movements of the protein backbone and through lateral chains. p-NBA is not stable in the active site as it moves in and out. RMSD translates these movements by showing a broad variability.

In comparison, the WT protein in complex with HMX shows similar stability (Figure 8.2). The protein exhibits a RMSD value of 2.00 ± 0.15 Å, whereas the FMNH2 gets a higher RMSD value at 3.25 ± 1.50 Å. However, HMX goes out of the active site in one simulation, but not in the two others. In the first one, HMX exhibits a RMSD value of 18.73 ± 1.19 Å. It seems to be stable in the first few ns but finally goes out, to the same extent as p-NBA. HMX gets lowers RMSD values for the two other simulations (respectively 8.52 ± 3.57 Å and 7.21 ± 1.19 Å). HMX behavior is not similar among the three MDs. In the first one, residues F124 and K123 of H6 do not manage to lock HMX in the pocket. As a result, the molecule leaves the active site. In the two other ones, due to a slightly different orientation of their lateral chain, residues F124 and Y123 prevent HMX from escaping the active site. As a consequence, HMX manages to establish transient H bonds with K41. However, these H bonds are not sufficient to stabilize HMX, which rolls in the pocket.

Our design project aimed to create an NR able to stabilize and thus make possible the reduction of HMX nitro groups. The designed NR displays comparable stability (Figure 8.3) when compared to the WT NR (2.25 ± 3.05 Å for the protein, 3.25 ± 0.75 Å for the cofactor). HMX reaches a stable state when complexed with the mutant enzyme. Indeed, even if some artifacts are observed due to the periodicity of the solvation box, the global RMSD is quite stable, regarding the complexed WT NR: 6.41 ± 1.20 Å, 4.35 ± 1.97 Å, and 5.11 ± 1.76 Å for the triplicate MD. In the mutant NR, HMX is more stable in the active site and is adequately positioned. The mutations Y123K and G120D (H6) stick HMX against the re face of the flavin, establishing H bonds with the nitro groups of HMX. Y123K mutation plays a crucial role in maintaining HMX in the active site. In the WT NR, Y123 did not manage to stabilize HMX through H bonds. As a consequence, HMX escapes or is pushed in the pocket without reaching a stable state. In the mutant NR, K123 catches and stabilizes HMX through H bonding when the molecule moves away from the active site. G120D mutation also provides better stabilization through H bonds between the nitro groups of HMX and its two oxygens. F124 prevents HMX from leaving to the same extent as in the WT NR. Moreover, this optimal support is reinforced by K41, which forms H bond bridges between the protein, the FMNH2 cofactor, and HMX. Finally, HMX is stabilized at the required distance to observe the hydride transfer.

As a final verification, an additional simulation was launched to verify whether the designed active site conserves its stability in the absence of HMX. As shown in Figure 8(4), both FMNH2 and the protein maintain their stability with an RMSD even more stable than those with HMX (respectively (3.88 ± 0.38 Å for the protein and 1.88 ± 1.23 Å for the cofactor). The flat RMSD means that the designed protein is stable, with and without HMX bound to the active site. The absence of high variations may be due to the lack of HMX above H6, which is known to show a high flexibility to accommodate different substrates.

2.2.3. Affinity Studies

Once the MD simulations of ligand recognition upon binding of HMX to the NR were performed, we also calculated the ligand-binding affinity.

The calculated binding free energies of each substrate for the WT and mutant NR were computed using the ensemble-average molecular mechanics energies combined with the generalized born and surface area continuum solvation (MM/GBSA) rescoring. MM/GBSA are popular methods used to estimate binding energies of small ligands to biological macromolecules. They are based on molecular dynamics simulations of the receptor–ligand complex. Each calculated binding free energy is averaged from snapshots extracted from the last four ns MD trajectories. The results are shown in Table 1.

We observed that the binding free affinity of the mutant NR and HMX complex is lower than those of the WT NR/HMX complex. This affinity value is also stronger than the WT NR/p-NBA one. These results tend to ensure that the designed enzyme complexed with HMX offers better stability compared to the WT NR. These calculations are confirmed by the structural study of the complex interaction.

2.2.4. HMX Behavior during the Simulation

The distance between each oxygen from the HMX nitro groups and the reductive N5 of FMNH2 has been calculated during the 130 ns simulation. As previously described, HMX does not stay bound in the active site of both WT and mutated NR. While HMX leaves the WT NR after several rolls in the active site, it remains trapped in the mutated NR active site. However, it does not keep a stable position and also rolls under FMNH2, exposing different nitro groups to the reductive N5.

A graph showing the medium distance between each oxygen from HMX (O1, O2 ... O8) for the complexes WT NR/HMX and mutant NR/HMX is shown in Figure 9. As previously mentioned, the possibility of hydride transfer depends on the distance between the flavin N5 atom and the atom donating or receiving the hydrogen, 3.8 Å being optimal [18].

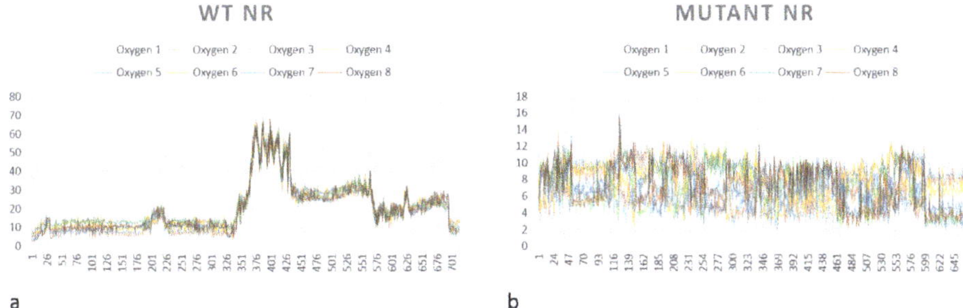

Figure 9. Average distance between each atom of oxygen from HMX and atom N5 of the FMNH2 in (**a**) the WT and (**b**) the mutant NR.

In the WT NR (Figure 9a), the distance to observe the reductive reaction is only inferior to 4 Å during the start of simulation. However, this distance is almost always inferior to this value for our mutant NR (Figure 9b).

These observations, combined with the affinity calculations, tend to prove that the mutant NR/HMX complex is more stable and that there is a higher probability of observing a hydride transfer between the mutant NR and HMX than between the WT NR and HMX.

2.3. Docking Studies

Our results show that the designed enzyme is able to better stabilize HMX than the WT NR, and that the affinity between the mutated NR and HMX is stronger than for the WT NR/HMX complex.

However, a specific point still needs to be clarified: is the mutant NR flavin still able to be reduced by NADPH? To answer this question, *in silico* docking of the NADPH co-factor to the predicted mutant NR structure was performed.

The last frame of the MD simulation for the mutant NR/HMX complex was extracted as .pdb. Prior to the docking, HMX, water molecules, and ions were removed. The docking zone was defined as follows: the FMN cofactor, H5 and H6 (residues 94 to 129), the loop between H3 and H4 (residues 67 to 74), the loop from amino acid 40 to 43, H7 (residues 138 to 142), the loop between H1 and H2 (residues nine to 22), and the loop between strand4 and H9 (residues 197 to 207). This zone spatially defines the active site. It also includes amino acids known to be involved in weak and dominant interactions with the NAPDH analog nicotinic acid adenine dinucleotide (NAAD). Indeed, Pitsawong et al, 2017 have shown that the nicotinic acid ring of NAAD stacks against the re face of the flavin over the uracyl and diazabenzene. They have also shown that weak interactions with the backbone of Gly120 and Thr67, and with the side chain of Asn71, engage the ribose hydroxyls. The dominant interactions stabilizing the phosphates are observed with the conserved side chains of Lys14 and Lys74. Also, the side chain of Phe124.B interacts with the nicotinamide.

A total of 150 docking poses were generated using the triangle match placement method associated to the London dG scoring function. Then, 25 top scoring poses were refined using the induced fit refinement scoring function GBVI/WSA dG with the generalized born solvation model (GBVI).

The top five scoring poses were evaluated. Three conformations of NADPH bound into the active site of the protein and interacting with FMN were observed. The ranking was based on the GBVI/WSA score and the maximum number of favorable non-bonded interactions between the mutant protein and NADPH. Configuration number 4 had the maximum number of protein–substrate–cofactor interactions and was hence considered to mimic the actual protein–ligand complex: carboxyl group of His11 forms one H-bond bridge between the protein, the phosphate group on NADPH, and the cofactor FMN atom 02 (Figure 10). The nicotinamide is not involved in pi stacking interactions with FMN, but it sits deep in the pocket, which contains the flavin rings, and which is stacked against the re face of the flavin over the uracyl and diazabenzene rings. Our observations are in good agreement with the crystal structure 5J8D of the NR bound in a complex with NAAD, an analog of NADPH [26]. The distance between the transferable hydrogen and the N5 of the flavin is 2.88 Å, allowing for a hydride transfer.

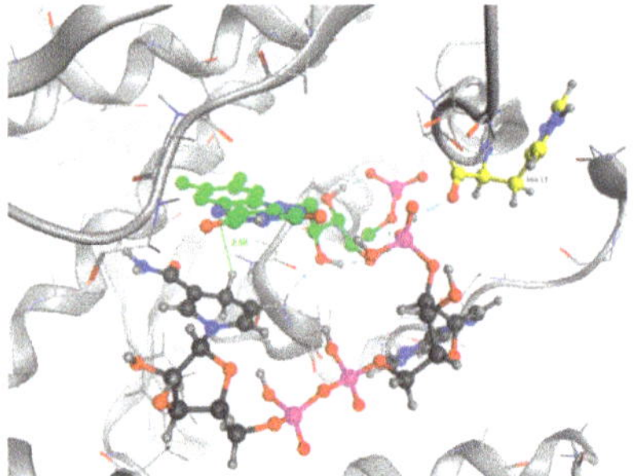

Figure 10. NADPH bound into the active site of the mutant NR. Protein backbone are shown in grey. Residue interacting with NADPH is shown as sticks. FMN is shown in green. The distance between the N5 of the flavin and the transferable H of NADPH is depicted.

3. Discussion

In this study, we designed the NR from *Enterobacter cloacae* to facilitate the reduction of a specific substrate: the high energy molecule HMX. This design is more likely a redesign of the active site, since only a few positions were mutated to allow a better stabilization of the ligand in the pocket. Indeed, in vitro studies have shown that HMX does not undergo full reduction. Different causes are possible—HMX has a very low solubility, and the NR is a soluble enzyme. Also, nitroaromatics are more likely to be stabilized in the active site, thanks to the pi stacking interactions with FMN, which is not the case of HMX.

The challenge here was to use a different way to stabilize HMX. The design allowed for the creation of various H-bonds with HMX (Figure 6), and the ligand fits more homogeneously in the mutant active site than in the WT NR. However, the model has been designed using a relatively rigid structure as a starting point. This is why we used MD simulations to allow for the relaxation of the complex and affinity calculations. RMSD calculations showed that the designed NR has a stability comparable to the WT NR. However, HMX reaches a more stable state when complexed with the mutant enzyme (Figure 9). Affinity calculations also indicate an improved affinity. As shown in Table 1, HMX and p-NBA contain oxygens with partially negative charges, suggesting electrostatic interactions, which should be a critical factor in the binding affinity. HMX has four nitro groups. Each of them has two H-bonds acceptors. As H-bonds and pi stacking are weak interactions, the presence of eight oxygens could compensate for the absence of the aromatic group and the subsequent absence of pi stacking for HMX. This also could explain the remaining of HMX in the WT NR active site during the start of the simulation. The escape of the ligand outside the pocket could be due to the size of HMX and the nature of the molecule, which presents inner movements, resulting in a larger occupied space than the one held by p-NBA.

This work also allowed us to show that the mutated NR is able to use NADPH in conjunction with oxidized FMN. In this regard, we performed a docking study. The optimal pose was selected among the top five scoring poses, according to the maximum number of protein-substrate-cofactor interactions. It has been shown that Asn71, Lys14, and Lys74 make polar contacts with the sugar-phosphate portion of the bound nicotinic acid adenine dinucleotide (NAAD), a precursor of NAD and NADPH [26]. It has also been shown that the NAAD binds the NR in an extended conformation [32] with its nicotinic ring against the face of the flavin. This is consistent with observations in other flavoenzymes [33]. These residues, identical or similar in NR homologs, suggest that NADH binding is similar among the family members. As none of the three conserved residues have been mutated during the design phase, we suppose that NADPH could adopt a stable conformation in the designed active site. However, our selected docking pose does not show the same interactions. NADPH is stabilized by a bond bridge between His 11, the phosphate group of NADPH, and the FMN atom 02. This configuration locks NADPH in the pocket. Also, we did not observe the interaction between the nicotinamide ring of NADPH and the backbone of Phe124, which was identified as conferring NR activity [34]. However, we observe the same position of the nicotinamide stacked against the re face of the flavin. Regarding the reaction, the hydride transfer is possible if the C4 of NADPH is 3.0Å from the flavin N5. This position promotes orbital overlap between the nicotinamide C4 hydrogen of NADH and the N5 of the isoalloxazine ring [18]. The selected pose of our study shows a distance of 2.88 Å between the FMN N5 and the hydrogen donor, which allows for the hydride transfer. Consequently, MD simulations would be relevant, as the structure, even minimized, could be trapped in local energy minimum. Indeed, docking was performed with an induced fit model could observe a rearrangement of the NADPH after a relaxation phase.

One last question still needs to be answered: does the mutant NR accommodate intermediate HMX derivative structures? Our work aimed to design an enzyme able to reduce the nitro groups of the explosive HMX specifically. In the process, the design was performed with HMX as a starting substrate. However, HMX contains four nitro groups, and the affinity of the intermediate structures, shown in Table 2, must be studied. The point is to understand how far the reduction could go, and how much we could detoxify the molecule. In 2014, Pitsawong et al. [14] showed that the WT NR does not generate p-aminobenzoic acid from p-NBA and therefore appears to not reduce nitro groups into the corresponding amine. The chemical explanation of this limitation was brought by McCormick et al. [35], who showed that the reduction rate of the nitro group increases with the groups present in the para position (with the following priorities: NH2 <OH <H <CH3 <COOH <NO2). It could be interesting to evaluate whether our mutant NR could overcome this condition.

The interest in going beyond the hydroxylamine state lies in the fact that HMX derivatives are also considered toxic. Indeed, we understand that the toxicity of HEMs is highly connected to the presence of the nitro groups. However, the hydroxylamine derivative is also known to interact with biomolecules, including DNA, and thus causing toxic and mutagenic effects. The toxic effects are related to the electrophilic character of these derivatives, whereas the mutagenic effects are mainly due to the formation of hydroxylamine moiety adducts through esterification with guanine [36]. This is why it would be interesting to perform a docking study of the HMX derivatives on the designed NR to investigate if the docking poses could promote hydride transfer. Also, these results would have to be confirmed by MD simulations and affinity calculations to confirm the stability of these molecules in the active site.

Table 2. HMX derivative molecules.

HMX Derivative Molecule	Name
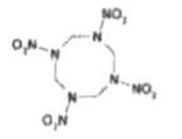	1,3,5,7-tetranitro-1,3,5,7-tetrazocane
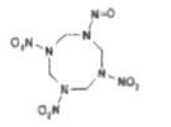	1,3,5-trinitro-7-nitroso-1,3,5,7-tetrazocane
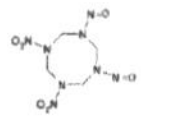	1,3-ditrinitro-5,7-dinitroso-1,3,5,7-tetrazocane
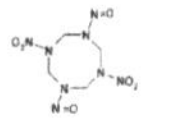	1,5-ditrinitro-3,7-dinitroso-1,3,5,7-tetrazocane
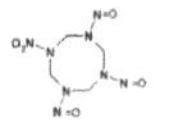	1-Nitro-3,5,7-trinitroso-1,3,5,7-tetrazocane
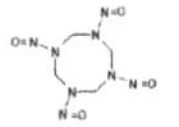	1,3,5,7-tetranitroso-1,3,5,7-tetrazocane
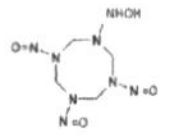	N-(3,5,7-trinitroso-1,3,5,7-tetrazocan-1-yl) hydroxylamine
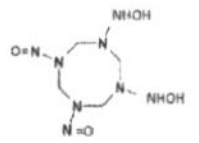	N-[3-(hydroxyamino)-5,7-dinitroso-1,3,5,7-tetrazocan-1-yl] hydroxylamine
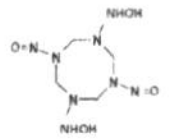	N-[5-(hydroxyamino)-3,7-dinitroso-1,3,5,7-tetrazocan-1-yl] hydroxylamine
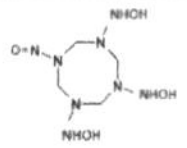	N-[(3,5-bis(hydroxyamino)-7-nitroso-1,3,5,7-tetrazocan-1-yl] hydroxylamine
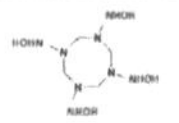	N-[3,5,7-tris(hydroxyamino)-1,3,5,7-tetrazocan-1-yl] hydroxylamine
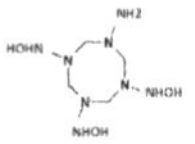	N1,N3,N5-trihydroxy-1,3,5,7-tetrazocane-1,3,5,7-tetramine
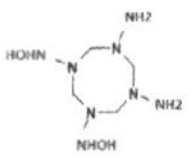	N1,N3-dihydroxy-1,3,5,7-tetrazocane-1,3,5,7-tetramine
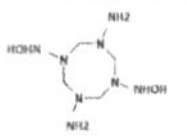	N1,N5-dihydroxy-1,3,5,7-tetrazocane-1,3,5,7-tetramine
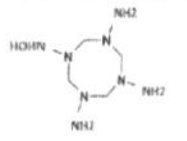	N1-hydroxy-1,3,5,7-tetrazocane-1,3,5,7-tetramine

From a technical point of view, it is important to address two points. Regarding the design phase, the choice of the algorithm depends on the protein design problem: given a desired structure, can we design an amino acid sequence capable of assuming a target structure? The goal of a protein design algorithm is to search for all the possible conformations of a sequence that could match a target fold. Then it must rank the sequences accordingly to the lowest energy conformation of each one, as determined by a protein design energy function. This ranking is highly connected to the optimization problem consisting of finding the conformation of minimum energy. A bunch of algorithms have been developed to solve the protein design problem. They are divided into two broad classes—exact algorithms, such as dead-end elimination (DEE), that lack runtime guarantees but guarantee the quality of the solution; and heuristic algorithms, such as Monte Carlo, that required fewer computational resources than exact algorithms, but have no guarantees for the optimality of the results. Regarding our starting protein, a NR able to metabolize, even incompletely, nitro compounds with experimental validation, the limited number of positions to design, and our computational resources, we chose to use a heuristic algorithm, and to enforce the resulting design results by MD simulation, allowing us to solve Newton's equations of motion and thus to gather dynamical information.

Second, the MD conclusion was based on the RMSD method. RMSD has the advantage of quickly translating into the stability of a protein. Nevertheless, it is essential to take this data with some hindsight. First of all, the main problem of the RMSD is closely related to the amplitude of error. Indeed, two identical structures could not be perfectly superimposed due to the movements of a single loop or a flexible terminus. In this case, such structures have a large global backbone RMSD and cannot be effectively overlapped by any algorithm that optimizes the global RMSD. The variations observed between our structure could be related to the high flexibility of H6, which is known to accommodate various substrates, in spite of their size. H6 is situated under FMNH2 and is connected to the rest of the protein through two loops. This unique situation allows, despite the mechanical rigidity of the helix, for a relative flexibility of the overall structure [37].

A similar approach has been performed in 2003. Loren et al. [38] computationally designed a receptor and a sensor protein with novel functions by using a DDE algorithm to construct efficient soluble receptors that binds TNT with high selectivity affinity. These designed receptors illustrate potential application of computational design and validate our approach.

Nonetheless, we only provided *in silico* results here. It would be interesting to produce the mutant and to proceed in vitro tests and calculations to get a better feedback on our work.

4. Materials and Methods

4.1. Workflow

The protocol aims to generate mutants and to computationally validate our model by performing MD simulations aiming to provide information on the free binding energy and on the stability of our newly designed structure. Each run was repeated iteratively until we get satisfying results in terms of structure, stability, and affinity (Figure 3).

4.2. Computational protein engineering

4.2.1. Protein Preparation

A 1.9 Å protein crystal structure (PDB ID: 5J8G [26]) of the bacterial NAD(P)H NR from *Enterobacter cloacae* in a ligand-bound complex with acid para nitrobenzoic (p-NBA) was obtained from the PDB [39] for use as a starting structure for computational modeling. This structure was examined and prepared for manipulation using the Structure Preparation feature in the Molecular Operating Environment software (MOE) [40]. Acid para nitro benzoic was removed and the structure was minimized using the Amber14 force field [41] to reach an energetically favorable conformation. Then all the unbound water and cofactor FMN molecules were removed from the structures.

4.2.2. Cofactor Preparation

FMNH2 structure file was built from 4PU0 [27] crystal structure and then converted into the appropriate conformer with MOE confsearch feature. FMNH2 was then superimposed to the oxidized

FMN into the 5J8G crystal structure and the coordinates were converted into .mol files. The .mol files were then parameterized by Rosetta31 (molfile_to_params.py script) to produce a parameter file and a new .pdb file.

4.2.3. Ligand Preparation

HMX structure was downloaded from Chemspider [42]. Then, HMX was minimized and placed into the active site of the protein structure 5J8G, in superposition of the acid para nitro benzoic position and in a way that the nitro group to be reduced was at a distance of 3.8 Å from the N5 atom of the flavin. HMX was then converted into a .mol file using MOE. Partial charges were corrected using MOE features. The .mol file was then parameterized by Rosetta (molfile_to_params.py script) to produce a parameters file and a new .pdb file.

4.2.4. Generation of HMX Rotamer Library

An HMX rotamer resource file was generated by random sampling using the searchconf function in MOE. This rotamer library was completed with crystal structure of HMX retrieved from the Cambridge database [43]. The library was then used for all subsequent Coupled Moves protocol simulations that used HMX as substrate ligand.

4.2.5. Resfile Generation

A 4.5 Å space around HMX was then determined. All the amino acids with a lateral chain not facing the active site were removed. Amino acid involved in the stabilization of the HMX cofactor or in the catalytic activity of the protein were also removed. The remaining amino acid were put in a text file named "resfile." This resfile gives information on which positions we want to design. Each target residue was allowed to mutate into rotamers of every amino acid ("ALLAA").

4.2.6. Design Method

The computational protocol used in this study redesigns enzyme active site. Coupled Moves algorithm primarily focuses on the optimization of side-chains for ligand binding and allows for a certain backbone flexibility [44].

The NAD(P)H NR protein structure prepared in the Structure Preparation section was split into separate .pdb files, consisting of the apo-protein, the HMX substrate and the FMNH2 cofactor. Small molecule structure files (HMX and FMNH2) were converted into .mol files. The .mol files were then parameterized by Rosetta (molfile_to_params.py script) to produce a parameters file and a new .pdb file for each structure. Then, we combined the protein, HMX, and FMNH2 structure files into a single .pdb input file for coupled moves command.

The substrate parameters files were edited to include the path to the HMX rotamer library file described in the "Generation of HMX rotamer library" section. The resfile including all the positions to be designed was created. FMNH2 and the apo-protein were allowed to use sampling rotamers of their current identity. HMX was supplemented with the rotamer library previously generated. The design.sh script was used to call all the prepared input files. It contains the variables and instructions for the Coupled Moves method simulation run.

4.2.7. Model Evaluation (Data Analysis)

Low-energy sequences generated by the Coupled Moves protocol were discarded to remove redundant sequences across multiple protocol runs, with the lowest energy rotamer conformations saved for each unique sequence. Then, all the sequences were aligned, and the results were compiled to form a logo sequence. For each designed position, the overall height of the stack indicates the sequence conservation at that position, while the height of symbols within the stack indicates the relative frequency of each amino or nucleic acid at that position. The sequence with the most conserved amino acid at each designed position were retrieved and pushed through a molecular dynamics simulation.

4.3. Molecular Dynamics Simulations

4.3.1. MD Methods

All the simulations were performed using the molecular dynamics program AMBER [41] using the Amber ff14SB force field for proteins, and the TIP3P model for all the water molecules in the system. The force field parameters for FMNH2 and HMX were provided by the General Amber Force Field (GAFF) [45]. Simulations were performed with the ligand bound to the protein with a cap comprising two layers of water molecules (TIP3PBOX) surrounding the complex within a distance up to 10 Å. The simulations systems were kept under isothermal/isobaric (NPT) conditions except for the heating phase.

Energy minimization was performed to obtain a low energy starting conformation for the subsequent MD simulation. The solvated complexes were minimized for a total of 5000 cycles, using the steepest descent method for 2500 cycles, followed by 2500 cycles of conjugate gradient. Then, a 1ns heating phase was performed from 0 to 300K at constant pressure and temperature. The equilibration/production was performed for 100 ns. Finally, the sampling phase was carried out at 300K for 30 ns. The time step of the simulations was 0.002 ps. Each MD simulation was performed in triplicate.

4.3.2. Trajectory Analysis

The VMD software [46] was used to visualize trajectories generated during the simulation. Root Mean Square Deviation (RMSD) was used to determine structure stability. RMSDs were calculated for every simulation.

4.3.3. Binding Free Energy Calculation

An ensemble average molecular mechanics energies combined with the generalized born and surface area continuum solvation (MM/GBSA) binding free-energy calculation was performed on the snapshots from the MD simulation to compare the binding affinity of HMX for the mutant and the WT NR. A total number of 200 snapshots were taken from the last 4 ns of the MD trajectory with an interval of 20 ps (only for all the trials where the ligand was successfully kept within the active state). The calculations were rendered by the MMPBSA.py [47] module of AMBER14. The MMGBSA method can be conceptually summarized as

$$\Delta G_{MM/GBSA} = G_{complex} - G_{receptor} - G_{ligand} = \Delta E_{MM} + \Delta G_{GB} + \Delta G_{NP} - T\Delta S$$

where ΔE_{MM} is the molecular mechanics interaction energy between the protein and the inhibitor, ΔG_{GB} and ΔG_{NP} are the electrostatic and nonpolar contributions to desolvation upon inhibitor binding, respectively, and $-T\Delta S$ is the conformational entropy change.

4.4. Docking Studies

Docking studies were performed with the MOE software. The last frame of the MD simulation involving the mutant NR complexed with HMX was extracted as .pdb. HMX, water molecules and ions were removed. The NADPH 3D conformer was downloaded from PubChem (PubChem CID:5884) and minimized in MOE using the AMBER14ff. A collection of poses was then generated from the pool of ligand conformations using one of the Triangle matcher placement methods. A total of 150 poses were generated. Each of the generated pose was attributed a London dG score. Poses generated by the placement methodology were then refined using the induced fit refinement scoring function

GBVI/WSA dG with the generalized born solvation model (GBVI). Twenty-five top scoring poses were sorted out. The top five scores were then evaluated.

5. Conclusions

Explosives contamination has become a major environmental issue during the past few years. The pollution with energetic material started with WWII and is still going on, due to manufacturing industries, conflicts, military operations, armed forces training activities, dumping of munitions, etc. Different strategies have been studied to remediate contaminated sites: burning, bursting, or chemical destruction. However, such methods are either costly or environmentally damaging. Bioremediation has recently emerged as an alternative way to detoxify soils from HEMs by using bacteria or plant metabolic pathways. Still, the rate of detoxification is highly variable from one HEM to another. The high energy molecules HMX is particularly problematic, as its solubility in water is lower than the one of other HEMs like TNT or RDX. Degradation of HMX by bacteria has been observed for a few strains: *Methylobacterium, K. pneumonia, C. Bifermentans*, and *Phanerchaete chrysosporium*. However, the degradation of the molecule is either incomplete, questioning the toxicity of these intermediates of degradation or occurs at meager rates.

To overcome these limitations, we rationally designed an enzyme known for its ability to reduce a broad range of nitro substituted compounds—the NR from *Enterobacter cloacae*. From structural data, we redesigned the active site specifically around HMX with the coupled moves algorithm of Rosetta. The mutated NR was then studied through MD simulations. Stability and affinity were calculated. HMX fits the designed active site in a better way than in the WT NR. The molecule makes more H-bonds, stabilizing the molecule, and exposing its nitro groups at a distance allowing for a hydride transfer from the FMNH2 cofactor. The distance remains acceptable for hydride transfer, and thus for the nitro reduction all along the 130 ns of the dynamics. Even if HMX is mobile in the active site, one of its eight oxygen atoms always remains close enough to the reductive N5 of the flavin to allow the hydride transfer. These results are encouraging, but further investigations need to be done. The basic functionality of the protein has to remain intact. Also, the total reduction reaction implies numerous HMX derivatives. Each of them has to be tested to check how far the reaction could go. Finally, it would be interesting to perform hydride quantum mechanics/molecular mechanics (QM/MM) studies to validate the reaction on an extended atomistic level.

Supplementary Materials: Supplementary materials can be found at http://www.mdpi.com/1422-0067/20/22/5556/s1.

Author Contributions: Investigation S.A.; methodology S.A., R.T.; supervision R.T.; writing original S.A.; writing_review and editing S.A., R.T.

Funding: This research received no external funding.

Acknowledgments: The authors would like to thank Dr. Simon Megy for his careful review of the manuscript.

Conflicts of Interest: The authors declare no conflict of interest.

Abbreviations

HEM	high energy molecule
WWII	world war II
NAD(P)	oxidized form of nicotinamide adenine dinucleotide phosphate
NAD(P)H	reduced form of nicotinamide adenine dinucleotide phosphate
NR	nitroreductase
FMN	oxidized form of flavin mononucleotide
FMNH2	reduced form of flavin mononucleotide
TNT	2,4,6-TriNitroToluene
RDX	hexahydro-1,3,5-trinitro-1,3,5-triazine
HMX	octahydro-1,3,5,7-tetranitro- 1,3,5,7-tetrazocine
p-NBA	acid para nitro benzoic
NAAD	nicotinic acid adenine dinucleotide
MD	molecular dynamics
MMGBSA	molecular mechanics energies combined with the generalized born and surface area continuum solvation
QM/MM	quantum-mechanics/molecular-mechanics
RMSD	root-mean-square deviation
WT	wild type
DEE	dead end elimination
ALLAA	All amino acids
PDB	Protein data bank
MOE	molecular operating environment
AMBER	Assisted Model Building with Energy Refinement

References

1. Stierstorfer, J.; Klapötke, T.M. High Energy Materials. Propellants, Explosives and Pyrotechnics. By Jai Prakash Agrawal. *Angew. Chem. Int. Ed.* **2010**, *49*, 6253. Available online: https://onlinelibrary.wiley.com/doi/abs/10.1002/anie.201003666 (accessed on 7 August 2019). [CrossRef]
2. Chatterjee, S.; Deb, U.; Datta, S.; Walther, C.; Gupta, D.K. Common explosives (TNT, RDX, HMX) and their fate in the environment: Emphasizing bioremediation. *Chemosphere* **2017**, *184*, 438–451. [CrossRef] [PubMed]
3. Rylott, E.L.; Bruce, N.C. Right on target: Using plants and microbes to remediate explosives. *Int. J. Phytoremediation* **2019**, 1–14. [CrossRef] [PubMed]
4. Lewis, T.A.; Newcombe, D.A.; Crawford, R.L. Bioremediation of soils contaminated with explosives. *J. Environ. Manag.* **2004**, *70*, 291–307. [CrossRef]
5. UN Environment. *The Fog of War*; UN Environment: Nairobi, Kenya, 2017; Available online: http://www.unenvironment.org/news-and-stories/story/fog-war (accessed on 19 August 2019).
6. Singh, S.N. (Ed.) *Biological Remediation of Explosive Residues*; Springer International Publishing: Berlin/Heidelberg, Germany, 2014.
7. Zhang, L.; Rylott, E.L.; Bruce, N.C.; Strand, S.E. Phytodetoxification of TNT by transplastomic tobacco (Nicotiana tabacum) expressing a bacterial nitroreductase. *Plant Mol. Biol.* **2017**, *95*, 99–109. [CrossRef]
8. Zhu, B.; Han, H.; Fu, X.; Li, Z.; Gao, J.; Yao, Q. Degradation of trinitrotoluene by transgenic nitroreductase in Arabidopsis plants. *Plant Soil Environ.* **2018**, *64*, 379–385. [CrossRef]
9. Zhang, L.; Rylott, E.L.; Bruce, N.C.; Strand, S.E. Genetic modification of western wheatgrass (Pascopyrum smithii) for the phytoremediation of RDX and TNT. *Planta* **2019**, *249*, 1007–1015. [CrossRef]
10. Beck, A.J.; Gledhill, M.; Schlosser, C.; Stamer, B.; Böttcher, C.; Sternheim, J.; Greinert, J.; Achterberg, E.P. Spread, Behavior, and Ecosystem Consequences of Conventional Munitions Compounds in Coastal Marine Waters. *Front. Mar. Sci.* **2018**, *5*. [CrossRef]
11. Nagar, S.; Shaw, A.K.; Anand, S.; Celin, S.M.; Rai, P.K. Aerobic biodegradation of HMX by Planomicrobium flavidum. *3 Biotech* **2018**, *8*, 455. [CrossRef]
12. Caballero, A.; Lazaro, J.J.; Ramos, J.L.; Esteve-Nunez, A. PnrA, a new nitroreductase-family enzyme in the TNT-degrading strain Pseudomonas putida JLR11. *Environ. Microbiol.* **2005**, *7*, 1211–1219. [CrossRef]

13. Kitts, C.L.; Green, C.E.; Otley, R.A.; Alvarez, M.A.; Unkefer, P.J. Type I nitroreductases in soil enterobacteria reduce TNT (2,4,6-trinitrotoluene) and RDX (hexahydro-1,3,5-trinitro-1,3,5-triazine). *Can. J. Microbiol.* **2000**, *46*, 8. [CrossRef] [PubMed]

14. Pitsawong, W.; Hoben, J.P.; Miller, A.-F. Understanding the Broad Substrate Repertoire of Nitroreductase Based on Its Kinetic Mechanism. *J. Biol. Chem.* **2014**, *289*, 15203–15214. [CrossRef] [PubMed]

15. Haynes, C.A.; Koder, R.L.; Miller, A.-F.; Rodgers, D.W. Structures of nitroreductase in three states: Effects of inhibitor binding and reduction. *J. Biol. Chem.* **2002**, *277*, 11513–11520. [CrossRef] [PubMed]

16. Fagan, R.L.; Palfey, B.A. 7.03—Flavin-Dependent Enzymes. In *Comprehensive Natural Products II*; Liu, H.-W., Mander, L., Eds.; Elsevier: Oxford, UK, 2010; pp. 37–113. [CrossRef]

17. Isayev, O.; Crespo-Hernández, C.E.; Gorb, L.; Hill, F.C.; Leszczynski, J. In silico structure–function analysis of E. cloacae nitroreductase. *Proteins Struct. Funct. Bioinform.* **2012**, *80*, 2728–2741. [CrossRef]

18. Fraaije, M.W.; Mattevi, A. Flavoenzymes: Diverse catalysts with recurrent features. *Trends Biochem. Sci.* **2000**, *25*, 126–132. [CrossRef]

19. Iuliano, J.N.; French, J.B.; Tonge, P.J. Chapter Eight—Vibrational spectroscopy of flavoproteins. In *Methods in Enzymology*; Palfey, B.A., Ed.; Academic Press: Cambridge, MA, USA, 2019; pp. 189–214. [CrossRef]

20. Koder, R.L.; Miller, A.-F. Steady-state kinetic mechanism, stereospecificity, substrate and inhibitor specificity of Enterobacter cloacae nitroreductase1This work was supported by PRF Grant ACS-PRF 28379 (A.F.M.) and a National Science Foundation Graduate Research Fellowship (R.L.K.).1. *Biochim. Biophys. Acta (BBA) Protein Struct. Mol. Enzymol.* **1998**, *1387*, 395–405. [CrossRef]

21. Race, P.R.; Lovering, A.L.; Green, R.M.; Ossor, A.; White, S.A.; Searle, P.F.; Wrighton, C.J.; Hyde, E.I. Structural and Mechanistic Studies of Escherichia coli Nitroreductase with the Antibiotic Nitrofurazone REVERSED BINDING ORIENTATIONS IN DIFFERENT REDOX STATES OF THE ENZYME. *J. Biol. Chem.* **2005**, *280*, 13256–13264. [CrossRef]

22. Chaignon, P.; Cortial, S.; Ventura, A.P.; Lopes, P.; Halgand, F.; Laprevote, O.; Ouazzani, J. Purification and identification of a Bacillus nitroreductase: Potential use in 3,5-DNBTF biosensoring system. *Enzym. Microb. Technol.* **2006**, *39*, 1499–1506. [CrossRef]

23. Yang, Y.; Lin, J.; Wei, D. Heterologous Overexpression and Biochemical Characterization of a Nitroreductase from Gluconobacter oxydans 621H. *Mol. Biotechnol.* **2016**, *58*, 428–440. [CrossRef]

24. Miller, A.-F.; Park, J.T.; Ferguson, K.L.; Pitsawong, W.; Bommarius, A.S. Informing Efforts to Develop Nitroreductase for Amine Production. *Molecules* **2018**, *23*, 211. [CrossRef]

25. Bryant, C.; DeLuca, M. Purification and characterization of an oxygen-insensitive NAD(P)H nitroreductase from Enterobacter cloacae. *J. Biol. Chem.* **1991**, *266*, 4119–4125. [PubMed]

26. Pitsawong, W.; Haynes, C.A.; Koder, R.L.; Rodgers, D.W.; Miller, A.-F. Mechanism-Informed Refinement Reveals Altered Substrate-Binding Mode for Catalytically Competent Nitroreductase. *Structure* **2017**, *25*, 978–987. [CrossRef] [PubMed]

27. Driggers, C.M.; Dayal, P.V.; Ellis, H.R.; Karplus, P.A. Crystal structure of Escherichia coli SsuE: Defining a general catalytic cycle for FMN reductases of the flavodoxin-like superfamily. *Biochemistry* **2014**, *53*, 3509–3519. [CrossRef] [PubMed]

28. Lovering, A.L.; Hyde, E.I.; Searle, P.F.; White, S.A. The structure of Escherichia coli nitroreductase complexed with nicotinic acid: Three crystal forms at 1.7 A, 1.8 A and 2.4 A resolution. *J. Mol. Biol.* **2001**, *309*, 203–213. [CrossRef]

29. Grove, J.I.; Lovering, A.L.; Guise, C.; Race, P.R.; Wrighton, C.J.; White, S.A.; Hyde, E.I.; Searle, P.F. Generation of *Escherichia Coli* Nitroreductase Mutants Conferring Improved Cell Sensitization to the Prodrug CB1954. *Cancer Res.* **2003**, *63*, 5532.

30. Johansson, E.; Parkinson, G.N.; Denny, W.A.; Neidle, S. Studies on the Nitroreductase Prodrug-Activating System. Crystal Structures of Complexes with the Inhibitor Dicoumarol and Dinitrobenzamide Prodrugs and of the Enzyme Active Form. *J. Med. Chem.* **2003**, *46*, 4009–4020. [CrossRef]

31. Ndibe, T.O.; Benjamin, B.; Eugene, W.C.; Usman, J.J. A Review on Biodegradation and Biotransformation of Explosive Chemicals. *Eur. J. Eng. Res. Sci.* **2018**, *3*, 58–65. [CrossRef]

32. Berrisford, J.M.; Sazanov, L.A. Structural basis for the mechanism of respiratory complex I. *J. Biol. Chem.* **2009**, *284*, 29773–29783. [CrossRef]

33. Pai, E.F.; Karplus, P.A.; Schulz, G.E. Crystallographic analysis of the binding of NADPH, NADPH fragments, and NADPH analogues to glutathione reductase. *Biochemistry* **1988**, *27*, 4465–4474. [CrossRef]

34. Zenno, S.; Koike, H.; Kumar, A.N.; Jayaraman, R.; Tanokura, M.; Saigo, K. Biochemical characterization of NfsA, the Escherichia coli major nitroreductase exhibiting a high amino acid sequence homology to Frp, a Vibrio harveyi flavin oxidoreductase. *J. Bacteriol.* **1996**, *178*, 4508–4514. [CrossRef]
35. McCormick, N.G.; Feeherry, F.E.; Levinson, H.S. Microbial transformation of 2,4,6-trinitrotoluene and other nitroaromatic compounds. *Appl. Environ. Microbiol.* **1976**, *31*, 949–958. [PubMed]
36. Corbett, M.D.; Corbett, B.R. Bioorganic Chemistry of the Arylhydroxylamine and Nitrosoarene Functional Groups. In *Biodegradation of Nitroaromatic Compounds*; Spain, J.C., Ed.; Springer: Boston, MA, USA, 1995; pp. 151–182. [CrossRef]
37. Kufareva, I.; Abagyan, R. Methods of protein structure comparison. *Methods Mol. Biol.* **2012**, *857*, 231–257. [CrossRef] [PubMed]
38. Looger, L.L.; Dwyer, M.A.; Smith, J.J.; Hellinga, H.W. Computational design of receptor and sensor proteins with novel functions. *Nature* **2003**, *423*, 185–190. [CrossRef] [PubMed]
39. Announcing the worldwide Protein Data Bank|Nature Structural & Molecular Biology. Available online: https://www.nature.com/articles/nsb1203-980 (accessed on 7 August 2019).
40. Chemical Computing Group. *Molecular Operating Environment (MOE)*; GTF Databanks Bulletin: Montreal, QC, Canada, 1994.
41. Case, D.A.; Ben-Shalom, I.Y.; Brozell, S.R.; Cerutti, D.S.; Cheatham, T.E., III; Cruzeiro, V.W.D.; Darden, T.A.; Duke, R.E.; Ghoreishi, D.; Gilson, M.K.; et al. *AMBER 2018*; University of California: San Francisco, CA, USA, 2018.
42. ChemSpider|Search and Share Chemistry. Available online: http://www.chemspider.com/ (accessed on 14 August 2019).
43. Cambridge Structural Database (CSD). Available online: www.ccdc.cam.ac.uk (accessed on 4 March 2019).
44. Ollikainen, N.; de Jong, R.M.; Kortemme, T. Coupling Protein Side-Chain and Backbone Flexibility Improves the Re-design of Protein-Ligand Specificity. *PLoS Comput. Biol.* **2015**, *11*, e1004335. [CrossRef]
45. Wang, J.; Wang, W.; Kollman, P.A.; Case, D.A. Automatic atom type and bond type perception in molecular mechanical calculations. *J. Mol. Graph. Model.* **2006**, *25*, 247–260. Available online: https://www.ncbi.nlm.nih.gov/pubmed/16458552 (accessed on 7 August 2019). [CrossRef] [PubMed]
46. VMD—Visual Molecular Dynamics. Available online: https://www.ks.uiuc.edu/Research/vmd/ (accessed on 14 August 2019).
47. Miller, B.R.; McGee, T.D.; Swails, J.M.; Homeyer, N.; Gohlke, H.; Roitberg, A.E. MMPBSA.py: An Efficient Program for End-State Free Energy Calculations. *J. Chem. Theory Comput.* **2012**, *8*, 3314–3321. [CrossRef]

Article

The Major Cat Allergen Fel d 1 Binds Steroid and Fatty Acid Semiochemicals: A Combined In Silico and In Vitro Study

Cécile Bienboire-Frosini [1,*], Rajesh Durairaj [1], Paolo Pelosi [2] and Patrick Pageat [3]

[1] Department of Molecular Biology and Chemical Communication (D-BMCC), Research Institute in Semiochemistry and Applied Ethology (IRSEA), Quartier Salignan, 84400 Apt, France; r.durairaj@group-irsea.com

[2] Austrian Institute of Technology GmbH, Biosensor Technologies, Konrad-Lorenzstraße, 3430 Tulln, Austria; ppelosi.obp@gmail.com

[3] Department of Chemical Ecology (D-EC), Research Institute in Semiochemistry and Applied Ethology (IRSEA), Quartier Salignan, 84400 Apt, France; p.pageat@group-irsea.com

* Correspondence: c.frosini@group-irsea.com; Tel.: +33-490-75-57-00

Received: 28 January 2020; Accepted: 13 February 2020; Published: 18 February 2020

Abstract: The major cat allergen Fel d 1 is a tetrameric glycoprotein of the secretoglobin superfamily. Structural aspects and allergenic properties of this protein have been investigated, but its physiological function remains unclear. Fel d 1 is assumed to bind lipids and steroids like the mouse androgen-binding protein, which is involved in chemical communication, either as a semiochemical carrier or a semiochemical itself. This study focused on the binding activity of a recombinant model of Fel d 1 (rFel d 1) towards semiochemical analogs, i.e., fatty acids and steroids, using both in silico calculations and fluorescence measurements. In silico analyses were first adopted to model the interactions of potential ligands, which were then tested in binding assays using the fluorescent reporter N-phenyl-1-naphthylamine. Good ligands were fatty acids, such as the lauric, oleic, linoleic, and myristic fatty acids, as well as steroids like androstenone, pregnenolone, and progesterone, that were predicted by in silico molecular models to bind into the central and surface cavities of rFel d 1, respectively. The lowest dissociation constants were shown by lauric acid (2.6 µM) and androstenone (2.4 µM). The specific affinity of rFel d 1 to semiochemicals supports a function of the protein in cat's chemical communication, and highlights a putative role of secretoglobins in protein semiochemistry.

Keywords: secretoglobin; odorant-binding protein; chemical communication; pheromone; N-phenyl-1-naphthylamine; in silico docking; molecular modeling; protein–ligand interactions; 2D interaction maps; ligand-binding assays

---

1. Introduction

The major cat allergen Fel d 1 is a secreted globular protein belonging to the family of secretoglobins. It is produced in large amounts in various anatomical areas of cats, such as the salivary, lacrimal, and sebaceous glands from the facial area, skin, and anal sacs [1–4]. The secretion of Fel d 1 is under androgen control [5]. Fel d 1 is a 35–38 kDa tetrameric glycoprotein composed of two heterodimers with a dimerization interface. Each heterodimer consists of two polypeptide chains encoded by independent genes and linked by three disulfide bridges. Chain 1 is made of 70 residues, and chain 2 of 90 or 92 residues [4,6,7]. Chain 2 contains an N-linked oligosaccharide composed of triantennary glycans [8].

Structural and immunological Fel d 1 polymorphisms have long been described in samples from various origins [8–10]. Fel d 1 is a resistant protein easily airborne that is abundantly found in different indoor environments [11,12]. Despite its high abundance and the serious health issues associated with this protein, the biological function of Fel d 1 remains unclear [4].

For other members of the secretoglobin family, different biological roles have been suggested, mainly related to immunoregulation [13–16], but also in chemical signaling [17–19]. Also for Fel d 1, a role in intra-species chemical communication has been proposed based on the fact that the protein is produced in the same areas known to release the cat semiochemicals, including the facial area, the podial complex, and the perianal zone, which contain glands that secrete chemical cues involved in cat territorial marking and/or social communication [20,21]. Besides, Fel d 1 immunological features have been linked to cat sex and behavior [22]. From a structural perspective, Fel d 1 also displays interesting features regarding ligand binding capabilities due to the presence of two internal cavities [23]. Structural similarities between Fel d 1 and another secretoglobin involved in mice mate selection and communication, the mouse salivary androgen-binding protein (ABP) [18], have been previously described [24,25]. Binding of some steroids to members of the secretoglobin family was previously reported, involving interactions with their central hydrophobic cavity [19,23,26]. In particular, a recent paper extensively describes the evolutionary divergence, functional sites, and surface structural resemblance between Fel d 1 and ABP, suggesting that the first protein could be involved in semiochemical transport/processing in intra-species communication [25]. However, so far, no experimental evidence has been provided on the capability of Fel d 1 to bind semiochemicals.

Production of recombinant Fel d 1 (rFel d 1) has been challenging in the past since the two chains are encoded by two different genes, and attempts to refold them in a correct (i.e., with retained disulfide formations) and stable way failed [4,27]. Hence, some authors proposed a rFel d 1 construct made of chain 1 linked to chain 2 via a flexible peptide linker of the (GGGGS)n type [28], which minimizes the steric hindrance between the two fusion partners, since the small size of these amino acids provides flexibility, and allows for mobility of the connecting functional domains [29]. The rFel d 1 displayed similar biological and structural properties (notably the disulfide pairing) to its natural counterparts [8,28].

In the current study, we have investigated the binding properties of this recombinant form of Fel d 1 produced in a *Pichia pastoris* clone with the N-glycosylation site N103 mutated and commercially available (INDOOR Biotechnologies) [30]. As a first step to verify the hypothesis of a role of Fel d 1 in chemical communication, we focused on putative ligands that had already been described as semiochemicals in the domestic cat [21,31], i.e., some fatty acids and their derivatives found in the composition of the feline facial pheromone F3 and the maternal cat appeasing pheromone. The feline facial pheromone F3 has been shown to promote calmness and reduce stress with its related undesirable consequences in cats, such as urine spraying and marking behavior [32–35]. The maternal cat appeasing pheromone has been shown to have appeasing effects and to facilitate social interactions in cats [36,37]. We have also tested some steroids since several secretoglobins have been experimentally shown to bind steroid hormones, including pig pheromaxein, rabbit uteroglobin, mouse salivary ABP, and rat prostatein [19,38–40]. To determine the affinity of these putative ligands and structurally characterize their interactions with rFel d 1, we used a double approach combining in silico analysis (molecular docking) with in vitro fluorescence binding assays.

2. Results

2.1. In Silico Molecular Docking of Fel d 1 with Putative Ligands

As a first approach to evaluate the binding properties of rFel d 1, we performed docking simulations of flexible ligands into the binding pocket of a rigid binding protein, represented as a grid box [41]. The collected data are reported in Table 1.

Table 1. In silico study of putative molecular residual interactions between recombinant Fel d 1 (rFel d 1) and the compounds and assessment of their capabilities to displace N-phenyl-1-naphthylamine (1-NPN).

n°	Compound Names	Estimated Free Energy of Binding (kcal/mol)	Total Intermolecular Energy (kcal/mol)	Frequency	H-Bond Residue	Hydrophobic Residue (Alkyl/Pi-Alkyl/Pi-Sigma)	In Silico Screening [a]	1-NPN Displacement Screening
				Fatty Acids and Other Derivatives				
1	Isobutyric acid	−2.96	−3.26	27%		A88, Y119, L129	No	ND
2	Capric Acid	−5.16	−7.4	23%		L61, F80, V83, F84	No	No
3	Lauric Acid	−5.84	−8.58	60%	Y119	L61, F80, V83, F84	Yes	Yes
4	Myristic Acid	−3.35	−7.02	36%	F84	F13, V133, M134, I137	Yes	Yes
5	Palmitic Acid	−2.33	−5.88	16%	F84	V133, M134	Yes	Yes
6	Oleic Acid	−2.82	−7.05	50%	M134	I64, F80, V83	Yes	Yes
7	Linoleic Acid	−2.95	−6.88	40%	P78	A88, Y119, L129	Yes	Yes
8	Dodecanal	−4.88	−7.42	2%		F84, M134	No	No
9	Dodecanol	−3.93	−7.02	2%		F84, V133, M134	No	No
10	Tetradecanol	−3.97	−7.89	6%		P78, Y81	No	No
11	Ethyl Laurate	−4.7	−8.02	12%	F84	L61, I64, F80, V83, V133, M134, I137	Yes	No
12	Methyl Palmitate	−2.53	−6.67	20%	T76		Yes	No
13	Nonanamide	−4.53	−6.51	4%		L61, I64, V83, F80	No	ND
14	Hexadecanamide	−2.84	−6.3	18%	T135	Y81	Yes	ND
15	Octadecanamide	−2.81	−6.96	6%	G131	Y81, F85	Yes	ND

Int. J. Mol. Sci. **2020**, 21, 1365

Table 1. *Cont.*

n°	Compound Names	Estimated Free Energy of Binding (kcal/mol)	Total Intermolecular Energy (kcal/mol)	Frequency	H-Bond Residue	Hydrophobic Residue (Alkyl/Pi-Alkyl/Pi-Sigma)	In Silico Screening [a]	1-NPN Displacement Screening
				Steroids				
1	Androstenone	−5.84	−5.84	65%	S138	P78, Y81, F85	Yes	Yes
2	Androstenedione	−5.83	−5.83	44%		Y81, F85	Yes	Yes
3	Androstenol	−5.06	−5.36	22%		Y81, F85	No	No
4	Progesterone	−5.74	−6.04	62%	T76	Y81, F85	Yes	Yes
5	Hydroxyprogesterone	−5.14	−5.54	39%	Y81	F85	Yes	Yes
6	Pregnenolone	−5.59	−6.17	58%	T76	Y81, F85	Yes	Yes
7	Estradiol	−4.94	−5.54	26%	T76	Y81, F85	Yes	Yes
8	Testosterone	−5.6	−5.9	35%	T76	Y81, F85	Yes	Yes
9	Dihydrotestosterone	−5.06	−5.35	12%		Y81, F85	No	No
10	Estrone	−3.56	−3.86	10%	D82, G131	F85	Yes	Yes
11	Dehydroepiandrosterone (DHEA)	−4.64	−4.94	30%		Y81, F85	Yes	Yes
12	Corticosterone	−5.35	−6.38	30%	T76, N89	Y81, F85	Yes	No
13	Deoxycorticosterone	−4.99	−5.29	12%		Y81, F85	No	No
				Fluorescent Probe				
1	1-NPN (Central)	−6.7	−7.41	50%	Y119	L14, L61, M112	Yes	/
2	1-NPN (Surface)	−4.74	−5.45	30%		Y81, F85	Yes	/

ND: Not determined because of the fluorescence increase, probably due to non-specific hydrophobic interactions [42]. [a] The in silico screenings were considered to result in positive outcomes ("yes") if the following were predicted: (1) minimum one H-bond interaction irrespective of the binding frequency or (2) ≥30% of binding frequency without H-bond. This threshold value of binding frequency (≥30%) was selected from the minimum binding frequency of the fluorescent probe (1-NPN) with rFel d 1.

Among the 15 fatty acids tested with rFel d 1, lauric, myristic, oleic, and linoleic acids were the best ligands based on their H-bond interactions, docking energy values, and binding frequency. In particular, lauric acid showed the highest frequency of binding with a free energy of −5.84 kcal/mol. The same compound also exhibited the lowest total intermolecular energy of −8.58 kcal/mol. Myristic, linoleic, and oleic acids were moderate ligands with free binding energies of −3.35, −2.95, and −2.82 kcal/mol, respectively. Furthermore, we observed non-bonded interactions (van der Waals and electrostatic), and pi-interactions with all the fatty acids tested.

The second series of chemicals tested includes several steroids. Among these, androstenone showed the maximum frequency of binding as well as the best free binding energy (−5.84 kcal/mol) with one H-bond interaction (S138) in rFel d 1. The behavior of androstenedione was very similar, with a binding energy of −5.83 kcal/mol, but this ligand exhibited a lower frequency of binding without H-bond interaction. On the other hand, progesterone and pregnenolone showed approximately 60% of the binding frequency, with binding energies comparable to those of androstenone and androstenedione. Pregnenolone and progesterone exhibited similar H-bond interactions (Thr76) but different from those of androstenone (S138). Furthermore, Tyr81 and Phe85 were often present as alkyl/pi-alkyl interactions in the steroid compounds.

Finally, our docking simulation predicted high binding activity of the fluorescent probe N-phenyl-1-naphthylamine (1-NPN) in the same range as those for fatty acids and steroids. Specifically, this compound has two potential binding localizations, i.e., in the central and in the surface binding cavities of rFel d 1. Conversely, some fatty acids and structurally related compounds (long-chain alcohols, aldehydes, ester, and amides), as well as few steroids, did not qualify as good ligands in docking simulations and fluorescent probe displacement (Table 1).

Overall, in silico screening indicated as the best potential ligands for the protein some fatty acids and steroids, which were further tested in fluorescence competitive binding assays.

2.2. Fluorescence Binding Studies

The rFel d 1 binds the fluorescent probe 1-NPN, producing a blueshift in the emission spectrum. Similarly to odorant-binding proteins (OBPs) and chemosensory proteins (CSPs) [42], the emission maximum occurs at 407 nm and is accompanied by a strong increase in fluorescence intensity. Figure 1 reports the actual emission spectra obtained with a rFel d 1 concentration of 1 μM and the relative binding curve obtained after processing the data with the GraphPad Software, Inc., giving a dissociation constant of 5.8 μM. Scatchard analysis confirmed the presence of a single binding site on the protein without any cooperativity effect and yielded a dissociation constant $K_{1\text{-NPN}}$ value of 4.8 μM. We also tested other fluorescent probes (2-NPN, 1-AMA (1-aminoanthracene), 1,8-ANS (8-anilinonaphtalene sulfonic acid), but none proved to perform better than 1-NPN (data not shown).

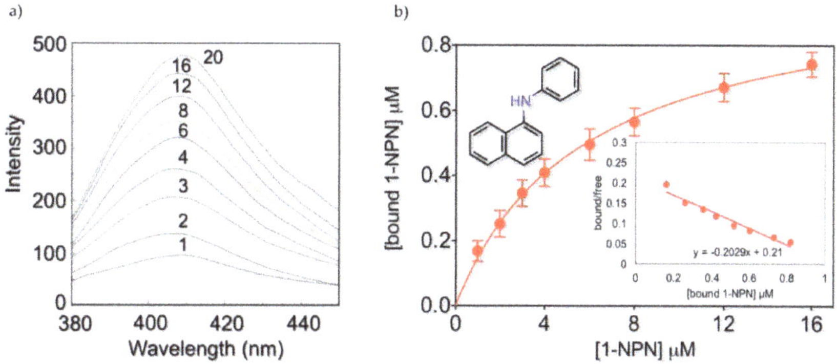

Figure 1. 1-NPN binding to rFel d 1. To a 1-μM solution of the protein in 50 mM Tris-HCl, pH 7.4, aliquots of 1 mM solution of 1-NPN in methanol were added to final concentrations of 1–20 μM. (**a**) The

representative emission curves experimentally obtained. No significant fluorescence emission was recorded in the same conditions with the protein alone (not shown). (**b**) The saturation binding curve obtained from the average of three experiments. Data were analyzed with GraphPad software and gave a value of 5.8 µM for the binding constant (SD 0.62). The relative Scatchard plot (inset) shows a linear behavior, apparently indicating the presence of a single binding site without cooperativity effects.

Among the 28 putative ligands, 5 fatty acids and 9 steroids were predicted to possibly interact with rFel d 1 based on the initial 1-NPN displacement screening (Table 1). These compounds were therefore tested in competitive binding experiments with 1-NPN and their displacement curves are reported in Figure 2. Table 2 lists the IC50 values for the best ligands, together with their dissociation constants. These were calculated using the value for 1-NPN (K_D 5.8 µM; SD 0.62), obtained with GraphPad software, more reliable than that evaluated from the Scatchard plot. Among the fatty acids, lauric acid exhibited the best affinity to rFel d 1 (Kd = 2.6 µM), while oleic, linoleic, and myristic acids displayed only moderate to low affinities, and palmitic acid proved to be the weakest ligand. Among the steroids, the strongest ligand was androstenone (Kd = 2.4 µM), followed by progesterone and pregnenolone. These results are in agreement with the in silico docking predictions.

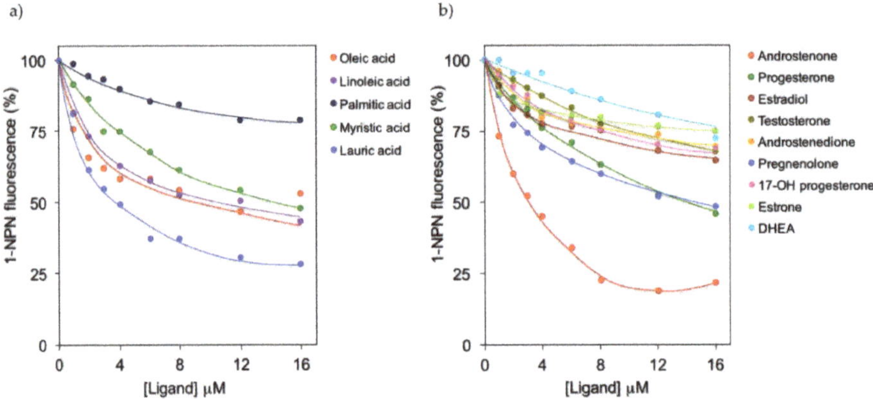

Figure 2. Competitive binding of selected fatty acids (**a**) and steroids (**b**) to rFel d 1. Fluorescence emission spectra were recorded at 25 °C in the presence of 1 µM of rFel d 1 and 2 µM of 1-NPN; excitation and emission wavelengths were 337 and 407 nm, respectively. Fluorescence of probe-protein complexes in the absence of a competitor was normalized to 100%.

Table 2. Affinities of different ligands to rFel d 1, evaluated in competitive binding assays.

Ligand	(IC$_{50}$) (µM)	K_d (µM)
Lauric acid	3.3	2.6
Oleic acid	10.0	7.7
Linoleic acid	10.1	7.8
Myristic acid	14.4	11.1
Androstenone	3.1	2.4
Pregnenolone	13.1	10.1
Progesterone	13.6	10.5

2.3. Visualization of Molecular Interactions

To visualize the possible binding modes of the best ligands to rFel d 1, molecular models and 2D molecular interaction maps were built and are shown in Figure 3. Lauric acid is predicted to bind in the central hydrophobic cavity of Fel d 1, where the strongest H-bond interaction occurs between the phenolic hydrogen of Tyr119 and the oxygen of lauric acid (Figure 3a). Androstenone, instead, is predicted to bind on the surface binding cavity of Fel d 1 and shows an H-bond between the Ser138 OH and the carbonyl group of the ligand (Figure 3b).

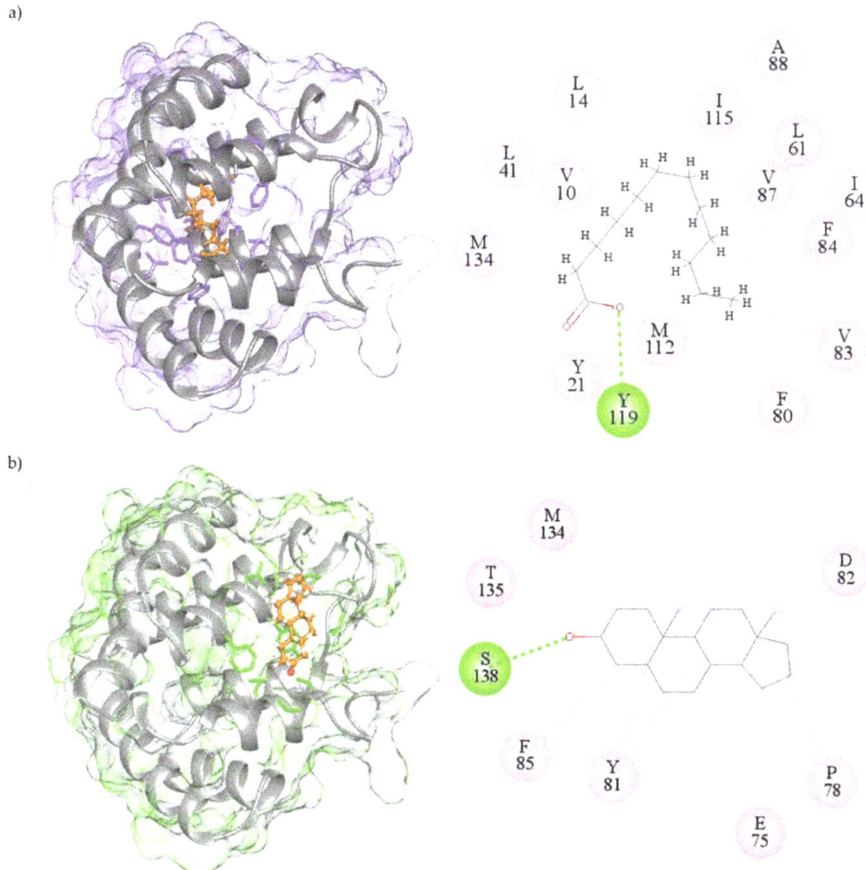

Figure 3. Molecular residue interactions of Fel d 1 with the best ligands, lauric acid (**a**) and androstenone (**b**). The interactions are shown in molecular ligand binding view (surface mesh) with a 2D-interaction map of the selective best-fitting ligands to the central and surface binding cavities of Fel d 1. The 2D map reports H-bond interactions in green color and hydrophobic interactions (van der Waals and alkyl/pi-alkyl) in pink color. All the amino acid residue interactions within 4 Å from the ligand are shown.

3. Discussion

On the basis of ligand-binding experiments, using the displacement of a fluorescent probe, and in silico docking simulations, we have shown that a recombinant form of Fel d 1 binds with good affinities some fatty acids and steroids, the best ligands being lauric acid and androstenone (Kd = 2.6 and 2.4 µM, respectively). Lauric acid is a component of the mixture of fatty acids described as the cat appeasing pheromone having effects on cats' social interactions [36,37], together with oleic, linoleic, and myristic acids, which also showed some affinity to rFel d 1. Androstenone is a volatile steroid pheromone found in high concentrations in the saliva of male pigs and triggers attraction/standing responses in estrous females [43]. Interestingly, some authors have also characterized the binding of isoforms from both native and recombinant pig OBP to fatty acids with appeasing effects and some steroids, indicating the biological relevance of these ligands in chemical communication [44,45]. Although the data here presented were obtained with a recombinant form of Fel d 1, they still support a role of this protein in the cat's chemical communication, probably as a semiochemicals carrier, similar in its function to OBPs [46].

From a structural perspective, molecular docking suggests that, among good ligands, fatty acids, except for linoleic acid, bind in the internal/central cavity of rFel d 1, while steroids bind in the cavity at the surface of the protein. 1-NPN, however, is predicted to fit into both cavities. This last observation could explain how both fatty acids and steroids can displace the fluorescent probe. The same fact might also account for the observation that lauric acid and androstenone, the two best ligands, fail to completely quench 1-NPN fluorescence, showing asymptotic behavior at concentrations much higher than zero. The same phenomenon might occur with other ligands but would not be clearly visible due to their much lower affinities. The presence of two binding sites for 1-NPN might contrast with the linear Scatchard plot. However, if the two sites present similar affinities for 1-NPN and there is no cooperativity effect, the Scatchard analysis would still produce a linear behavior. Incidentally, it is worth noting that, to the best of our knowledge, this is the first report of using a hydrophobic fluorescent probe (1-NPN) to monitor the binding activity of a secretoglobin family member. This probe, therefore, represents a useful tool for monitoring ligand binding properties with other proteins of the family and investigating their putative involvement in chemical signaling [47].

Looking more closely at the residual interactions, the in silico predictions revealed that fatty acids would mainly interact with the hydrophobic residues Val10, Phe13, Leu14, Tyr21, Phe80, Phe84, Val87, Met112, Tyr119, Asp130, and Met134. In the same way, the amino acids Glu75, Thr76, Pro78, Tyr81, Asp82, Phe85, Gly131, Thr135, and Ser138 (all corresponding to only chain 2 residues of the natural Fel d 1 [48]) displayed predicted hydrophobic interactions with the steroids. The present results are in agreement with the few steroid interactions previously described in Fel d 1 [23]. In particular, Tyr21 was previously reported to be highly conserved in several secretoglobins [25] and possibly involved in ligand binding [49,50]. Phe6 was also predicted to interact with ligands [50]. These previous reports suggested that both these amino acids could be important for a function of the protein in chemical communication. Likewise, in the present study, we predicted that Tyr21, Phe84, and Tyr119 could interact with fatty acids, while Tyr81 and Phe85 could interact with steroids.

A limitation of the in silico study is that we used the docking protocol, which is a static or quasi-static method, to obtain the structure of the various Fel d 1-ligand complexes. Using a scoring function that is meant to reproduce the binding affinity in terms of free binding energy, these structures are ranked to reveal the best-fit ligands in a way comparable to the rank based on experimental data [51]. Although the molecular docking free energy differences estimations are fast, simple, and useful for the screening of ligands, they are not the most precise ones (compared to the free binding energies determined by molecular dynamics simulations for instance) due to the absence of mobility or the absence of an explicit solvation of the system [51]. Nevertheless, here, we also considered other computational factors like binding frequencies and residue interactions before concluding about the results of the in silico screening displayed in Table 1. Moreover, these results were further confirmed by in vitro experiments.

A limitation of the in vitro study is that we used a recombinant model of the native Fel d 1, in which a peptide segment was introduced as a linker between the two subunits in the place of disulfide bridges. However, the recombinant and native Fel d 1 secondary structures were found to be similar based on circular dichroism [52]. Most importantly, the disulfide pairing of recombinant Fel d 1 corresponds with that of the native Fel d 1 [8,52]. Therefore, the peptide link in rFel d 1 seems not to introduce major differences in the overall folding of the protein. Whereas the overall structures of native Fel d 1 and of its recombinant are reasonably similar, differences in the flexibility and residual conformations can still exist. Even minor changes may affect the binding activity of a protein: for instance, several authors have shown that post-translational modifications, such as phosphorylation and O-glycosylation, influence the binding profiles of pig OBP isoforms, and phosphorylation can even enhance the binding affinities for some compounds in native OBPs compared to their recombinant counterparts [45,53,54]. It was also hypothesized that the glycosylation pattern of Fel d 1 might affect its structural features, notably by reducing its cavity size, thus possibly altering/modulating its ligand-binding properties [23]. Therefore, we cannot exclude that differences between natural and recombinant forms of Fel d 1 may affect the binding properties of the protein. Confirming our results with the native Fel d 1 would be necessary to definitely assess its putative function as semiochemical carrier.

The proteins that participate in chemical communication have complex roles, such as solubilizing, transporting, serving as reservoirs, assisting in the controlled release of semiochemicals, or even acting themselves as chemical messages (e.g., MUPs) [55,56]. The binding and controlled release of volatile chemical cues via proteins are of particular interest for Felidae, which are mostly solitary carnivores and use scent marks to delimit their territories of variable sizes according to ecological resources [57]. Domestic cats vary greatly in spatial organization, from being solitary in well-dispersed populations at densities of a single individual per square km or lower to living in highly populated groups [58]. Whatever the cats' social organization is, chemical communication mediated by scent marks is essential to assess social and territorial relationships [59]. The chemical composition of the marks can also provide physiological information in some cases, such as sex or sexual status [60]. Interestingly, other Felidae species also secrete proteins similar to Fel d 1 [61], which might as well have the function of extending the persistence of chemical cues in their environment. Because territory marking involves high energy costs [62], it is important to keep the chemical message as long as possible in general and specifically for Felidae [63].

In mammals, OBPs, sometimes referred to as pheromone-binding proteins (PBPs), are the main proteins that have been reported to mediate chemical communication. These proteins belong to the large family of lipocalins and bind semiochemicals and odorants representing various chemical classes [46,64]. The cat lipocalin Fel d 4 was shown to be involved in chemical communication as a kairomone by eliciting defensive behavior in mice [65]. The structure of secretoglobins (α-helix bundles assembled in a boomerang configuration, creating a central hydrophobic pocket), to which Fel d 1 belongs, is completely different from that of lipocalins (barrel of β-strands with a central apolar cavity) [64,66]. However, the binding data collected with a structural model of Fel d 1 suggest that a function of semiochemical carrier could be considered also for secretoglobins. More experimental evidence is needed, such as studying the expression of Fel d 1 in cat chemosensory organs, confirming its binding activity with the native protein, and perhaps identifying its natural ligands. We hope that our work can stimulate more research in the field of secretoglobins and confirm their putative role in mammalian chemical communication.

Unveiling the ligand-binding properties of Fel d 1 towards semiochemical compounds supports a function of this protein as a semiochemical carrier. As Fel d 1 is one of the most important aeroallergens [4], it is possible that lipid binding might also affect the allergenicity of this protein. Indeed, some authors have shown that another version of recombinant Fel d 1 was able to bind lipopolysaccharides (LPS), enhance lipid cellular signaling through Toll-like receptors, and potentiate the production of the pro-inflammatory cytokine TNF-α (Tumor Necrosis Factor-α), which could eventually influence the allergic sensitization process [67]. In this respect, ligand binding characteristics of Fel d 1 might help to understand the allergenic effects of the protein itself compared to that of its complexes with ligands [68]. Besides, the binding of a ligand to Fel d 1 might affect the allergen recognition by Immunoglobulin E (IgE) if the epitopes are altered through B-cell epitope conformational changes induced by the ligand or if the amino acid residues involved in IgE binding are obscured in the ligand-protein complex. Then, elucidating the ligand binding properties of Fel d 1 might provide valuable insights into this putative phenomenon of ligand-induced epitope masking. Along the same line, several approaches aiming at decreasing or controlling the cat production of immunologically active Fel d 1 have recently been investigated in order to alleviate the symptoms suffered by allergic cat owners [69]. In particular, the use of a diet supplemented with anti-Fel d 1 avian IgY [70] or the immunization of cats with a modified form of recombinant Fel d 1 to stimulate the production of neutralizing antibodies [71] have been proposed. However, as the results of this study suggest that Fel d 1 could play an important role in the cat's chemical communication, our opinion is that any attempt to alter the production of Fel d 1 should consider possible consequences that might affect the cat's biology.

4. Materials and Methods

4.1. System Configuration

All the computational analyses were carried out in a high-performance GPU workstation with Cent OS V.7.6 Linux and the Windows OS. The hardware specifications of the workstation (Model: LVX-1 × RTX-2080Ti) include a powerful Intel Core i9-9920X processor with 1GPU Nvidia RTX-2080Ti, 32GB RAM, running with a superfast boot-home 1 × M2-1TB NVME SSD and 2 × 8TB independent hard disks. The workstation has passed all the validation tests by the Linuxvixion GPU certified system.

4.2. Collection and Structure Conversion of Ligands

Molecular structures of the 28 putative ligands (15 fatty acids and their derivatives (FA) and 13 steroids) and N-phenyl-1-naphthylamine (1-NPN) (fluorescent probe) were collected from PubChem (https://pubchem.ncbi.nlm.nih.gov/). All the 2D structures of the ligands were converted into the corresponding three-dimensional (3D) coordinates (sdf to mol2 format) using OpenBabelGUI tools V.2.3.1 (http://openbabel.org). The selected compounds were used to obtain a drug-likeness score from the Lipinski rule of five (RO5) webserver (http://www.scfbio-iitd.res.in/software/drugdesign/lipinski.jsp) [72].

4.3. Physio-Chemical Properties Analysis

The physico-chemical properties of all putative ligands and 1-NPN were collected from various chemical databases such as PubChem and ChemSpider. The compound properties were classified as the chemical formula, molecular weight, H-bond donor, acceptor, topological polar surface area, and RO5 (Table 3).

4.4. Molecular Docking Analysis

4.4.1. Ligand Optimization

The retrieved molecular structures of the putative ligands and 1-NPN (.mol) were energy minimized using the geometry optimization method (MMFF94 force field) with pH 7.0. The Gasteiger partial charge was added to the ligand atoms and the MMFF94 energies were found to differ between all the compounds. All the nonpolar atoms were merged, and rotatable bonds were defined.

4.4.2. Protein Grid Parameters

The 3D crystal structure of rFel d 1 (PDB ID: 2EJN) was retrieved from the Protein Data Bank (PDB) (https://www.rcsb.org/). The protein dimer and ligand dataset were uploaded to the DockingServer (https://www.dockingserver.com; Virtua Drug, Hungary), a web-based interface module consisting of Gasteiger and PM6 semiempirical quantum-mechanical partial charge calculations to enhance the accuracy of docking output utilizing the AutoDock 4 method [73]. The essential hydrogen atoms, Kollman united atom-type charges, and solvation parameters were added to the 3D structure of rFel d 1. The Gasteiger charge calculation method was selected for the protein clean step. The 3D dimensional grid box was constructed for permitting ligands to interact in the binding sites of Fel d 1. The affinity grid parameters (nx = 23; ny = 23; nz = 23 and cx = −0.48; cy = 0.81; cz = 0.22) and 0.375 Å spacing were generated using the Autogrid program [74]. The total Gasteiger charge of rFel d 1 was −6.959 kcal/mol. After completion of this step, the rFel d 1 structure was prepared for the docking simulation analysis.

Table 3. Molecular structural properties of all putative semiochemical compounds and the fluorescent probe N-phenyl-1-naphthylamine (1-NPN).

n°	Compounds	PubChem Compound ID (CID)	Chemical Formula	Molecular Weight (g/mol)	H-Bond Donor	H-Bond Acceptor	Topological Polar Surface Area (Å^2)	Lipinski Rule of Five (RO5)
			Fatty Acids and Their Derivatives					
1	Isobutyric acid	CID_6590	$C_4H_8O_2$	88.106	1	2	37.3	0
2	Capric acid	CID_2969	$C_{10}H_{20}O_2$	172.268	1	2	37.3	0
3	Lauric acid	CID_3893	$C_{12}H_{24}O_2$	200.322	1	2	37.3	0
4	Myristic acid	CID_11005	$C_{14}H_{28}O_2$	228.376	1	2	37.3	0
5	Palmitic acid	CID_985	$C_{16}H_{32}O_2$	256.43	1	2	37.3	1
6	Oleic acid	CID_445639	$C_{18}H_{34}O_2$	282.468	1	2	37.3	1
7	Linoleic acid	CID_5280450	$C_{18}H_{32}O_2$	280.442	1	2	37.3	1
8	Dodecanal	CID_8194	$C_{12}H_{24}O$	184.323	0	1	17.1	0
9	Dodecanol	CID_8193	$C_{12}H_{26}O$	186.339	1	1	20.2	0
10	Tetradecanol	CID_8209	$C_{14}H_{30}O$	214.393	1	1	20.2	0
11	Ethyl Laurate	CID_7800	$C_{14}H_{28}O_2$	228.376	0	2	26.3	0
12	Methyl palmitate	CID_8181	$C_{17}H_{34}O_2$	270.457	0	2	26.3	1
13	Nonanamide	CID_70709	$C_9H_{19}NO$	157.257	1	1	43.1	0
14	Hexadecanamide	CID_69421	$C_{16}H_{33}NO$	255.446	1	1	43.1	0
15	Octadecanamide	CID_31292	$C_{18}H_{37}NO$	283.5	1	1	43.1	1
			Steroids					
1	Androstenone	CID_6852393	$C_{19}H_{28}O$	272.432	0	1	17.1	1
2	Androstenedione	CID_6128	$C_{19}H_{26}O_2$	286.415	0	2	34.1	0
3	Androstenol	CID_101989	$C_{19}H_{30}O$	274.448	1	1	20.2	1
4	Progesterone	CID_5994	$C_{21}H_{30}O_2$	314.469	0	2	34.1	0
5	Hydroxyprogesterone	CID_6238	$C_{21}H_{30}O_3$	330.468	1	3	54.4	0
6	Pregnenolone	CID_8955	$C_{21}H_{32}O_2$	316.485	1	2	37.3	0
7	Estradiol	CID_5757	$C_{18}H_{24}O_2$	272.388	2	2	40.5	0
8	Testosterone	CID_6013	$C_{19}H_{28}O_2$	288.431	1	2	37.3	0
9	Dihydrotestosterone	CID_10635	$C_{19}H_{30}O_2$	290.447	1	2	37.3	0
10	Estrone	CID_5870	$C_{18}H_{22}O_2$	270.372	1	2	37.3	0
11	Dehydroepiandrosterone (DHEA)	CID_5881	$C_{19}H_{28}O_2$	288.431	1	2	37.3	0
12	Corticosterone	CID_5753	$C_{21}H_{30}O_4$	346.467	2	4	74.6	0
13	Deoxycorticosterone	CID_6166	$C_{21}H_{30}O_3$	330.468	1	3	54.4	0
			Fluorescent Probe					
1	N-phenyl-1-naphthylamine (1-NPN)	CID_7013	$C_{16}H_{13}N$	219.287	1	1	12	0

4.4.3. Semi-Empirical Calculations

The docking simulation was performed using the Lamarckian genetic algorithm (LGA) and the Solis and Wets local search method to determine the optimum complex [75] in the AutoDock method. The AutoDock parameter set- and distance-dependent dielectric functions were used in the calculation of the van der Waals and the electrostatic terms, respectively. The initial position, orientation, and torsion of the ligand molecules were set randomly, and all rotatable torsions were released during docking. Each docking calculation was derived from 100 runs, which were set to terminate after a maximum of 2,500,000 energy calculations (540,000 for a generation with a population size of 150). A translational step of 0.2 Å, quaternion, and torsion steps of five were employed as parameters for the docking analyses. The AutoDock algorithms calculate the free binding energy to assess the orientation of a ligand binding pose to a protein while forming a stable complex. The protein–ligand complex was analyzed, and the molecular interaction poses of each compound were selected for the ranking of the best-fit ligands according to the docking score with several docking parameters. The estimation of the binding free energy was selected from the best- docked conformation of the protein–ligand complex using docking simulation.

4.4.4. Molecular Visualization

The protein–ligand interactions were visualized using Discovery studio visualizer DSV 4.5 (Accelrys, San Diego, CA, USA), USCF Chimera (https://www.cgl.ucsf.edu/chimera/) and the LigPlot V.4.5.3 program. The evaluation of semi-empirical docking values was computed regarding the score of lowest binding energy, hydrogen bonding (H-bonding), and polar and steric interactions.

4.5. Fluorescence Measurement and Binding Assays

N-Phenyl-1-naphthylamine (1-NPN) was used as a non-polar fluorescent probe in competitive binding experiments with the ligands (Sigma, France) to investigate binding efficiency of semiochemical analogs with pure rFel d 1 (INDOOR Biotechnologies, UK) [30]. The fluorescence experiments were performed on an FP-750 spectrofluorometer (JASCO, Japan) instrument at 25 °C in a right-angle configuration with a 1 cm light path fluorimeter quartz cuvette and 5-nm slits for both excitation and emission. The probe 1-NPN was excited at 337 nm and emission spectra were recorded between 380 and 450 nm, at 25 °C. The protein was dissolved in 50 mM Tris-HCl, pH 7.4, and ligands were added as 1 mM methanol solutions.

The rFel d 1 intrinsic fluorescence was expected to be negligible since no tryptophan is present in the sequences of both Fel d 1 chains [48], yet it was verified. The binding of 1-NPN to rFel d 1 was tested at two protein concentrations (1 μM and 2 μM) by titrating the protein solution with aliquots of a 1-mM solution of 1-NPN in methanol to final concentrations of 1–20 μM. The bound ligand was evaluated from the values of fluorescence intensity assuming that the protein was 100% active, with a stoichiometry of 1:1 protein: ligand. Dose–responses curves were performed in triplicate and linearized using Scatchard plots to calculate the 1-NPN dissociation constant (Kd $_{\text{1-NPN}}$).

Semiochemicals were first screened for their capabilities to bind rFel d 1 using 1 μM of rFel d 1, 1 μM of 1-NPN, and 2 μM of a competitive ligand. Active compounds were then used to measure their affinity to the protein, using a concentration range of 0–16 μM. The dissociation constants of the competitor ligands (Kd) were calculated from the respective IC50 values (IC50: competitor's concentration halving the initial fluorescence), using the equation:

$$\text{Kd} = [\text{IC50}]/(1 + [1 - \text{NPN}]/K_{1 - \text{NPN}})$$

where [1-NPN] is the free concentration of 1-NPN and $K_{\text{1-NPN}}$ is the dissociation constant of the complex rFel d 1/1-NPN. IC50 was graphically determined from the dose–response curve of each competitor ligand.

Author Contributions: C.B.-F. and P.P. (Patrick Pageat) conceived the study; C.B.-F. and P.P. (Paolo Pelosi) conceived, designed, and performed the in vitro experiments and analyzed the data; R.D. conceived, designed, and performed the in silico experiments and analyzed the data; P.P. (Patrick Pageat) and P.P. (Paolo Pelosi) contributed reagents/materials/analysis tools; C.B.-F. and P.P. (Patrick Pageat) supervised and administered the project; C.B.-F. and R.D. wrote the original draft preparation; All authors have read and agreed to the published version of the manuscript.

Funding: This research received no external funding.

Acknowledgments: The authors thank the American Journal Expert for the English editing of this manuscript.

Conflicts of Interest: Patrick Pageat is the inventor of the patent "Properties of cat's facial pheromone" n° WO1996023414A1 and the patent "Cat Appeasing Pheromone" n° WO2015140631A1.

References

1. Van Milligen, F.J.; Vroom, T.M.; Aalberse, R.C. Presence of Felis domesticus Allergen I in the Cat's Salivary and Lacrimal Glands. *Int. Arch. Allergy Appl. Immunol.* **1990**, *92*, 375–378. [CrossRef] [PubMed]
2. Charpin, C.; Mata, P.; Charpin, D.; Lavaut, M.N.; Allasia, C.; Vervloet, D. Fel d I allergen distribution in cat fur and skin. *J. Allergy Clin. Immunol.* **1991**, *88*, 77–82. [CrossRef]
3. De Andrade, A.D.; Birnbaum, J.; Magalon, C.; Magnol, J.P.; Lanteaume, A.; Charpin, D.; Vervloet, D. Fel d I levels in cat anal glands. *Clin. Exp. Allergy* **1996**, *26*, 178–180. [CrossRef]
4. Bonnet, B.; Messaoudi, K.; Jacomet, F.; Michaud, E.; Fauquert, J.L.; Caillaud, D.; Evrard, B. An update on molecular cat allergens: Fel d 1 and what else? Chapter 1: Fel d 1, the major cat allergen. *Allergy Asthma Clin. Immunol.* **2018**, *14*, 1–9. [CrossRef] [PubMed]
5. Zielonka, T.M.; Charpin, D.; Berbis, P.; Lucciani, P.; Casanova, D.; Vervloet, D. Effects of castration and testosterone on Fel dI production by sebaceous glands of male cats: I–Immunological assessment. *Clin. Exp. Allergy* **1994**, *24*, 1169–1173. [CrossRef] [PubMed]
6. Kaiser, L.; Grönlund, H.; Sandalova, T.; Ljunggren, H.G.; van Hage-Hamsten, M.; Achour, A.; Schneider, G. The crystal structure of the major cat allergen Fel d 1, a member of the secretoglobin family. *J. Biol. Chem.* **2003**, *278*, 37730–37735. [CrossRef] [PubMed]
7. Kaiser, L.; Velickovic, T.C.; Badia-Martinez, D.; Adedoyin, J.; Thunberg, S.; Hallen, D.; Berndt, K.; Grönlund, H.; Gafvelin, G.; Van Hage, M.; et al. Structural characterization of the tetrameric form of the major cat allergen Fel d 1. *J. Mol. Biol.* **2007**, *370*, 714–727. [CrossRef]
8. Kristensen, A.K.; Schou, C.; Roepstorff, P. Determination of isoforms, N-linked glycan structure and disulfide bond linkages of the major cat allergen Fel d1 by a mass spectrometric approach. *Biol. Chem.* **1997**, *378*, 899–908. [CrossRef]
9. Bienboire-Frosini, C.; Lebrun, R.; Vervloet, D.; Pageat, P.; Ronin, C. Distribution of core fragments from the major cat allergen Fel d 1 is maintained among the main anatomical sites of production. *Int. Arch. Allergy Immunol.* **2010**, *152*, 197–206. [CrossRef]
10. Bienboire-Frosini, C.; Lebrun, R.; Vervloet, D.; Pageat, P.; Ronin, C. Variable content of Fel d 1 variants in house dust and cat extracts may have an impact on allergen measurement. *J. Investig. Allergol. Clin. Immunol.* **2012**, *22*, 270–279.
11. Liccardi, G.; D'Amato, G.; Russo, M.; Canonica, G.W.; D'Amato, L.; De Martino, M.; Passalacqua, G. Focus on cat allergen (Fel d 1): Immunological and aerodynamic characteristics, modality of airway sensitization and avoidance strategies. *Int. Arch. Allergy Immunol.* **2003**, *132*, 1–12. [CrossRef] [PubMed]
12. Zahradnik, E.; Raulf, M. Animal allergens and their presence in the environment. *Front. Immunol.* **2014**, *5*, 76. [CrossRef] [PubMed]
13. Mukherjee, A.B.; Zhang, Z.; Chilton, B.S. Uteroglobin: A steroid-inducible immunomodulatory protein that founded the Secretoglobin superfamily. *Endocr. Rev.* **2007**, *28*, 707–725. [CrossRef] [PubMed]
14. Yokoyama, S.; Cai, Y.; Murata, M.; Tomita, T.; Yoneda, M.; Xu, L.; Pilon, A.L.; Cachau, R.E.; Kimura, S. A novel pathway of LPS uptake through syndecan-1 leading to pyroptotic cell death. *Elife* **2018**, *7*, 1–25. [CrossRef] [PubMed]
15. Chiba, Y.; Kurotani, R.; Kusakabe, T.; Miura, T.; Link, B.W.; Misawa, M.; Kimura, S. Uteroglobin-related protein 1 expression suppresses allergic airway inflammation in mice. *Am. J. Respir. Crit. Care Med.* **2006**, *173*, 958–964. [CrossRef]

16. Maccioni, M.; Riera, C.M.; Rivero, V.E. Identification of rat prostatic steroid binding protein (PSBP) as an immunosuppressive factor. *J. Reprod. Immunol.* **2001**, *50*, 133–149. [CrossRef]

17. Karn, R.C. Evolution of Rodent Pheromones: A Review of the ABPs with Comparison to the ESPs and the MUPs. *Int. J. Biochem. Res. Rev.* **2013**, *3*, 328–363. [CrossRef]

18. Chung, A.G.; Belone, P.M.; Bímová, B.V.; Karn, R.C.; Laukaitis, C.M. Studies of an Androgen-Binding Protein Knockout Corroborate a Role for Salivary ABP in Mouse Communication. *Genetics* **2017**, *205*, 1517–1527. [CrossRef]

19. Austin, C.J.; Emberson, L.; Nicholls, P. Purification and characterization of pheromaxein, the porcine steroid-binding protein. A member of the secretoglobin superfamily. *Eur. J. Biochem.* **2004**, *271*, 2593–2606. [CrossRef]

20. Carayol, N.; Birnbaum, J.; Magnan, A.; Ramadour, M.; Lanteaume, A.; Vervloet, D.; Tessier, Y.; Pageat, P. Fel d 1 production in the cat skin varies according to anatomical sites. *Allergy* **2000**, *55*, 570–573. [CrossRef]

21. Pageat, P.; Gaultier, E. Current research in canine and feline pheromones. *Vet. Clin. N. Am. Small Anim. Pract.* **2003**, *33*, 187–211. [CrossRef]

22. Bienboire-Frosini, C.; Cozzi, A.; Lafont-Lecuelle, C.; Vervloet, D.; Ronin, C.; Pageat, P. Immunological differences in the global release of the major cat allergen Fel d 1 are influenced by sex and behaviour. *Vet. J.* **2012**, *193*, 162–167. [CrossRef] [PubMed]

23. Ligabue-Braun, R.; Sachett, L.G.; Pol-Fachin, L.; Verli, H. The Calcium Goes Meow: Effects of Ions and Glycosylation on Fel d 1, the Major Cat Allergen. *PLoS ONE* **2015**, *10*, e0132311. [CrossRef] [PubMed]

24. Karn, R.C. The mouse salivary androgen-binding protein (ABP) alpha subunit closely resembles chain 1 of the cat allergen Fel dI. *Biochem. Genet.* **1994**, *32*, 271–277. [CrossRef]

25. Durairaj, R.; Pageat, P.; Bienboire-Frosini, C. Another cat and mouse game: Deciphering the evolution of the SCGB superfamily and exploring the molecular similarity of major cat allergen Fel d 1 and mouse ABP using computational approaches. *PLoS ONE* **2018**, *13*, e0197618. [CrossRef]

26. Karn, R.C.; Clements, M.A. A comparison of the structures of the alpha:beta and alpha:gamma dimers of mouse salivary androgen-binding protein (ABP) and their differential steroid binding. *Biochem. Genet.* **1999**, *37*, 187–199. [CrossRef]

27. Chapman, M.D.; Smith, A.M.; Vailes, L.D.; Arruda, L.K.; Dhanaraj, V.; Pomés, A. Recombinant allergens for diagnosis and therapy of allergic disease. *J. Allergy Clin. Immunol.* **2000**, *106*, 409–418. [CrossRef]

28. Vailes, L.D.; Sun, A.W.; Ichikawa, K.; Wu, Z.; Sulahian, T.H.; Chapman, M.D.; Guyre, P.M. High-level expression of immunoreactive recombinant cat allergen (Fel d 1): Targeting to antigen-presenting cells. *J. Allergy Clin. Immunol.* **2002**, *110*, 757–762. [CrossRef]

29. Chen, X.; Zaro, J.L.; Shen, W.C. Fusion protein linkers: Property, design and functionality. *Adv. Drug Deliv. Rev.* **2013**, *65*, 1357–1369. [CrossRef]

30. Wuenschmann, S.; Vailes, L.D.; King, E.; Aalberse, R.C.; Chapman, M.D. Expression of a Deglycosylated Recombinant Fel d 1 in Pichia pastoris. *J. Allergy Clin. Immunol.* **2008**, *121*, S214. [CrossRef]

31. Vitale Shreve, K.R.; Udell, M.A.R. Stress, security, and scent: The influence of chemical signals on the social lives of domestic cats and implications for applied settings. *Appl. Anim. Behav. Sci.* **2017**, *187*, 69–76. [CrossRef]

32. Mills, D.S.; White, J.C. Long-term follow up of the effect of a pheromone therapy on feline spraying behaviour. *Vet. Rec.* **2000**, *147*, 746–747. [PubMed]

33. Mills, D.S.; Mills, C. Evaluation of a novel method for delivering a synthetic analogue of feline facial pheromone to control urine spraying by cats. *Vet. Rec.* **2001**, *149*, 197–199. [CrossRef]

34. Kronen, P.W.; Ludders, J.W.; Erb, H.N.; Moon, P.F.; Gleed, R.D.; Koski, S. A synthetic fraction of feline facial pheromones calms but does not reduce struggling in cats before venous catheterization. *Vet. Anaesth. Analg.* **2006**, *33*, 258–265. [CrossRef]

35. Pereira, J.S.; Fragoso, S.; Beck, A.; Lavigne, S.; Varejão, A.S.; da Graça Pereira, G. Improving the feline veterinary consultation: The usefulness of Feliway spray in reducing cats' stress. *J. Feline Med. Surg.* **2016**, *18*, 959–964. [CrossRef]

36. Cozzi, A.; Monneret, P.; Lafont-Lecuelle, C.; Bougrat, L.; Gaultier, E.; Pageat, P. The maternal cat appeasing pheromone: Exploratory study of the effects on aggressive and affiliative interactions in cats. *J. Vet. Behav. Clin. Appl. Res.* **2010**, *5*, 37–38. [CrossRef]

37. DePorter, T.L.; Bledsoe, D.L.; Beck, A.; Ollivier, E. Evaluation of the efficacy of an appeasing pheromone diffuser product vs placebo for management of feline aggression in multi-cat households: A pilot study. *J. Feline Med. Surg.* **2019**, *21*, 293–305. [CrossRef]

38. Fridlansky, F.; Milgrom, E. Interaction of uteroglobin with progesterone, 5αpregnane-3, 20-dione and estrogens. *Endocrinology* **1976**, *99*, 1244–1251. [CrossRef]

39. Karn, R.C. Steroid binding by mouse salivary proteins. *Biochem. Genet.* **1998**, *36*, 105–117. [CrossRef]

40. Chen, C.; Schilling, K.; Hiipakka, R.A.; Huang, I.Y.; Liao, S. Prostate α-protein. Isolation and characterization of the polypeptide components and cholesterol binding. *J. Biol. Chem.* **1982**, *257*, 116–121.

41. Taylor, R.D.; Jewsbury, P.J.; Essex, J.W. A review of protein-small molecule docking methods. *J. Comput. Aided Mol. Des.* **2002**, *16*, 151–166. [CrossRef]

42. Pelosi, P.; Zhu, J.; Knoll, W. From radioactive ligands to biosensors: Binding methods with olfactory proteins. *Appl. Microbiol. Biotechnol.* **2018**, *102*, 8213–8227. [CrossRef]

43. Dorries, K.M.; Adkins-Regan, E.; Halpern, B.P. Sensitivity and behavioral responses to the pheromone androstenone are not mediated by the vomeronasal organ in domestic pigs. *Brain Behav. Evol.* **1997**, *49*, 53–62. [CrossRef]

44. Guiraudie-Capraz, G.; Pageat, P.; Cain, A.H.; Madec, I.; Nagnan-Le Meillour, P. Functional characterization of olfactory binding proteins for appeasing compounds and molecular cloning in the vomeronasal organ of pre-pubertal pigs. *Chem. Senses* **2003**, *28*, 609–619. [CrossRef]

45. Nagnan-Le Meillour, P.; Joly, A.; Le Danvic, C.; Marie, A.; Zirah, S.; Cornard, J.P. Binding specificity of native odorant-binding protein isoforms is driven by phosphorylation and O-N-acetylglucosaminylation in the pig Sus scrofa. *Front. Endocrinol.* **2019**, *9*, 816. [CrossRef]

46. Pelosi, P. The role of perireceptor events in vertebrate olfaction. *Cell. Mol. Life Sci.* **2001**, *58*, 503–509. [CrossRef]

47. Stopkova, R.; Klempt, P.; Kuntova, B.; Stopka, P. On the tear proteome of the house mouse (*Mus musculus musculus*) in relation to chemical signalling. *PeerJ* **2017**, *5*, e3541. [CrossRef]

48. Morgenstern, J.P.; Griffith, I.J.; Brauer, A.W.; Rogers, B.L.; Bond, J.F.; Chapman, M.D.; Kuo, M.-C. Amino acid sequence of Fel dI, the major allergen of the domestic cat: Protein sequence analysis and cDNA cloning. *Proc. Natl. Acad. Sci. USA* **1991**, *88*, 9690–9694. [CrossRef]

49. Emes, R.D.; Riley, M.C.; Laukaitis, C.M.; Goodstadt, L.; Karn, R.C.; Ponting, C.P. Comparative evolutionary genomics of androgen-binding protein genes. *Genome Res.* **2004**, *14*, 1516–1529. [CrossRef]

50. Jackson, B.C.; Thompson, D.C.; Wright, M.W.; McAndrews, M.; Bernard, A.; Nebert, D.W.; Vasiliou, V. Update of the human secretoglobin (SCGB) gene superfamily and an example of "evolutionary bloom" of androgen-binding protein genes within the mouse Scgb gene superfamily. *Hum. Genomics* **2011**, *5*, 691–702. [CrossRef]

51. Golebiowski, J.; Topin, J.; Charlier, L.; Briand, L. Interaction between odorants and proteins involved in the perception of smell: The case of odorant-binding proteins probed by molecular modelling and biophysical data. *Flavour Fragr. J.* **2012**, *27*, 445–453. [CrossRef]

52. Grönlund, H.; Bergman, T.; Sandström, K.; Alvelius, G.; Reininger, R.; Verdino, P.; Hauswirth, A.; Liderot, K.; Valent, P.; Spitzauer, S.; et al. Formation of disulfide bonds and homodimers of the major cat allergen Fel d 1 equivalent to the natural allergen by expression in Escherichia coli. *J. Biol. Chem.* **2003**, *278*, 40144–40151. [CrossRef] [PubMed]

53. Le Danvic, C.; Guiraudie-Capraz, G.; Abderrahmani, D.; Zanetta, J.P.; Nagnan-Le Meillour, P. Natural ligands of porcine olfactory binding proteins. *J. Chem. Ecol.* **2009**, *35*, 741–751. [CrossRef] [PubMed]

54. Brimau, F.; Cornard, J.P.; Le Danvic, C.; Lagant, P.; Vergoten, G.; Grebert, D.; Pajot, E.; Nagnan-Le Meillour, P. Binding specificity of recombinant odorant-binding protein isoforms is driven by phosphorylation. *J. Chem. Ecol.* **2010**, *36*, 801–813. [CrossRef] [PubMed]

55. Burger, B.V. Mammalian semiochemicals. In *The Chemistry of Pheromones and Other Semiochemicals II*; Springer: Berlin/Heidelberg, Germany, 2005; pp. 231–278.

56. Beynon, R.J.; Armstrong, S.D.; Gómez-Baena, G.; Lee, V.; Simpson, D.; Unsworth, J.; Hurst, J.L. The complexity of protein semiochemistry in mammals. *Biochem. Soc. Trans.* **2014**, *42*, 837–845. [CrossRef] [PubMed]

57. Bradshaw, J.W.S.; Cameron-Beaumont, C. The signalling repertoire of the domestic cat and its undomesticated relatives. In *The Domestic Cat, The Biology of Its Behaviour*; Turner, D.C., Bateson, P., Eds.; Cambridge University Press: Cambridge, UK, 2000; pp. 68–93.

58. Natoli, E.; De Vito, E. Agonistic behaviour, dominance rank and copulatory success in a large multi-male feral cat, *Felis catus* L., colony in central Rome. *Anim. Behav.* **1991**, *42*, 227–241. [CrossRef]

59. Natoli, E. Behavioural Responses of Urban Feral Cats to Different Types of Urine Marks. *Behaviour* **1985**, *94*, 234–243. [CrossRef]

60. Smith, J.L.D.; McDougal, C.; Miquelle, D. Scent marking in free-ranging tigers, *Panthera tigris. Anim. Behav.* **1989**, *37*, 1–10. [CrossRef]

61. De Groot, H.; van Swieten, P.; Aalberse, R.C. Evidence for a Fel d I-like molecule in the "big cats" (*Felidae* species). *J. Allergy Clin. Immunol.* **1990**, *86*, 107–116. [CrossRef]

62. Burgos, T.; Virgós, E.; Valero, E.S.; Arenas-Rojas, R.; Rodríguez-Siles, J.; Recio, M.R. Prey density determines the faecal-marking behaviour of a solitary predator, the Iberian lynx (*Lynx pardinus*). *Ethol. Ecol. Evol.* **2019**, *31*, 219–230. [CrossRef]

63. Darden, S.K.; Steffensen, L.K.; Dabelsteen, T. Information transfer among widely spaced individuals: Latrines as a basis for communication networks in the swift fox? *Anim. Behav.* **2008**, *75*, 425–432. [CrossRef]

64. Tegoni, M.; Pelosi, P.; Vincent, F.; Spinelli, S.; Grolli, S.; Ramoni, R.; Cambillau, C. Mammalian odorant binding proteins. *Biochim. Biophys. Acta* **2000**, *1482*, 229–240. [CrossRef]

65. Papes, F.; Logan, D.W.; Stowers, L. The Vomeronasal Organ Mediates Interspecies Defensive Behaviors through Detection of Protein Pheromone Homologs. *Cell* **2010**, *141*, 692–703. [CrossRef]

66. Callebaut, I.; Poupon, A.; Bally, R.; Demaret, J.P.; Housset, D.; Delettre, J.; Hossenlopp, P.; Mornon, J.P. The uteroglobin fold. *Ann. N. Y. Acad. Sci.* **2000**, *923*, 90–112. [CrossRef]

67. Herre, J.; Grönlund, H.; Brooks, H.; Hopkins, L.; Waggoner, L.; Murton, B.; Gangloff, M.; Opaleye, O.; Chilvers, E.R.; Fitzgerald, K.; et al. Allergens as immunomodulatory proteins: The cat dander protein Fel d 1 enhances TLR activation by lipid ligands. *J. Immunol.* **2013**, *191*, 1529–1535. [CrossRef]

68. Bublin, M.; Eiwegger, T.; Breiteneder, H. Do lipids influence the allergic sensitization process? *J. Allergy Clin. Immunol.* **2014**, *134*, 521–529. [CrossRef]

69. Satyaraj, E.; Wedner, H.J.; Bousquet, J. Keep the cat, change the care pathway: A transformational approach to managing Fel d 1, the major cat allergen. *Allergy Eur. J. Allergy Clin. Immunol.* **2019**, *74*, 5–17. [CrossRef]

70. Satyaraj, E.; Gardner, C.; Filipi, I.; Cramer, K.; Sherrill, S. Reduction of active Fel d1 from cats using an antiFel d1 egg IgY antibody. *Immun. Inflamm. Dis.* **2019**, *7*, 68–73. [CrossRef]

71. Thoms, F.; Jennings, G.T.; Maudrich, M.; Vogel, M.; Haas, S.; Zeltins, A.; Hofmann-Lehmann, R.; Riond, B.; Grossmann, J.; Hunziker, P.; et al. Immunization of cats to induce neutralizing antibodies against Fel d 1, the major feline allergen in human subjects. *J. Allergy Clin. Immunol.* **2019**, *144*, 193–203. [CrossRef]

72. Jayaram, B.; Singh, T.; Mukherjee, G.; Mathur, A.; Shekhar, S.; Shekhar, V. Sanjeevini: A freely accessible web-server for target directed lead molecule discovery. *BMC Bioinform.* **2012**, *13*, S7. [CrossRef]

73. Bikadi, Z.; Hazai, E. Application of the PM6 semi-empirical method to modeling proteins enhances docking accuracy of AutoDock. *J. Cheminform.* **2009**, *1*, 15. [CrossRef]

74. Morris, G.M.; Huey, R.; Lindstrom, W.; Sanner, M.F.; Belew, R.K.; Goodsell, D.S.; Olson, A.J. AutoDock4 and AutoDockTools4: Automated Docking with Selective Receptor Flexibility. *J. Comput. Chem.* **2009**, *30*, 2785–2791. [CrossRef]

75. Solis, F.J.; Wets, R.J.-B. Minimization by Random Search Techniques. *Math. Oper. Res.* **1981**, *6*, 19–30. [CrossRef]

International Journal of
Molecular Sciences

Article

Enhancement of Binding Affinity of Folate to Its Receptor by Peptide Conjugation

Roopa Dharmatti [1,2], Hideyuki Miyatake [1,*], Avanashiappan Nandakumar [3], Motoki Ueda [1,3], Kenya Kobayashi [1], Daisuke Kiga [2,4], Masayuki Yamamura [2] and Yoshihiro Ito [1,2,3,*]

[1]   Nano Medical Engineering Laboratory, RIKEN Cluster for Pioneering Research, 2-1 Hirosawa, Wako, Saitama 351-0198, Japan; roopa.dharmatti@riken.jp (R.D.); motoki.ueda@riken.jp (M.U.); kenya.kobayashi@alpsalpine.com (K.K.)

[2]   Department of Computer Science, School of Computing, Tokyo Institute of Technology, 4259 Nagatsuta-cho, Midori-ku, Yokohama 226-8503, Japan; kiga@waseda.jp (D.K.); my@c.titech.ac.jp (M.Y.)

[3]   Emergent Bioengineering Materials Research Team, RIKEN Center for Emergent Matter Science, 2-1 Hirosawa, Wako, Saitama 351-0198, Japan; nandakumar.avanashiappan@riken.jp

[4]   Department of Electrical Engineering and Bioscience, Waseda University, 2-2 Wakamatsu Cho, Shinjyuku-ku, Tokyo 162-8480, Japan

*   Correspondence: miyatake@riken.jp (H.M.); y-ito@riken.jp (Y.I.); Tel.: +81-48-467-4979 (Y.I.); Fax: +81-48-467-9300 (Y.I.)

Received: 11 April 2019; Accepted: 25 April 2019; Published: 30 April 2019

Abstract: (1) Background: The folate receptor (FR) is a target for cancer treatment and detection. Expression of the FR is restricted in normal cells but overexpressed in many types of tumors. Folate was conjugated with peptides for enhancing binding affinity to the FR. (2) Materials and Methods: For conjugation, folate was coupled with propargyl or dibenzocyclooctyne, and 4-azidophenylalanine was introduced in peptides for "click" reactions. We measured binding kinetics including the rate constants of association (k_a) and dissociation (k_d) of folate-peptide conjugates with purified FR by biolayer interferometry. After optimization of the conditions for the click reaction, we successfully conjugated folate with designed peptides. (3) Results: The binding affinity, indicated by the equilibrium dissociation constant (K_D), of folate toward the FR was enhanced by peptide conjugation. The enhanced FR binding affinity by peptide conjugation is a result of an increase in the number of interaction sites. (4) Conclusion: Such peptide-ligand conjugates will be important in the design of ligands with higher affinity. These high affinity ligands can be useful for targeted drug delivery system.

Keywords: folate; folate receptor; peptide conjugation; click reaction; biolayer interferometry

1. Introduction

Traditional cancer therapy involves removal of tumor cells by surgery, radiation and non-selective types of chemotherapy [1,2]. Surgery and radiation are often effective with tumors that are primary or localized and have not metastasized to multiple sites throughout the body [3]. Chemotherapy is effective in the treatment of metastatic cancers because typical chemotherapeutic agents focus on rapidly growing tissues, which is a property common to cancer cells. Nonetheless, chemotherapy also often has a high incidence of unwanted and damaging side effects in normal tissues because these tissues are also undergoing growth [4,5]. Therefore, monoclonal antibodies against cellular targets that are unique to cancer cells have been developed [4,6], and antibody-drug conjugates (ADCs) have also been developed [6]. Targeted treatments exert their anticancer effects through multiple mechanisms, including proliferation inhibition [6], apoptosis induction [7], metastasis suppression [8], immune function regulation [9] and multidrug resistance reversal [5,10]. A few ADCs have been used successfully in clinical trials [5,10,11]. However, there are several points to consider when using an antibody as a drug-transporter that targets tumors. Limitations owing to poor therapeutic efficacy of ADCs include: (i) manufacturing procedures that create heterogeneous mixtures of ADCs with a number of drug molecules conjugated inconsistently; (ii) the synthesis costs are extremely high with difficulties in quality control; and (iii) the larger size of ADCs hampers penetration of ADCs into tumor tissue [12]. Small molecules or peptides are potential therapeutic molecules that overcome these problems [2]. In contrast to antibodies, these agents provide advantages such as reduced immunogenicity, quick clearance, increased diffusion and tissue penetration, chemical stability and ease of synthesis [2,6].

Due to the remarkable expression of the folate receptor (FR) on the surface of tumor cells, the FR can be exploited as a cancer diagnostic and therapeutic target [13]. Folate is an intrinsic ligand of the FR, consisting of a pterin ring, a central *p*-amino benzoic acid and an L-glutamic acid tail [4,14,15], and has been conjugated with anti-cancer drugs [4,16] and drug carriers [17–23] for targeted delivery of drugs to tumor cells. For example, a peptide that binds to the α isoform of the FR, which is a subtype of FRs, was selected by phage display; however, the affinity of this peptide was low when compared with that of folate [24].

In this report, we conjugate folate with peptides to enhance binding affinity toward the FR. Previously, Li and Roberts [25] prepared a penicillin-peptide conjugate that has at least 100-fold higher activity than penicillin. Wang et al. introduced aminophenylalanine coupled with purvalanol into peptides to enhance the inhibitory activity of purvalanol against kinases [26,27]. Peptide conjugation should increase the affinity between the target protein and ligand by increasing the number of interaction sites, as shown in Figure 1.

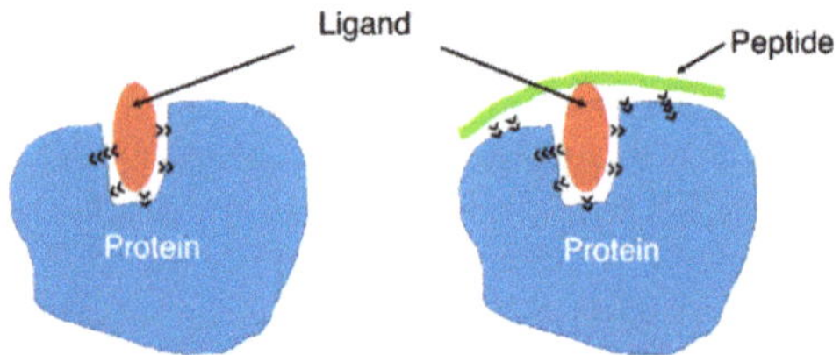

Figure 1. Schematic showing how a peptide conjugated to a ligand (folate) enhances the affinity of the ligand toward the target protein (folate receptor). The black arrowheads indicate molecular interactions.

For conjugation, we added propargyl or dibenzocyclooctyne (DBCO) to folate and 4-azidophenylalanine (AzPhe) in the peptide for the "click" reaction, as shown in Figure 2, because it is possible to introduce the azidophenyl groups into proteins by bio-orthogonal approaches [28,29]. The binding assay of the synthesized folate-peptide conjugates with FR was performed by biolayer interferometry (BLI), and the association rate constant (k_a) and dissociation rate constant (k_d) were determined. The study demonstrated that the conjugation of folate with peptides enhanced the affinity of folate toward the FR.

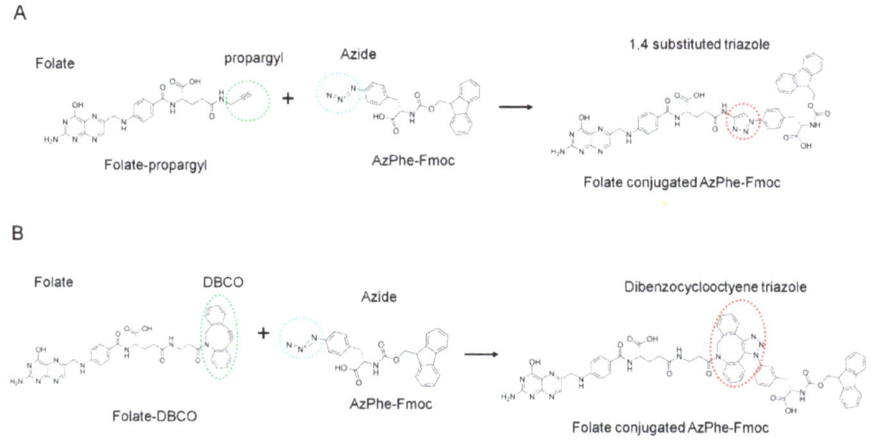

Figure 2. Schemes showing the synthesis of the folate-conjugated AzPhe-Fmoc. (**A**) The Cu(I)-catalyzed alkyne-azide cycloaddition (CuAAC) click reaction between the propargyl group (green dotted circle) and the azide group (cyan dotted circle) to conjugate folate via the triazole ring (red dotted circle). (**B**) The strain-promoted alkyne-azide cycloaddition (SPAAC) click reaction between DBCO (green dotted circle) and the azide group (cyan dotted circle) to conjugate folate via dibenzocyclooctyne triazole (red dotted circle).

2. Results and Discussion

2.1. Folate-Phe Conjugation by Click Reactions

Two types of folate analogues were prepared by addition of the propargyl group (Figure 2A) and DBCO (Figure 2B), and both were adjacent to the γ-carboxyl group of folate. The additions enabled confirmation of the click reaction between folate analogues and AzPhe-Fmoc. Folate-propargyl was used for the Cu(I)-catalyzed alkyne-azide cycloaddition (CuAAC) click reaction with AzPhe-Fmoc. To promote the CuAAC reaction, Cu(I) stabilizing ligands such as Tris (2-benzimidazoylmethyl) amine (BimH$_3$) and microwaves were also employed at 50 °C. However, absorbance from the triazole ring on the target compound was not detected under the conditions shown in Table 1.

Currently, some groups have reported success of the CuAAC reaction between folate-propargyl and polymers containing an azido group [30–34]. However, their folate-propargyl conjugates were a mixture of propargyl groups bound to the C$^\alpha$ and C$^\gamma$ of the glutamic acid part of folate. The present conjugate is the first example of a folate-propargyl with the propargyl group specifically linked to the C$^\gamma$ of folate. The results in Table 1 indicate that the C$^\gamma$-binding propargyl group shows low reactivity in the CuAAC reaction. The other possibility is that coordination by the -N and -NH groups of the folate-propargyl with Cu(I) interferes with alkyne-Cu(I) complexation.

In contrast, the strain-promoted azide-alkyne cycloaddition (SPAAC) "click" reaction between folate-DBCO and AzPhe-Fmoc was successful (Table 1). The yield increased up to 88% by using twice the molar ratio of folate-DBCO against AzPhe-Fmoc, and the reaction temperature did not affect yields noticeably. Golas et al. [35] studied the substituent effect on azide reactivity in CuAAC using various azide compounds with propargyl alcohol. The electronic properties and steric congestion near end groups are major determinants for the reactivity of azide compounds. Azide with electron withdrawing groups, such as ethyl azido-acetate, methyl 2-azidopropionate and azidoacetonitrile, react faster than similar compounds with a neighboring aromatic ring (benzyl azide and 1-phenylethyl azide). In addition, primary azides such as benzyl azide and ethyl azido-acetate react faster than their secondary analogues, 1-phenylethyl azide and methyl 2-azidopropionate, respectively. In this case, AzPhe is less reactive because the electron-withdrawing is affected by the aromatic ring. Nonetheless,

AzPhe can be more reactive through SPAAC because DBCO enhances the reactivity by its resonant structure [36].

Table 1. Reaction conditions of folate-propargyl or folate-DBCO with AzPhe-Fmoc.

Folate-alkyne [a]	Molar Ratio of Folate-alkyne: Azide	Reaction Conditions	Yield (%) [b]
9	1:1	CuCl (0.1 mM), BimH$_3$ (0.1 mM), Na ascorbate (0.1 mM), 11% (*v/v*) DMSO + 89% (*v/v*) H$_2$O, room temperature (RT), 12 h	N.D.
9	1:1	CuCl (0.2 mM), BimH$_3$ (0.1 mM), Na ascorbate (0.2 mM), 11% (*v/v*) DMSO + 89% (*v/v*) H$_2$O, 50 °C, 10 h	N.D.
9	1:1	CuSO$_4$ (0.1mM), BimH$_3$ (0.1 mM), Na ascorbate (0.6 mM), 11% (*v/v*) DMSO + 89% (*v/v*) H$_2$O, MW [c], 1h	N.D.
16	1:1	10% (*v/v*) DMF + 10% (*v/v*) H$_2$O + 80% (*v/v*) MeOH, RT, 16 h	60
16	1:1	10% (*v/v*) DMF + 10% (*v/v*) H$_2$O + 80% (*v/v*) MeOH, 50 °C, 16 h	56
16	2:1	20% (*v/v*) DMF + 10% (*v/v*) H$_2$O + 70% (*v/v*) MeOH, RT, 16 h	88

[a] The number corresponds the compound number in Figure 6. [b] High performance liquid chromatography (HPLC) yields; [c] Microwave conditions; N.D. Not detected.

2.2. Preparation of Folate-Peptide Conjugates by the SPAAC Click Reaction

Since folate-DBCO was demonstrated to conjugate efficiently to AzPhe by the SPAAC reaction, the preparation of folate-peptide conjugates was performed by this reaction (Figure 3). Three peptide sequences, GF[AzPhe]IQ, SE[AzPhe]KA and DSE[AzPhe]KAY, were synthesized. The folate-peptide conjugates were designed by the program ICM-Pro (Molsoft L.L.C., San Diego, CA, USA). After successful conjugation of folate with AzPhe by SPAAC, we considered the folate-conjugated AzPhe as one unit and increased the length of the peptide by adding amino acids at N-terminal and C-terminal of the AzPhe. This length was increased by trial and error procedure. The peptides were synthesized by a conventional solid phase synthesis method. For BLI measurements, in which a biotin group binds to streptavidin bound to coated sensor chips, the N-terminus of the peptides was modified with biotin-(PEG$_{24}$)-NHS. The coupling was performed before release from the solid phase synthesis resin (Figure 3A) [37,38]. The polyethylene glycol (PEG) linker functions as a spacer between the immobilized and interaction sites and as a solubilizer of the folate-peptide conjugates in aqueous solutions.

The same coupling reaction conditions were used for peptide conjugation. After the click reaction and purification, each folate-peptide conjugate was identified by matrix assisted laser desorption/ionization-time of flight mass spectrometry (MALDI-TOF MS). From the mass spectra, folate was confirmed to bind successfully to the side chain of AzPhe in the peptides.

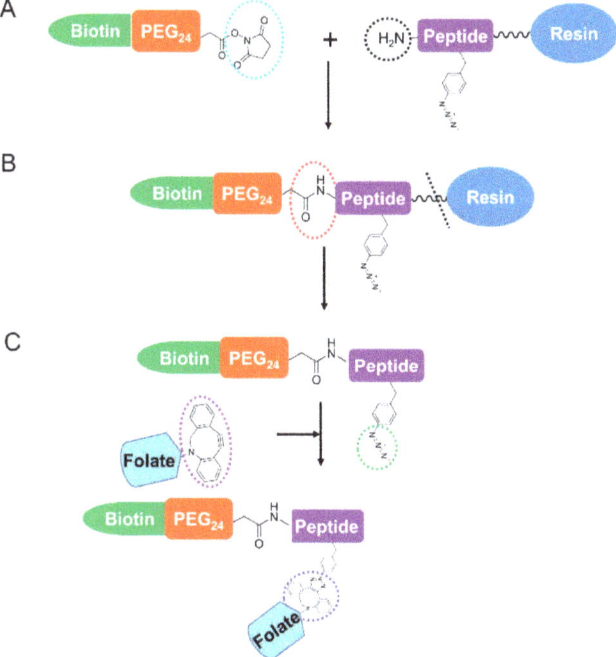

Figure 3. Synthesis procedure for peptide conjugates. (**A**) *N*-terminal peptide modification with biotin-PEG$_{24}$ was achieved by reacting ester NHS (cyan dotted circle) and the NH$_2$ group of the peptide (black dotted circle).The black wavy line between resin and peptide indicated various peptide lengths. (**B**) After the *N*-terminal modification with biotin-PEG$_{24}$, the amide bond (red dotted circle) was formed. Next, peptide was cleaved from resin (black dotted line). (**C**) The folate-DBCO-AzPhe containing peptide was achieved by the SPAAC click reaction between DBCO (purple dotted circle) of the folate and azide groups (green dotted circle) of the AzPhe in the peptide to form the folate-peptide conjugate via the dibenzocyclooctyne triazole (blue dotted circle).

2.3. BLI Measurement

Table 2 and Figure 4 show the results of the BLI measurements to evaluate the affinities of the folate-peptide conjugates toward folate receptor alpha (FRα).

Commercially available folate-PEG$_8$-biotin was used as a control for BLI analysis. The equilibrium dissociation constant (K_D) between FRα and folate was 1.14 nM. Wibowo et al. [39] and Chen et al. [14] used a radiolabeled ligand assay and isothermal calorimetry for measurement of the K_D of folate with FRα and yielded values of ~10 pM and ~190 pM, respectively. Combined with our results, the differences in K_D values indicate that the method used to measure the K_D has a strong influence on the outcome.

Table 2. BLI results for the binding affinity of folate and folate-peptide conjugates.

Ligands	K_D (nM)	k_a (M^{-1} s^{-1})	k_d (s^{-1})
Folate	1.14	6.74×10^6	7.69×10^{-3}
GFZIQ	0.18	4.11×10^5	7.53×10^{-5}
SEZKA	0.90	8.91×10^4	8.01×10^{-5}
DSEZKAY	0.24	1.10×10^6	2.65×10^{-4}

Z = folate-conjugated AzPhe.

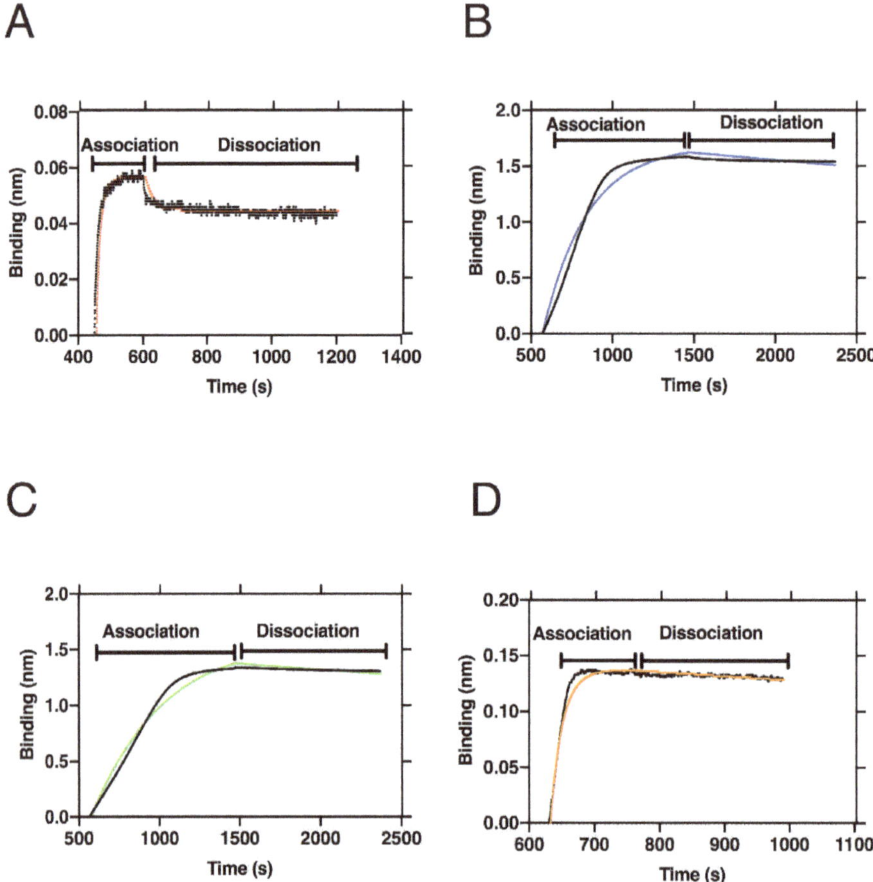

Figure 4. BLI data for binding of (**A**) folate-PEG$_8$-biotin, (**B**) GFZIQ, (**C**) SEZKA and (**D**) DSEZKAY with FRα. In all cases, analyte only data was kept as a reference and 1:1 local analysis was used. The black curve lines are run data and colored curve lines are fitting data.

An advantage of BLI is evaluation of the k_a and k_d. The binding mode of folate to FRα shows a non-equilibrium binding mode, in which the k_d (7.69 × 10^{-3} s^{-1}) was ~10^3 times slower than that of the association rate (6.74 × 10^6 M^{-1} s^{-1}). This difference between the k_a and k_d corresponds well with the scenario previously proposed for folate binding to FRα [39]. In crystallographic work that compared the apo- and folate binding forms of FRs, large conformational changes around the folate binding pocket upon folate binding were observed, i.e., from the relaxed (open) to tight (closed) forms. In the closed form, the inhibitory loop, basic loop and switching helix around the binding pocket cooperatively undergo conformational changes to bind the folate tightly. The bound folate in the FRs then dissociates from the receptors after endocytosis of the FRs into cells, which is triggered by the acidic environment of the cells. Such a non-equilibrium-binding mode promotes efficient uptake of folate into cells. Thus, our BLI data provide the first indication that the proposed trafficking mechanism of folate is valid by revealing the asymmetric binding kinetics of FRs.

2.4. Interaction of Folate-Peptide Conjugates with FRα

By conjugation with peptides, the affinity increased to sub-nanomolar ($\sim10^{-10}$ M) K_D values (Table 2). The peptide-conjugates showed slower k_a values that ranged from 8.91×10^4 to 1.10×10^6 ($M^{-1}\ s^{-1}$). Results presented in Figure 4B–D show significantly slow dissociation even after incubation in buffer. As a result, the k_d slows from 7.53×10^{-5} to $2.65 \times 10^{-4}\ s^{-1}$, which increases the K_D values. These observations suggest that peptide modification further stabilizes the complex formed between the peptide-conjugates and FRs, most probably by increasing the number of interaction sites between them.

In the peptide-conjugates, SEZKA and DSEZKAY share the common SEZKA sequence. Addition of aspartic acid (D) at the N-terminus and tyrosine (Y) at the C-terminus leads to a 12-fold faster association constant and 3-fold faster dissociation constant for the DSEZKAY peptide-conjugate, resulting in a 4-fold lower K_D. This increase in affinity occurs by lengthening SEZKA to DSEZKAY. This result indicates that we can alter the affinity of peptide-conjugate compounds by increasing the length of the peptides at both the N- and C-termini. This may provide a way to manipulate binding properties of peptide-conjugated compounds by increasing the length of the peptide part of the conjugates, which may increase the number of interaction contacts with the target protein.

Figure 5 shows the results of the docking simulation, which demonstrates the interaction mode of DSEZKAY with FRα. As expected in Figure 1, the structure of the complex shows an increase in the number of interactions to FRα from the peptide portion around the folate-binding pocket. Previous reports have demonstrated greater than 100-fold increases in binding affinity by peptide conjugates [25–27], whereas the present result was lower than these previous increases in affinity. However, the present investigation also revealed that peptide conjugation is a useful tool to enhance the binding affinity to the target molecule. Future efforts will focus on using the folate-peptide conjugate to target anti-cancer drug delivery.

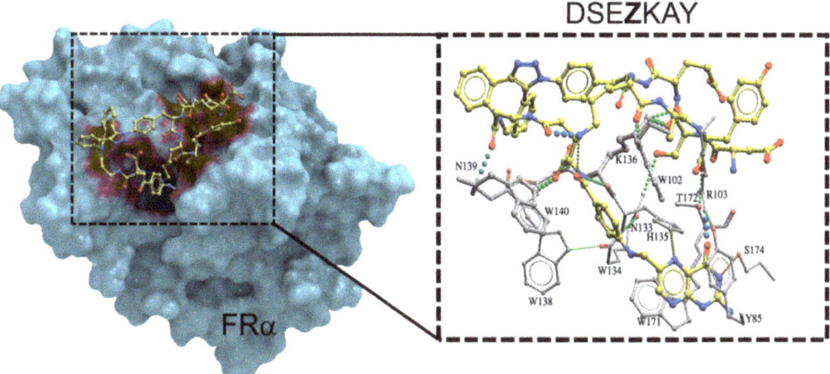

Figure 5. Docking model of DSEZKAY (yellow) with the surface of the FRα (blue). The gradation of yellow and magenta colors on the surface of the FRα indicated the interaction between the ligand and surface of the FRα. This figure was prepared by the program ICM-Pro. The left dotted box area of interaction between DSEZKAY with FRα is zoomed in right dotted box. All the dotted lines in right dotted box indicate an "increased" interaction of DSEZKAY with the FRα (except for the interaction of folate with the FRα).

3. Materials and Methods

3.1. Materials

Fmoc-Phe(4-N$_3$)-OH (AzPhe-Fmoc) was purchased from Watanabe Chemical Industries, Ltd. (Hiroshima, Japan) to incorporate non-natural amino acids during the solid phase peptide synthesis procedure. BimH$_3$ was purchased from Tokyo Chemical Industry Co., Ltd. (Tokyo, Japan). Biotin-PEG$_{24}$-

NHS was purchased from Thermo Fisher Scientific (Waltham, MA, USA) for biotin-PEG$_{24}$ modification at the N-terminus of the folate-peptide conjugates. For the activity assay, streptavidin (SA) biosensors were purchased from ForteBio (Fremont, CA, USA). Folate-PEG$_8$-biotin was purchased from Nanocs (New York, NY, USA). Reagents used for reversed-phase high performance liquid chromatography (RP-HPLC) were of HPLC grade. All other chemicals used were of biochemical research grade. MALDI-TOF MS (Microflex, Bruker Daltonics, Billerica, MA, USA.) was employed for molecular weight measurement.

3.2. Synthesis of Folate-Propargyl and Folate-DBCO

Synthetic schemes of folate derivatives are presented in Figure 6. Each compound was synthesized and confirmed as follows.

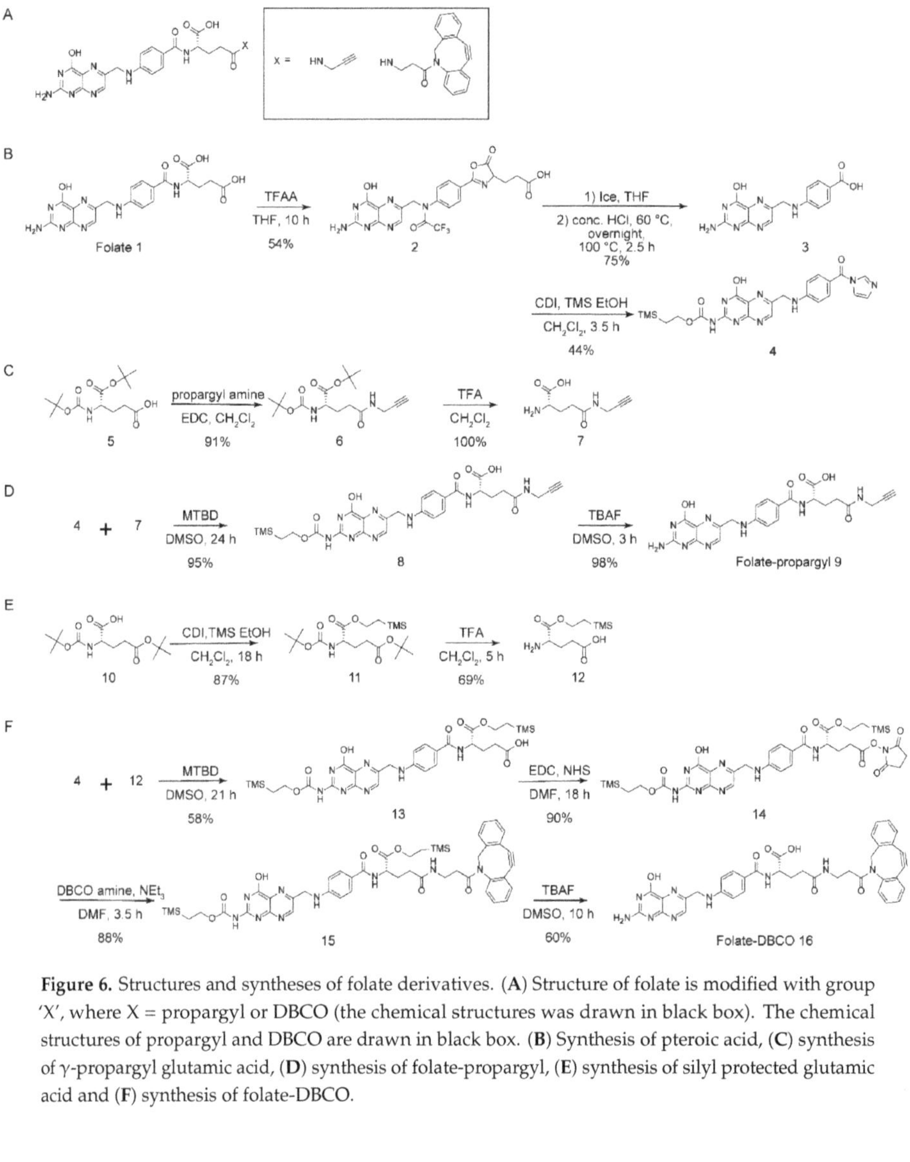

Figure 6. Structures and syntheses of folate derivatives. (**A**) Structure of folate is modified with group 'X', where X = propargyl or DBCO (the chemical structures was drawn in black box). The chemical structures of propargyl and DBCO are drawn in black box. (**B**) Synthesis of pteroic acid, (**C**) synthesis of γ-propargyl glutamic acid, (**D**) synthesis of folate-propargyl, (**E**) synthesis of silyl protected glutamic acid and (**F**) synthesis of folate-DBCO.

3.2.1. Compound 2

To a solution of folate **1** (10 g, 0.022 mol) and 100 mL anhydrous tetrahydrofuran (THF) in a three-neck flask, 24 mL, 0.176 mol trifluoroacetic anhydride [(CF$_3$CO)$_2$O] was slowly added at 0 °C for 30 min. The dark brown homogeneous mixture was stirred at room temperature (RT). After 10 h, the reaction mixture was filtered through a pad of celite to remove the small amount of solid residue. The filtrate was concentrated under reduced pressure and the viscous liquid was dissolved with a minimum amount of THF (5 mL), which was slowly transformed into a flask of well-stirred diethyl ether (Et$_2$O). The yellow precipitate formed in Et$_2$O was collected by filtration and washed with Et$_2$O (25 mL ×2) to yield the crude compound **2**.

3.2.2. Compound 3

The crude compound **2** (6 g) was dissolved in THF (50 mL) followed by the addition of ice (~10 g) with stirring for 5 h. The mixture was slowly transferred into stirred Et$_2$O (200 mL). The yellowish precipitate was collected by filtration, washed with Et$_2$O (200 mL ×3) and dried for 24 h under vacuum. To the suspension of yellowish precipitate, conc. HCl (60 mL) was added and refluxed at 60 °C overnight and then 100 °C for 2.5 h. The reaction mixture was poured into water (100 mL). The precipitate formed in the solution was collected by filtration and washed with Et$_2$O to afford compound **3** (75%). ^{1}H NMR (DMSO-d$_6$, 400 MHz): 4.60 [singlet (s), 2 H], 6.66 [doublet (d), *J* = 8.8 Hz, 2 H], 7.66 (d, *J* = 8.8 Hz, 2 H), 8.68 [broad singlet (brs), 2 H], 8.78 (s, 1 H).

3.2.3. Compound 4

Compound **3** (3.0 g, 9.6 mmol), Et$_3$N (5.36 mL, 38.0 mmol), and 1,1′-carbonyldiimidazole (CDI) (6.2 g, 38.0 mmol) in 30 mL dimethyl sulfoxide (DMSO) was stirred at RT for 3.5 h. To the resulting solution, 2-(trimethylsilyl) ethanol (11 mL, 76.8 mmol) was added. After 5 h stirring at RT, the reaction mixture was poured into a mixture of water (330 mL), 9.6 mL acetic acid (AcOH) and Et$_2$O (192 mL). The resulting yellow precipitate was collected by filtration and purified on a silica gel column with 10% (*v*/*v*) methanol (MeOH) in CHCl$_3$ to give a yellow solid, which was further washed with Et$_2$O to give compound **4** (2.14 g, 44%). ^{1}H NMR (DMSO-d$_6$, 400 MHz): 0.06 (s, 9H), 1.03–1.07 [multiplet (m), 2H], 4.28–4.32 (m, 2H), 4.66 (d, *J* = 6.4 Hz, 2H), 6.79 (d, *J* = 8.8 Hz, 2H), 7.10 (s, 1H), 7.61–7.67 (m, 4H), 8.15 (s, 1H), 8.89 (s, 1H).

3.2.4. Compound 6

To compound **5** (800 mg, 2.6 mmol) in THF (20 mL), propargylamine hydrochloride (275 mg, 3.0 mmol), 1-ethyl-3-(3-dimethylaminopropyl) carbodiimide hydrochloride (EDC) (575 mg, 3.0 mmol) and Et$_3$N (575 mL) were added, and subsequently CH$_2$Cl$_2$ (30 mL) and MeOH (10 mL) were added to dissolve solids completely. After stirring at RT for 5 h the solvent was evaporated. The residue was purified on a silica gel column to give compound **6** (620 mg, 91%). ^{1}H NMR (CDCl$_3$, 400 MHz): 1.46 (s, 18 H), 1.86 (s, CC*H*, 1 H), 2.12–2.31 (m, 4 H), 4.06 [quartet (q), 2 H], 4.16 (m, $^\alpha$C*H*, 1 H), 5.24 (s, N*H*, 1 H), 6.52 (s, N*H*, 1 H).

3.2.5. Compound 7

Compound **6** (620 mg, 1.8 mmol) was dissolved in CH$_2$Cl$_2$ (5.4 mL) and cooled to 0 °C. To the solution, 12.6 mL trifluoroacetic acid (TFA) was added while stirring. After stirring at RT for 4 h, the solvent was evaporated under reduced vacuum. MeOH was added to dissolve the crude powder and then solidification was performed by the addition of Et$_2$O. The solvent was evaporated and the precipitate dried to give compound **7** (340 mg, 100%). ^{1}H NMR (D$_2$O, 400 MHz): 2.00 (dd, 2 H), 2.30 (dd, 2 H), 2.45 (s, CC*H*, 1 H), 3.66 (t, $^\alpha$C*H*, 1 H), 3.81 (s, 2 H).

3.2.6. Compound **8**

Compound **7** (107 mg, 0.58 mmol), compound **4** (224 mg, 0.44 mmol) and 7-methyl-1,5,7-triazabicyclo [4.4.0] dec-5-ene (MTBD) (0.2 mL) were dissolved in DMSO (5 mL) and the mixture was stirred for 24 h under a nitrogen atmosphere. The solution was drop-wise added into a mixture of 1 M AcOH (30 mL), MeOH (15 mL) and $CHCl_3$ (30 mL). The solution was then washed with a AcOH:MeOH (1:1, *v/v*) mixture once and with a H_2O:MeOH (2:1, *v/v*) mixture twice. The organic layer was then dried with $MgSO_4$ and evaporated. The solid was dissolved in a minimum volume of $CHCl_3$ and was solidified by the addition of Et_2O. The precipitate was collected by decantation and the solvent evaporated, and the precipitate dried under vacuum to yield compound **8** (263 mg, 95%). ^{1}H NMR (DMSO-d_6, 400 MHz): 0.05 (s, 9 H), 1.05 (m, 2 H), 1.91 (m, 2 H), 2.20 (m, 2 H), 3.07 (s, C*CH*, 1 H), 3.83 (s, 2 H), 4.30 (m, 3 H), 4.60 (d, 2 H), 6.66 (d, 2 H), 7.03 [triplet (t), folate amine, 1 H], 7.65 (d, 2 H), 8.17 (br, folate amide, 1 H), 8.30 (br, amide, folate amide, 2 H), 8.84 (s, 1 H), 11.9 (br, folate O*H*, 2 H). MALDI-TOF MS calculated for $C_{28}H_{34}N_8O_7Si$ [M + H]$^+$ 623.239; obtained [M+H]$^+$ 623.403.

3.2.7. Compound **9** (Folate-Propargyl)

To a solution of compound **8** (70 mg, 0.11 mmol) in DMSO (1 mL), 1 M tetrabutylammonium fluoride (TBAF) in THF (1 mL) was added and then stirred. After 3 h stirring at RT, the mixture was solidified by the addition of H_2O:AcOH (2:1, *v/v*), and the material purified by centrifugation and decantation. This procedure was performed three times. The compound was solidified by Et_2O and centrifuged once. The orange powder of compound **9** (53 mg, 98%) was obtained by drying *in vacuo*. Figure S1 shows ^{1}H NMR data of compound **9**. ^{1}H NMR (DMSO-d_6, 400 MHz): 1.94 (m, 2 H), 2.21 (t, 2 H), 3.07 (s, C*CH*, 1 H), 3.82 (s, 2 H), 4.26 (m, 1 H), 4.48 (d, 2 H), 6.64 (d, 2 H), 6.93 (t, folate amine, 1 H), 7.65 (d, 2 H), 8.13 (d, folate amide, 1 H), 8.29 (t, folate amide, 1 H), 8.65 (s, 1 H), 12.0 (br, folate O*H*, 2 H). MALDI-TOF MS calculated for $C_{22}H_{23}N_8O_5$ [M + H]$^+$ 479.179; obtained [M + H]$^+$ 479.310.

3.2.8. Compound **11**

A mixture of compound **10** (3 g, 9.9 mmol) and CDI (1.60 g, 9.9 mmol) in CH_2Cl_2 (30 mL) was stirred at RT for 1 h, followed by the addition of 1.46 mL of 9.9 mmol tetramethylsilane ethanol (TMS EtOH), and this sample was stirred for a further 18 h. H_2O (150 mL) was added to the reaction mixture and the resulting mixture was partitioned. The organic layer was dried with anhydrous Na_2SO_4 and the solvent evaporated under reduced pressure. The residue was purified on a silica gel column with 25% (*v/v*) ethyl acetate in hexane to give a colorless oil 11 (3.46 g, 87%). ^{1}H NMR (DMSO-d_6, 400 MHz): 0.05 (s, 9 H), 0.99–1.04 (m, 2 H), 1.44–1.45 (m, 18 H), 1.85–1.95 (m, 1 H), 2.07–2.16 (m, 1 H), 2.24–2.38 (m, 2 H), 4.20–4.30 (m, 3 H), 5.09 (d, *J* = 8.4 Hz, 1 H).

3.2.9. Compound **12**

A mixture of compound **11** (2 g, 4.9 mmol) and TFA:CH_2Cl_2 (1:2, *v/v*) (15 mL) was stirred at 0 °C for 30 min. The reaction mixture was then allowed to acquire at RT for 4.5 h. while stirring. The solvent of the reaction mixture was evaporated and the material purified on a silica gel column with 20–35% (*v/v*) MeOH in $CHCl_3$ to give compound **12** (0.842 g, 69%, as a colorless semisolid). ^{1}H NMR (DMSO-d_6, 400 MHz): 0.05 (s, 9 H), 0.99–1.03 (m, 2 H), 1.94–2.05 (m, 2 H), 2.36–2.48 (m, 2 H), 4.03 (t, *J* = 6.4 Hz, 1 H), 4.22–4.26 (m, 2 H).

3.2.10. Compound **13**

Compound **12** (1.75 g, 3.4 mmol), compound **4** (1.28 g, 5.2 mmol) and MTBD (1.48 mL, 10 mmol) in DMSO (15 mL) were stirred at RT in a 100 mL two neck round bottom flask. After 21 h, the resulting mixture was poured into a mixture of aqueous AcOH (1 M, 600 mL), MeOH (250 mL) and $CHCl_3$ (600 mL). The organic layer was then washed with 1 M AcOH:MeOH (1/1, *v/v*) (400 mL) and H_2O:MeOH (2/1) (600 mL ×2). The resulting organic solution was dried with anhydrous Na_2SO_4 and evaporated under reduced pressure. The crude mixture was purified on a silica gel column with CHCl$_3$:MeOH:ethyl acetate:AcOH (17:1:2:0.08, *v/v/v/v*) and then CHCl$_3$:MeOH:AcOH (9:1:0.025, *v/v/v*) to afford a yellow solid compound **13** (1.38 g, 58%). ^{1}H NMR (DMSO-d$_6$, 400 MHz): 0.01 (s, 9 H), 0.06 (s, 9 H), 0.91–0.95 (m, 2 H), 1.03–1.07 (m, 2 H), 1.86–1.95 (m, 1 H), 1.99–2.06 (m, 1 H), 2.30–2.34 (m, 2 H), 4.09–4.13 (m, 2 H), 4.28–4.34 (m, 3 H), 4.59 (d, *J* = 6 Hz, 2 H), 6.65 (d, *J* = 8.8 Hz, 2 H), 7.02 (t, *J* = 6.4 Hz, 1 H), 7.65 (d, *J* = 8.4 Hz, 2 H), 8.22 (d, *J* = 7.6 Hz, 1 H), 8.84 (s, 1 H), 11.88 (br, 4 H).

3.2.11. Compound **14**

To a solution of compound **13** (1 g, 1.5 mmol) and 5 mL *N,N*-dimethylformamide (DMF), NHS (202 mg, 1.7 mmol) and EDC (279 mg, 1.5 mmol) were added. The resulting mixture was stirred at RT for 18 h. The reaction mixture was poured into water (300 mL) and the yellow precipitate was collected by filtration to afford compound **14** (1.03 g, 90%). ^{1}H NMR (DMSO-d$_6$, 400 MHz): 0.00 (s, 9 H), 0.06 (s, 9 H), 0.91–0.95 (m, 2 H), 1.02–1.07 (m, 2 H), 2.04–2.16 (m, 2 H), 2.76–2.84 (m, 6 H), 4.09–4.15 (m, 2 H), 4.28–4.32 (m, 3 H), 4.59 (s, 2 H), 6.66 (d, *J* = 8.8 Hz, 2 H), 7.66 (d, *J* = 9.2 Hz, 2 H), 8.32 (d, *J* = 7.6 Hz, 1 H), 8.84 (s, 1 H), 11.70 (br, 2 H).

3.2.12. Compound **15**

Compound **14** (142 mg, 0.18 mmol), DBCO amine (50 mg, 0.18 mmol) and triethylamine (Et$_3$N) (0.04 mL, 0.29 mmol) in 3 mL CH$_2$Cl$_2$ were added and stirred at RT for 3.5 h in a 20 mL round bottom flask. The reaction mixture was diluted with CHCl$_3$ (25 mL) and washed with water (25 mL ×2). The organic layer was dried with anhydrous Na$_2$SO$_4$, evaporated and the sample purified on a Sephadex LH-20 column with CHCl$_3$:MeOH = 1:1 (*v/v*) to afford compound **15** (0.150 mg; 88%). ^{1}H NMR (DMSO-d$_6$, 400 MHz): −0.007 (s, 9H), 0.054 (s, 9 H), 0.89–0.93 (m, 2 H), 1.02–1.06 (m, 2 H), 1.75–2.08 (m, 5 H), 2.36–2.43 (m, 1 H), 2.90–2.95 (m, 1 H), 3.04–3.12 (m, 1 H), 3.60 (dd, *J* = 13.6 Hz, *J* = 6.8 Hz, 1 H), 4.07–4.11 (m, 2 H), 4.21–4.31 (m, 3 H), 4.59 (d, *J* = 5.6 Hz, 2 H), 5.01 (t, *J* = 14.4 Hz, 1 H), 6.65 (dd, *J* = 8.8 Hz, *J* = 4.4 Hz, 2 H), 7.05 (q, *J* = 6 Hz, 1 H), 7.24–7.46 (m, 6 H), 7.52–7.66 (m, 5 H), 8.26 (d, *J* = 7.2 Hz, 1 H), 8.84 (s, 1 H), 11,69 (br, 1 H). ^{13}C NMR (DMSO-d$_6$, 100 MHz): −1.5, 16.8, 17.0, 26.2, 31.6, 34.1, 35.0, 46.0, 52.4, 54.8, 62.4, 64.6, 108.0, 111.2, 114.3, 121.3, 121.4, 122.4, 125.2, 126.8, 127.7, 128.0, 128.2, 128.9, 129.0, 129.5, 130.0, 132.3, 132.4, 148.3, 149.2, 150.7, 151.4, 152.1, 155.0, 159.5, 166.3, 170.1, 171.3, 172.3.

3.2.13. Compound **16** (Folate-DBCO)

To a solution of compound **15** (100 mg, 0.1 mmol) in DMSO (1 mL), TBAF [1.14 mL of 1 M in anhydrous THF, 10 equivalent (eq.)] was added and then stirred at RT. After 10 h stirring, AcOH (1.25 mL) was added and the mixture was poured into a mixture of CHCl$_3$ and ethyl acetate (4:1, 25mL). The yellowish precipitate formed in the solution was collected by filtration and then recrystallized in a mixture EtOH:MeOH to give the yellow solid compound **16** (folate-DBCO). ^{1}H NMR data is displayed in Figure S2A. ^{1}H NMR (DMSO-d$_6$, 400 MHz): 1.77–2.04 (m, 5 H), 2.33–2.40 (m, 1 H), 2.88–2.92 (m, 1 H), 3.05–3.11 (m, 1 H), 3.61 (dd, *J* = 14 Hz, *J* = 3.6 Hz, 1 H), 4.10–4.14 (m, 1 H), 4.47 (d, *J* = 6 Hz, 2 H), 5.01 (dd, *J* = 14.4 Hz, *J* = 8.4 Hz, 1 H), 6.63 (dd, *J* = 8.8 Hz, *J* = 2.8 Hz, 2 H), 6.91–6.94 (m, 1 H), 7.07 (br, 1 H), 7.27–7.48 (m, 6 H), 7.55–7.66 (m, 5 H), 7.94 (br, 1 H), 8.63 (s, 1 H). ^{13}C NMR data is displayed in Figure S2B. ^{13}C NMR (DMSO-d$_6$, 100 MHz): 27.4, 31.9, 34.1, 35.0, 46.0, 52.8, 54.8, 108.1, 111.3, 114.3, 121.4, 121.8, 122.5, 125.3, 126.8, 127.7, 128.0, 128.2, 128.7, 129.0, 129.5, 132.4, 148.3, 148.5, 150.6, 151.4, 154.3, 161.5, 165.7, 170.2, 171.7, 174.4. HRMS data is displayed in Figure S2C. HRMS (QSTAR Elite, AB SCIEX, Framingham, MA, USA) calculated for C$_{37}$H$_{33}$N$_9$NaO$_6$ [M + Na]$^+$ 722.2446; obtained [M + Na]$^+$ 722.2445.

3.3. *Click Reaction of Folate-Propargyl or Folate-DBCO with AzPhe-Fmoc*

Reaction schemes for the click reactions of folate-propargyl and folate-DBCO with AzPhe-Fmoc are shown in Figure 2. A 1 mM stock of folate-propargyl and a 10 mM stock of BimH$_3$ were prepared in DMSO for CuAAC. A 10 mM stock of AzPhe, sodium ascorbate, 2 mM stock of copper (II) sulfate ($CuSO_4$) and copper (I) chloride (CuCl) were prepared in H_2O.

A 10 mM folate-DBCO stock was prepared in DMF and a 10 mM stock of AzPhe-Fmoc was dissolved in H_2O for SPAAC. Several trials were performed for both compounds listed in Table 1. RP-HPLC using an Inertsil ODS-3 column (Nacalai tesque Inc., Kyoto, Japan) at 25 °C for 55 min was performed with H_2O containing 0.1 % (*v/v*) TFA (solvent A) and acetonitrile containing 0.1 % (*v/v*) TFA (solvent B) as a solvent system with a gradient from 0–0.10 min at 90% A, 5–40 min at 90–50% A, 40–45 min at 50–0% A and 45–47 min at 0–90% A, and flow rate of 1 mL/min. In some cases, a gradient from 0–10 min at 90% A, 5–40 min at 90–30% A, 40–43 min at 30–0% A and 43–45 min at 0–90% A was used, and a flow rate of 1 mL/min.

3.4. Synthesis and Purification of Peptides with N-terminal Biotin-PEG$_{24}$

The folate-peptide conjugates were synthesized by conventional Fmoc based solid-phase synthesis methods using a high purity single peptide synthesizer MultiPep CF and micro-column (INTAVIS Co. Ltd., Cologne, Germany). During synthesis, coupling and deprotection steps were carried out in the peptide synthesizer. All peptides were synthesized at the 10 μmol scale. Peptide synthesis is as follows: Preloaded 0.21 mmol/g of fmoc-Gln(Trt)-NovaSyn TGA (Novabiochem, Darmstadt, Germany), 0.19 mmol/g of fmoc-Ala-NovaSyn TGA (Novabiochem) or 0.24 mmol/g of fmoc-Tyr (tBu)-NovaSyn (Novabiochem) was used for synthesis of GF[AzPhe]IQ, SE[AzPhe]KA and DSE[AzPhe]KAY, respectively. Fmoc deprotection was performed by using 20% (*v/v*) piperidine in *N*-methyl-2-pyrrolidone (NMP) or 1% (*v/v*) formic acid + 20% (*v/v*) piperidine in NMP, depending on the amino acid content of the peptide. For the coupling step, the corresponding amino acid (5 times mol with respect to resin) was incubated with the resin for 30 min in the presence of NMP (8 μL), 0.5 M (2-(1H-benzotriazol-1-yl)-1,1,3,3-tetramethyluronium hexafluorophosphate (150 μL) and 4 M N-methylmorpholine (45 μL). AzPhe (65 ng) was used during each synthesis. After confirming the mass of the synthesized peptides by MALDI-TOF MS, the beads were incubated overnight on a shaker with a mixture of biotin-PEG$_{24}$-NHS (34 mg, 1.5 mol eq.), hydroxybenzotriazole (8.3 mg, 0.6 mol eq.) with the addition of NMP (300 μL). The reaction scheme for *N*-terminal peptide modification with biotin-PEG$_{24}$ modification is shown in Figure 3A. After confirming the mass of the product with MALDI-TOF MS, the peptides were cleaved from the resin using a cleavage cocktail [95.0% (*v/v*) TFA, 2.5% (*v/v*) triisopropylsilane and 2.5% (*v/v*) H_2O]. Depending on the amino acid content of each peptide, resins were incubated with the cleavage cocktail for 2–4 h in a light protected container with intermittent shaking. The cleavage mixture was then filtered to remove the beads and peptides were precipitated using cold Et_2O. The resulting precipitate was centrifuged and washed three times with Et_2O. Et_2O was removed by overnight vacuum lyophilization and peptides were obtained in powder form. The products were further purified by RP-HPLC using the Inertsil ODS-3 column at 25 °C for GFZIQ and SEZKA: 50 min with a gradient of 1–51% (*v/v*) acetonitrile in water containing 0.1% (*v/v*) TFA. An Inert Sustain C18 column (Nacalai tesque Inc.) at 50 °C was used to further purify the DSEZKAY peptide. The gradient was 25–55% (*v/v*) acetonitrile in water containing 0.1 % (*v/v*) TFA for 30 min. Peptides were purified as shown in Figure S3 and analyzed by MALDI-TOF MS (Figure S4A–C) and the results are summarized in Table S1. The purified peptides were lyophilized and stored until required.

3.5. SPAAC Click Chemistry to Conjugate Folate into Peptides

Folate-DBCO (3.55 mg) was dissolved in 1 mM DMSO (2.5 mL). Folate-DBCO was then diluted to 0.1 mM with MeOH [total DMSO = 10% (*v/v*)]. A 1 mM stock of ~50% purified peptide-PEG$_{24}$-biotin was prepared in MeOH. 1 mol eq., 0.1 mM (1 mL) biotin-PEG$_{24}$-peptide was mixed with 2 mol eq., 0.1 mM (2 mL) folate-DBCO in 7.14% (*v/v*) DMSO and 92.86% MeOH [40]. The mixture was constantly rotated at 5 rpm and 25 °C on a rotator for 16 h under dark conditions. After the reaction, mixtures were purified by RP-HPLC using the Inertsil ODS-3 column at 25 °C for 50 min with a gradient of 1–51% (*v/v*) acetonitrile in water containing 0.1% (*v/v*) TFA. Folate-peptide conjugates were analyzed by MALDI-TOF MS analysis.

3.6. Purification and Refolding of FRα

All steps performed for purification and refolding of FRα were carried out according to our previous report [41]. However, some different reagents were used. For cell body washing, we used 4 M urea instead of Triton X-114. The inclusion bodies were solubilized and purified with 8 M urea instead of 6 M guanidine HCl. Purification and refolding data are shown in Figure S5, and Tables S2 and S3.

3.7. BLI Measurements

The binding affinity of refolded FRα toward folate and folate conjugated peptide aptamers was measured by biolayer interferometry at 25 °C using a BLItz system (ForteBio) with kinetics buffer [10 mM PBS, pH 7.4, 0.5% (w/v) BSA and 0.01% (v/v) Tween 20]. The measurement procedure has been reported previously [41]. Streptavidin-coated biosensors (SA sensors were hydrolyzed for 2 h in 250 μL kinetics buffer and then soaked with 250 μL folate-PEG_8-biotin (2.5 μM), or a variety of concentrations of 250 μL folate-peptide-PEG_{24}-biotin conjugates at a stirring speed of 2200 rpm. Two baselines were measured for each sensor in kinetics buffer for 30 and 300 s prior to the immobilization and association step, respectively. Folate-PEG_8-biotin or folate-peptide-PEG_{24}-biotin conjugates immobilized to SA biosensors were dipped into FRα solutions for the association step. Dissociation was monitored in 250 μL immune assay kinetics buffer. To eliminate errors from non-specific binding of the analyte (FRα) on the SA biosensor chips, reference data with the same concentrations of analyte were also measured.

The obtained binding data were analyzed using a 1:1 local analysis mode applied with association and dissociation step corrections by the BLItz Pro1.2 software (ForteBio). The reference measurements were subtracted during data analysis to determine k_a, k_d and K_D.

4. Conclusions

By conjugation with peptides the affinities of folate to the receptor were enhanced. The conjugation with designed peptides will be useful for enhancement of ligands affinities through the increase of binding sites.

Supplementary Materials: Supplementary materials can be found at http://www.mdpi.com/1422-0067/20/9/2152/s1.

Author Contributions: Y.I., H.M., D.K. and M.Y. conceived the project. A.N., M.U. and K.K. synthesized the organic compounds. H.M. designed the peptide sequences in the conjugates. R.D. and H.M. constructed the protein overexpression systems and purified the proteins. R.D. and H.M. carried out the click reactions and the subsequent HPLC assays. R.D. and H.M. measured the binding kinetics. R.D., H.M., M.U. and Y.I. wrote the manuscript.

Funding: This project was funded by the Incentive Research Program of RIKEN (FY2016, H.M.) and JSPS-Turkey (I2016652, Y.I.). R.D. and A.N. were supported by the International Program Associate of RIKEN (IPA number: 151022) and SPDR (Grant number: 201801061156), respectively.

Acknowledgments: We thank the Support Unit for Bio-material Analysis, RIKEN CBS Research Resources Center for peptide synthesis, *N*-terminal modification with biotin-PEG_{24} and HPLC purification. We thank Kenji Suzuki to help the computational design of the compounds. We are thankful to Katsunori Tanaka, Masashi Ueki and Yun Heo for their input in CuAAC reaction. We are also thankful to Primetech Corp. for help with analysis of the BLI data. We thank the Edanz Group (www.edanzediting.com/ac) for editing a draft of this manuscript.

Conflicts of Interest: The authors declare no conflict of interest.

Abbreviations

AcOH	Acetic acid
ADC	Antibody-drug conjugate
AzPhe	4-Azido phenylalanine
BimH$_3$	Tris-(2-benzimidazolylmethyl) amine
BLI	Biolayer interferometry
BSA	Bovine serum albumin
brs	Broad singlet
CDI	1,1′-carbonyldiimidazole
CuAAC	Cu(I)-catalyzed alkyne-azide cycloaddition
d	Duplet
DBCO	Dibenzylcyclooctyne
DMF	*N,N*-Dimethylformamide
DMSO	Dimethyl sulfoxide
EDC	1-Ethyl-3-(3-dimethylaminopropyl) carbodiimide hydrochloride
eq	Equivalent
EtOH	Ethanol
Et$_2$O	Diethyl ether
Et$_3$N	Triethylamine
Fmoc	9-Fluorenylmethyloxycarbonyl group
FR	Folate receptor
FRα	Folate receptor alpha
k_a	Association rate constant
K_D	Equilibrium dissociation constant
k_d	Dissociation rate constant
m	Multiplet
MALDI-TOF MS	Matrix assisted laser desorption/ionization-time of flight mass spectrometry
MeOH	Methanol
MTBD	7-Methyl-1,5,7-triazabicyclo [4.4.0] dec-5-ene
NHS	*N*-hydroxysuccinimide
NMP	*N*-methyl-2-pyrrolidone
PBS	Phosphate buffered saline
PEG	Polyethylene glycol
q	Quartet
RT	Room temperature
s	Singlet
SA	Streptavidin
SPAAC	Strain-promoted alkyne-azide cycloaddition
TBAF	Tetrabutylammonium fluoride
TFA	Trifluoroacetic acid
THF	Tetrahydrofuran
TMS	Tetramethylsilane

References

1. Bahrami, B.; Mohammadnia-Afrouzi, M.; Bakhshaei, P.; Yazdani, Y.; Ghalamfarsa, G.; Yousefi, M.; Sadreddini, S.; Jadidi-Niaragh, F.; Hojjat-Farsangi, M. Folate-conjugated nanoparticles as a potent therapeutic approach in targeted cancer therapy. *Tumour Biol.* **2015**, *36*, 5727–5742. [CrossRef]
2. Vrettos, E.I.; Mezo, G.; Tzakos, A.G. On the design principles of peptide-drug conjugates for targeted drug delivery to the malignant tumor site. *Beilstein J. Org. Chem.* **2018**, *14*, 930–954. [CrossRef]
3. Kato, T.; Jin, C.S.; Ujiie, H.; Lee, D.; Fujino, K.; Wada, H.; Hu, H.P.; Weersink, R.A.; Chen, J.; Kaji, M.; et al. Nanoparticle targeted folate receptor 1-enhanced photodynamic therapy for lung cancer. *Lung Cancer* **2017**, *113*, 59–68. [CrossRef] [PubMed]

4. Fernandez, M.; Javaid, F.; Chudasama, V. Advances in targeting the folate receptor in the treatment/imaging of cancers. *Chem. Sci.* **2018**, *9*, 790–810. [CrossRef] [PubMed]

5. Ak, G.; Yilmaz, H.; Gunes, A.; Hamarat Sanlier, S. In vitro and in vivo evaluation of folate receptor-targeted a novel magnetic drug delivery system for ovarian cancer therapy. *Artif. Cells Nanomed. Biotechnol.* **2018**, *46*, 926–937. [CrossRef]

6. Cheung, A.; Bax, H.J.; Josephs, D.H.; Ilieva, K.M.; Pellizzari, G.; Opzoomer, J.; Bloomfield, J.; Fittall, M.; Grigoriadis, A.; Figini, M.; et al. Targeting folate receptor alpha for cancer treatment. *Oncotarget* **2016**, *7*, 52553–52574. [CrossRef] [PubMed]

7. Hassan, M.; Watari, H.; AbuAlmaaty, A.; Ohba, Y.; Sakuragi, N. Apoptosis and molecular targeting therapy in cancer. *Biomed. Res. Int.* **2014**, *2014*, 150845. [CrossRef] [PubMed]

8. Doi, S.; Zou, Y.; Togao, O.; Pastor, J.V.; John, G.B.; Wang, L.; Shiizaki, K.; Gotschall, R.; Schiavi, S.; Yorioka, N.; et al. Klotho inhibits transforming growth factor-beta1 (TGF-beta1) signaling and suppresses renal fibrosis and cancer metastasis in mice. *J. Biol. Chem.* **2011**, *286*, 8655–8665. [CrossRef]

9. Kim, M.; Pyo, S.; Kang, C.H.; Lee, C.O.; Lee, H.K.; Choi, S.U.; Park, C.H. Folate receptor 1 (FOLR1) targeted chimeric antigen receptor (CAR) T cells for the treatment of gastric cancer. *PLoS ONE* **2018**, *13*, e0198347. [CrossRef] [PubMed]

10. Mitra, A.; Renukuntla, J.; Shah, S.; Boddu, S.H.S.; Vadlapudi, A.D.; Vadlapatla, R.K.; Pal, D. Functional characterization and expression of folate receptor-α in T47D human breast cancer cells. *Drug Dev. Ther.* **2015**, *6*, 52–61. [CrossRef]

11. Parasassi, T.; Giusti, A.M.; Raimondi, M.; Ravagnan, G.; Sapora, O.; Gratton, E. Cholesterol protects the phospholipid bilayer from oxidative damage. *Free Radic. Biol. Med.* **1995**, *19*, 511–516. [CrossRef]

12. Reichert, J.M. Antibody-based therapeutics to watch in 2011. *mAbs* **2014**, *3*, 76–99. [CrossRef]

13. Elnakat, H.; Ratnam, M. Distribution, functionality and gene regulation of folate receptor isoforms: Implications in targeted therapy. *Adv. Drug Deliv. Rev.* **2004**, *56*, 1067–1084. [CrossRef] [PubMed]

14. Chen, C.; Ke, J.; Zhou, X.E.; Yi, W.; Brunzelle, J.S.; Li, J.; Yong, E.L.; Xu, H.E.; Melcher, K. Structural basis for molecular recognition of folic acid by folate receptors. *Nature* **2013**, *500*, 486–489. [CrossRef]

15. Vlahov, I.R.; Leamon, C.P. Engineering folate-drug conjugates to target cancer: From chemistry to clinic. *Bioconjug. Chem.* **2012**, *23*, 1357–1369. [CrossRef]

16. Kam, N.W.; O'Connell, M.; Wisdom, J.A.; Dai, H. Carbon nanotubes as multifunctional biological transporters and near-infrared agents for selective cancer cell destruction. *Proc. Natl. Acad. Sci. USA* **2005**, *102*, 11600–11605. [CrossRef]

17. Zhu, L.; Dong, D.; Yu, Z.-L.; Zhao, Y.-F.; Pang, D.-W.; Zhang, Z.-L. Folate-Engineered Microvesicles for Enhanced Target and Synergistic Therapy toward Breast Cancer. *ACS Appl. Mater. Interfaces* **2017**, *9*, 5100–5108. [CrossRef] [PubMed]

18. Sun, X.; Du, R.; Zhang, L.; Zhang, G.; Zheng, X.; Qian, J.; Tian, X.; Zhou, J.; He, J.; Wang, Y.; et al. A pH-Responsive Yolk-Like Nanoplatform for Tumor Targeted Dual-Mode Magnetic Resonance Imaging and Chemotherapy. *ACS Nano* **2017**, *11*, 7049–7059. [CrossRef] [PubMed]

19. Wang, H.; Sun, S.; Zhang, Y.; Wang, J.; Zhang, S.; Yao, X.; Chen, L.; Gao, Z.; Xie, B. Improved drug targeting to liver tumor by sorafenib-loaded folate-decorated bovine serum albumin nanoparticles. *Drug Deliv.* **2019**, *26*, 89–97. [CrossRef]

20. Sudimack, J.; Lee, R.J. Targeted drug delivery via the folate receptor. *Adv. Drug Deliv. Rev.* **2000**, *41*, 147–162. [CrossRef]

21. Lu, Y.; Low, P.S. Folate-mediated delivery of macromolecular anticancer therapeutic agents. *Adv. Drug Deliv. Rev.* **2002**, *54*, 675–693. [CrossRef]

22. Bae, Y.; Jang, W.-D.; Nishiyama, N.; Fukushima, S.; Kataoka, K. Multifunctional polymeric micelles with folate-mediated cancer cell targeting and pH-triggered drug releasing properties for active intracellular drug delivery. *Mol. Biosyst.* **2005**, *1*, 242–250. [CrossRef]

23. Bae, Y.; Nishiyama, N.; Kataoka, K. In Vivo Antitumor Activity of the Folate-Conjugated pH-Sensitive Polymeric Micelle Selectively Releasing Adriamycin in the Intracellular Acidic Compartments. *Bioconjug. Chem.* **2007**, *18*, 1131–1139. [CrossRef]

24. Xing, L.; Xu, Y.; Sun, K.; Wang, H.; Zhang, F.; Zhou, Z.; Zhang, J.; Zhang, F.; Caliskan, B.; Qiu, Z.; et al. Identification of a peptide for folate receptor alpha by phage display and its tumor targeting activity in ovary cancer xenograft. *Sci. Rep.* **2018**, *8*, 8426. [CrossRef]

25. Li, S.; Roberts, R.W. A Novel Strategy for In Vitro Selection of Peptide-Drug Conjugates. *Chem. Biol.* **2003**, *10*, 233–239. [CrossRef]
26. Wang, W.; Hirano, Y.; Uzawa, T.; Liu, M.Z.; Taiji, M.; Ito, Y. In vitro selection of a peptide aptamer that potentiates inhibition of cyclin-dependent kinase 2 by purvalanol. *Medchemcomm* **2014**, *5*, 1400–1403. [CrossRef]
27. Wang, W.; Hirano, Y.; Uzawa, T.; Taiji, M.; Ito, Y. Peptide-Assisted Enhancement of Inhibitory Effects of Small Molecular Inhibitors for Kinases. *Bull. Chem. Soc. Jpn.* **2016**, *89*, 444–446. [CrossRef]
28. Bazewicz, C.G.; Liskov, M.T.; Hines, K.J.; Brewer, S.H. Sensitive, site-specific, and stable vibrational probe of local protein environments: 4-azidomethyl-L-phenylalanine. *J. Phys. Chem. B* **2013**, *117*, 8987–8993. [CrossRef] [PubMed]
29. Link, A.J.; Mock, M.L.; Tirrell, D.A. Non-canonical amino acids in protein engineering. *Curr. Opin. Biotech.* **2003**, *14*, 603–609. [CrossRef] [PubMed]
30. Huang, L.; Li, Z.; Zhao, Y.; Zhang, Y.; Wu, S.; Zhao, J.; Han, G. Ultralow-Power Near Infrared Lamp Light Operable Targeted Organic Nanoparticle Photodynamic Therapy. *J. Am. Chem. Soc.* **2016**, *138*, 14586–14591. [CrossRef]
31. Consoli, G.M.L.; Granata, G.; Fragassi, G.; Grossi, M.; Sallese, M.; Geraci, C. Design and synthesis of a multivalent fluorescent folate–calix[4]arene conjugate: Cancer cell penetration and intracellular localization. *Org. Biomol. Chem.* **2015**, *13*, 3298–3307. [CrossRef]
32. De, P.; Gondi, S.R.; Sumerlin, B.S. Folate-Conjugated Thermoresponsive Block Copolymers: Highly Efficient Conjugation and Solution Self-Assembly. *Biomacromolecules* **2008**, *9*, 1064–1070. [CrossRef] [PubMed]
33. White, B.M.; Zhao, Y.; Kawashima, T.E.; Branchaud, B.P.; Pluth, M.D.; Jasti, R. Expanding the Chemical Space of Biocompatible Fluorophores: Nanohoops in Cells. *ACS Cent. Sci.* **2018**, *4*, 1173–1178. [CrossRef]
34. Lehner, R.; Liu, K.; Wang, X.; Hunziker, P. Efficient Receptor Mediated siRNA Delivery in Vitro by Folic Acid Targeted Pentablock Copolymer-Based Micelleplexes. *Biomacromolecules* **2017**, *18*, 2654–2662. [CrossRef] [PubMed]
35. Golas, P.L.; Tsarevsky, N.V.; Matyjaszewski, K. Structure-re activity correlation in "Click" chemistry: Substituent effect on azide reactivity. *Macromol. Rapid Comm.* **2008**, *29*, 1167–1171. [CrossRef]
36. Long, N.; Wong, W.T. *The Chemistry of Molecular Imaging*; Wiley: London, UK, 2014; pp. 25–54.
37. Li, H.; Aneja, R.; Chaiken, I. Click chemistry in peptide-based drug design. *Molecules* **2013**, *18*, 9797–9817. [CrossRef]
38. McKay, C.S.; Finn, M.G. Click chemistry in complex mixtures: Bioorthogonal bioconjugation. *Chem. Biol.* **2014**, *21*, 1075–1101. [CrossRef] [PubMed]
39. Wibowo, A.S.; Singh, M.; Reeder, K.M.; Carter, J.J.; Kovach, A.R.; Meng, W.; Ratnam, M.; Zhang, F.; Dann, C.E., 3rd. Structures of human folate receptors reveal biological trafficking states and diversity in folate and antifolate recognition. *Proc. Natl. Acad. Sci. USA* **2013**, *110*, 15180–15188. [CrossRef]
40. Dommerholt, J.; Rutjes, F.; van Delft, F.L. Strain-Promoted 1,3-Dipolar Cycloaddition of Cycloalkynes and Organic Azides. *Top. Curr. Chem. (Cham)* **2016**, *374*, 16. [CrossRef]
41. Dharmatti, R.; Miyatake, H.; Zhang, C.; Ren, X.; Yumoto, A.; Kiga, D.; Yamamura, M.; Ito, Y. Escherichia coli expression, purification, and refolding of human folate receptor alpha (hFRalpha) and beta (hFRbeta). *Protein Expr. Purif.* **2018**, *149*, 17–22. [CrossRef]

International Journal of
Molecular Sciences

Article

Insight into Structural Characteristics of Protein-Substrate Interaction in Pimaricin Thioesterase

Shuobing Fan, Rufan Wang, Chen Li, Linquan Bai, Yi-Lei Zhao and Ting Shi *

State Key Laboratory of Microbial Metabolism, Joint International Research Laboratory of Metabolic and
Developmental Sciences, School of Life Sciences and Biotechnology, Shanghai Jiao Tong University,
Shanghai 200240, China; bigyellowjar@sjtu.edu.cn (S.F.); wangrufan@sjtu.edu.cn (R.W.);
ll105309790@sjtu.edu.cn (C.L.); bailq@sjtu.edu.cn (L.B.); yileizhao@sjtu.edu.cn (Y.-L.Z.)
* Correspondence: tshi@sjtu.edu.cn; Tel./Fax: +86-21-34207347

Received: 21 January 2019; Accepted: 6 February 2019; Published: 18 February 2019

Abstract: As a polyene antibiotic of great pharmaceutical significance, pimaricin has been extensively studied to enhance its productivity and effectiveness. In our previous studies, pre-reaction state (PRS) has been validated as one of the significant conformational categories before macrocyclization, and is critical to mutual recognition and catalytic preparation in thioesterase (TE)-catalyzed systems. In our study, molecular dynamics (MD) simulations were conducted on pimaricin TE-polyketide complex and PRS, as well as pre-organization state (POS), a molecular conformation possessing a pivotal intra-molecular hydrogen bond, were detected. Conformational transition between POS and PRS was observed in one of the simulations, and POS was calculated to be energetically more stable than PRS by 4.58 kcal/mol. The structural characteristics of PRS and POS-based hydrogen-bonding, and hydrophobic interactions were uncovered, and additional simulations were carried out to rationalize the functions of several key residues (Q29, M210, and R186). Binding energies, obtained from MM/PBSA calculations, were further decomposed to residues, in order to reveal their roles in product release. Our study advanced a comprehensive understanding of pimaricin TE-catalyzed macrocyclization from the perspectives of conformational change, protein-polyketide recognition, and product release, and provided potential residues for rational modification of pimaricin TE.

Keywords: pimaricin thioesterase; protein-substrate interaction; macrocyclization; molecular dynamics (MD) simulation; pre-reaction state

1. Introduction

Produced predominantly by the genus *Streptomyces*, polyene polyketides consist of a large family of natural compounds [1,2], including pimaricin [3], amphotericin B [4], nystatin A1 [5], and candicidin/FR008 [6]. The members of this class are widely used in clinical medicine for their broad spectral antifungal properties [7]. With constant progress in scientific research, their potential pharmaceutical values on antiviral, antiprotozoal, antiprion, and anticancer activity have been progressively clarified [8,9].

As a potent polyene antibiotic produced by *Streptomyces natalensis*, pimaricin (i.e. natamycin) primarily functions in the treatment of fungal infections caused by *Candida*, *Fusarium*, *Penicillium*, and *Aspergillus* organisms [10]. It is also known as an additive in food industry [11]. Pimaricin was approved by the Food and Drug Administration (FDA) as a drug for fungal keratitis in 1978 [12]. Ergosterol constitutes a major component in fungal and trypanosomatids plasma membrane, while absent in animal cells [13]. Pimaricin serves to bind specifically to ergosterol, downregulate vesicle trafficking, suppress membrane protein transport, and interfere with endocytosis, as well as exocytosis without permeabilizing the membrane [14–16]. Its strong performance in clinical trials makes pimaricin appealing to researchers, and its biosynthetic pathway modification and drug design have become new science hotspots [17].

Pimaricin is synthesized by type I polyketide synthases (PKSs), which consists of several covalently-connected multi-domain "modules." Each module contains a set pattern of domains, with acyltransferase (AT) adding acyl-CoA building blocks, acyl carrier protein (ACP) carrying the polyketide between modules, and ketosynthase (KS) accepting the growing chain from ACP [18]. An extra combination of domains, such as ketoreductase (KR), dehydratase (DH), methyltransferase (MT) were responsible for the production of distinctive macrolactones [19–21]. Situated in the last module, the thioesterase (TE) domain off-loads the ACP-tethered polyketide from PKS via macrocyclization or hydrolysis.

Consistent with serine hydrolase, a two-step mechanism has been proposed for TE-mediated catalysis of macrocyclic polyketides [22]. The first step is acylation of a universally conserved serine residue in the catalytic triad, generating an acyl-enzyme intermediate and stabilized for a considerable time [23]. The second step takes place with an intra-molecular hydroxyl group on polyketide which initiates a nucleophilic attack and leads to cyclization, or hydrolysis of the acyl-enzyme intermediate with no efficient intra-molecular nucleophile present.

In our previous work concerning 6-deoxyerythronolide B synthase (DEBS)-TE [24], a hydrogen bond emerged between the polyketide chain terminal hydroxyl group O_{11} and carbonyl oxygen O (Figure 1), as accompanied by the swing of C_{11} ethyl of substrate. This structure has been reported in Trauger's work in 2001, where it was referred to as the "pre-organization state" (POS). According to Trauger [25], the "product-like" conformation might contribute to pre-organization of the substrate for cyclization. The conformation with a hydrogen bond, forming between the O_{11} and Nε of His259 in the catalytic triad, was defined as an *active state* in our study. This conformation maintained for ~100 ns in our simulations with considerable steadiness. However, the distance of O_{11}-C_1 for nucleophilic attack was larger than 6 Å in *active state*. Finally, an advantageous conformer (the pre-reaction state, PRS) was found [24] after ~270 ns MD simulation, which possessed both hydrogen bond O_{11}-Nε_{H259} and an appropriate distance between O_{11} and C_1 to facilitate nucleophilic attack. Critical to mutual recognition and catalytic preparation between TE and substrate, the PRS seemed decisive in the occurrence of macrocyclization.

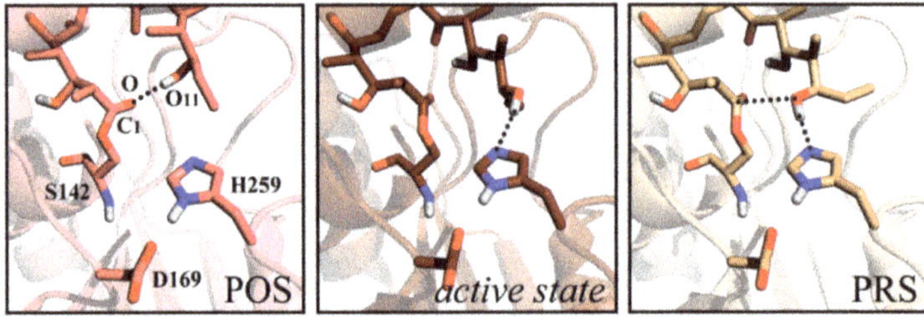

Figure 1. Structures of pre-organization state (POS), active state and the pre-reaction state (PRS) of 6-deoxyerythronolide B synthase (DEBS) thioesterase (TE) system.

To understand the molecular basis of pimaricin-TE (pima-TE) catalyzed macrocyclization, MD simulations were employed on enzyme-substrate complex. POS, *active state*, and PRS were discovered during 5 × 50 ns molecular dynamics (MD) simulations, and the conformational transition between POS and PRS was explored. The structural characteristics of POS and PRS were uncovered by conducting analyses of hydrogen-bonding and hydrophobic interactions. Additional simulations on several mutants (including Q29A, M210G, R186F, R186Y and S138C) were carried out to validate the functions of several key residues in substrate recognition and product release. At last, the binding energies of enzyme-product complex were obtained through MM/PBSA calculations, and with critical residues highlighted. We also provided an explanation on the departure of product from the active site.

2. Results and Discussion

2.1. Key Structural Conformations in MD Simulations

Intrigued by the recognition mechanism of pima-TE, 5×50 ns MD simulations were performed on constructed complex. Root-mean-square deviation (RMSD) analysis revealed that all five trajectories attained equilibrium (Figure S1). The root-mean-squared fluctuation (RMSF) values highlighted six parts on pima-TE. Firstly, the lid region was violently jacked up by the erected polyketide (Figure 2). As a polyene molecule with 26-atom skeleton, pimaricin's accommodation would require a larger space, compared with pikromycin, a 14-membered ring. It was conceivable that the relaxation of the substrate would incur close contact with the lid. Next, as components of the channel, αL2, as well as loop *l1* & *l2*, presented evident structural dynamism and various coiling states, while αL3 exhibited negligible fluctuation. Helix αL2 was proposed to wield a larger influence on protein-intermediate recognition than αL3, and *l1* and *l2* were responsible for the exit and entrance size. At last, RMSF indicated that loop *l3* adopted larger fluctuations than loop *l1* and *l2*, and the b-factor calculation [26,27] disclosed an inherent mobility of loop *l3*.

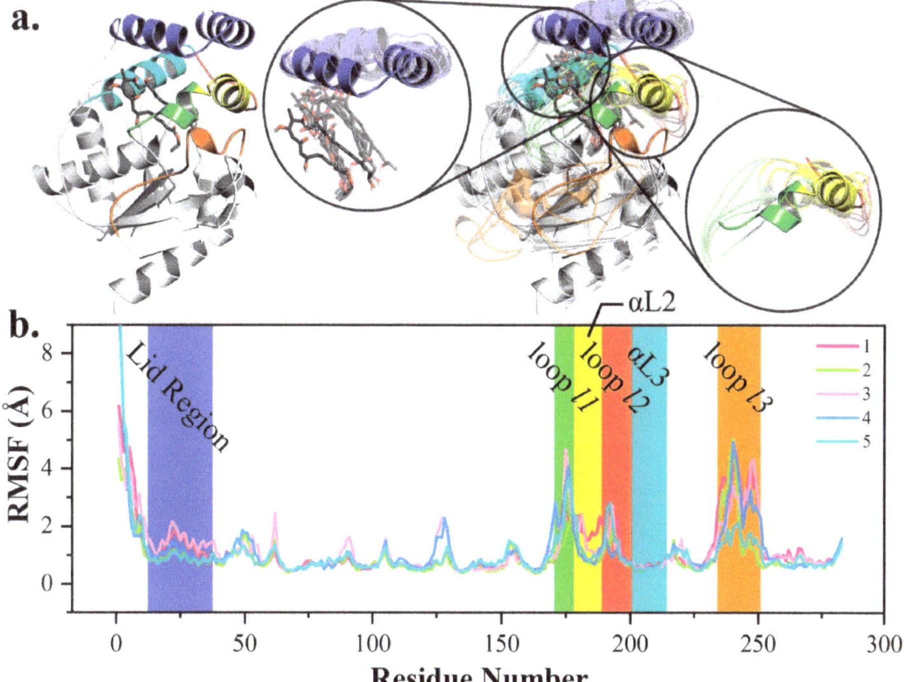

Figure 2. Conformational change of pima-TE system during molecular dynamics (MD) simulations. (**a**) Structural variations between post (opaque) and pre-simulation (transparent) complexes, with lid region, polyketide chain, α-helix αL2, αL3 and loop *l1*, *l2* & *l3* colored in tv_blue, gray, yellow, cyan, green, red and orange. (**b**) Root-mean-squared fluctuation (RMSF) of five trajectories with key structural elements highlighted.

Next, conformational variations at active site in each trajectory were carefully studied. The entire 250 ns trajectory was divided into three categories, based on distance measurement. With a hydrogen bond coming into being between terminal hydroxyl O_7 and $N\varepsilon_{H261}$ (distance $(O_7\text{-}N\varepsilon_{H261}) \leq 3.0$ Å), the intermediate was regarded ready to be de-protonated by H261, namely, an *active state*. The *active state* was observed in all five trajectories (8.7, 3.1, 17.1, 79.5, and 23.4%, respectively), with the highest proportion in md4 (Figure 3). Moreover, the terminal O_7 was proposed to be conducive for nucleophilic attack onto carboxyl C_1 with distance (O_7-C_1) $\leq$ 4.5 Å. The PRS was defined as both criteria were met, and was present in md4 for 4700 frames (18.8%, Figure S2); in other trajectories, PRS appeared with a significantly lower frequency, testifying to its unsteadiness as a transient state.

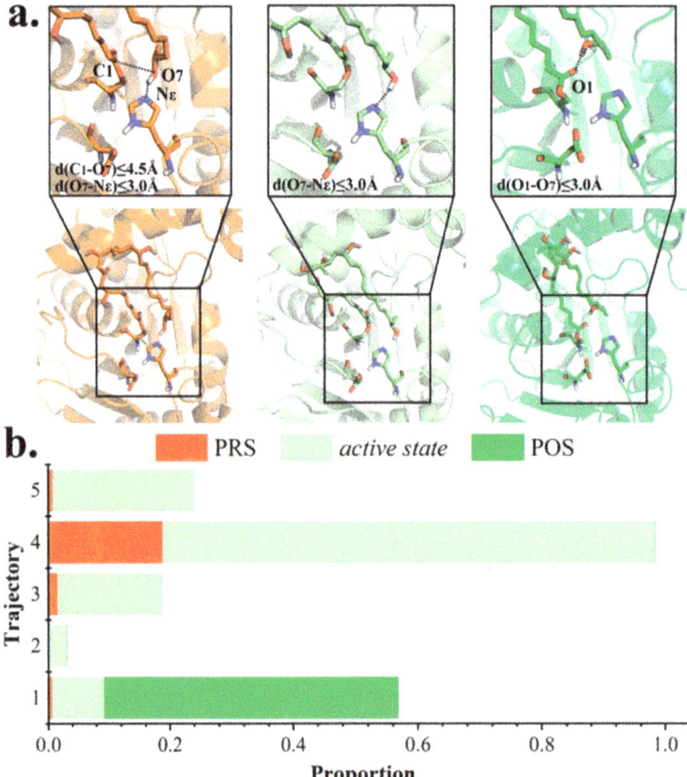

Figure 3. Classification of trajectory frames based on polyketide chain conformation. (**a**) Representative structures of PRS, *active state* and POS extracted from clustering analysis. (**b**) Proportion of PRS, *active state*, and POS in each trajectory.

Distinguished from PRS and *active state* that ultimately lead to macrocyclization, inactive conformations are susceptible to hydrolysis. Notably, among inactive conformations, the POS, which is characterized by a hydrogen bond between O_7 and carbonyl oxygen O_1 of substrate, was also observed within md1 for 11896 frames (47.6%, Figure S2), whereas it was nearly absent in others (3, 2, 20 and 0 frames in md2–5).

2.2. Conformational Transition between POS and PRS

Next, the transformation between POS and PRS was studied using dihedral angle C_α-C_β-C_γ-O_7 as an indicator of polyketide terminal rotation. In PRS, bond C_α-C_β ran anti-parallel against C_γ-O_7 ($-180°$), but in POS, the dihedral angle was altered to an acute angle fluctuating between ($-30°$, $-70°$). The conformational flip took place in 18–22 ns of md1 trajectory, with conformation altering progressively from PRS ($-180°$) to POS ($-60°$). As presented in Figure 4, terminal hydroxyl O_7 firstly swung up and disassociated from H261, followed by C_β-C_γ twisting clockwise and terminal methyl oriented towards the entrance (I→II). Further, the intermediate swelled to diminish distance O_7-O_1. After quick adjustment, POS came into being and maintained for rest of the trajectory (II→III).

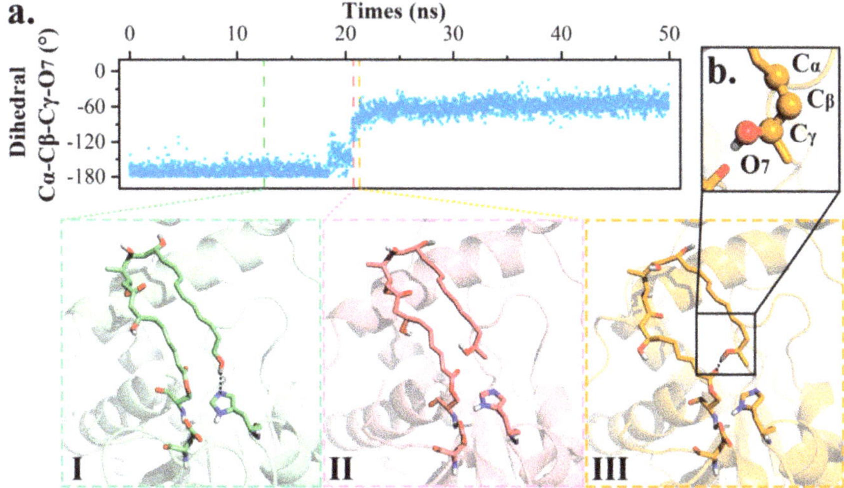

Figure 4. Conformational transformation between PRS and POS. (**a**) Dihedral value representation of md1 along with intermediate conformation change. (**b**) Presentation of dihedral angle C_α-C_β-C_γ-O_7.

An energy calculation was also conducted to investigate the structural stability of aforementioned conformations. As expected, POS harbored a lower energy than PRS by 4.58 kcal/mol, indicating the steadiness of the O_7-O_1 intra-molecular hydrogen bond. On the other hand, the *active state* was calculated to be 0.18 kcal/mol less stable than PRS. The slight difference prompted that conformational transition between *active state* and PRS would easily achieve through C_β-C_γ bond rotation.

In conclusion, a conformational transformation between POS and PRS was accomplished through dihedral flip and conformation adjustment, and the energies on POS, *active state*, and PRS were computed to understand the reaction process.

2.3. Hydrophilic and Hydrophobic Interactions in Pima-TE System

Based on MD simulations, hydrogen bonding and hydrophobic interactions between pima-TE and substrate were studied. As exhibited in Figure 5, in PRS, loop *l1* (residue 170–177) played a crucial part in fastening the substrate. The atoms O_2, O_3, O_4 or O_5 were anchored by H172 (13.35%), T177 (15.20%) and Q174 (4.55%) without fixed pattern. Residue Q29, stretching downward from the lid region, served as a crane to lift up O_6 and gave rise to an erected molecule (39.26%). The main chain of M210 fixed O_4 at the center of the molecule (28.56%), while its side chain laid parallel to the hydrophobic area of conjugated polyene (99.92%). The hydrophobic segments of T73, L183 and Y180 were also oriented towards the polyketide chain and worked jointly to conserve a water-free sub-environment with frequencies of 94.88, 89.27 and 78.50%, respectively.

Under the circumstance of POS, Q29 (19.38%) and H172 (10.71%) assisted the intermediate erection, except for this time the molecule slighted twisted to stabilize the intra-molecular hydrogen bond, which enabled S33 from the lid region to contribute in the hydrogen bonding network (8.90%). Identical to PRS, M210 again assumed the role of both a hydrophilic stake (56.98%) and a hydrophobic driving force (99.28%). Besides Y180 (36.17%) and L183 (57.06%) as in PRS, L213 also participated in hydrophobicity maintenance (25.43%).

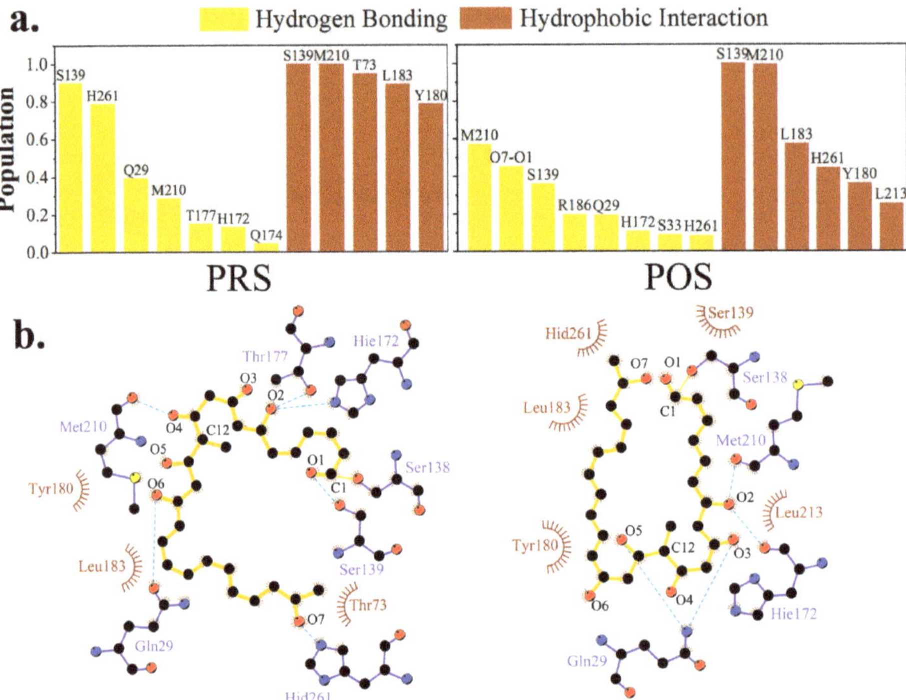

Figure 5. Hydrogen-bonding and hydrophobic interaction network variations in PRS and POS. (**a**) Proportion of top-ranked hydrogen bonding interactions. (**b**) Diagram of protein-substrate interaction produced by ligplot⁺ [28]. Backbone of the polyketide substrate was colored in yellow, and residues providing hydrogen bonding and hydrophobic interactions in slate_blue and brown.

In a word, binding modes of pimaricin polyketide with TE shared considerable similarity between PRS and POS, with Q29, M210 and residues on loop *l1* interacting with the chain via hydrogen bonding, and M210, Y180, and L183 contributing to hydrophobic network.

2.4. Key Residues Analyzed Via Mutant Simulations

2.4.1. Mutation 1-Q29A

According to the analyses of wild type simulations, Q29, located at lid region of pima-TE, could mediate the distance O_7-$N\varepsilon_{H261}$ within a favorable range through bonding with O_6 of substrate. When mutated to Ala, with Q29's side chain shortened and hydrogen bond abolished, it was speculated that the substrate would fall off from its original position. Here, the distance between O_6 and $C\alpha$ of Q/A29 (designated as O_6-$C_{Q/A29}$) was utilized to depict the substrate's spatial displacement. As shown in Figure 6, the distance O_6-$C_{Q/A29}$ fluctuated acutely in Q29A trajectory md2 and md3, with the substrate either overlength (**2₁**, **3₁**), hydrogen bonding to other residues (**3ₙ**), or drifting aimlessly (**2ₙ**, **3ₘ**). The erratic change also decreased PRS formation by a large margin (4.37% vs. 0.54%, Table S1). On the other hand, POS was observed in Q29A md1 with a frequency of 80.84% (**1₁**, **1ₙ**).

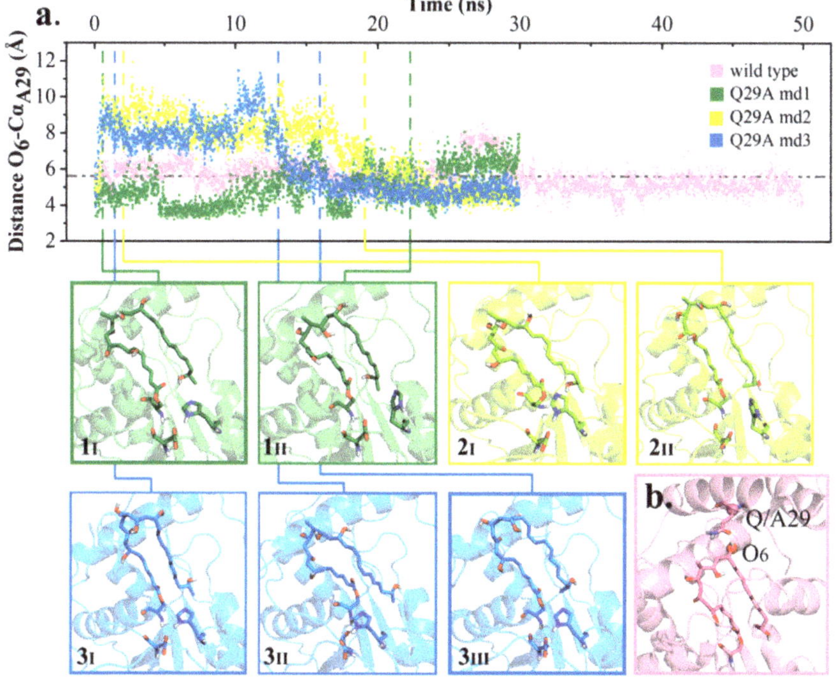

Figure 6. Instability of polyketide conformation. (**a**) Distance O_6-$C_{Q/A29}$ in wild type md4 and Q29A md1–3 along with conformation transformation. (**b**) Diagram of distance O_6-$C_{Q/A29}$.

From our perspective, Q29 could regulate hydrogen bond O_7-$N\varepsilon_{H261}$ and PRS formation by binding the substrate position with a hydrogen bond, while having little effect on POS.

2.4.2. Mutation 2-M210G

To validate M210's function in hydrophobic interaction network, M210 was mutated into Gly. The substrate backbone's distance RMSD (dRMSD) was calculated with the first frame as a reference. As seen in Figure 7, the larger dRMSD signified variation in substrate conformation, and its irregularity suggested volatility. Furthermore, new patterns of hydrogen bonding were observed in mutant (Figure 7): In md1, the polyketide chain leaned towards αL3 and interacted with N214 (23.08%); in md2 and md3, the substrate slightly rotated and bonded with D179 on αL2 (76.35% and 87.10%). Having lost M210 as a hydrophobic barrier, the polyketide chain would adjust its position, and M206 from the neighboring cycle of αL3 exhibited hydrophobicity. Owing to the altered interaction network, it was hard for the substrate to attain PRS as in wild type complex (4.37% vs. 0.72%).

Specifically, the aforementioned conformational change of substrate also produced a similar effect on loop *l1* seeking hydrogen bonding, and gave rise to a shrinking channel exit, while a bigger exit would be favored in product release. Taken together, M210 was crucial in maintaining the polyketide chain in between αL2 and αL3, which was conduced to protein-substrate interaction and an advantageous channel exit shape.

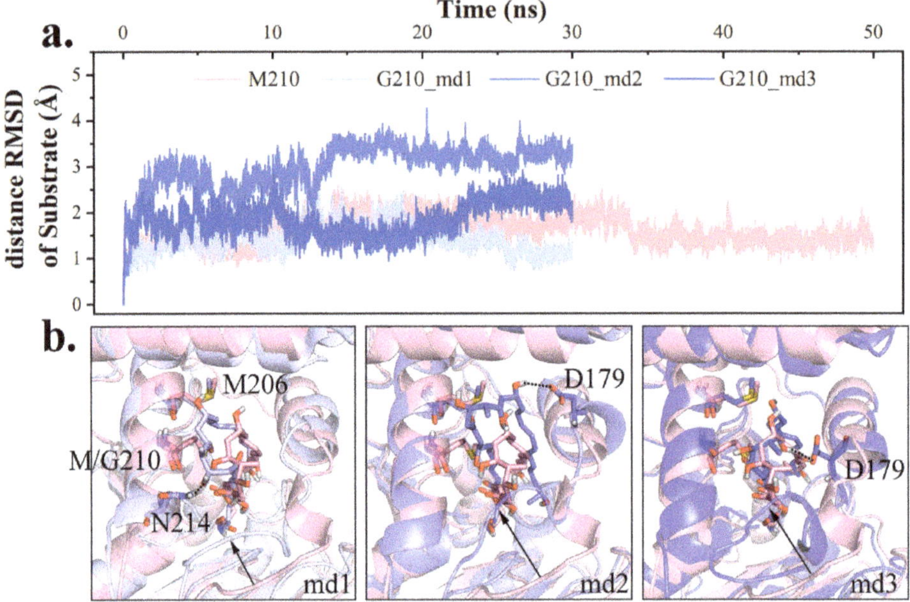

Figure 7. Conformational change of intermediate change upon M210 mutation. (**a**) Distance RMSD (dRMSD) value of substrate backbone in wild type md4 and M210G (lightpink for wild type, lightblue, slate and tv_blue for M210G md1–3). (**b**) Diagram of the dominant structure in each trajectory.

2.4.3. Mutation 3-R186F & R186Y

As a residue containing multiple hydrophilic groups, R186 bonded with O_7 for a rather high probability in wild type complex simulations. As seen in Figure 8a, in md1–3, high frequency of O_7-N_{R186} bonding could account for the scarce existence of O_7-$N\varepsilon_{H261}$ interaction. To promote PRS formation, R186 was firstly mutated to Phe.

To our disappointment, the frequency of PRS formation did not improve (4.37% vs. 1.64%). It was determined that E80 was coupled with R186 and R266, to pose a spatial barrier at the entrance and prevent the admission of other substrate, while functionally maintaining the closure and hydrophobicity of substrate pocket. Nevertheless, when R186 was mutated to Phe, a crack appeared (Figure 8b), and frequency of E80-R266 interaction was lowered as well (Table S2). Worse still, lacking the tying force, the distance $C\alpha_{F186}$-$C\alpha_{E80}$ also increased, implying a larger entrance (Figure 8c).

Based on our findings, pima-TE was re-modified into R186Y mutant. This time, we endowed a hydroxyl to the side chain of mutated residue to bond with E80, while the remainder stayed hydrophobic. We were more than pleased to find a significant rise in PRS ratio (4.37% vs. 18.14%), with Y186-E80 bonding partly restored (Table S3). Of particular note, a close-to-reaction PRS conformation appeared in md3 and maintained for over 10 ns, with the terminal methyl oriented towards the entrance and forcing O_7 closer to C_1 (Figure 8d). We thus regard R186Y as a promising modification towards pimaricin productivity advancement.

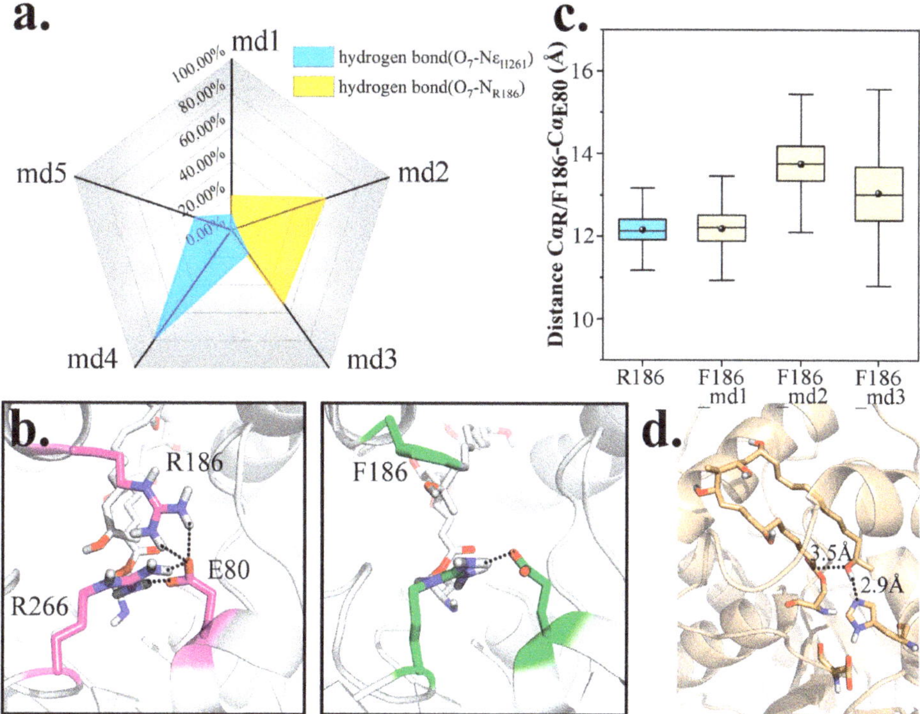

Figure 8. Mutational trials on R186. (**a**) Radar chart indicating the proportion of hydrogen bond O_7-$N\varepsilon_{H261}$ and O_7-N_{R186} formation within 5 wild type simulations. (**b**) Larger entrance of pima-TE after R186F mutation. (**c**) Coupling and non-coupling states of three entrance residues (R/F186, E80 and R266) located on different structure elements. (**d**) A favorable PRS emerged in R186Y md3.

2.4.4. Mutation 4-S138C

As presented by Koch et al. [29], compared with pikromycin synthase (PICS)-TE$_{WT}$, PICS-TE$_{S148C}$ could promote macrocyclization efficiency by over 300%. Therefore, pima-TE$_{S138C}$ mutant were subjected to MD simulations to study whether the superiority of Cys over Ser applied in pima-TE as well. After the clustering analysis, the dominant polyketide structure of each S138C trajectory demonstrated unbelievable similarity (Figure S3). As seen in Figure 9, S138C frames were significantly more concentrated in the conducive range for reaction compared to wild type ones, suggesting potential catalytic advantage. However, due to O→S atomic radius enlargement, bond length involving O/S increased by 0.3 Å, and 0.5 Å, respectively, and distance O_7-C_1 would mostly gather around 5.5 Å in mutated complex. The density of advantageous conformations in S138C system strongly suggested the favorability of this mutation.

2.5. Study on TE's Effect on the Release of Pimaricin Product

The binding energy between pima-TE and polyketide product was calculated with MM/PBSA program (Figure 10a). In study of product (i.e., MOL) movement across the channel, distance between its mass center and $C\alpha_{S138}$ was measured (Figure 10b). As a result, the product migrated towards the exit for approximately 4 Å in md1, while it hardly moved in others. Therefore, md1 was regarded to have a tendency of product release, and the other two disclosed the stabilization effect produced by the protein. Energy decomposition revealed residues around the exit to play key parts in protein-product interactions (Figure 10c), and to assume important roles in product release. (Figure S4)

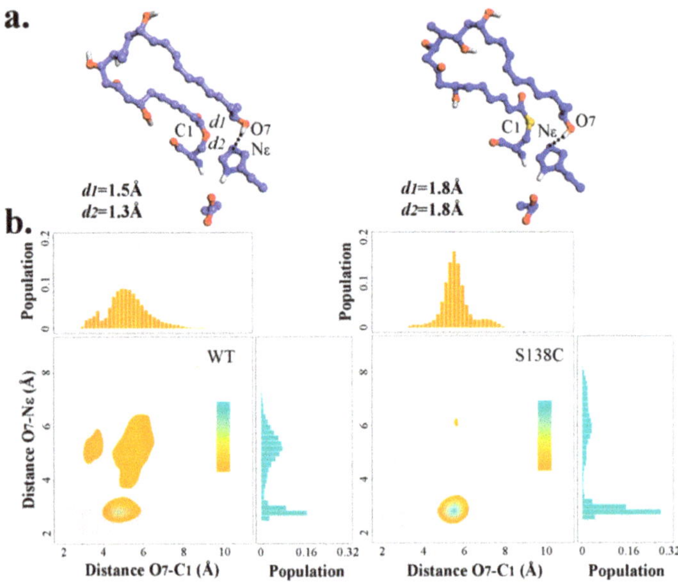

Figure 9. (**a**) Diagram of representative PRS conformations in wild type (**left**) and S138C (**right**) trajectories. (**b**) Density map and marginal histogram indicating the distribution of all frames on the basis of distances O_7-$N\varepsilon_{H261}$ and O_7-C_1 in wild type and S138C trajectories. The rectangles in light-coral, slate-blue, and thistle highlight points with distance (O_7-$N\varepsilon_{H261}$) $\leq$ 3.0 Å, distance (O_7-C_1) $\leq$ 4.5 Å and 4.5 Å $\leq$ distance (O_7-C_1) $\leq$ 6.0 Å, respectively.

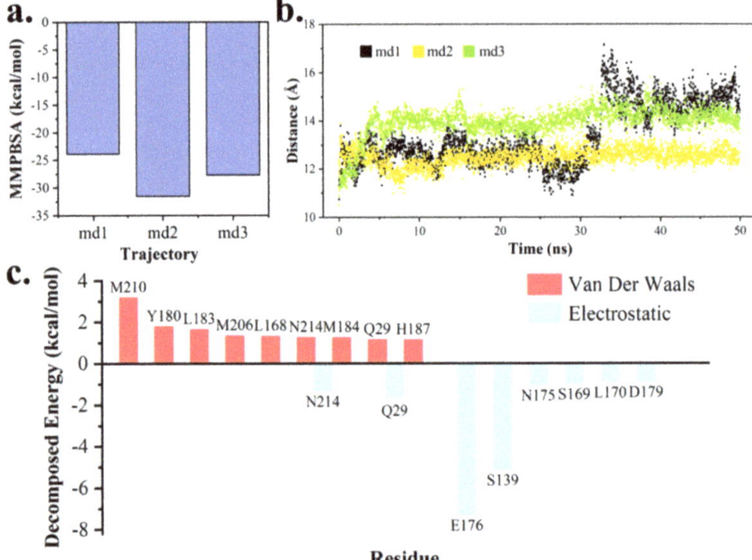

Figure 10. MM/PBSA analysis on the outward trend of macro-lactone product. (**a**) Binding energy between pima-TE and product ring. (**b**) Distance between ring mass center and $C\alpha_{S138}$. (**c**) Residues with top-ranked van der Waals (VDW) and electrostatic contributions to binding free energy.

Next, a careful analysis was conducted on the disengagement of product in md1, and three patterns of hydrogen bonding between Q29 and product were generalized (Figure 11). For the first 25 ns, the product remained its original state and O_6 from the product ring continued bonding to Q29 (I). Afterwards, in cooperation with Q29's side chain turnover (II), the ring lied down a little and interacted with Q29 from the right (III). Then, the free hydroxyl on Q29 (OE_1) grasped O_2 from the other side of the molecule, further enabling the molecule to lie flat (IV). In a word, steps II-III and III-IV played decisive roles in altering the product's layout and pulling the product farther away from the active site. Due to distinguished distribution of hydrophilic and hydrophobic areas on protein, rotation of the product might partly attenuate its interaction with peripheral residues, and impel the product's departure.

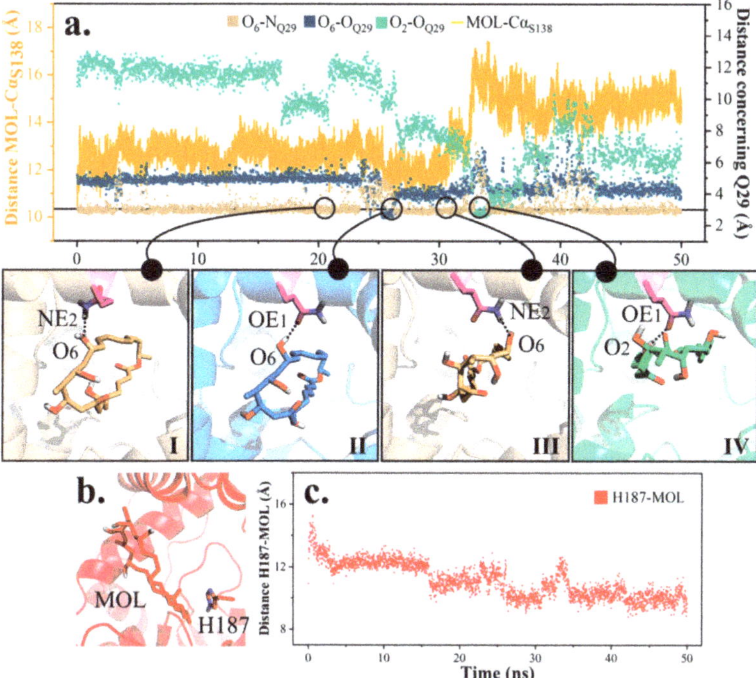

Figure 11. Diagram of product movement in substrate channel in md1. (**a**) Three hydrogen bonding modes between Q29 and MOL and their transformation. (**b**) Spatial location of H187 with respect to MOL. (**c**) Distance H187-MOL in md1.

On the other hand, H187 seemed to provide thrust towards the release of product (Figure 11). Distance between the mass center of H187 and product ring shrank along the simulation, revealing established hydrophobic interaction between the imidazole of H187 and terminal methyl on the product (Figure 11).

Taken together, after cyclization, the product would stay in the vicinity of active site for a while due to van der Waals (VDW) and electrostatic interactions from peripheral residues. Later on, the product layout was altered by molecule rotation, varied hydrogen bonding, etc., which impaired the spatial constraint, and caused the ring to gradually migrate towards the exit, with Q29 hydrogen bonding as a driving force and H187 as a rear helper.

3. Discussion

Due to the limitation of experimental instruments, present computational strategies combining homology modeling, molecular docking, MD simulation, and QM/MM calculation have been extensively utilized to provide insight into atomistic details in protein-substrate recognition and catalytic mechanism. Over recent years, packages and software [30–32] to study protein- substrate interaction have sprung up relentlessly, and MD simulation has become a regular routine herein [33–35].

In this work, MD simulations were carried out on pima-TE-substrate/product complexes. Residues playing critical roles in product recognition, assembly, and release were uncovered through hydrogen bonding and hydrophobic interaction network analysis, which could be obtained from representative conformations of trajectories, as well as decomposition of MM/PBSA binding energy. Q29 and M210 might contribute to tight binding effect, and the structural correlation between protein and substrate was reduced once they were eliminated. R186 was uncovered to maintain pocket hydrophobicity yet distract the substrate from a proper position, and its mutation to Tyr could benefit macrocyclization by raising the proportion of advantageous conformation. The computer-aided methods could provide theoretical basis to enzyme clarification.

Since the transition states of enzyme catalysis were hard to obtain in silico, we chose pre-reaction state (PRS) as an evaluation indicator. According to our previous research [26,36], PRS was the very prior stage of macrocyclization in terms of both structure and energy, and its formation was decisive to TE cyclization. The proportion of PRS was regarded as the degree of reaction readiness. Besides, PRS proportion was used in mutated systems as well to help elucidate the functions of these residues and speculate their effect on TE activity. However, a more accurate account of the mutation required explanation in energy and experimental verification as well.

To conclude, the study approach applied in our work involved protein-substrate interaction, residue targeting, and mutation analyses with PRS occurrence as an indicator. The strategy could provide structural rationale for TE-substrate complex and guide future experiments on design of efficient protein mutants or novel compounds.

4. Materials and Methods

4.1. System Preparation

Given the unavailability of pima-TE crystallization data from Protein Data Bank (PDB), initial structure of pima-TE was produced through homology modeling with PICS-TE [37] (PDB: 2H7Y) as a template (sequence similarity: 48.1%). Twenty pima-TE models were generated in the discovery studio 3.5 [38]. The one with the lowest total energy was selected, and its stereochemical quality was further validated by Procheck [39], with 93.7% of its residues falling in the most favored region.

Considering the extensively-acknowledged catalytic process of pimaricin PKS, the mature pimaricin product was disconnected at carbonyl C_1, and the lactonic ring as well as exocyclic mycosamine were removed. Furthermore, carboxyl on C_{12} was also substituted by a methyl. The precursor was optimized with Gaussian09 [40] AM1 method [41], after which the buckled conformation still sustained. The energetically-stabilized substrate was then covalently bonded to active site Ser138 on pima-TE model, with a hydrogen bond forming between its terminal hydroxyl and Nε of active site His261. Protonation state of His261 was altered to HID to facilitate PRS formation. The polyketide-bound acyl-enzyme intermediate was utilized as the initial structure of MD simulations.

During the preparation of the system parameters, an N-terminal cap (-CO-CH$_3$) and a C-terminal cap (-NH-CH$_3$) were firstly added onto the Ser138 to block its ends. Conformational optimization at the level of HF/6-31G(d) was then employed on the intermediate, and its electrostatic surface potential (ESP) charge was computed. Afterwards, a two-step restrained electrostatic potential (RESP) model was applied to determine charge distribution on the substrate. Finally, two prior-added caps were removed, and parameters for the intermediate were generated by the Antechamber package, on the basis of which topology files for protein-substrate complexes were prepared with *tleap* module in

AMBER 14. Through *tleap*, pima TE-substrate system was placed in an octahedral TIP3P water box [42], with 12 sodium ions added to maintain charge neutralization.

4.2. Molecular Dynamics Simulation

Starting from the solvated polyketide-bound acyl-enzyme intermediate, classical molecular dynamics simulations were carried out utilizing AMBER14 [43] ff03.r1 force field [44]. The system was firstly subjected to 10,000 steps of steepest descent energy minimization followed by 1000 cycles of conjugate gradient minimization with bonds involving hydrogen constrained by SHAKE algorithm [45], and then another 10,000 steps of steepest descent energy minimization followed by 5000 cycles of conjugate gradient minimization with no constraint exerted. The system was then gradually heated from 0 to 300 K through 25000 iterations. After a 200ps-equilibrium in NPT ensemble, five 50-ns simulations (300 K, 1 atm) with different random seeds were conducted. The VDW interactions were cut off at 10 Å and long-range electrostatic interactions were calculated with particle mesh Ewald (PME) method [46]. Analyses of trajectories were performed using *cpptraj* in Ambertools14.

4.3. Quantum Mechanics/Molecular Mechanics) Calculation

Quantum mechanics/molecular mechanics (QM/MM) calculations were performed with a two-layered ONIOM method [47,48] in Gaussian09 program. Geometrical snapshots from the dominant MD cluster were extracted as PRS, *active state*, and POS, and were further subject to geometry optimization. The quantum mechanical (QM) layer consisted of side chains of active site triad (Ser138, Asp166 and His261) and the polyketide chain, which added up to 96 atoms and bore one negative charge. The optimization process was carried out under M06-2X [49] functional and basis set 6-31G(d) [50].

4.4. Simulation of Site Mutation Proteins

Based on the analyses, M210 and Q29 were selected as key residues in protein-substrate interaction. Considering the optimization of pima-TE, R186 was mutated to Phe and Tyr in succession to reduce its interference against PRS formation. In accordance with a previously published article of Koch et al. [29], S138 was also mutated to Cys to examine pima-TE$_{S138C}$'s effectiveness. Single site mutation was employed directly on the initial structure of wild type pima-TE, and all mutants (M210G, Q29A, R186F, R186Y, S138C) went through 30–50 ns simulations following identical procedures as mentioned in Section 4.2.

4.5. Free Energy Calculation and Conformational Stability Analysis

The polyketide chain, which was extracted from dominant structure in wild type md4, was manually rang up and subjected to conformational optimization with Gaussian 09 AM1 method. The optimized product was then docked into the channel with C_1 adjacent to the active site. After the model construction, 3×50 ns MD simulations was carried out with AMBER14 program.

After clustering analysis in *cpptraj*, a 20 ns segment with dominant conformation was extracted from each trajectory, and was further subject to a molecular mechanics Poisson-Boltzmann surface area [51] (MM/PBSA) calculation to estimate the free energy difference (ΔG^{tot}) between bound and detached states of product-protein complexes in solution. The MMPBSA.py program in AMBER14 was performed, and the free energy discrepancy was decomposed to peripheral residues in terms of hydrophobic and electrostatic forces. Table 1 lists the number and duration of all MD simulations utilized in the study.

Table 1. List of MD Runs Performed.

Substrate Type	Name	System	No. of Runs Per Complex	Length Per Run (ns)
Polyketide Chain	wild type	pima-TE$_{WT}$ + polyketide chain	5	50
	M210G	pima-TE$_{M210G}$ + polyketide chain	3	30
	Q29A	pima-TE$_{Q29A}$ + polyketide chain	3	30
	R186F	pima-TE$_{R186F}$ + polyketide chain	3	30
	R186Y	pima-TE$_{R186Y}$ + polyketide chain	3	30
	S138C	pima-TE$_{S138C}$ + polyketide chain	3	50
Product	ring	pima-TE$_{WT}$ + product	3	50

5. Conclusions

In this paper, MD simulations were utilized as a primary tool to explore pimaricin TE catalysis on an atomic level. Firstly, 5 × 50 ns trajectories on polyketide were conducted in search of pre-reaction states (PRS), and transformation between POS and PRS were examined. POS was found to bear lower energy, yet less mature conformation in comparison with PRS. Protein-polyketide hydrogen bonding and hydrophobic interactions were deciphered, with several key residues subjected to mutations. As discovered, Q29 was responsible for holding a polyketide hydroxyl and controlling the substrate position, and M210 contributed to favorable protein-ligand interaction by virtue of its hydrophobicity. R186Y might promote productivity by reducing the interference on PRS formation, and S138C could effectively enhance the proportion of required conformations. Ultimately, the MM/PBSA program was employed to unveil residues mediating product release, and the postulation of a mechanism of polyketide product departure from the active site was proposed. We gave a comprehensive overview on pima-TE catalysis, with computational methods, and offered opinions for protein engineering.

Supplementary Materials: Supplementary materials can be found at http://www.mdpi.com/1422-0067/20/4/877/s1.

Author Contributions: Conceptualization, L.B., Y.-L.Z., and T.S.; formal analysis, S.F., R.W., and C.L.; funding acquisition, L.B. and Y.-L.Z.; investigation, S.F., R.W., and C.L.; methodology, Y.-L.Z. and T.S.; writing-original draft, S.F. and T.S.; writing-review and editing, S.F. and T.S.

Funding: This research was funded by the National Science Foundation of China, grant number 21377085 and 31770070.

Acknowledgments: The authors thank SJTU-YG2016MS42, SJTU-ZH2018ZDA26, SJTU-YG2016MS33 for financial supports and SJTU-HPC computing facility award for computational hours.

Conflicts of Interest: The authors declare no conflict of interest.

Abbreviations

PKS	Polyketide Synthase
TE	Thioesterase
MD	Molecular Dynamics
PRS	Pre-reaction State
POS	Pre-organization State

References

1. Gil, J.A.; Martin, J.F. *Biotechnology of Antibiotics*, 2nd ed.; W. Strohl, M., Ed.; Dekker: New York, NY, USA, 1997.

2. Aparicio, J.F.; Mendes, M.V.; Anton, N.; Recio, E.; Martin, J.F. Polyene macrolide antibiotic biosynthesis. *Curr. Med. Chem.* **2004**, *11*, 1645–1656. [CrossRef]

3. Szlinder-Richert, J.; Mazerski, J.; Cybulska, B.; Grzybowska, J.; Borowski, E. MFAME, *N*-methyl-*N*-D-fructosyl amphotericin B methyl ester, a new amphotericin B derivative of low toxicity: Relationship between self-association and effects on red blood cells. *Biochim. Biophys. Acta Gen. Sub.* **2001**, *1528*, 15–24. [CrossRef]

4. Ogasawara, Y.; Katayama, K.; Minami, A.; Otsuka, M.; Eguchi, T.; Kakinuma, K. Cloning, Sequencing, and Functional Analysis of the Biosynthetic Gene Cluster of Macrolactam Antibiotic Vicenistatin in Streptomyces halstedii. *Chem. Biol.* **2006**, *11*, 79–86. [CrossRef] [PubMed]

5. Cereghetti, D.M.; Carreira, E.M. Amphotericin B: 50 Years of Chemistry and Biochemistry. *Synthesis* **2006**, *37*, 914–942. [CrossRef]

6. Carmody, M.; Murphy, B.; Byrne, B.; Power, P.; Rai, D.; Rawlings, B.; Caffrey, P. Biosynthesis of Amphotericin Derivatives Lacking Exocyclic Carboxyl Groups. *J. Biol. Chem.* **2005**, *280*, 34420–34426. [CrossRef] [PubMed]

7. Gantt, R.W.; Peltierpain, P.; Thorson, J.S. Enzymatic methods for glyco (diversification/randomization) of drugs and small molecules. *Nat. Prod. Rep.* **2011**, *28*, 1811–1853. [CrossRef]

8. Zotchev, S.B. Polyene Macrolide Antibiotics and their Applications in Human Therapy. *Curr. Med. Chem.* **2003**, *10*, 211–223. [CrossRef]

9. Baginski, M.; Czub, J.; Sternal, K. Interaction of amphotericin B and its selected derivatives with membranes: Molecular modeling studies. *Chem. Rec.* **2010**, *6*, 320–332. [CrossRef]

10. Atta, H.M.; Selim, S.M.; Zayed, M.S. Natamycin antibiotic produced by Streptomyces sp.: Fermentation, purification and biological activities. *J. Ame. Sci.* **2012**, *8*, 469–475.

11. Stark, J. Natamycin: An effective fungicide for food and beverages. *Nat. Antimicrobials Minim. Process. Foods* **2003**, 82–97. [CrossRef]

12. Austin, A.; Lietman, T.; Rose-nussbaumer, J. Update on the management of infectious keratitis. *Ophthalmology* **2017**, *124*, 1678–1689. [CrossRef] [PubMed]

13. Priya, A.B.; Kalyan, M. In vitro leishmanicidal effects of the anti-fungal drug natamycin are mediated through disruption of calcium homeostasis and mitochondrial dysfunction. *Apoptosis* **2018**, *23*, 420–435. [CrossRef]

14. Te Welscher, Y.M.; Jones, L.; Van Leeuwen, M.R.; Dijksterhuis, J.; de Kruijff, B.; Eitzen, G.; Breukink, E. Natamycin Inhibits Vacuole Fusion at the Priming Phase via a Specific Interaction with Ergosterol. *Antimicrob. Agents Chemother.* **2010**, *54*, 2618–2625. [CrossRef] [PubMed]

15. Tanner, W. Membrane transport inhibition as mode of action of polyene antimycotics: Recent data supported by old ones. *Food Technol. Biotechnol.* **2014**, *52*, 8–12. [CrossRef]

16. Van Leeuwen, M.R.; Golovina, E.A.; Dijksterhuis, J. The polyene antimycotics nystatin and filipin disrupt the plasma membrane, whereas natamycin inhibits endocytosis in germinating conidia of Penicillium discolor. *J. Appl. Microbiol.* **2009**, *106*, 1908–1918. [CrossRef] [PubMed]

17. Mccall, L.I.; Aroussi, A.E.; Choi, J.Y.; Vieira, D.F.; Muylder, G.D.; Johnston, J.B.; Chen, S.; Kellar, D.; Siqueira-Neto, J.L.; Roush, W.R.; et al. Targeting Ergosterol Biosynthesis in Leishmania donovani: Essentiality of Sterol 14 alpha-demethylase. *PLoS Negl. Trop. Dis.* **2015**, *9*, 1–17. [CrossRef] [PubMed]

18. Dutta, S.; Whicher, J.R.; Hansen, D.A.; Hansen, W.A.; Chelmer, J.A.; Congdon, G.R.; Alison, R.H.N.; Kristina, H.; Sherman, D.H.; Smith, J.L.; Skiniotis, G. Structure of a modular polyketide synthase. *Nature* **2014**, *510*, 512–517. [CrossRef] [PubMed]

19. Skiba, M.A.; Sikkema, A.P.; Fiers, W.D.; Gerwick, W.H.; Sherman, D.H.; Aldrich, C.C.; Smith, J.L. Domain Organization and Active Site Architecture of a Polyketide Synthase C-methyltransferase. *ACS Chem. Biol.* **2016**, *11*, 3319–3327. [CrossRef]

20. Curran, S.C.; Hagen, A.; Poust, S.; Chan, L.J.G.; Garabedian, B.M.; Rond, T.; Baluyot, M.J.; Vu, J.T.; Lau, A.K.; Yuzawa, S.; et al. Probing the flexibility of an iterative modular polyketide synthase with non-native substrates in vitro. *ACS Chem. Biol.* **2018**, *13*, 2261–2268. [CrossRef]

21. Rittner, A.; Paithankar, K.S.; Vu, K.H.; Grininger, M. Characterization of the polyspecific transferase of murine type I fatty acid synthase (FAS) and implications for polyketide synthase (PKS) engineering. *ACS Chem. Biol.* **2018**, *13*, 723–732. [CrossRef]

22. Ferscht, A. *Enzyme Structure and Mechanism,* 2nd ed.; W. H. Freeman and Company: New York, NY, USA, 1985. [CrossRef]

23. Kormana, T.P.; Crawfordb, J.M.; Labonteb, J.W.; Newmanb, A.G.; Wongc, J.; Townsendb, C.A.; Tsaia, S.C. Structure and function of an iterative polyketide synthase thioesterase domain catalyzing Claisen cyclization in aflatoxin biosynthesis. *Proc. Natl. Acad. Sci. USA* **2010**, *107*, 6246–6251. [CrossRef] [PubMed]

24. Chen, X.P.; Shi, T.; Wang, X.L.; Wang, J.T.; Chen, Q.H.; Bai, L.Q.; Zhao, Y.L. Theoretical studies on the Mechanism of Thioesterase-catalyzed Macrocyclization in Erythromycin Biosynthesis. *ACS Catal.* **2016**, *6*, 4369–4378. [CrossRef]

25. Trauger, J.W.; Kohli, R.M.; Walsh, C.T. Cyclization of Backbone-Substituted Peptides Catalyzed by the Thioesterase Domain from the Tyrocidine Nonribosomal Peptide Synthetase. *Biochemistry* **2001**, *40*, 7092–7098. [CrossRef] [PubMed]

26. Parthasarathy, S.; Murthy, M.R.N. Analysis of temperature factor distribution in high-resolution protein structures. *Protein Sci.* **1997**, *6*, 2561–2567. [CrossRef] [PubMed]

27. Pan, X.Y.; Shen, H.B. Robust Prediction of B-Factor Profile from Sequence Using Two-Stage SVR Based on Random Forest Feature Selection. *Protein Peptide Lett.* **2009**, *16*, 1447–1454. [CrossRef]

28. Laskowski, R.A.; Swindells, M.B. LigPlot+: Multiple ligand-protein interaction diagrams for drug discovery. *J. Chem. Inf. Model.* **2011**, *51*, 2778–2786. [CrossRef]

29. Koch, A.A.; Hansen, D.A.; Shende, V.V.; Furan, L.R.; Houk, K.N.; Gonzalo Jiménez-Osés, G.; Sherman., D.H. A Single Active Site Mutation in the Pikromycin Thioesterase Generates a More Effective Macrocyclization Catalyst. *J. Am. Chem. Soc.* **2017**, *139*, 13456–13465. [CrossRef]

30. Lu, T.; Chen, F.W. Multiwfn: A multifunctional wavefunction analyzer. *J. Comput. Chem.* **2012**, *33*, 580–592. [CrossRef]

31. Szklarczyk, D.; Morris, J.H.; Cook, H.; Kuhn, M.; Wyder, S.; Simonovic, M.; Santos, A.; Doncheva, N.T.; Roth, A.; Bork, P.; et al. The STRING database in 2017: Quality-controlled protein-protein association networks, made broadly accessible. *Nucleic Acids Res.* **2017**, *45*, D362–D368. [CrossRef]

32. Onur, S.; Pemra, O. gRINN: A tool for calculation of residue interaction energies and protein energy network analysis of molecular dynamics simulations. *Nucleic Acids Res.* **2018**, *46*, W554–W562. [CrossRef]

33. Li, J.; Sun, R.; Wu, Y.H.; Song, M.Z.; Li, J.; Yang, Q.Y.; Chen, X.Y.; Bao, J.K.; Zhao, Q. L1198F Mutation Resensitizes Crizotinib to ALK by Altering the Conformation of Inhibitor and ATP Binding Sites. *Int. J. Mol. Sci.* **2017**, *18*, 482. [CrossRef] [PubMed]

34. Lee, J.; Gokey, T.; Ting, D.; He, Z.H.; Guliaev, A.B. Dimerization misalignment in human glutamate-oxaloacetate transaminase variants is the primary factor for PLP release. *PLOS ONE* **2018**, *13*, e0203889. [CrossRef] [PubMed]

35. Liu, W.P.; Liu, G.J.; Zhou, H.Y.; Fang, X.; Fang, Y.; Wu, J.H. Computer prediction of paratope on antithrombotic antibody 10B12 and epitope on platelet glycoprotein VI via molecular dynamics simulation. *BioMed. Eng. OnLine* **2016**, *15*, 647–658. [CrossRef] [PubMed]

36. Shi, T.; Liu, L.X.; Tao, W.T.; Luo, S.G.; Fan, S.B.; Wang, X.L.; Bai, L.Q.; Zhao, Y.L. Theoretical studies on the Catalytic Mechanism and Substrate Diversity for Macrocyclization of Pikromycin Thioesterase. *ACS Catal.* **2018**, *8*, 4323–4332. [CrossRef]

37. Giraldes, J.W.; Akey, D.L.; Kittendorf, J.D.; Sherman, D.H.; Smith, J.L.; Fecik, R.A. Structural and mechanistic insights into polyketide macrolactonization from polyketide-based affinity labels. *Nat. Chem. Biol.* **2006**, *2*, 531–536. [CrossRef] [PubMed]

38. *Accelrys Discovery Studio Visualizer 3.5*; Accelrys Software Inc.: San Diego, CA, USA, 2005.

39. Laskowski, R.A.; MacArthur, M.W.; Moss, D.S.; Thornton, J.M. PROCHECK: A program to check the stereochemical quality of protein structures. *J. Appl. Cryst.* **1993**, *26*, 283–291. [CrossRef]

40. *Amber 2014*; Univeristy of California: San Francisco, CA, USA, 2014.

41. Jakalian, A.; Bush, B.L.; Jack, D.B.; Bayly, C.I. Fast, efficient generation of high-quality atomic Charges. AM1-BCC model: I. Method. *J. Comput. Chem.* **2000**, *21*, 132–146. [CrossRef]

42. Jorgensen, W.L.; Chandrasekhar, J.; Madura, J.D.; Impey, R.W.; Klein, M.L. Comparison of simple potential functions for simulating liquid water. *J. Chem. Phys.* **1983**, *79*, 926–935. [CrossRef]

43. *Gaussian 09*; Revision A.02; Gaussian Inc.: Wallingford, CT, USA, 2009.

44. Duan, Y.; Wu, C.; Chowdhury, S.; Lee, M.C.; Xiong, G.M.; Zhang, W.; Yang, R.; Cieplak, P.; Luo, R.; Lee, T.; et al. A point-charge force field for molecular mechanics simulations of proteins based on condensed-phase quantum mechanical calculations. *J. Comput. Chem.* **2003**, *24*, 1999–2012. [CrossRef]

45. Ryckaert, J.P.; Ciccotti, G.; Berendsen, H.J.C. Numerical integration of the Cartesian equations of motion of a system with constraints: Molecular dynamics of *N*-alkanes. *J. Chem. Phys.* **1977**, *23*, 327–341. [CrossRef]

46. Darden, T.; York, D.; Pedersen, L. Particle mesh Ewald: An N.log(N) method for Ewald sums in large systems. *J. Chem. Phys.* **1993**, *98*, 10089–10092. [CrossRef]

47. Vreven, T.; Byun, K.S.; Komáromi, I.; Dapprich, S.; Montgomery, J.A., Jr.; Morokuma, K.; Frisch, M.J. Combining quantum mechanics methods with molecular mechanics methods in ONIOM. *J. Chem. Theory Comput.* **2006**, *2*, 815–826. [CrossRef] [PubMed]

48. Vreven, T.; Frisch, M.; Kudin, K.; Schlegel, H.; Morokuma, K. Geometry optimization with QM/MM methods II: Explicit quadratic coupling. *Mol. Phys.* **2006**, *104*, 701–714. [CrossRef]
49. Zhao, Y.; Truhlar, D.G. The M06 suite of density functionals for main group thermochemistry, thermochemical kinetics, noncovalent interactions, excited states, and transition elements: Two new functionals and systematic testing of four M06-class functionals and 12 other functionals. *Theor. Chem. Acc.* **2008**, *120*, 215–241. [CrossRef]
50. Rassolov, V.A.; Ratner, M.A.; Pople, J.A.; Redfern, P.C.; Curtiss, L.A. 6-31G* Basis Set for Third-Row Atoms. *J. Comput. Chem.* **2001**, *22*, 976–984. [CrossRef]
51. Swanson, J.M.J.; Henchman, R.H.; McCammon, J.A. Revisiting free energy calculations: A theoretical connection to MM/PBSA and direct calculation of the association free energy. *Biophys. J.* **2004**, *86*, 67–74. [CrossRef]

International Journal of
Molecular Sciences

Article

Preparation of Biphenyl-Conjugated Bromotyrosine for Inhibition of PD-1/PD-L1 Immune Checkpoint Interactions

Eun-Hye Kim [1,2], Masuki Kawamoto [1,3,*], Roopa Dharmatti [1], Eiry Kobatake [2], Yoshihiro Ito [1,3] and Hideyuki Miyatake [1,*]

[1] Nano Medical Engineering Laboratory, RIKEN Cluster of Pioneering Research, 2-1 Hirosawa, Wako, Saitama 351-0198, Japan; eunhye.kim@riken.jp (E.-H.K.); roopa.dharmatti@riken.jp (R.D.); y-ito@riken.jp (Y.I.)

[2] Department of Life Science and Technology, School of Life Science and Technology, Tokyo Institute of Technology, 4259 Nagatsuta-cho, Midori-ku, Yokohama 226-8503, Japan; kobatake.e.aa@m.titech.ac.jp

[3] Emergent Bioengineering Materials Research Team, RIKEN Center for Emergent Matter Science, 2-1 Hirosawa, Wako, Saitama 351-0198, Japan

* Correspondence: mkawamot@riken.jp (M.K.); miyatake@riken.jp (H.M.); Tel.: +81-48-467-2752 (M.K.); +81-48-467-4979 (H.M.); Fax: +81-48-467-9300 (M.K.); +81-48-467-9300 (H.M.)

Received: 7 April 2020; Accepted: 17 May 2020; Published: 21 May 2020

Abstract: Cancer immunotherapy has been revolutionized by the development of monoclonal antibodies (mAbs) that inhibit interactions between immune checkpoint molecules, such as programmed cell-death 1 (PD-1), and its ligand PD-L1. However, mAb-based drugs have some drawbacks, including poor tumor penetration and high production costs, which could potentially be overcome by small molecule drugs. **BMS-8**, one of the potent small molecule drugs, induces homodimerization of PD-L1, thereby inhibiting its binding to PD-1. Our assay system revealed that **BMS-8** inhibited the PD-1/PD-L1 interaction with IC_{50} of 7.2 μM. To improve the IC_{50} value, we designed and synthesized a small molecule based on the molecular structure of **BMS-8** by in silico simulation. As a result, we successfully prepared a biphenyl-conjugated bromotyrosine (**X**) with IC_{50} of 1.5 μM, which was about five times improved from **BMS-8**. We further prepared amino acid conjugates of **X** (**amino-X**), to elucidate a correlation between the docking modes of the **amino-X**s and IC_{50} values. The results suggested that the displacement of **amino-X**s from the **BMS-8** in the pocket of PD-L1 homodimer correlated with IC_{50} values. This observation provides us a further insight how to derivatize **X** for better inhibitory effect.

Keywords: PD-1/PD-L1; immune checkpoint inhibitors; biphenyl-conjugated bromotyrosine; amino acid conjugation; **amino-X**; in silico simulation; IC_{50}

---

1. Introduction

Immunotherapy has recently emerged as a fourth modality for cancer therapy, together with surgery, chemotherapy, and radiation therapy [1–4]. The immunotherapy promotes T-cells to kill cancer cells by the blockade of immune checkpoint pathways [5,6]. One of the major immune checkpoint pathways is inactivated by the binding of programmed cell-death 1 (PD-1) [7], which is largely expressed on T cells, and its ligand PD-L1 [3,8,9], which is mainly expressed on antigen-presenting cells under physiological conditions but is upregulated on cancer cells [10]. PD-L1 binding to PD-1 suppresses T-cell function, including cytolytic activity, leading to downregulation of the anti-tumor immune response [2,5]. Another immune checkpoint is mediated by binding of the ligands B7-1/2

(CD80, CD86) on activated antigen-presenting cells or cancer cells to cytotoxic T-lymphocyte-associated protein 4 (CTLA-4) on T cells, which also suppresses T-cell activity [11,12]. Identification of these immunosuppressive pathways led to the development of monoclonal antibody (mAb)-based cancer therapies that inhibit PD-1/PD-L1 or CTLA-4/B7 pathways, thereby reinvigorating the host anti-tumor immune response [2,13–17]. Among the therapies currently approved for clinical use are the anti-CTLA-4 mAb ipilimumab (Yervoy®), which was the first immune checkpoint inhibitor to demonstrate an anti-cancer effect [18,19], and the anti-PD-1 mAb nivolumab (Opdivo®) [20]. In addition to these and other approved mAb-based immune checkpoint inhibitors [21], many others are currently in clinical trials for various cancers and immune-based diseases [22–25].

Protein-based drugs such as mAbs have some important drawbacks, such as high production costs associated with the preparation of biologicals [26], poor tumor penetration due to their large molecular weights (~150 kDa) [27], and unexpected post-translational glycosylation patterns [28]. Small molecule drugs, which are generally orally active and can overcome many of the challenges associated with protein drugs, are therefore being pursued as attractive alternative immune checkpoint inhibitors [28,29].

Until now, Bristol-Myers Squibb (BMS) has disclosed the patent claim [30] with structures of a number of BMS compounds, which are the potential inhibitors of the PD-1/PD-L1 pathway. Previous works have shown that one of the BMS compounds, **BMS-8**, binds directly to PD-L1 and induces formation of PD-L1 homodimers, which in turn prevents the interaction with PD-1 [31]. In the patent claims, the homogenous time-resolved fluorescence (HTRF) assay report that **BMS-8** has a sub μM order of IC$_{50}$, 0.146 μM [30], with other BMS compounds [32]. In this study, however, our amplified luminescence proximity homogeneous assay (Alpha) measured the IC$_{50}$ of **BMS-8** as 7.2 μM. Therefore, we aimed to prepare higher affinity compounds by taking the advantage of the complex structure of **BMS-8**/PD-L1 [31] with in silico simulation [33–35]. Figure 1 shows our strategies to improve the affinity of **BMS-8**. We used fragmented structures of 3-hydroxymethyl-2-methylbiphenyl (**1**) and 3-bromotyrosine (**2**). After conjugation of **1** and **2**, a biphenyl-conjugated bromotyrosine (denoted as **X**) was synthesized. Because an amino and carboxyl group included in **X**, it could be conjugated to various amino acids. [36,37]. During the procedures, we employed in silico simulation and IC$_{50}$ assay to reveal molecular mechanism of the inhibition.

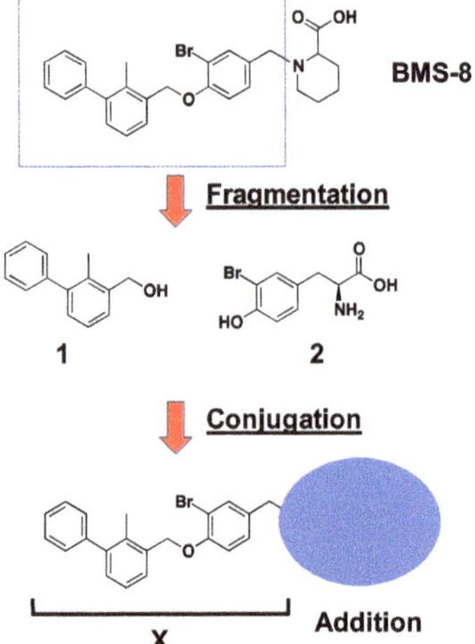

Figure 1. Strategies to improve inhibitory effect of **BMS-8**. 3-hydroxymethyl-2-methylbiphenyl (**1**) and 3-bromotyrosine (**2**) were selected as fragmented structures. A biphenyl-conjugated bromotyrosine **X** was synthesized after conjugation of **1** and **2**. We conjugated a variety of amino acids as additions, to the amino- and carboxyl-groups of **X** to reveal molecular mechanism of the inhibition.

2. Results

2.1. In Silico Docking Simulation and Organic Chemistry Synthesis of a Biphenyl-Conjugated Bromotyrosine

We designed a biphenyl-conjugated bromotyrosine (denoted as **X**), based on the **BMS-8**. We docked **X** into the crystal structure of **BMS-8**/PD-L1$_{AB}$ complex (PDB ID: 5J8O) [31] using ICM 3.8-7 software (Molsoft L.L.C., San Diego, CA, USA) [33–35], without guidance and induced fitting to avoid over-fitting. We obtained the docking score of −42.96 for **X**, which was the same order of **BMS-8**, −49.5 (Table 1). Based on the scores, we confirmed the potential of **X** for inhibition. Therefore, we synthesized **X** by the organic chemistry procedures. Scheme 1 shows the synthetic route for a biphenyl-bromotyrosine **6**. Full synthesis details are provided in Materials and Methods. The C- and N-terminals of 3-bromotyrosine (**2**) were first protected by *tert*-butyl and fluorenylmethyloxycarbonyl (Fmoc) groups, respectively, to produce the amino acid **4**, which was then reacted with 3-hydroxymethyl-2-methylbiphenyl (**1**) through the Mitsunobu reaction to yield compound **5**. Deprotection of the *tert*-butyl group in compound **5** produced the Fmoc-protected amino acid **6**. Deprotection of the Fmoc group in **6** yielded the compound **X**. Peptide conjugates were obtained by solid-state peptide synthesis using compound **6**. ^{1}H NMR spectra of the compounds are shown in Figures S1–S4. A summary of the analytical data for the synthesized compounds is given in Table S1. The analytical data indicate the successful synthesis of **X** and 29 **amino-X** derivatives consisting of 2-mers (**GX, XG, XS, XR, XA, XW**), 3-mers (**YXC, WXG, QXQ, CXA, RXN, SXR, NXR, CXR, GXG, XNL, XNH, XHP, XGG**), 4-mers (**XCSE, XGGG**), 5-mers (**WRXNN, ERXNK, WRXNQ, XRRRR, XGGGG**), 6-mer (**XGGGGG**), and 7-mers (**CERXNKM, FWRXNNI**).

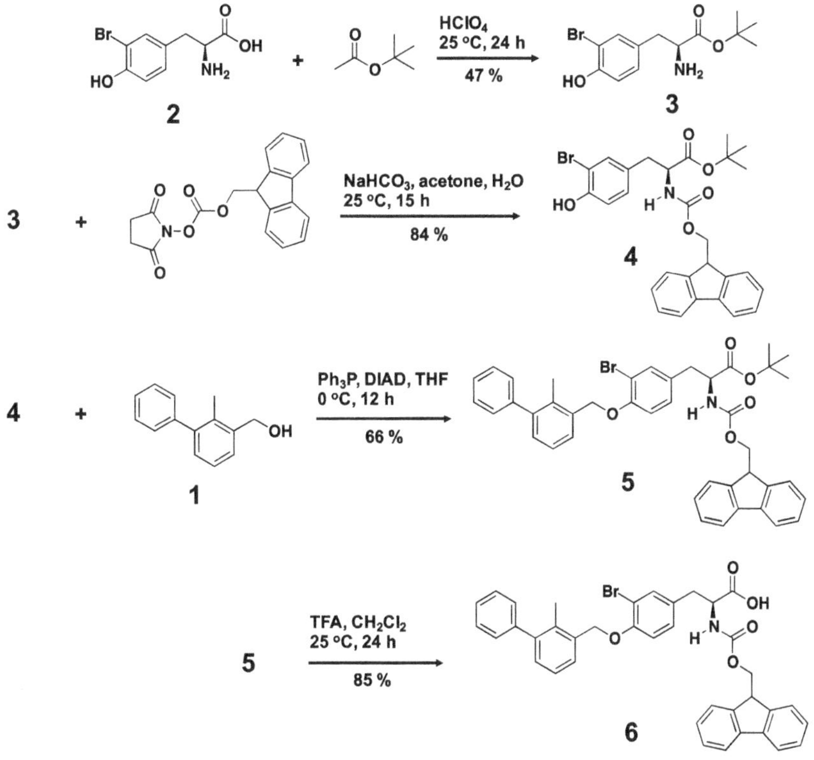

Scheme 1. Synthetic scheme for the biphenyl-conjugated bromotyrosine **6**.

Table 1. Docking simulation and IC_{50} measurements of **BMS-8** and **amino-X**s.

Amino Acid Length	Sequence	Score	RMSD (Å)	IC_{50} (µM)
-	**BMS-8**	−49.5	-	7.2
1	**X**	−42.96	0.40	1.5
2	**GX**	−41.0	0.52	448.5
	XG	−46.9	0.28	2.1
	XS	−42.1	0.60	2655.0
	XR	−45.7	0.37	892.0
	XA	−43.1	0.47	22.3
	XW	−43.3	0.51	845.0
3	**YXC**	−37.1	0.46	465.0
	WXG	−50.6	0.48	404.8
	QXQ	−37.7	0.73	1961.0
	CXA	−42.0	0.48	665.0
	RXN	−40.3	0.63	405.3
	SXR	−36.7	0.58	796.0
	NXR	−50.3	0.46	982.0
	CXR	−41.5	0.54	550.0
	GXG	−43.6	0.39	676.0
	XNL	−43.0	0.58	855.0
	XNH	−40.7	0.50	313.0
	XHP	−33.5	0.55	359.0
4	**XGG**	−36.1	0.57	6505.0
	XCSE	−32.6	0.45	1555.0
	XGGG	−53.3	0.51	6766.0
5	**WRXNN**	−38.1	0.38	157.4
	ERXNK	−21.3	0.48	15.6
	WRXNQ	−19.4	0.49	163.2
	XRRRR	−28.3	0.45	435.6
	XGGGG	−41.8	0.75	647.5
6	**XGGGGG**	−45.3	0.48	846.0
7	**CERXNKM**	4.65	1.80	308.2
	FWRXNNI	−7.30	0.41	311.8

2.2. Inhibition Assays of PD-1/PD-L1 Binding by BMS-8 and X

To evaluate the binding affinities of the compounds for PD-L1, we used the amplified luminescence proximity homogeneous assay (Alpha) by using the AlphaLISA® assay kit [38]. This assay is based on photoinduced energy transfer between donor and acceptor beads conjugated to PD-1 and PD-L1, respectively (Figure S6).

The AlphaLISA® assay revealed that the intermediates of **X**, compounds **1–6**, showed a few hundred µM or weaker IC_{50} values (Figure 3). **BMS-8** inhibited the PD-1/PD-L1 interaction with IC_{50} of 7.2 µM (Figure 2), which was weaker than that previously reported, IC_{50} of 0.146 µM [30]. On the other hand, nivolumab showed nano-molar order of inhibition (IC_{50} = 5.1 nM, Figure 2), corresponding to the previously reported value [39], which suggests the validity of our assay system.

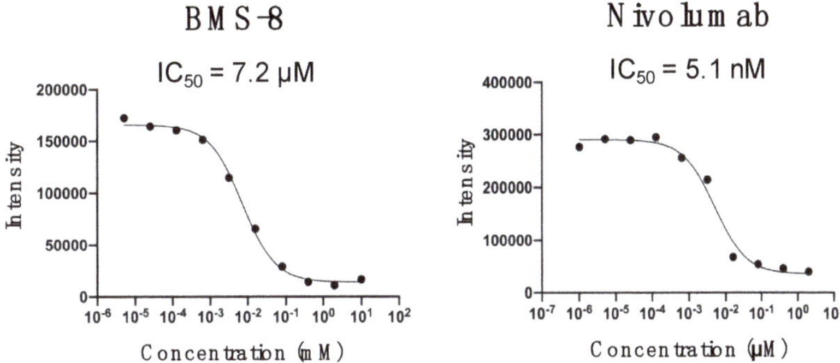

Figure 2. Inhibition of PD-1/PD-L1 interaction by **BMS-8** and nivolumab measured by the AlphaLISA® assay.

2.3. Fragmentation of BMS-8 and Conjugation of Compounds to Prepare X

To prepare higher affinity compounds based on **BMS-8**, we first considered a scenario that smaller groups of **BMS-8**, compounds **1–6** (Scheme 1), showed better inhibitory effect for PD-1/PD-L1 PPI. The docking scores of the compounds, however, were larger than that of **BMS-8** (−49.5), suggesting pooper inhibition effect. Actually, AlphaLISA assay revealed that the IC_{50} values were a few hundred µM, which were much weaker than that of **BMS-8** (7.2 µM) (Figure 3).

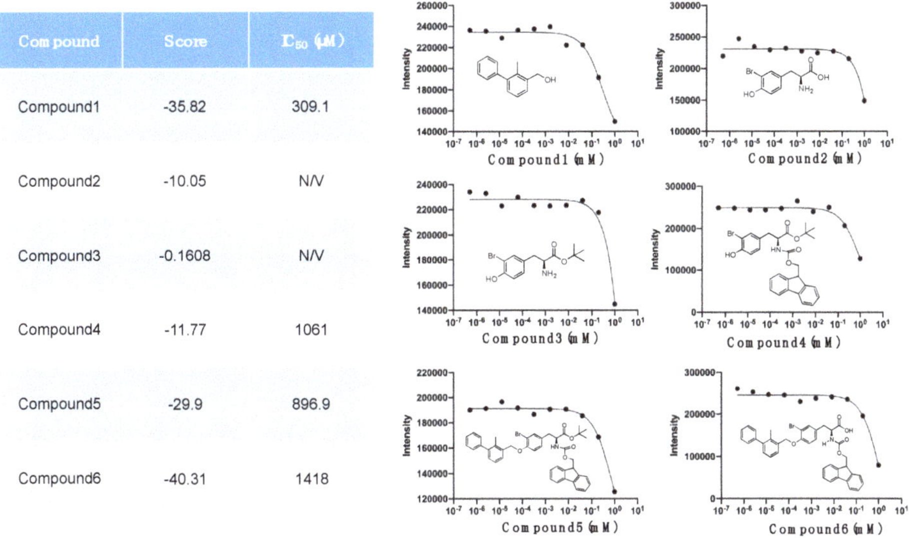

Compound	Score	IC_{50} (µM)
Compound1	-35.82	309.1
Compound2	-10.05	N/V
Compound3	-0.1608	N/V
Compound4	-11.77	1061
Compound5	-29.9	896.9
Compound6	-40.31	1418

Figure 3. Docking scores, IC_{50} values and measurements of compounds **1–6**. All compounds showed larger scores than that of **X** (score = −42.96) with a few hundred µM of IC_{50} values.

Therefore, we considered the next scenario of conjugation of compounds; we conjugated compound **4** and compound **1** to prepare biphenyl-bromotyrosine (**X**), which resembled **BMS-8** except the terminal amino- and carboxyl-groups. In turn, **X** showed a docking score of −42.96, comparable to that of **BMS-8** (−49.5). In fact, **X** inhibited PD-1/PD-L1 PPI with IC_{50} = 1.5 µM (Figure 4), which was five times better than that of **BMS-8** (7.2 µM).

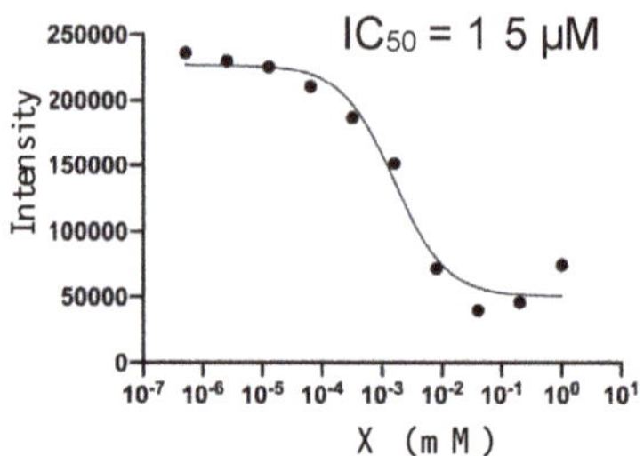

Figure 4. AlphaLISA assay of **X**. **X** shows IC_{50} = 1.5 µM with docking score = −42.96.

2.4. Docking Simulation and Inhibition Assay of Amino-Xs

The binding mode of the BMS compounds and derivatives to PD-L1 has previously been revealed by X-ray crystallography [31,40–42]. BMS compounds induces transient homodimerization of PD-L1$_{AB}$ on the binding, which masks the binding site for PD-1 located in the homodimerization interface. We docked **amino-X**s to the crystal structure of **BMS-8**/PD-L1$_{AB}$ complex (PDB ID: 5J8O) [31], using ICM 3.8-7 software (Molsoft L.L.C., San Diego, CA, USA) [33–35], without guidance and induced fitting to avoid over-fitting. After the docking, we calculated the root mean square deviation (RMSD) of distances between atoms in compound **BMS-8** and **X**, excluding Cα, NH$_2$, and COOH atoms (Figure 5).

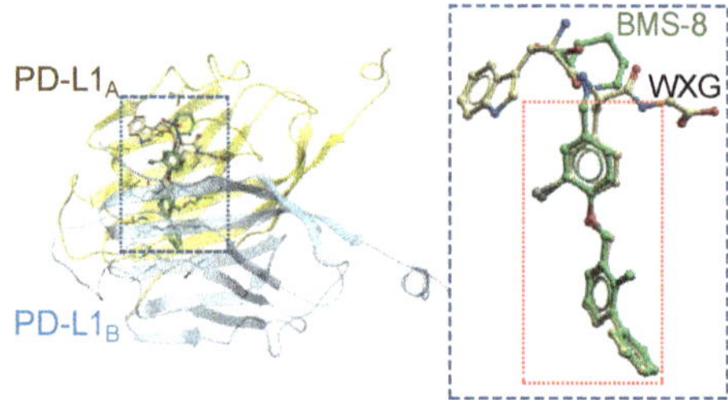

Figure 5. Root mean square deviation (RMSD) calculation between amino-X and **BMS-8** bound to PD-L1$_{AB}$ homodimer. After docking of amino-X (in this case, WXG), we calculated RMSD between a part of **WXG** (excluding Cα, amino-group and carboxyl-group) and the corresponding part of **BMS-8**, as shown by the red dotted-rectangle.

Table 1 shows the docking scores and RMSD values for **amino-X**s docked to PD-L1$_{AB}$. Also, the IC_{50} values for the **amino-X**s are listed in Table 1. As a result, they suggested some positive correlations. The IC_{50} values of the 1–2-mer **amino-X**s showed moderate correlations with both the RMSDs (CC 0.67, Table 2) and the scores (CC 0.40, Table 2). However, these correlations weakened as the number of conjugated amino acids increased (RMSD from 0.67 to 0 and CC 0.40 to −0.20, Table 2). These results suggest that the current in silico docking worked better for **amino-X**s conjugated with shorter amino acids.

Table 2. Correlation coefficients (CC) for IC_{50} vs. Score and IC_{50} vs. RMSD.

Length	CC of Score/IC_{50}	CC of RMSD/IC_{50}
1-2	0.40	0.67
1-3	0.35	0.37
1-4	0	0.28
1-7	−0.20	0

CC values were calculated by the Microsoft Excel.

To discuss the correlations further, we compared the docking structures of **X** (IC_{50} = 1.5 µM), **XG** (IC_{50} = 2.1 µM), and **GX** (IC_{50} = 448.5 µM).

We compared the binding modes of **BMS-8** and **X** in the pocket of PD-L1$_{AB}$ homodimer (Figure 6). **BMS-8**, with IC_{50} of 7.2 µM (Figure 2), binds the pocket with a hydrogen bind to Q66$_A$ and a hydrophobic interaction with V68$_A$ (Figure 6A), respectively. On the other hand, **X** forms a hydrogen bond with the hydroxy group of the side chain of Y56$_A$, which stabilizes the binding (Figure 6A), with IC_{50} of 1.5 µM (Figure 4). The superposition of **X** onto **BMS-8** showed an RMSD displacement of 0.40 Å (Figure 6B) We conclude that binding of **X** would not markedly impede PD-L1 homodimerization, which is consistent with its relatively low IC_{50} value of 1.5 µM (Figure 4). These results suggest that we can improve an IC_{50} value by substituting the six-membered group of **BMS-8** with some proper groups, leading to rearrangement of interactions around it. Besides, smaller displacement of biphenyl-bromotyrosine portion shown by RMSD is preferable for higher affinity.

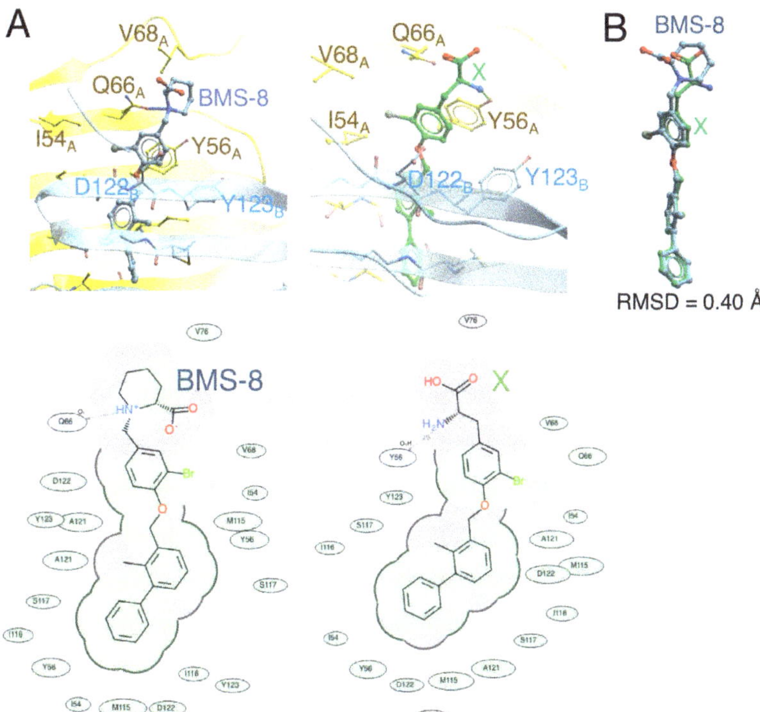

Figure 6. Docking conformations of **BMS-8** and **X**. (**A**) The docking modes of **BMS-8** and **X** were revealed by the X-ray crystallography and in silico docking simulation, respectively, which the 2D binding pictures. The 2D figures show that biphenyl portions of the ligands bind into the pocket by hydrophobic interactions shown in light-green color. In contrast, the amino cation at the six-membered ring of **BMS-8** makes a hydrogen bond with the sidechain of Q66$_A$ in cyan color. In addition, the six-membered ring makes hydrophobic interaction with V68$_A$. On the other hand, amino-group of bromo-tyrosine in **X** makes a hydrogen bonding to the hydroxyl group of Y56$_A$ colored in cyan, without other hydrophobic interaction, as shown in the 2D picture. (**B**) **BMS-8** and **X** without Cα, NH$_2$, and COOH superposed each other with RMSD of 0.40 Å.

Modeling of **XG** identified two potential hydrogen bonds between the N-terminal of **XG** and the side chain of Q66$_A$ and between the carboxyl group of Gly and R125$_B$ in the side chain (Figure 7A). The RMSD between **XG** and **BMS-8** was 0.28 Å (Figure 7B), which suggested that the IC_{50} value of

XG would be similar to that of **X**. Indeed, **XG** had a measured IC_{50} for PD-1/PD-L1 binding of 2.1 µM (Figure 7C). **X** and **XG** potentially have the inhibitory effect for PD-1/PD-L1 interaction because K_D between PD-1 and PD-L1 are reported as 6.4 µM [43].

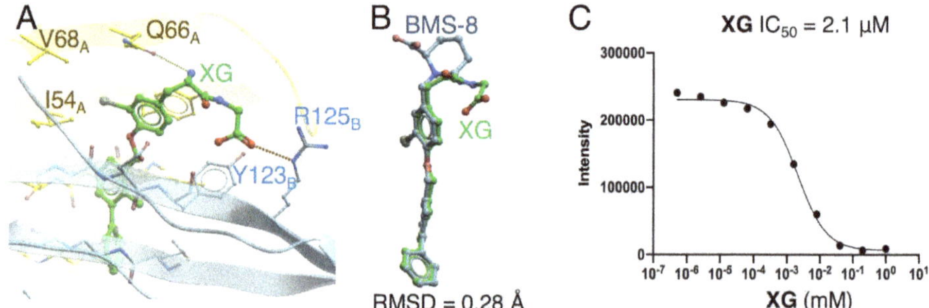

Figure 7. In silico binding mode of **XG**. (**A**) Behavior of **XG** in the binding pocket of the PD-L1$_{AB}$ homodimer. (**B**) Superposition of **XG** onto **BMS-8**. The RMSD for displacement was 0.28 Å. (**C**) IC_{50} of **XG** for PD-1/PD-L1 binding.

GX docking into the binding pocket of the PD-L1 homodimer revealed two hydrogen bonds formed between **GX** amino groups and carbonyl group of Y123$_B$ (Figure 8A). As a result, the calculated RMSD between **GX** and **BMS-8** was 0.52 Å (Figure 8B), which was larger than the RMSD of **X** and **XG**. This observation suggests that **GX** binding might sterically hinder PD-L1 homodimerization, leading to poorer inhibition of PD-1/PD-L1 binding. Consistent with this, the measured IC_{50} for **GX** was 448.5 µM (Figure 8C), which was several hundred times higher than those for **X** and **XG** (Figure 4; Figure 7C). It is possible that the larger displacement of **X** of **GX** caused to deform the pocket of the PD-L1 homodimer, leading to the weaker inhibition of **GX** than those of **X** and **XG**.

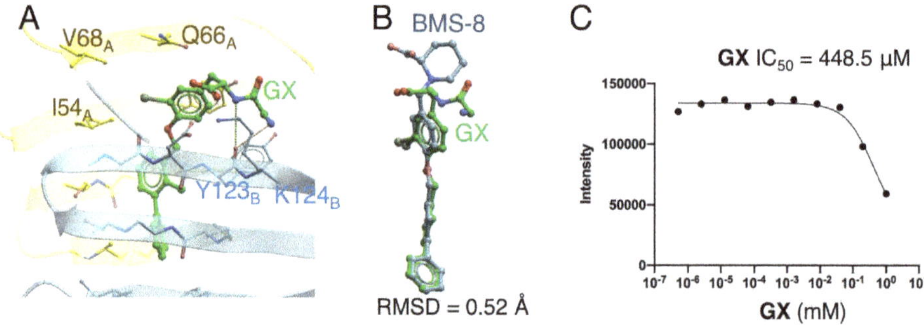

Figure 8. In silico binding mode of **GX**. (**A**) Behavior of **GX** in the binding pocket of the PD-L1$_{AB}$ homodimer. (**B**) Superposition of **GX** onto **BMS-8**. The RMSD for displacement was 0.52 Å RMSD. (**C**) IC_{50} of **GX** for PD-1/PD-L1 binding.

The **X** portion of **BMS-8** without Cα, NH_2, COOH atoms formed hydrophobic interactions in the crystal structure (PDB ID: 5J8O), with residues I54$_A$, Y56$_A$, V68$_A$, M115$_A$, I116$_A$, S117$_A$, A121$_A$, D122$_A$, I54$_B$, Y56$_B$, M115$_B$, I116$_B$, S117$_B$, A121$_B$, D122$_B$, and Y123$_B$ of the PD-L1 homodimer (Figure 9A). The space-filling representation of **X** shows the adherent interaction mode to the binding pocket (Figure 9B,C). The intermediate compounds of **BMS-8**, compounds **1–6** (Scheme 1) showed a poor ability to inhibit PD-1/PD-L1 binding (Figure 3), which was probably due to insufficient hydrophobic filling of the compounds in the binding pocket.

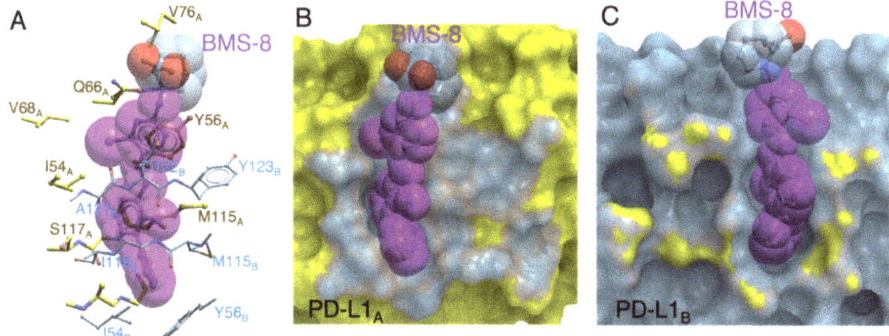

Figure 9. Schematic drawing and space-filling representation of **BMS-8** binding in the binding pocket of the PD-L1 homodimer. In (**A–C**), violet represents **X** without Cα, NH$_2$, COOH atoms. (**A**) Binding mode of **BMS-8** in the pocket of the PD-L1$_{AB}$ homodimer (PDB ID:5J8O). Yellow and cyan represent PD-L1$_A$ and PD-L1$_B$ side chains, respectively. (**B**) Space-filling representation of **BMS-8** bound to the surface of PD-L1$_A$ (yellow) and contact area with PD-L1$_B$ (cyan). (**C**) Space-filling representation of **BMS-8** bound to the surface of PD-L1$_B$ (cyan) and contact area with PD-L1$_A$ (yellow).

Collectively, our results suggest that the larger displacement of **amino-Xs** from **BMS-8** prevents PD-L1$_A$/PD-L1$_B$ homodimer formation. The docking simulations suggest that **X** and **GX** promote homodimerization of PD-L1, resulting in low IC$_{50}$ values, whereas the larger displacement of **amino-Xs** prevents PD-L1 homodimer formation and increase the IC$_{50}$ values.

The results of this study advance our understanding of how small molecule compounds could be rationally designed to inhibit PD-1/PD-L1 interactions with high affinity. In silico docking simulations have typically shown that target proteins have stable binding pockets during ligand binding, even allowing for some local flexibility of the side chains within the pockets [37,44]. In that scenario, binding scores generally correlate well with experimentally determined inhibitor activity [45]. However, binding of **X** and **amino-X** in the PD-L1 pocket occurs through strict interactions, indicating that even a slight displacement of the **X** conformation leads to deformation of the PD-L1 homodimer, which deceases the inhibitory effect. Consistent with this, the **amino-Xs** with shorter amino acid conjugates showed moderate positive correlations between the measured IC$_{50}$ values and RMSDs in the no template/flexible docking mode, whereas the correlation was weakened by further amino acid addition.

3. Materials and Methods

3.1. Materials for Organic Chemistry Synthesis

Sodium chloride (NaCl), lysozyme, monosodium phosphate (NaH$_2$PO$_4$), imidazole, glycerol, reduced glutathione, oxidized glutathione, methanol, dimethyl sulfoxide (DMSO), trifluoroacetic acid (TFA), tert-butyl acetate, perchloric acid (HClO$_4$), hydrochloric acid (HCl), sodium carbonate, ethyl acetate, sodium sulfate, hexane, sodium hydrogen carbonate (NaHCO$_3$), acetone, triphenyl phosphine (Ph$_3$P), anhydrous dichloromethane (CH$_2$Cl$_2$), and anhydrous tetrahydrofuran (THF) were purchased from Wako Pure Chemical Industries Ltd. (Osaka, Japan). 3-Bromo-tyrosine, 3-hydroxymethyl-2-methylbiphenyl, and diisopropyl azodicarboxylate (DIAD; 40% in toluene, approximately 1.9 mol L^{-1}) were purchased from Tokyo Chemical Industry Co., Ltd. (Tokyo, Japan). Magnesium sulfate and CH$_2$Cl$_2$ were purchased from Junsei Chemical Co., Ltd. (Tokyo, Japan). Deuterochloroform (CDCl$_3$) was purchased from Isotec, Inc. (Miamisburg, OH, USA), and *N*-[(9H-fluoren-9-ylmethoxy) carbonyloxy] succinimide (Fmoc-Osu) was purchased from Watanabe Chemical Industries, Ltd. (Hiroshima, Japan).

3.2. Synthesis of a Biphenyl-Conjugated Bromotyrosine

(S)-tert-Butyl 2-amino-3-(3-bromo-4-hydroxyphenyl) propanoate (3). A suspension of 3-bromotyrosine (2; 1.0 g, 3.9 mmol) in tert-butyl acetate (16 mL, 92 mmol) was cooled to 0 °C, and stirred for 30 min. $HClO_4$ (0.5 mL, 7.7 mmol) was then slowly added to the suspension at 0 °C, and the reaction mixture was warmed to 25 °C and stirred for 24 h. The mixture was washed with water and 1N HCl, and the aqueous phase was brought to pH 9 using sodium carbonate and then extracted with ethyl acetate. The resulting organic phase was washed with water and dried with sodium sulfate. The solvent was evaporated under reduced pressure, yielding an oily compound. This crude product was washed with cold hexane and then dried under reduced pressure to yield compound **3** (0.57 g, 47%). ^{1}H-NMR (400 MHz, $CDCl_3$): δ = 1.41 (s, 9H, $–OC(CH_3)_3$), 2.73 (dd, 1H, J = 14.4, 8.0 Hz, $HOPh(Br)–CH_2CH(NH_2)–$), 2.93 (dd, 1H, J = 13.6, 5.2 Hz, $HOPh(Br)–CH_2CH(NH_2)–$), 3.57 (dd, 1H, J = 7.2, 5.6 Hz, $HOPh(Br)–CH_2CH(NH_2)–$), 3.70 (m, 3H, $HOPh(Br)–CH_2CH(NH_2)–$), 6.70 (d, 1H, J = 8.0 Hz, aromatic ring), 6.94 (dd, 1H, J = 8.4, 2.0 Hz, aromatic ring), 7.26 (d, 1H, J = 1.6 Hz, aromatic ring).

(S)-tert-Butyl 2-({[(9H-fluoren-9-yl)methoxy]carbonyl}amino)-3-(3-bromo-4-hydroxyphenyl) propanoate (4). A suspension of **3** (0.5 g, 1.6 mmol) and $NaHCO_3$ (0.27 g, 3.2 mmol) in water (20 mL) was cooled to 0 °C. Fmoc-Osu (1.1 g, 3.2 mmol) in acetone (40 mL) was added to the suspension slowly, and the reaction mixture was then stirred at 25 °C for 15 h. The solvent was removed and washed with 1N HCl and water. After drying under vacuum, the crude product was purified by column chromatography on silica gel (eluent: ethyl acetate/hexane = 1:3 v/v) to yield compound **4** (0.71 g, 84%). ^{1}H-NMR (400 MHz, $CDCl_3$): δ = 1.42 (s, 9H, $–OC(CH_3)_3$), 3.00 (d, 2H, J = 5.6 Hz, $HOPh(Br)–CH_2CH(NHCOOCH_2CH–)–$), 4.21 (t, 1H, J = 7.2 Hz, $HOPh(Br)–CH_2CH(NHCOOCH_2CH–)–$), 4.33 (dd, 1H, J = 10.4, 7.2 Hz, $HOPh(Br)–CH_2CH(NHCOOCH_2CH–)–$), 4.43–5.00 (m, 2H, $HOPh(Br)–CH_2CH(NHCOOCH_2CH–)–$), 5.29 (d, 1H, J = 8.0 Hz, $HOPh(Br)–CH_2CH(NHCOOCH_2CH–)–$), 5.43 (s, 1H, $HOPh(Br)–CH_2CH(NHCOOCH_2CH–)–$), 6.91 (d, 1H, J = 8.4 Hz, aromatic ring), 6.96 (d, 1H, J = 9.2 Hz, aromatic ring), 7.26–7.33 (m, 3H, aromatic ring), 7.40 (dd, 2H, J = 7.4, 7.4 Hz, aromatic ring), 7.57 (dd, 2H, J = 6.2, 6.2 Hz, aromatic ring), 7.76 (d, 2H, J = 7.2 Hz, aromatic ring); high resolution mass spectrometry (HRMS) calculated for $C_{28}H_{28}BrNO_5$ ($[M + H]^+$): 538.1224, found: 538.1224.

4 +

Ph$_3$P, DIAD, THF
0 °C, 12 h

66 %

1

5

(S)-tert-Butyl 2-({[(9H-fluoren-9-yl)methoxy]carbonyl}amino)-3-{3-bromo-4-[(2-methyl-1,1′-biphenyl-3-yl)methoxy]phenyl}propanoate (5). To a solution of **4** (0.1 g, 0.19 mmol), 3-hydroxymethyl-2-methylbiphenyl (**1**; 39 mg, 0.20 mmol), and triphenyl phosphine (57 mg, 0.20 mmol) in anhydrous THF (10 mL), DIAD (0.1 mL, 0.22 mmol) was added at 0 °C under argon, and the reaction mixture was stirred at 0 °C for 12 h under argon. The organic phase was extracted with CH$_2$Cl$_2$ and dried over anhydrous magnesium sulfate. The solvent was then evaporated under reduced pressure, with the temperature kept below 30 °C. The crude product was purified by column chromatography on silica gel (eluent: ethyl acetate/hexane = 1:4 v/v) to yield compound **5** (0.09 g, 66%). ^{1}H-NMR (400 MHz, CDCl$_3$): δ = 1.46 (s, 9H, –OC(CH$_3$)$_3$), 2.27 (s, 3H, Biphenyl(CH$_3$)–CH$_2$OPh(Br)–CH$_2$CH(NHCOOCH$_2$CH–)–), 3.05 (d, 2H, J = 5.6 Hz, Biphenyl(CH$_3$)–CH$_2$OPh(Br)–CH$_2$CH(NHCOOCH$_2$CH–)–), 4.23 (t, 1H, J = 7.6 Hz, Biphenyl(CH$_3$)–CH$_2$OPh(Br)–CH$_2$CH(NHCOOCH$_2$CH–)–), 4.34 (dd, 1H, J = 6.8, 6.8 Hz, Biphenyl(CH$_3$)–CH$_2$OPh(Br)–CH$_2$CH(NHCOOCH$_2$CH–)–), 4.46-4.56 (m, 2H, Biphenyl(CH$_3$)–CH$_2$OPh(Br)–CH$_2$CH(NHCOOCH$_2$CH–)–), 5.13 (s, 2H, Biphenyl(CH$_3$)-CH$_2$OPh(Br)-CH$_2$CH(NHCOOCH$_2$CH–)–), 5.37 (d, 1H, J = 7.6 Hz, Biphenyl(CH$_3$)–CH$_2$OPh(Br)-CH$_2$CH(NHCOOCH$_2$CH–)–), 6.92 (d, 1H, J = 8.0 Hz, aromatic ring), 7.05 (d, 1H, J = 7.2 Hz, aromatic ring), 7.19–7.45 (m, 14H, aromatic ring), 7.53 (d, 1H, J = 6.8 Hz, aromatic ring), 7.60 (dd, 2H, J = 6.4, 6.4 Hz, aromatic ring), 7.77 (d, 2H, J = 7.2 Hz, aromatic ring); HRMS calculated for C$_{42}$H$_{40}$BrNO$_5$ ([M + H]$^+$): 718.2163, found: 718.2164.

5

TFA, CH$_2$Cl$_2$
25 °C, 24 h

85 %

6

(S)-2-({[(9H-fluoren-9-yl)methoxy]carbonyl}amino)-3-{3-bromo-4-[(2-methyl-1,1′-biphenyl-3-yl)methoxy]phenyl}propanoic acid (6). A solution of **5** (3.9 g, 5.42 mmol) in anhydrous CH$_2$Cl$_2$ (36 mL) was stirred at 0 °C under argon for 15 min. TFA (1.3 mL, 16.6 mmol) was added dropwise to the solution at 0 °C, and the reaction mixture was stirred at 25 °C under argon. After 6 h, TFA (1.5 mL, 19.5 mmol) was added to the reaction mixture, which was then stirred at 25 °C for 18 h under argon. The solvent was removed under reduced pressure, with the temperature kept below 40 °C. The crude product was purified by column chromatography on silica gel (eluent: CH$_2$Cl$_2$/methanol = 97:3 v/v) to yield compound **6** (3.2 g, 85%). ^{1}H-NMR (400 MHz, CDCl$_3$): δ = 2.24 (s, 3H, Biphenyl(CH$_3$)–CH$_2$OPh(Br)–CH$_2$CH(NHCOOCH$_2$CH–)–), 3.05 (dd, 1H, J = 14.0, 6.0 Hz, Biphenyl(CH$_3$)–CH$_2$OPh(Br)–CH$_2$CH(NHCOOCH$_2$CH–)–), 3.15 (dd, 1H, J = 14.8, 5.2 Hz, Biphenyl(CH$_3$)–CH$_2$OPh(Br)–CH$_2$CH(NHCOOCH$_2$CH–)–), 4.21 (t, 1H, J = 6.8 Hz, Biphenyl(CH$_3$)–CH$_2$OPh(Br)–CH$_2$CH(NHCOOCH$_2$CH–)–), 4.36 (dd, 1H, J = 6.8, 6.8 Hz, Biphenyl(CH$_3$)–CH$_2$OPh(Br)–CH$_2$CH(NHCOOCH$_2$CH–)–), 4.46 (dd, 1H, J = 10.0, 7.2 Hz, Biphenyl(CH$_3$)–CH$_2$OPh(Br)–CH$_2$CH(NHCOOCH$_2$CH–)–), 4.66

(dd, 1H, J = 13.2, 6.0 Hz, Biphenyl(CH$_3$)–CH$_2$OPh(Br)–CH$_2$CH(NHCOOCH$_2$CH–)–), 5.09 (s, 2H, Biphenyl(CH$_3$)-CH$_2$OPh(Br)–CH$_2$CH(NHCOOCH$_2$CH–)–), 5.23 (d, 1H, J = 8.4 Hz, Biphenyl(CH$_3$)-CH$_2$OPh(Br)–CH$_2$CH(NHCOOCH$_2$CH–)–), 6.91 (d, 1H, J = 8.8 Hz, aromatic ring), 7.03 (d, 1H, J = 7.6 Hz, aromatic ring), 7.21–7.55 (m, 15H, aromatic ring), 7.74 (d, 2H, J = 7.2 Hz, aromatic ring); HRMS calculated for C$_{38}$H$_{32}$BrNO$_5$ ([M + H]$^+$): 662.1537, found: 662.1520.

3.3. Solid-State Peptide Synthesis

Amino-Xs were synthesized using an automated peptide synthesizer (MultiPep CF, INTAVIS Bioanalytical Instruments AG, Cologne, Germany). The synthetic protocol for glycine-conjugated peptide **XG** was as follows: Fmoc-protected glycine attached to a polystyrene resin (Fmoc-Gly NovaSyn TGT, Merck KGaA, Darmstadt, Germany) was deprotected by piperidine (20% in *N*-methylpyrrolidone (NMP). The resulting resin was reacted with **6** (99 mg, 0.14 mmol), 1-[bis(dimethylamino)methylene]-1H-benzotriazolium 3-oxide hexafluorophosphate (HBTU; 150 µL, 0.5 M in *N,N*-dimethylformamide (DMF), *N*-methylmorpholine (45 µL, 4.0 M in DMF) in NMP (8 µL) for 45 min. After washing, the *N*-α-protecting group of Fmoc in compound **6** was deprotected by piperidine (20% in NMP). Finally, the obtained peptide was cleaved from the resin using TFA (95% in water), yielding **XG**. Other peptides were synthesized using a similar method. (*S*)-2-amino-3-[3-bromo-4-{(2-methyl-1,1'-biphenyl-3-yl)methoxy}phenyl]propanoic acid (**X**) was obtained by deprotection of Fmoc in **6** using piperidine (20% in NMP).

3.4. Characterization

The synthesized compounds were identified using ^{1}H NMR spectroscopy (JNM–ECZ400R, JEOL Ltd., Tokyo, Japan) and HRMS (QSTAR Elite, AB SCIEX, Framingham, MA, USA).

3.5. Determination of the IC$_{50}$ Value by AlphaLISA®

3.5.1. Principle of the Competitive Binding Assay

The binding affinity of the inhibitors to PD-L1 were measured using the AlphaLISA® assay kit (AL356 HV/C/F, PerkinElmer) according to the manufacturer's instructions, with the anti-PD-1 mAb nivolumab (Selleck Chemicals, Houston, TX, USA) included as a positive control [41]. In this assay, direct binding of an inhibitor to PD-L1 is detected by photoinduced energy transfer (Figure S6). Biotin-conjugated PD-1 is attached to streptavidin-coated donor beads and histidine (His)-tagged PD-L1 is attached to anti-His-conjugated acceptor beads. Photoexcitation of the donor beads at 680 nm yields singlet oxygen. If PD-L1–PD-1 binding is successful, energy is transferred through singlet oxygen, leading to an increase in fluorescence intensity at 615 nm (Figure S6).

3.5.2. Preparation of Samples

BMS-8 was purchased from AA Blocks LLC (San Diego, CA, USA). Stock solutions of inhibitors in DMSO (stock solution A, 5 mM) were serially diluted (Figure S5A) to obtain 10 assay solutions (1–10) with concentrations ranging from 5.0 mM to 2.6 nM (Table S2). An aliquot of solution 1–10 (2 µL) was mixed with His-tagged PD-L1 (25 nM, 2 µL), biotin-conjugated PD-1 (25 nM, 2 µL), anti-His acceptor beads (0.55 g L^{-1}, 2 µL), and streptavidin-coated donor beads (1.1 g L^{-1}, 2 µL) (Figure S5B) in a final volume of 10 µL and incubated at 25 °C for 90 min. Positive and negative technical controls were included in parallel. Positive controls contained buffer (2 µL) in place of solution 1–10, and negative controls contained only the beads (2 µL each) and buffer (6 µL).

3.5.3. AlphaLISA® Measurement and Analysis

The reaction samples (10 µM) were placed in a 384-well microplate and photoirradiated at 680 nm from the top. Fluorescence at 615 nm was detected using an EnSpire multimode plate reader

(Perkin Elmer, Waltham, MA, USA). IC_{50} values were estimated from a sigmoidal curve of fluorescence intensity vs. inhibitor concentration using a relative weighting method ($1/Y^2$ weighting) with GraphPad Prism 8 (GraphPad Software Inc., San Diego, CA, USA).

3.6. Docking Simulation of Compounds

The docking simulation software ICM 3.8-7 [33] was used to investigate the binding modes of **X** and **amino-X**s to the PD-L1 homodimer complexed with **BMS-8** (PDB ID: 5J8O) [31]. We performed docking without template docking [37] or introducing flexibility [37] to avoid over-fitting of the ligands into the pocket. The docking simulation supposed Monte Carlo pseudo-Brownian motion [46]. In the simulation, the score suggests goodness of docking, defined as follows [45]:

$$\text{Score} = \Delta E_{\text{IntFF}} + T\Delta S_{\text{Tor}} + \alpha_1 \Delta E_{\text{HBond}} + \alpha_2 \Delta E_{\text{HBDesol}} + \alpha_3 \Delta E_{\text{solEl}} + \alpha_4 \Delta E_{\text{HPob}} + \alpha_5 Q_{\text{size}} \tag{1}$$

where $\alpha_1 - \alpha_5$ = weight, ΔE_{IntFF} = ligand–target van der Waals interactions and internal force field energy of the ligand, $T\Delta S_{\text{Tor}}$ = free energy changes due to conformational energy loss upon ligand binding, ΔE_{HBond} = hydrogen bonding interactions, $\Delta E_{\text{HBDesol}}$ = hydrogen bond donor–acceptor desolvation energy, ΔE_{solEl} = solvation electrostatic energy upon ligand binding, ΔE_{HPob} = hydrophobic free energy gain, and Q_{size} = a size correction term proportional to the number of ligand atoms [45,47,48]. We calculated RMSD values by using CORREL function in the Microsoft Excel.

4. Conclusions

This study reports that we prepared the new biphenyl-conjugated bromotyrosine, which inhibits the PD-1/PD-L1 interaction with better effect than that of **BMS-8**. In addition, the **amino-X**s, which are conjugates of **X** with a variety of amino acids, provide the molecular mechanism how amino acid modifications of **X** affects inhibition of PD-1/PD-L1 interactions. Binding of the **X** without the $C\alpha$, NH_2, and COOH atoms portion of **amino-X**s into the PD-L1 binding pocket is required to promote transient homodimerization of PD-L1$_A$/PD-L1$_B$, leading to formation of a stable ternary complex composed of **X** and PD-L1$_{AB}$. Amino acid conjugation, however, alters the **X** docking conformation in the PD-L1 pocket, reducing the IC_{50} values dramatically. We conclude that improper interactions between amino acids conjugated to **X** and those in the binding pocket induced displacement of the compounds, thereby reducing inhibitory effect. In the future, we plan to design conjugates with amino acids that do not disturb the conformation of **X** in the PD-L1 binding pocket.

Supplementary Materials: The following are available online at http://www.mdpi.com/1422-0067/21/10/3639/s1.

Author Contributions: H.M., M.K., Y.I., and E.K. conceived the project. E.-H.K. and M.K. conducted the organic syntheses. H.M. and E.-H.K. performed the in silico simulations. E.-H.K., R.D., and H.M. measured the IC_{50} values by the AlphaLISA®. E.-H.K., M.K., and H.M. wrote the manuscript. All of the authors discussed the results and approved the content of the final manuscript. All authors have read and agreed to the published version of the manuscript.

Funding: This study was partly supported by the Incentive Research Fund of RIKEN (FY2017-2018, H.M. and M.K.) and the Japan Agency for Medical Research and Development (AMED: FY2018-2020, Y.I.). E.-H.K. was financially supported by the Junior Research Associate Program at RIKEN.

Acknowledgments: We thank the Bio-material Analysis Support Unit, RIKEN Center for Brain Science for synthesis of biphenyl-conjugated amino acids. We thank Anne M. O'Rourke, from Edanz Group (www.edanzediting.com/ac) for editing a draft of this manuscript.

Conflicts of Interest: The authors declare no conflicts of interest.

Abbreviations

PD-1	Programmed cell death 1
PD-L1	Programmed cell death-ligand 1
PD-L1$_A$	PD-L1 chain A
PD-L1$_B$	PD-L1 chain B
PD-L1$_{AB}$	Homodimer of PD-L1$_A$/PD-L1$_B$ chains
Alpha	Amplified Luminescence Proximity Homogeneous Assay
X	biphenyl-conjugated bromotyrosine
Amino-X	Amino acid conjugated-**X**
MALDI-TOF MS	Matrix assisted laser desorption/ionization-time of flight mass spectrometry
RMSD	Root mean square deviation
K$_D$	Equilibrium dissociation constant
IC$_{50}$	50% maximal inhibitory concentration
CC	Correlation coefficient
HTRF	Homogenous Time-Resolved Fluorescence

References

1. Rius, M.; Lyko, F. Epigenetic cancer therapy: Rationales, targets and drugs. *Oncogene* **2011**, *31*, 4257–4265. [CrossRef] [PubMed]
2. Postow, M.A.; Callahan, M.K.; Wolchok, J.D. Immune checkpoint blockade in cancer therapy. *J. Clin. Oncol.* **2015**, *33*, 1974–1982. [CrossRef]
3. Lee, L.; Gupta, M.; Sahasranaman, S. Immune Checkpoint inhibitors: An introduction to the next-generation cancer immunotherapy. *J. Clin. Pharmacol.* **2016**, *56*, 157–169. [CrossRef] [PubMed]
4. Hoos, A. Development of immuno-oncology drugs-from CTLA4 to PD1 to the next generations. *Nat. Rev. Drug Discov.* **2016**, *15*, 235–247. [CrossRef] [PubMed]
5. Iwai, Y.; Ishida, M.; Tanaka, Y.; Okazaki, T.; Honjo, T.; Minato, N. Involvement of PD-L1 on tumor cells in the escape from host immune system and tumor immunotherapy by PD-L1 blockade. *Proc. Natl. Acad. Sci. USA* **2002**, *99*, 12293–12297. [CrossRef] [PubMed]
6. Mellman, I.; Coukos, G.; Dranoff, G. Cancer immunotherapy comes of age. *Nature* **2011**, *480*, 480–489. [CrossRef]
7. Ishida, Y.; Agata, Y.; Shibahara, K.; Honjo, T. Induced expression of PD-1, a novel member of the immunoglobulin gene superfamily, upon programmed cell death. *EMBO J.* **1992**, *11*, 3887–3895. [CrossRef]
8. Dong, H.; Zhu, G.; Tamada, K.; Chen, L. B7-H1, a third member of the B7 family, co-stimulates T-cell proliferation and interleukin-10 secretion. *Nat. Med.* **1999**, *5*, 1365–1369. [CrossRef]
9. Freeman, G.J.; Long, A.J.; Iwai, Y.; Bourque, K.; Chernova, T.; Nishimura, H.; Fitz, L.J.; Malenkovich, N.; Okazaki, T.; Byrne, M.C.; et al. Engagement of the PD-1 immunoinhibitory receptor by a novel B7 family member leads to negative regulation of lymphocyte activation. *J. Exp. Med.* **2000**, *192*, 1027–1034. [CrossRef]
10. Dong, H.D.; Strome, S.E.; Salomao, D.R.; Tamura, H.; Hirano, F.; Flies, D.B.; Roche, P.C.; Lu, J.; Zhu, G.F.; Tamada, K.; et al. Tumor-associated B7-H1 promotes T-cell apoptosis: A potential mechanism of immune evasion. *Nat. Med.* **2002**, *8*, 793–800. [CrossRef]
11. Lee, K.M.; Chuang, E.; Griffin, M.; Khattri, R.; Hong, D.K.; Zhang, W.; Straus, D.; Samelson, L.E.; Thompson, C.B.; Bluestone, J.A. Molecular basis of T cell inactivation by CTLA-4. *Science* **1998**, *282*, 2263–2266. [CrossRef] [PubMed]
12. Stamper, C.C.; Zhang, Y.; Tobin, J.F.; Erbe, D.V.; Ikemizu, S.; Davis, S.J.; Stahl, M.L.; Seehra, J.; Somers, W.S.; Mosyak, L. Crystal structure of the B7-1/CTLA-4 complex that inhibits human immune responses. *Nature* **2001**, *410*, 608–611. [CrossRef] [PubMed]
13. Drake, C.G. Basic overview of current immunotherapy approaches in urologic malignancy. *Urol. Oncol. Semin. Orig. Investig.* **2006**, *24*, 413–418. [CrossRef]
14. Weiner, L.M.; Surana, R.; Wang, S. Monoclonal antibodies: Versatile platforms for cancer immunotherapy. *Nat. Rev. Immunol.* **2010**, *10*, 317–327. [CrossRef] [PubMed]
15. Guo, Y.-T.; Hou, Q.-Y.; Wang, N. Monoclonal antibodies in cancer therapy. *Clin. Oncol. Cancer Res.* **2011**, *8*, 215–219. [CrossRef]

16. Pento, J.T. Monoclonal Antibodies for the Treatment of Cancer. *Anticancer Res.* **2017**, *37*, 5935–5939. [CrossRef]
17. Reichert, J.M. Monoclonal Antibodies as Innovative Therapeutics. *Curr. Pharm. Biotechnol.* **2008**, *9*, 423–430. [CrossRef]
18. Leach, D.R.; Krummel, M.F.; Allison, J.P. Enhancement of antitumor immunity by CTLA-4 blockade. *Science* **1996**, *271*, 1734–1736. [CrossRef]
19. Hodi, F.S.; O'Day, S.J.; McDermott, D.F.; Weber, R.W.; Sosman, J.A.; Haanen, J.B.; Gonzalez, R.; Robert, C.; Schadendorf, D.; Hassel, J.C.; et al. Improved survival with ipilimumab in patients with metastatic melanoma. *New Engl. J. Med.* **2010**, *363*, 711–723. [CrossRef]
20. Wang, C.Y.; Thudium, K.B.; Han, M.H.; Wang, X.T.; Huang, H.C.; Feingersh, D.; Garcia, C.; Wu, Y.; Kuhne, M.; Srinivasan, M.; et al. In vitro characterization of the anti-PD-1 antibody nivolumab, BMS-936558, and in vivo toxicology in non-human primates. *Cancer Immunol. Res.* **2014**, *2*, 846–856. [CrossRef]
21. Kimiz-Gebologlu, I.; Gulce-Iz, S.; Biray-Avci, C. Monoclonal antibodies in cancer immunotherapy. *Mol. Boil. Rep.* **2018**, *45*, 2935–2940. [CrossRef] [PubMed]
22. Suzuki, S.; Ishida, T.; Yoshikawa, K.; Ueda, R. Current status of immunotherapy. *Jpn. J. Clin. Oncol.* **2016**, *46*, 191–203. [CrossRef] [PubMed]
23. Zhang, H.; Chen, J. Current status and future directions of cancer immunotherapy. *J. Cancer* **2018**, *9*, 1773–1781. [CrossRef] [PubMed]
24. Sharma, P.; Hu-Lieskovan, S.; Wargo, J.A.; Ribas, A. Primary, adaptive, and acquired resistance to cancer immunotherapy. *Cell* **2017**, *168*, 707–723. [CrossRef] [PubMed]
25. El-Osta, H.; Shahid, K.; Mills, G.M.; Peddi, P. Immune checkpoint inhibitors: The new frontier in non-small-cell lung cancer treatment. *Onco Targets Ther.* **2016**, *9*, 5101–5116. [CrossRef]
26. Shukla, A.A.; Thommes, J. Recent advances in large-scale production of monoclonal antibodies and related proteins. *Trends Biotechnol.* **2010**, *28*, 253–261. [CrossRef]
27. Chames, P.; Van Regenmortel, M.; Weiss, E.; Baty, D. Therapeutic antibodies: Successes, limitations and hopes for the future. *Br. J. Pharmacol.* **2009**, *157*, 220–233. [CrossRef]
28. Bojadzic, D.; Buchwald, P. Toward small-molecule inhibition of protein-protein interactions: general aspects and recent progress in targeting costimulatory and coinhibitory (immune checkpoint) interactions. *Curr. Top. Med. Chem.* **2018**, *18*, 674–699. [CrossRef]
29. Golani, L.K.; Wallace-Povirk, A.; Deis, S.M.; Wong, J.; Ke, J.; Gu, X.; Raghavan, S.; Wilson, M.R.; Li, X.; Polin, L.; et al. Tumor targeting with novel 6-substituted pyrrolo [2,3-d] pyrimidine antifolates with heteroatom bridge substitutions via cellular uptake by folate receptor alpha and the proton-coupled folate transporter and inhibition of de novo purine nucleotide biosynthesis. *J. Med. Chem.* **2016**, *59*, 7856–7876. [CrossRef]
30. Chupak, L.S.; Zheng, X. Compounds Useful as Immunomodulators. Bristol-Myers Squibb Company. WO2015034820 A1, 12 March 2015.
31. Zak, K.M.; Grudnik, P.; Guzik, K.; Zieba, B.J.; Musielak, B.; Dömling, A.S.S.; Dubin, G.; Holak, T.A. Holak. Structural basis for small molecule targeting of the programmed death ligand 1 (PD-L1). *Oncotarget* **2016**, *7*, 30323–30335. [CrossRef]
32. Musielak, B.; Kocik, J.; Skalniak, L.; Magiera-Mularz, K.; Sala, D.; Czub, M.; Stec, M.; Siedlar, M.; Holak, T.A.; Plewka, J. CA-170-A potent small-molecule PD-L1 inhibitor or not? *Molecules* **2019**, *24*, 2804. [CrossRef] [PubMed]
33. Abagyan, R.; Totrov, M.; Kuznetsov, D. ICM–A New method for protein modeling and design-applications to docking and structure prediction from the distorted native conformation. *J. Comput. Chem.* **1994**, *15*, 488–506. [CrossRef]
34. Bottegoni, G.; Kufareva, I.; Totrov, M.; Abagyan, R. Four-dimensional docking: A fast and accurate account of discrete receptor flexibility in ligand docking. *J. Med. Chem.* **2009**, *52*, 397–406. [CrossRef] [PubMed]
35. Bottegoni, G.; Kufareva, I.; Totrov, M.; Abagyan, R. A new method for ligand docking to flexible receptors by dual alanine scanning and refinement (SCARE). *J. Comput. Aided. Mol. Des.* **2008**, *22*, 311–325. [CrossRef]
36. Wang, W.; Hirano, Y.; Uzawa, T.; Liu, M.; Taiji, M.; Ito, Y. In vitro selection of a peptide aptamer that potentiates inhibition of cyclin-dependent kinase 2 by purvalanol. *MedChemComm* **2014**, *5*, 1400–1403. [CrossRef]
37. Dharmatti, R.; Miyatake, H.; Nandakumar, A.; Ueda, M.; Kobayashi, K.; Kiga, D.; Yamamura, M.; Ito, Y. Enhancement of binding affinity of folate to its receptor by peptide conjugation. *Int. J. Mol. Sci.* **2019**, *20*, 2152. [CrossRef] [PubMed]

38. Eglen, R.M.; Reisine, T.; Roby, P.; Rouleau, N.; Illy, C.; Bosse, R.; Bielefeld, M. The use of AlphaScreen technology in HTS: Current status. *Curr. Chem. Genom.* **2008**, *1*, 2–10. [CrossRef]

39. Tan, S.; Zhang, H.; Chai, Y.; Song, H.; Tong, Z.; Wang, Q.; Qi, J.; Wong, G.; Zhu, X.; Liu, W.J.; et al. An unexpected N-terminal loop in PD-1 dominates binding by nivolumab. *Nat. Commun.* **2017**, *8*, 14369. [CrossRef]

40. Guzik, K.; Zak, K.M.; Grudnik, P.; Magiera, K.; Musielak, B.; Torner, R.; Skalniak, L.; Domling, A.; Dubin, G.; Holak, T.A. Small-molecule inhibitors of the programmed cell death-1/programmed death-ligand 1 (PD-1/PD-L1) interaction via transiently induced protein states and dimerization of PD-L1. *J. Med. Chem.* **2017**, *60*, 5857–5867. [CrossRef]

41. Skalniak, L.; Zak, K.M.; Guzik, K.; Magiera, K.; Musielak, B.; Pachota, M.; Szelazek, B.; Kocik, J.; Grudnik, P.; Tomala, M.; et al. Small-molecule inhibitors of PD-1/PD-L1 immune checkpoint alleviate the PD-L1-induced exhaustion of T-cells. *Oncotarget* **2017**, *8*, 72167–72181. [CrossRef]

42. Magiera-Mularz, K.; Skalniak, L.; Zak, K.M.; Musielak, B.; Rudzinska-Szostak, E.; Berlicki, L.; Kocik, J.; Grudnik, P.; Sala, D.; Zarganes-Tzitzikas, T.; et al. Bioactive macrocyclic inhibitors of the PD-1/PD-L1 immune checkpoint. *Angew. Chem. Int. Ed.* **2017**, *56*, 13732–13735. [CrossRef] [PubMed]

43. Lazar-Molnar, E.; Scandiuzzi, L.; Basu, I.; Quinn, T.; Sylvestre, E.; Palmieri, E.; Ramagopal, U.A.; Nathenson, S.G.; Guha, C.; Almo, S.C. Structure-guided development of a high-affinity human Programmed Cell Death-1: Implications for tumor immunotherapy. *EBioMedicine* **2017**, *17*, 30–44. [CrossRef]

44. Wang, W.; Hirano, Y.; Uzawa, T.; Taiji, M.; Ito, Y. Peptide-assisted enhancement of inhibitory effects of small molecular inhibitors for kinases. *Bull. Chem. Soc. Jpn.* **2016**, *89*, 444–446. [CrossRef]

45. Neves, M.A.; Totrov, M.; Abagyan, R. Docking and scoring with ICM: The benchmarking results and strategies for improvement. *J. Comput. Mol. Des.* **2012**, *26*, 675–686. [CrossRef] [PubMed]

46. Abagyan, R.; Totrov, M. Biased probability monte-carlo conformational searches and electrostatic calculations for peptides and proteins. *J. Mol. Biol.* **1994**, *235*, 983–1002. [CrossRef] [PubMed]

47. Totrov, M.; Abagyan, R. Flexible protein-ligand docking by global energy optimization in internal coordinates. *Proteins* **1997**, *29* (Suppl. 1), 215–220. [CrossRef]

48. Schapira, M.; Totrov, M.; Abagyan, R. Prediction of the binding energy for small molecules, peptides and proteins. *J. Mol. Recognit.* **1999**, *12*, 177–190. [CrossRef]

International Journal of
Molecular Sciences

Article

NMR Fragment-Based Screening against Tandem RNA Recognition Motifs of TDP-43

Gilbert Nshogoza [1], Yaqian Liu [1], Jia Gao [1], Mingqing Liu [1], Sayed Ala Moududee [1], Rongsheng Ma [1], Fudong Li [1], Jiahai Zhang [1], Jihui Wu [1], Yunyu Shi [1,2] and Ke Ruan [1,*]

[1] Hefei National Laboratory for Physical Sciences at the Microscale, School of Life Sciences, University of Science and Technology of China, Hefei 230027, China
[2] CAS, Center for Excellence in Biomacromolecules, Chinese Academy of Sciences, Beijing 100101, China
* Correspondence: kruan@ustc.edu.cn; Tel.: +86-551-6360-7767

Received: 28 April 2019; Accepted: 28 June 2019; Published: 30 June 2019

Abstract: The TDP-43 is originally a nuclear protein but translocates to the cytoplasm in the pathological condition. TDP-43, as an RNA-binding protein, consists of two RNA Recognition Motifs (RRM1 and RRM2). RRMs are known to involve both protein-nucleotide and protein-protein interactions and mediate the formation of stress granules. Thus, they assist the entire TDP-43 protein with participating in neurodegenerative and cancer diseases. Consequently, they are potential therapeutic targets. Protein-observed and ligand-observed nuclear magnetic resonance (NMR) spectroscopy were used to uncover the small molecule inhibitors against the tandem RRM of TDP-43. We identified three hits weakly binding the tandem RRMs using the ligand-observed NMR fragment-based screening. The binding topology of these hits is then depicted by chemical shift perturbations (CSP) of the ^{15}N-labeled tandem RRM and RRM2, respectively, and modeled by the CSP-guided High Ambiguity Driven biomolecular DOCKing (HADDOCK). These hits mainly bind to the RRM2 domain, which suggests the druggability of the RRM2 domain of TDP-43. These hits also facilitate further studies regarding the hit-to-lead evolution against the TDP-43 RRM domain.

Keywords: epigenetics; protein-RNA interaction; RRM domain inhibitor; NMR fragment-based screening; TDP-43

1. Introduction

RNA recognition motifs (RRMs) play diverse roles in post-transcriptional gene expression events such as RNA transport, localization, stability, and mRNA and rRNA processing. RRM is also known as the ribonucleoproteins (RNP) domain, as it contains the short and conserved elements RNP1 and RNP2, or RNA binding domain (RBD), that are abundantly distributed in higher vertebrates [1] and ubiquitously found in all kingdoms of life, including viruses and prokaryotes. In addition, they also participate in important functions such as microRNA biogenesis, apoptosis, and cell division [2,3]. RRMs are not only known to be involved in protein–nucleotide interactions, but also in protein–protein interactions [4].

The transactive response DNA-binding Protein 43kDa (TDP-43) is a RRM-containing protein, which plays important functions in mRNA metabolism regulation, including transcription repression, exon skipping, and RNA splicing [5,6]. TDP-43 is originally a nuclear protein, but translocates to the cytoplasm upon a pathological condition. It is a ubiquitously expressed, highly conserved, and multifunctional RNA and DNA-binding protein [7]. TDP-43 stabilizes the mRNA of human low-molecular-weight neurofilament (hNFL) [8]. Depletion of TDP-43 has important consequences in essential metabolic processes in human cells, like nuclear shape deformation, apoptosis, and misregulation of the cell cycle [9]. The disruption of TDP-43 auto-regulation impacts both localization of TDP-43 and its level, which results in TDP-43 accumulation in the cytoplasm. Based on its crucial roles in RNA processing, dysfunctional TDP-43 causes some abnormalities in alternative mRNA splicing, miRNA biogenesis, and RNA-rich granules formation [10].

The dysregulation of TDP-43 is hence associated with a variety of human diseases, especially neurodegenerative diseases, e.g., frontotemporal lobar degeneration (FTLD), amyotrophic lateral sclerosis (ALS), brain ischemia, aging, and Alzheimer's disease [11–13]. For instance, in cases of FTLD and ALS, TDP-43 is the main constituent of their ubiquitin inclusions [14]. During the stress conditions, TDP-43 is localized in the cytoplasm, with mRNA binding to its RRM and glycine-rich domain, and thus forms the isolated liquid compartment enriching the mRNA and proteins. Such stress granules (SGs) in cells and in pathological brain tissue play crucial roles in FTLD/ALS pathology [15,16]. Aggregate-prone TDP-43 variants or exposure to oxidative stress generates distinct TDP-43 inclusions devoid of SGs [17]. The toxicity of the TDP-43 overexpression requires the presence of functional RNA Recognition motifs [18–20]. Recently, the proteinopathy of both important mutations (D169G and K263E located at RRM1 and RRM2, respectively) was computationally explored and the mutants are more prone to aggregation, causing neurological disorders [21].

Apart from the TDP-43 involvement in neurodegenerative diseases, an accumulating amount of evidence suggests that TDP-43 is a cancer responsive factor. TDP-43 positively contributes to the anticancer activity for curcumin in MCF-7 cells [22] and as a tumor suppressor by partnering with the TRIM16 in inhibiting the viability and proliferation of neuroblastoma and breast cancer cells [23]. In addition, normal levels of TDP-43 might be a crucial protective factor for cells under apoptotic insult [24]. On the contrary, the TDP-43 inhibition suppressed cervical cancer cell growth and induced cell cycle arrest while its overexpression promoted cancer cell progression and drove the cell cycle [25]. TDP-43 may regulate melanoma cell proliferation and metastasis by modulating glucose metabolism [26]. TDP-43 also plays an oncogenic role in malignant glioma cell progression by stabilizing small nucleolar RNA host gene 12 (SNHG12) [27]. The findings demonstrated that TDP-43 regulates the MALAT1, a non-coding RNA overexpressed in non-small cell lung cancer (NSCLC), through direct binding to MALAT1 RNA at the 3′ region by RRM, whose participation is compulsory. This controls the growth, invasion, and migration of NSCLC cells [28]. Reduced tumor progression, including proliferation and metastasis, was observed upon the knockdown of TDP-43 in triple-negative breast cancer (TNBC) and RRM involvement is assured [29]. These studies suggest that targeting the TDP-43 RRM domains may, therefore, be an effective therapeutic approach for neurodegenerative diseases and cancers.

Although more is known about the TDP-43 biology and its association with neurodegenerative and cancer diseases, the development of treatments toward TDP-43 is mostly lagging behind those targeting other proteins involved in such diseases [30]. RRM and RNA complexes have long been attractive targets for small molecule inhibition targeting the RNA, not the protein [31,32]. Firstly, the aminoacridine derivative was discovered to interrupt the formation of RNA and U1A RRM1 complex [33]. Additionally, a high-throughput screening assay, based on AlphaScreen®, technology was used to characterize DNA and RNA oligonucleotides (bt-TAR-32 and bt-TG6, respectively) binding to TDP-43 and their interaction inhibition was assessed [34]. Later, that series of 4-aminoquinoline derivatives were characterized for their capacity to modulate TDP-43 metabolism and function, whereby they bind to TDP-43, reduce its interaction with the oligonucleotide, and stimulate caspase-mediated cleavage of TDP-43 [35], but information is still lacking on the binding topology. Furthermore, some medicinal treatment reduces the TDP-43 inclusions through the autophagy pathway were discussed [36]. However, no compounds directly targeting RRM domains of TDP-43 have been uncovered to our best knowledge.

NMR spectroscopy is a powerful approach which has been extensively used by the pioneers in fragment-based drug discovery for detecting molecular interactions between the target and the fragment libraries [37–39] and to facilitate structure-based drug design [40]. Consistently, the fragment-based screening approach has been fruitful for identifying hits for the challenging protein-protein interaction "hot-spots" [41–45]. We expect it shall be effective in the case of the shallow RNA binding pocket of TDP-43 tandem RRMs.

Here, we carried out automated NMR fragment-based screening [46] to identify three hits of the tandem RRMs of TDP-43. Chemical shift perturbations of the ^{15}N labeled TDP-43 tandem RRMs demonstrate that these hits bind to the same site, mainly on the RRM2 domain. It has also been validated by the chemical shift perturbation experiments for TDP-43 RRM2 alone. The CSP-driven HADDOCK was used to generate the protein-hits binding mode. Collectively, our work provides a class of compounds for further hit-to-lead evolution of the TDP-43 RRM domain and paves the path for targeting protein-RNA interactions using the fragment-based approach.

2. Results

Structurally, TDP-43 tandem RRMs are approximately 160 amino acids long and display a $\beta1\alpha1\beta2\beta3\alpha2\beta4$ arrangement of secondary structure, with an additional β-hairpin named $\beta3'\beta3''$ [47] or $\beta5$ [48,49] which is located between $\alpha2\beta4$, and extends the β-sheet surface to be accessible to binding by multiple RNA nucleotides. This leads to a rare RRMs orientation type ($\beta2\beta4$) and the 14-aa linker needs to connect four β-strands instead of two [2,47]. Diverse studies revealed that TDP-43 tandem RRMs can interact with both short and long single-stranded nucleic acids rich in UG/TG, either separately or collectively, to achieve high affinity and specificity [47–49]. Given the RNA recognition mode by tandem RRMs, TDP-43 RRMs are independent of each other in unbound form but they establish a rigid structure upon RNA binding on the flat surface β-sheet [47]. In general, this RNA-recognition pocket is much shallower than the ATP-binding sites of kinases. Hence, it poses a grand challenge for conventional high throughput screening aimed at discovering strong binders. Conversely, the fragment-based approach has proven fruitful for uncovering the initial hits, albeit at weak affinities.

NMR ligand-observed methods detect the weak protein-ligand binding by detecting changes in the characteristics of the ligand spectrum that occur upon binding to the protein. Using the ligand-based experiments, i.e., saturation transfer difference (STD) [50], water ligand observed via gradient spectroscopy (WaterLOGSY) [51], Carr–Purcell–Meiboom–Gill (CPMG) [52], and ligand-based 1D proton, we found 17 hits from the primary screening of 89 cocktails containing 10 compounds each (Figure 1a). The binders present signals while the non-binders present no signals in the STD spectra. Accordingly, the binders show inverted or a fast decay of signals in the WaterLOGSY and CPMG experiments, respectively. The combined output of these spectra enabled the identification of primary screening hits from cocktails. It is worth noting that the reference 1D proton spectra of each individual compound might be slightly different from the screening spectra as a different buffer was used to be better compatible with TDP-43 tandem RRMs. The primary screening hits were further validated by the secondary screening for individual hits using the same set of NMR experiments (Figure 1b and Figure S1). The aromatic peaks of the hit are depicted as they suffer less from the interference of buffer signals. The secondary screening eliminated 13 primary hits, probably due to sample aggregation in cocktails, ambiguous selection of hits with degenerated chemical shifts, and/or spectrometer instability. Among the remaining 4 hits, hit 2 demonstrated a distinct topology relative to hits 1 and 3 (Figure 1c).

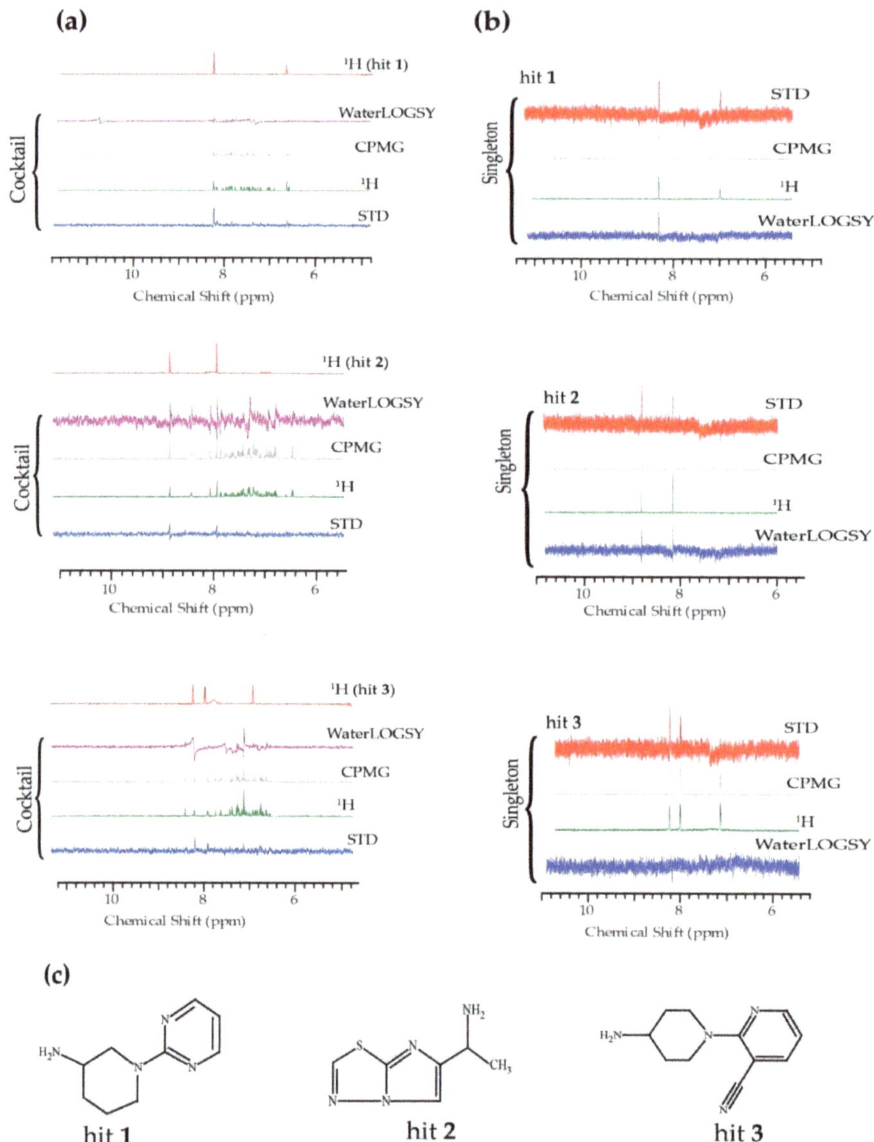

Figure 1. NMR fragment-based screening against the tandem RRM domain of TDP-43. (**a**) The primary screening WaterLOGSY, CPMG, ¹H and STD spectra for three representative cocktails. The ¹H reference spectrum of the respective hit is shown for comparison. (**b**) The secondary screening spectra for individual hit 1, 2, and 3, respectively. (**c**) The chemical structures of hits 1, 2, and 3.

The 4 secondary screening hits were then cross-validated using the chemical shift perturbations (CSPs) of the ^{15}N-labeled tandem RRMs of TDP-43 and 3 of them induced significant chemical shift changes of the tandem RRM (Figures 2 and 3). This approach has been extensively applied in the interrogation of protein-ligand interactions in an affinity ranging from nM to mM. As CSP is a sensitive indicator of chemical environment changes induced by ligand titration, it is particularly powerful in the detection of weak bindings. The linewidths of the amide signals of TDP-43 tandem RRM show almost no changes upon titration of hit 1 (Table S1), which suggests that hit 1 induces no protein aggregation. This is a useful approach to remove false positives, which are commonly found in drug screening because of protein aggregation [53]. Titration of hit 1 induces dose-dependent CSPs of residues G245, E246, H256, I257, S258 (Figure 2b and Figure S2). However, the curve does not reach the saturation point, as it is limited by the weak binding affinity and the low aqueous solubility of the hit. Hence, the binding affinity of those weak binders cannot be robustly estimated from CSPs. The disturbed residues were then mapped on the surface representation of the solution structure of TDP-43 tandem RRMs (PDB code: 4BS2) [47]. Residues H256, I257, S258 locate on the β4 strand, while residues G245 and E246 bridge the α2 and β3 (Figure 2c).

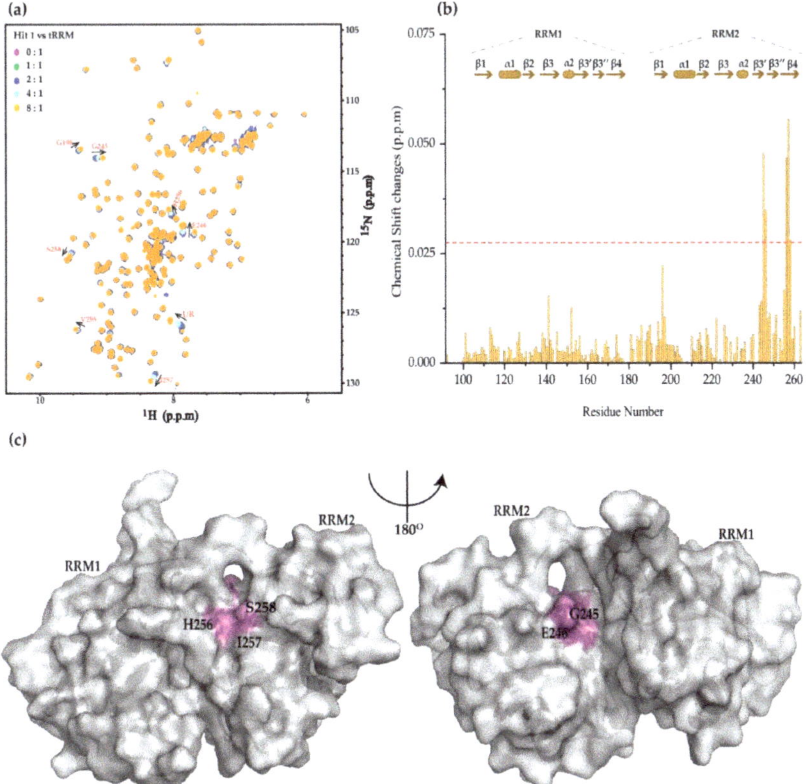

Figure 2. The binding topology of hit 1 on the tandem RRMs of TDP-43 using NMR chemical shift perturbations. (**a**) The chemical shift perturbations of ^{15}N-labeled tandem RRM domain of TDP-43 upon titration of hit 1. The ligand/protein molar ratios are annotated. The perturbed residues are labeled and the arrows indicate the direction of chemical shift changes. UR stands for unassigned residue. (**b**) Chemical shift changes of the TDP-43-tandem RRM domain are at the ligand protein molar ratio of 8:1. The red horizontal dashed line represents two standard deviations above the averaged chemical shift changes of residues. (**c**) Surface representation of TDP-43 tandem RRM domain (PDB code: 4BS2) showing the purple-colored residues with significant chemical shift changes.

Consistently, hits 2 and 3 titrations also point to the same binding topology in the tandem RRM of TDP-43 (Figure 3). For example, hit 2 perturbed residues G245, H256, and I257 (Figure 3a,c), while hit 3 induced significant CSPs for residues G245, E246, H256, and I257 (Figure 3b,d). The similarity of the binding pattern of the three hits suggests that weak but specific binders were successfully identified using the NMR fragment-based screening.

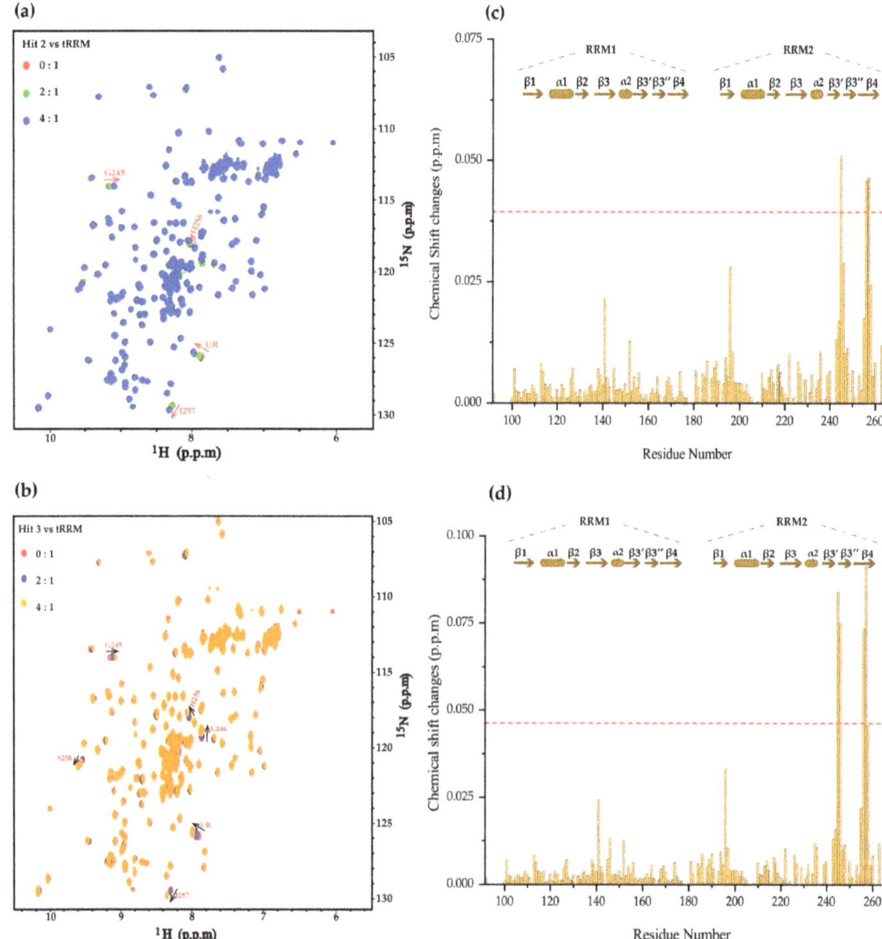

Figure 3. Chemical shift perturbations of tandem RRM upon binding of hit 2 and 3. (**a,b**) The chemical shift perturbations of TDP-43 tandem RRM domain induced by titration of hit 1 and 2, respectively. Annotated are the hits: Protein molar ratios. UR stands for unassigned residue. (**c,d**) Residue-by-residue chemical shift changes of tandem RRM at the hit/protein molar ratio of 8:1 for compound 2 and 3, respectively. The red dashed lines represent two standard deviations above the averaged chemical shift changes of residues.

Having confirmed that 3 different hits bind on the same site of the TDP-43 RRM2 domain, we further investigated whether RRM2 alone is sufficient for ligand binding. Hit 2 was thus titrated to the ^{15}N-labeled RRM2 domain of TDP-43 (Figure 4a). Consequently, the residues G245, on loop bridging the α2 and β3′, H256, and I257, located on β4-strand, were perturbed (Figure 4b). Those residues were mapped on the surface representation of the TDP-43 RRM2 [49] domain in complex with a single-stranded DNA (Figure 4c). The hit binds to the same sites of either TDP-43 tandem RRM or RRM2 alone. That is to say, TDP-43 RRM2 is the main contributor for ligand binding and should be considered as the target for follow-up hit-to-lead evolutions.

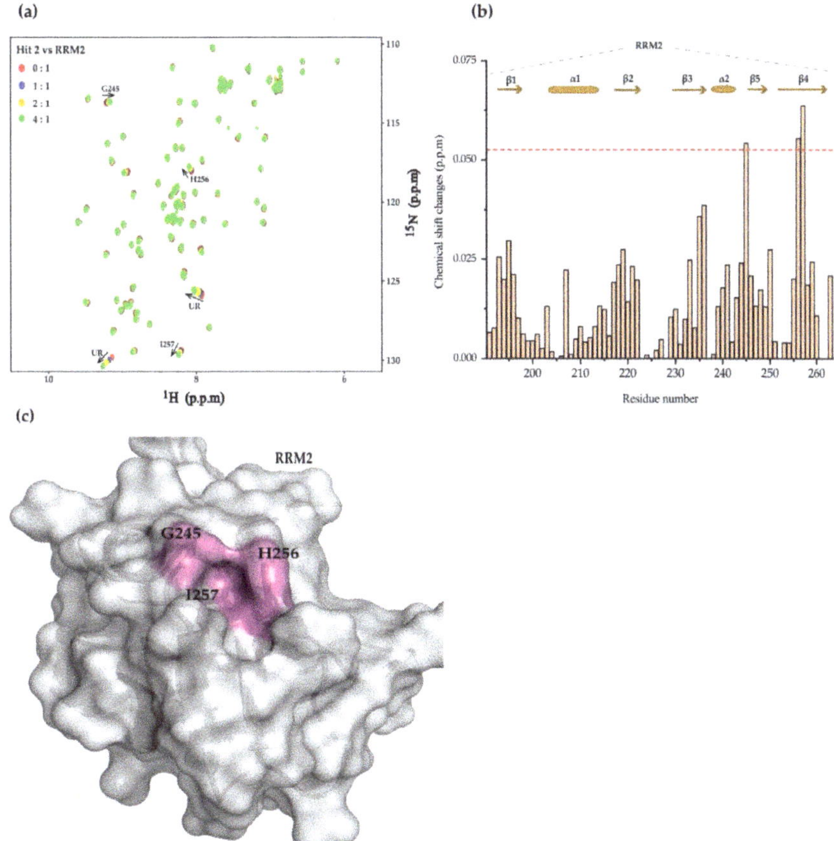

Figure 4. Chemical shift perturbations of the TDP-43-RRM2 domain upon hit 2 titration. (**a**) The chemical shift perturbations of the RRM2 domain of TDP-43 by hit 2 titration. (**b**) Chemical shift changes of the TDP-43 RRM2 domain residues at a hit 2; protein molar ratio of 4:1. The red dashed line represents two standard deviations above the averaged chemical shift changes of residues. (**c**) Residues (colored in purple) undergo significant chemical shift changes and are mapped on the surface representation of TDP-43-RRM2 domain (PDB code: 1WF0).

We further compared the small molecule binding topology with the nucleic acid recognition sites of the TDP-43 RRM domain. In TDP-43 tandem RRMs, 10 out of 12 nucleotides of the AUG12 RNA (GUGUGAAUGAAU) interact with RRM1 and RRM2 (PDB code: 4BS2) [47]. Among them, the first five ($G_1U_2G_3U_4G_5$) nucleotides are accommodated on the RRM1 β-sheet and the following two nucleotides (A_6A_7) act as a connector between two RRMs, while the next three nucleotides ($U_8G_9A_{10}$) lie on the RRM2. The U_8 nucleotide of RNA is recognized on S258 (β4) through hydrogen bonds, on the backbone carbonyl oxygen of N259 (β4), and the backbone amide of E261 from the C-terminus [47]. Comparatively, all three hits have perturbed some residues located on the β4-strand, hits 1 and 3 specifically disturbed S258 (β4). This also interacts with the U_8 nucleotide in tandem RRM (Figure 5a). Furthermore, the RRM2 residues D247 (loop α2-β3′) and I249 (β3′) are involved in inter-RRM interactions upon RNA binding on the tandem RRM of TDP-43. This study revealed that their nearby residues, G245 and E246 (loop α2-β3′), display higher chemical shift perturbations induced by the hits binding (Figures 2b and 3c,d).

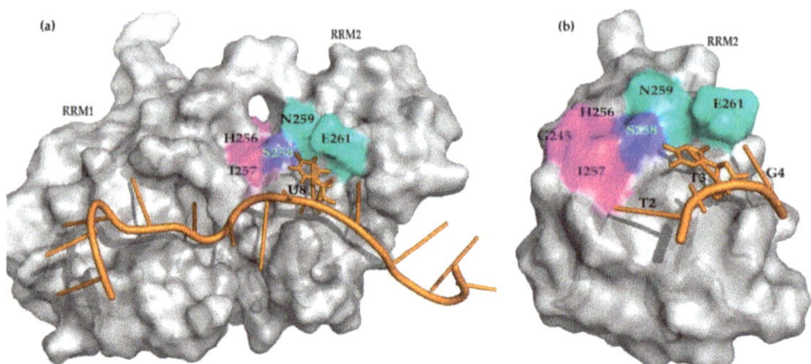

Figure 5. Comparison of binding sites of nucleic acids and hits on TDP-43. (**a**) Surface representation of TDP-43 tandem RRMs in complex with AUG12 RNA (orange cartoon), where residues interact with the U_8 nucleotide (stick) and hits are highlighted in cyan and magenta, respectively. Residue S258 (blue) interacts with both U_8 and hit 1. (**b**) Surface representation of TDP-43 RRM2 in complex with ssDNA (PDB code: 3D2W) using the same coloring scheme.

Accordingly, the crystal structure of TDP-43 RRM2 in complex with ss-DNA 5′-GTTGAGCGTT-3′ (PDB entry: 3D2W) reveals that only three 5′ end nucleotides (T2, T3, G4) make extensive contacts with β-sheet residues of RRM2, whereby T3 particularly contacts with S258, Asn259, and Glu261 through hydrogen bonds [49], while in our study the residues H256 and I257, nearby the S258 (β4), have been perturbed upon hit binding on the single RRM2 (Figure 5b). This suggests that the fragment screening hits bind to a proximal site for RNA/DNA recognition, thus new hits can be designed using a fragment grow strategy to block the DNA/RNA recognition capability of TDP-43 RRM2.

To further characterize the binding mode, a data-driven approach, HADDOCK [54], was used to model the tandem RRM-hit 1 complex structure. Residues G245, E246, H256, I257, and S258 were defined as active ones in the binding site. Among the docking poses generated by HADDOCK, the best-fit ones were filtered out based on CSP and STD restraints [41,55,56]. One representative docking pose (Figure 6) indicates that hit 1 forms a hydrogen bond with the side chain of S258 and the aromatic ring of hit 1 is proximal to residues G245, E246, H256, and I257. These docking poses pave the path for following structure-guided hit-to-lead evolution.

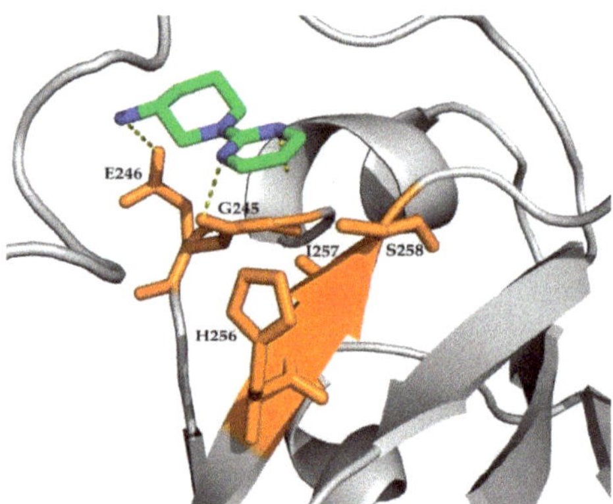

Figure 6. The representative docking model of hit 1 in consistency with experimental CSP and STD restraints. Hit 1 (green color) in the binding site of tandem RRM (PDB: 4bs2) where the carbonyl hydrogen is oriented toward G245, while the side chain hydrogen interacts with E246 residue of tandem RRM. Other active residues (orange sticks), H256, I257, and S258 are located in proximal of the hit 1.

3. Discussion

Proteins containing RRM domains function in important aspects of the posttranscriptional regulation of gene expression, mRNA maturation, and other RNA processing machinery. These proteins perform their diverse roles depending on the dual ability to recognize RNA and to interact with other proteins by using their RRM domain [31]. As TDP-43 is closely correlated with neurodegenerative and cancerous diseases [29,57], the RRM domain of TDP-43 becomes an attractive therapeutic target. However, there is no direct inhibitor targeting the RRM discovered to date.

We uncovered three small molecules binding to the tandem RRM domain of TDP-43 by using NMR fragment-based screening techniques. The NMR spectroscopy, one of a plethora of biophysical methods, is particularly powerful to detect even ultra-weak protein-ligand interactions. Accordingly, chemical shift perturbations observed in the heteronuclear single-quantum coherence (HSQC) spectra or the linewidth analysis of the small molecules allow the determination of binding affinity [58,59]. This is sometimes recalcitrant, as the titration to saturation point may be infeasible in case of weak binding affinities and low aqueous solubility of compounds.

NMR is extensively applied in fragment-based lead discovery [60]. The central idea is to screen a small library (500–2000 molecules) of low-molecular-weight compounds (110–250 Da), as their low complexity enhances the probability of matched interactions between the target and these fragment compounds. The reasonable hit rate indicates the druggability of the TDP-43 tandem RRM domain.

Although the 4-aminoquinolines molecules have been discovered through high throughput screening against the full-length TDP-43 [34], the enlightenment on binding site is still lacking. TDP-43 contains two RNA-binding RRM domains and the C-terminal low complexity domain, which may form liquid–liquid phase separation as a reservoir of mRNAs. Here, it is essential to determine the small molecule binding topology on TDP-43. The tandem RRM of TDP-43 is composed of a canonical RRM arrangement (β1α1β2β3α2β4), with an additional β-hairpin (β3′β3″ or β5) found between α2 and β4 which extends the β-sheet surface for RNA recognition [2,47,49]. The binding topology of our fragment screening hits and CSP-guided HADDOCK modeling reveal a ligand-binding "hot spot" of TDP-43 RRM2, proximal to H256, I257, and S258. Interestingly, these residues are also close to the RRM1 and RRM2 interface. The previous study proposed that both RRM domains are indispensable for achieving the greater binding affinity between the TDP-43 and nucleic acids [49]. Since this "hot spot" is partially overlapped with the RNA/DNA recognition site, it directs the following structure-guided hit-to-lead evolution against TDP-43 tandem RRM domains.

4. Materials and Methods

4.1. Cloning, Expression, and Protein Purification

The tandem RRM domain of TDP-43 (residues 101–269) was synthesized by GENEWIZ (Suzhou, China) and sub-cloned into the pET22b vector (GE Healthcare, Shanghai, China) with the His$_6$ tag. The RRM2 domain was amplified from the tandem RRM construct and then sub-cloned into the pET22b vector (GE Healthcare, Shanghai, China) with the His$_6$ tag. The constructs were transformed into *Escherichia coli* BL21 and cultivated in 1 L LB media, incubated at 37 °C. The proteins were expressed at 16 °C after induction by 0.5 mM isopropyl β-D-thiogalactosidase (IPTG) for 20 h. The bacteria were harvested by centrifugation (5000 rpm, 10 min), resuspended in lysis buffer (25 mM Tris, 500 mM NaCl at pH 7.5), and then lysed by sonication. The cell lysates were centrifuged (13,000 rpm, 30 min). The collected supernatant was purified on a column filled with Nickel-chelated resin (QIAGEN, Shanghai, China). The impurities were washed out using a buffer (25 mM Tris, 1 M NaCl at pH 7.5) containing a linear gradient of 20–40 mM imidazole, then the same buffer containing 500 mM imidazole was used to elute out the target proteins. All proteins were further purified by size exclusion chromatography using a HiLoad 16/600 Superdex 75 column (GE Healthcare, Shanghai, China). The target proteins were confirmed by SDS-PAGE.

For ^{15}N-labeled proteins, the cells were first cultured in 1 L LB media, harvested when A_{600} reached 1.0 and then transferred to 1 L M9 media containing ^{15}NH$_4$Cl. The cells were induced by 0.4 mM IPTG to express the proteins (tandem RRMs and RRM2 domains). The purified proteins were concentrated in PBS buffer plus the 5 mM DTT at pH 7.5.

4.2. NMR Fragment-Based Screening

All NMR fragment screening experiments were carried out at 25 °C using an Agilent 700 MHZ spectrometer equipped with a 96 well auto-sampler and a 5 mm cryoprobe. During the primary screening, the ligand-based NMR spectra (STD, WaterLOGSY, CPMG, and 1D ^{1}H) were acquired against the 890 fragments library (ChemBridge, San Diego, CA, USA) as described previously in detail [46,61]. Those fragments were distributed in 89 cocktails, composed of 10 compounds each, at a final concentration of 0.4 mM. These cocktails were incubated with protein (10 μM) in sodium phosphate (50 mM, pH 7.5), NaCl (200 mM), dithiothreitol (5 mM), and D$_2$O (50%). To further confirm the identified primary hits, secondary screening was individually carried out for single hits using the same buffer and NMR experimental settings. We then automatically processed and visualized the primary and secondary data with our ACD/Labs scripts, as previously described [46].

4.3. NMR Chemical Shift Perturbation

NMR HSQC spectra were acquired at 25 °C on either an Agilent 700MHZ spectrometer equipped with a cryoprobe or an Agilent 500MHz spectrometer equipped with a room temperature probe. The ^{15}N-labeled proteins (0.1 mM or 0.2 mM), in PBS buffer (50 mM, pH 7.5), containing NaCl (200 mM), dithiothreitol (5 mM), and D$_2$O (10%) were titrated by small molecules stocked in DMSO at a concentration of 200 mM, using a series of hit/protein molar ratios of 0.0, 0.5, 1.0, 2.0, 4.0, and 8.0 for TDP-43 tandem RRMs and 0.0, 0.5, 1.0, 2.0, and 4.0 for RRM2, respectively. Spectra were processed in NMRpipe and analyzed with Sparky. The chemical shift changes ($\Delta\delta$) relative to the free form of protein were defined as follows:

$$\Delta\delta = \sqrt{(\delta_{1_H})^2 + (0.2\delta_{15_N})^2}, \tag{1}$$

where $\delta^1{}_H$ and $\delta^{15}{}_N$ are the chemical shift differences of the ^{1}H and ^{15}N dimensions, respectively. We referred to the following chemical shift assignments previously deposited in the Biological Magnetic Resonance Data Bank: RRM1 (BMRB Entry 18765), RRM2 (BMRB Entry 19922), and tandem RRM (BMRB Entry 19290). All structures figures were prepared by Pymol (DeLano Scientific, LLC, Palo Alto, CA, USA).

4.4. Molecular Docking

HADDOCK is an information-driven docking technique used for modeling biomolecule structures by using experimental or predictive restraints [54,62]. The CSPs, obtained from the NMR HSQC titration data, were used both as HADDOCK restraints and for defining the protein active residues. The tandem RRM structure (PDB: 4bs2) served as the starting structure, while the hit **1** PDB file was generated by the PRODRG [63]. The docking calculations were done by the HADDOCK web server and clustered 186 structures in 16 clusters according to the RMSD threshold of 2 Å.

4.5. Linewidth Analysis

The NMR HSQC spectra at molar ratios of 0:1 and 8:1 (hit/protein) were processed using the same NMRpipe script, e.g., 2-fold zero-filling, Fourier transformation, and phase corrections. The spectra were then analyzed, with randomly selected peaks, using Sparky. After peak integration, the linewidth, i.e., the full width at half the peak height, was automatically estimated by Sparky.

Int. J. Mol. Sci. **2019**, *20*, 3230

Supplementary Materials: Supplementary materials can be found at http://www.mdpi.com/1422-0067/20/13/3230/s1.

Author Contributions: Conceptualization and formal analysis, G.N. and K.R.; resources, J.G., Y.L., M.L., S.A.M., R.M., F.L., J.Z., J.W., and, Y.S.; writing-original draft, G.N.; writing-review and editing, G.N. and K.R.; supervision, Y.S., and K.R.; funding acquisition, J.W., Y.S., and K.R.

Funding: We thank the financial support by grants from the Ministry of Science and Technology of China (2016YFA0500700), Strategic Priority Research Program of the Chinese Academy of Sciences (XDA12020355 and XDPB10), National Natural Science Foundation of China (21703254, 21807095, 21874123 and U1632153,), and Fundamental Research Funds for the Central Universities (WK2070080002 and WK2060190086).

Acknowledgments: Part of our NMR work was performed at the National Facility for Protein Sciences Shanghai and the High Magnetic Field Laboratory, Chinese Academy of Sciences. We thank Jiuyang Liu and Na Wang for their technical and experimental support. We also thank the University of Science and Technology of China which hosted Gilbert Nshogoza as a Ph.D. student financially supported by the Chinese Scholarship Council.

Conflicts of Interest: The authors declare no conflict of interest.

Abbreviations

ALS	Amyotrophic lateral sclerosis
ATG7	Autophagy-related protein 7
BMRB	Biological Magnetic Resonance Data Bank
CPMG	Carr–Purcell–Meiboom–Gill
CSPs	Chemical shift perturbations
FBS	Fragment-based screening
FTLD	Frontotemporal lobar degeneration
HADDOCK	High Ambiguity Driven biomolecular DOCKing
HDAC6	Histone deacetylase 6
hNFL	human low molecular weight neurofilament
HSQC	Heteronuclear single-quantum coherence
ITC	Isothermal titration calorimetry
NMR	Nuclear Magnetic Resonance
RNP	Ribonucleoproteins
RRM	RNA Recognition Motifs
SGs	Stress granules
SNHG12	Small nucleolar RNA host gene 12
SPR	Surface Plasmon Resonance
STD	Saturation Transfer difference
TDP-43	Transactive response DNA-binding Protein 43
TNBC	Triple-negative breast cancer
TRIM16	Tripartite motif-containing protein 16
WaterLOGSY	Water ligand observed via gradient spectroscopy

References

1. Venter, J.C. The sequence of the human genome. *Science* **2001**, *291*, 1304–1351. [CrossRef] [PubMed]
2. Afroz, T.; Cienikova, Z.; Clery, A.; Allain, F.H. One, Two, Three, Four! How Multiple RRMs Read the Genome Sequence. *Methods Enzym.* **2015**, *558*, 235–278. [CrossRef]
3. Clery, A.; Blatter, M.; Allain, F.H. RNA recognition motifs: Boring? Not quite. *Curr. Opin. Struct. Biol.* **2008**, *18*, 290–298. [CrossRef] [PubMed]
4. Kielkopf, C.L.; Lucke, S.; Green, M.R. U2AF homology motifs: Protein recognition in the RRM world. *Genes Dev.* **2004**, *18*, 1513–1526. [CrossRef] [PubMed]
5. Buratti, E.; Brindisi, A.; Giombi, M.; Tisminetzky, S.; Ayala, Y.M.; Baralle, F.E. TDP-43 binds heterogeneous nuclear ribonucleoprotein A/B through its C-terminal tail: An important region for the inhibition of cystic fibrosis transmembrane conductance regulator exon 9 splicing. *J. Biol. Chem.* **2005**, *280*, 37572–37584. [CrossRef] [PubMed]

6.   Mercado, P.A.; Ayala, Y.M.; Romano, M.; Buratti, E.; Baralle, F.E. Depletion of TDP 43 overrides the need for exonic and intronic splicing enhancers in the human apoA-II gene. *Nucleic Acids Res.* **2005**, *33*, 6000–6010. [CrossRef] [PubMed]

7.   Banks, G.T.; Kuta, A.; Isaacs, A.M.; Fisher, E.M. TDP-43 is a culprit in human neurodegeneration, and not just an innocent bystander. *Mamm. Genome* **2008**, *19*, 299–305. [CrossRef]

8.   Strong, M.J.; Volkening, K.; Hammond, R.; Yang, W.C.; Strong, W.; Leystra-Lantz, C.; Shoesmith, C. TDP43 is a human low molecular weight neurofilament (hNFL) mRNA-binding protein. *Mol. Cell. Neurosci.* **2007**, *35*, 320–327. [CrossRef]

9.   Ayala, Y.M.; Misteli, T.; Baralle, F.E. TDP-43 regulates retinoblastoma protein phosphorylation through the repression of cyclin-dependent kinase 6 expression. *Proc. Natl. Acad. Sci. USA* **2008**, *105*, 3785–3789. [CrossRef]

10. Weskamp, K.; Barmada, S.J. TDP43 and RNA instability in amyotrophic lateral sclerosis. *Brain Res.* **2018**, *1693*, 67–74. [CrossRef]

11. Wilson, C.A.; Dugger, B.; Dickson, D.W.; Wang, D.S. TDP-43 in aging and Alzheimer's disease. *Int. J. Clin. Exp. Pathol.* **2011**, *4*, 147–155. [PubMed]

12. Sun, M.; Yamashita, T.; Shang, J.; Liu, N.; Deguchi, K.; Liu, W.; Ikeda, Y.; Feng, J.; Abe, K. Acceleration of TDP43 and FUS/TLS protein expressions in the preconditioned hippocampus following repeated transient ischemia. *J. Neurosci. Res.* **2014**, *92*, 54–63. [CrossRef] [PubMed]

13. Josephs, K.A. TDP-43 is a key player in the clinical features associated with Alzheimer's disease. *Acta Neuropathol.* **2014**, *127*, 811–824. [CrossRef] [PubMed]

14. Neumann, M.; Sampathu, D.M.; Kwong, L.K.; Truax, A.C.; Micsenyi, M.C.; Chou, T.T.; Bruce, J.; Schuck, T.; Grossman, M.; Clark, C.M.; et al. Ubiquitinated TDP-43 in frontotemporal lobar degeneration and amyotrophic lateral sclerosis. *Science* **2006**, *314*, 130–133. [CrossRef] [PubMed]

15. Bentmann, E.; Neumann, M.; Tahirovic, S.; Rodde, R.; Dormann, D.; Haass, C. Requirements for stress granule recruitment of fused in sarcoma (FUS) and TAR DNA-binding protein of 43 kDa (TDP-43). *J. Biol. Chem.* **2012**, *287*, 23079–23094. [CrossRef] [PubMed]

16. Liu-Yesucevitz, L.; Bilgutay, A.; Zhang, Y.J.; Vanderweyde, T.; Citro, A.; Mehta, T.; Zaarur, N.; McKee, A.; Bowser, R.; Sherman, M.; et al. Tar DNA binding protein-43 (TDP-43) associates with stress granules: Analysis of cultured cells and pathological brain tissue. *PLoS ONE* **2010**, *5*, e13250. [CrossRef]

17. Chen, Y.; Cohen, T.J. Aggregation of the nucleic acid–binding protein TDP-43 °Ccurs via distinct routes that are coordinated with stress granule formation. *J. Biol. Chem.* **2019**.

18. Ash, P.E.; Zhang, Y.J.; Roberts, C.M.; Saldi, T.; Hutter, H.; Buratti, E.; Petrucelli, L.; Link, C.D. Neurotoxic effects of TDP-43 overexpression in C. elegans. *Hum. Mol. Genet.* **2010**, *19*, 3206–3218. [CrossRef]

19. Li, Y.; Ray, P.; Rao, E.J.; Shi, C.; Guo, W.; Chen, X.; Woodruff, E.A., 3rd; Fushimi, K.; Wu, J.Y. A Drosophila model for TDP-43 proteinopathy. *Proc. Natl. Acad. Sci. USA* **2010**, *107*, 3169–3174. [CrossRef]

20. Voigt, A.; Herholz, D.; Fiesel, F.C.; Kaur, K.; Muller, D.; Karsten, P.; Weber, S.S.; Kahle, P.J.; Marquardt, T.; Schulz, J.B. TDP-43-mediated neuron loss in vivo requires RNA-binding activity. *PLoS ONE* **2010**, *5*, e12247. [CrossRef]

21. Bhandare, V.V.; Ramaswamy, A. The proteinopathy of D169G and K263E mutants at the RNA Recognition Motif (RRM) domain of tar DNA-binding protein (tdp43) causing neurological disorders: A computational study. *J. Biomol. Struct. Dyn.* **2018**, *36*, 1075–1093. [CrossRef] [PubMed]

22. Fang, H.Y.; Chen, S.B.; Guo, D.J.; Pan, S.Y.; Yu, Z.L. Proteomic identification of differentially expressed proteins in curcumin-treated MCF-7 cells. *Phytomedicine* **2011**, *18*, 697–703. [CrossRef] [PubMed]

23. Kim, P.Y.; Tan, O.; Liu, B.; Trahair, T.; Liu, T.; Haber, M.; Norris, M.D.; Marshall, G.M.; Cheung, B.B. High TDP43 expression is required for TRIM16-induced inhibition of cancer cell growth and correlated with good prognosis of neuroblastoma and breast cancer patients. *Cancer Lett.* **2016**, *374*, 315–323. [CrossRef] [PubMed]

24. Nan, Y.N.; Zhu, J.Y.; Tan, Y.; Zhang, Q.; Jia, W.; Hua, Q. Staurosporine induced apoptosis rapidly downregulates TDP- 43 in glioma cells. *Asian Pac. J. Cancer Prev.* **2014**, *15*, 3575–3579. [CrossRef] [PubMed]

25. Zhou, Y.L.; Li, Y.D.; Guo, F.J. Expression of TDP43 in cervical cancer. *Int. J. Clin. Exp. Pathol.* **2016**, *9*, 1467–1473.

26. Zeng, Q.; Cao, K.; Liu, R.; Huang, J.; Xia, K.; Tang, J.; Chen, X.; Zhou, M.; Xie, H.; Zhou, J. Identification of TDP-43 as an oncogene in melanoma and its function during melanoma pathogenesis. *Cancer Biol.* **2017**, *18*, 8–15. [CrossRef] [PubMed]

27. Liu, X.; Zheng, J.; Xue, Y.; Qu, C.; Chen, J.; Wang, Z.; Li, Z.; Zhang, L.; Liu, Y. Inhibition of TDP43-Mediated SNHG12-miR-195-SOX5 Feedback Loop Impeded Malignant Biological Behaviors of Glioma Cells. *Mol. Nucleic Acids* **2018**, *10*, 142–158. [CrossRef]

28. Guo, F.; Jiao, F.; Song, Z.; Li, S.; Liu, B.; Yang, H.; Zhou, Q.; Li, Z. Regulation of MALAT1 expression by TDP43 controls the migration and invasion of non-small cell lung cancer cells in vitro. *Biochem. Biophys. Res. Commun.* **2015**, *465*, 293–298. [CrossRef]

29. Ke, H.; Zhao, L.; Zhang, H.; Feng, X.; Xu, H.; Hao, J.; Wang, S.; Yang, Q.; Zou, L.; Su, X.; et al. Loss of TDP43 inhibits progression of triple-negative breast cancer in coordination with SRSF3. *Proc. Natl. Acad. Sci. USA* **2018**, *115*, E3426–E3435. [CrossRef]

30. Boxer, A.L.; Gold, M.; Huey, E.; Gao, F.B.; Burton, E.A.; Chow, T.; Kao, A.; Leavitt, B.R.; Lamb, B.; Grether, M.; et al. Frontotemporal degeneration, the next therapeutic frontier: Molecules and animal models for frontotemporal degeneration drug development. *Alzheimers Dement.* **2013**, *9*, 176–188. [CrossRef]

31. Varani, G.; Nagai, K. RNA recognition by RNP proteins during RNA processing. *Annu. Rev. Biophys. Biomol. Struct.* **1998**, *27*, 407–445. [CrossRef] [PubMed]

32. Mei, H.Y.; Cui, M.; Heldsinger, A.; Lemrow, S.M.; Loo, J.A.; Sannes-Lowery, K.A.; Sharmeen, L.; Czarnik, A.W. Inhibitors of protein-RNA complexation that target the RNA: Specific recognition of human immunodeficiency virus type 1 TAR RNA by small organic molecules. *Biochem.-Us* **1998**, *37*, 14204–14212. [CrossRef] [PubMed]

33. Gayle, A.Y.; Baranger, A.M. Inhibition of the U1A–RNA Complex by an Aminoacridine Derivative. *Bioorg. Med. Chem. Lett.* **2002**, *12*, 2839–2842. [CrossRef]

34. Cassel, J.A.; Blass, B.E.; Reitz, A.B.; Pawlyk, A.C. Development of a novel nonradiometric assay for nucleic acid binding to TDP-43 suitable for high-throughput screening using AlphaScreen technology. *J. Biomol. Screen.* **2010**, *15*, 1099–1106. [CrossRef] [PubMed]

35. Cassel, J.A.; McDonnell, M.E.; Velvadapu, V.; Andrianov, V.; Reitz, A.B. Characterization of a series of 4-aminoquinolines that stimulate caspase-7 mediated cleavage of TDP-43 and inhibit its function. *Biochimie* **2012**, *94*, 1974–1981. [CrossRef] [PubMed]

36. Vidal, R.L.; Matus, S.; Bargsted, L.; Hetz, C. Targeting autophagy in neurodegenerative diseases. *Trends Pharm. Sci* **2014**, *35*, 583–591. [CrossRef] [PubMed]

37. Hajduk, P.J.; Bures, M.; Praestgaard, J.; Fesik, S.W. Privileged molecules for protein binding identified from NMR-based screening. *J. Med. Chem* **2000**, *43*, 3443–3447. [CrossRef] [PubMed]

38. Hajduk, P.J.; Meadows, R.P.; Fesik, S.W. NMR-based screening in drug discovery. *Q Rev. Biophys.* **1999**, *32*, 211–240. [CrossRef]

39. Shuker, S.B.; Hajduk, P.J.; Meadows, R.P.; Fesik, S.W. Discovering high-affinity ligands for proteins: SAR by NMR. *Science* **1996**, *274*, 1531–1534. [CrossRef]

40. Fesik, S.W. NMR structure-based drug design. *J. Biomol. NMR* **1993**, *3*, 261–269. [CrossRef]

41. Liu, J.; Gao, J.; Li, F.; Ma, R.; Wei, Q.; Wang, A.; Wu, J.; Ruan, K. NMR characterization of weak interactions between RhoGDI2 and fragment screening hits. *Biochim. Et Biophys. Acta (Bba)-Gen. Subj.* **2017**, *1861*, 3061–3070. [CrossRef] [PubMed]

42. Liu, J.; Li, F.; Bao, H.; Jiang, Y.; Zhang, S.; Ma, R.; Gao, J.; Wu, J.; Ruan, K. The polar warhead of a TRIM24 bromodomain inhibitor rearranges a water-mediated interaction network. *FEBS J.* **2017**, *284*, 1082–1095. [CrossRef] [PubMed]

43. Liu, J.; Zhang, S.; Liu, M.; Liu, Y.; Nshogoza, G.; Gao, J.; Ma, R.; Yang, Y.; Wu, J.; Zhang, J.; et al. Structural plasticity of the TDRD3 Tudor domain probed by a fragment screening hit. *FEBS J.* **2018**, *285*, 2091–2103. [CrossRef] [PubMed]

44. Ruan, K.; Gao, J.; Ma, R. NMR in Fragment Based Lead Discovery. *Chin. J. Magn. Reson.* **2012**, *29*, 163–181.

45. Tang, H.; Nshogoza, G.; Liu, M.; Liu, Y.; Ruan, K.; Ma, R.; Gao, J. Identification of Novel Hits of the NSD1 SET Domain by NMR Fragment-Based Screening. *Chin. J. Magn. Reson.* **2018**, *36*, 148–154. [CrossRef]

46. Gao, J.; Ma, R.; Wang, W.; Wang, N.; Sasaki, R.; Snyderman, D.; Wu, J.; Ruan, K. Automated NMR fragment based screening identified a novel interface blocker to the LARG/RhoA complex. *PLoS ONE* **2014**, *9*, e88098. [CrossRef] [PubMed]

47. Lukavsky, P.J.; Daujotyte, D.; Tollervey, J.R.; Ule, J.; Stuani, C.; Buratti, E.; Baralle, F.E.; Damberger, F.F.; Allain, F.H. Molecular basis of UG-rich RNA recognition by the human splicing factor TDP-43. *Nat. Struct. Mol. Biol.* **2013**, *20*, 1443–1449. [CrossRef]

48. Kuo, P.H.; Chiang, C.H.; Wang, Y.T.; Doudeva, L.G.; Yuan, H.S. The crystal structure of TDP-43 RRM1-DNA complex reveals the specific recognition for UG- and TG-rich nucleic acids. *Nucleic Acids Res.* **2014**, *42*, 4712–4722. [CrossRef]

49. Kuo, P.H.; Doudeva, L.G.; Wang, Y.T.; Shen, C.K.; Yuan, H.S. Structural insights into TDP-43 in nucleic-acid binding and domain interactions. *Nucleic Acids Res.* **2009**, *37*, 1799–1808. [CrossRef]

50. Viegas, A.; Manso, J.o.; Nobrega, F.L.; Cabrita, E.J. Saturation-Transfer Difference (STD) NMR: A Simple and Fast Method for Ligand Screening and Characterization of Protein Binding. *J. Chem. Educ.* **2011**, *88*, 990–994. [CrossRef]

51. Dalvit, C.; Fogliatto, G.; Stewart, A.; Veronesi, M.; Stockman, B. WaterLOGSY as a method for primary NMR screening: Practical aspects and range of applicability. *J. Biomol. NMR* **2001**, *21*, 349–359. [CrossRef] [PubMed]

52. Hajduk, P.J.; Olejniczak, E.T.; Fesik, S.W. One-dimensional relaxation- and diffusion-edited NMR methods for screening compounds that bind to macromolecules. *J. Am. Chem. Soc.* **1997**, *119*, 12257–12261. [CrossRef]

53. Zega, A. NMR Methods for Identification of False Positives in Biochemical Screens. *J. Med. Chem.* **2017**, *60*, 9437–9447. [CrossRef] [PubMed]

54. Van Zundert, G.C.P.; Rodrigues, J.P.G.L.M.; Trellet, M.; Schmitz, C.; Kastritis, P.L.; Karaca, E.; Melquiond, A.S.J.; van Dijk, M.; de Vries, S.J.; Bonvin, A.M.J.J. The HADDOCK2.2 webserver: User-friendly integrative modeling of biomolecular complexes. *J. Mol. Biol.* **2015**, *428*, 720–725. [CrossRef] [PubMed]

55. Aguirre, C.; ten Brink, T.; Cala, O.; Guichou, J.F.; Krimm, I. Protein–ligand structure guided by backbone and side-chain proton chemical shift perturbations. *J. Biomol. Nmr.* **2014**, *60*, 147–156. [CrossRef] [PubMed]

56. Aguirre, C.; ten Brink, T.; Guichou, J.F.; Cala, O.; Krimm, I. Comparing Binding Modes of Analogous Fragments Using NMR in Fragment-Based Drug Design: Application to PRDX5. *PLoS ONE* **2014**, *9*. [CrossRef] [PubMed]

57. Harrison, A.F.; Shorter, J. RNA-binding proteins with prion-like domains in health and disease. *Biochem. J.* **2017**, *474*, 1417–1438. [CrossRef]

58. Shortridge, M.D.; Hage, S.H.; Harbison, G.S.; Powers, R. Estimating Protein-Ligand Binding Affinity using High-Throughput Screening by NMR. *J. Comb. Chem.* **2008**, *10*, 948–958. [CrossRef]

59. Mashalidis, E.H.; Sledz, P.; Lang, S.; Abell, C. A three-stage biophysical screening cascade for fragment-based drug discovery. *Nat. Protoc.* **2013**, *8*, 2309–2324. [CrossRef]

60. Gossert, A.D.; Jahnke, W. NMR in drug discovery: A practical guide to identification and validation of ligands interacting with biological macromolecules. *Prog. Nucl. Magn. Reson. Spectrosc.* **2016**, *97*, 82–125. [CrossRef]

61. Wang, N.; Li, F.; Bao, H.; Li, J.; Wu, J.; Ruan, K. NMR Fragment Screening Hit Induces Plasticity of BRD7/9 Bromodomains. *ChemBioChem* **2016**, *17*, 1456–1463. [CrossRef] [PubMed]

62. Wassenaar, T.A.; Van Dijk, M.; Loureiro-Ferreira, N.; Van Der Schot, G.; De Vries, S.J.; Schmitz, C.; Van Der Zwan, J.; Boelens, R.; Giachetti, A.; Ferella, L.; et al. WeNMR: Structural Biology on the Grid. *J. Grid Comput.* **2012**, *10*, 743–767. [CrossRef]

63. Schuttelkopf, A.W.; van Aalten, D.M. PRODRG: A tool for high-throughput crystallography of protein-ligand complexes. *Acta Cryst. D Biol. Cryst.* **2004**, *60*, 1355–1363. [CrossRef] [PubMed]

International Journal of
Molecular Sciences

Article

A Galactoside-Binding Protein Tricked into Binding Unnatural Pyranose Derivatives: 3-Deoxy-3-Methyl-Gulosides Selectively Inhibit Galectin-1

Kumar Bhaskar Pal [1], Mukul Mahanti [1], Hakon Leffler [2] and **Ulf J. Nilsson [1],***

[1] Centre for Analysis and Synthesis, Department of Chemistry, Lund University, Box 124, SE-221 00 Lund, Sweden

[2] Section MIG, Department of Laboratory Medicine, Lund University, BMC-C1228b, Klinikgatan 28, SE-221 84 Lund, Sweden

* Correspondence: ulf.nilsson@chem.lu.se; Tel.: +46-46-222-8218; Fax: +46-46-222-8209

Received: 23 June 2019; Accepted: 22 July 2019; Published: 2 August 2019

Abstract: Galectins are a family of galactoside-recognizing proteins involved in different galectin-subtype-specific inflammatory and tumor-promoting processes, which motivates the development of inhibitors that are more selective galectin inhibitors than natural ligand fragments. Here, we describe the synthesis and evaluation of 3-C-methyl-gulopyranoside derivatives and their evaluation as galectin inhibitors. Methyl 3-deoxy-3-C-(hydroxymethyl)-β-D-gulopyranoside showed 7-fold better affinity for galectin-1 than the natural monosaccharide fragment analog methyl β-D-galactopyranoside, as well as a high selectivity over galectin-2, 3, 4, 7, 8, and 9. Derivatization of the 3-C-hydroxymethyl into amides gave gulosides with improved selectivities and affinities; methyl 3-deoxy-3-C-(methyl-2,3,4,5,6-pentafluorobenzamide)-β-D-gulopyranoside had K_d 700 μM for galectin-1, while not binding any other galectin.

Keywords: galectin-1; gulopyranosides; fluorescence polarization; benzamide; selective

1. Introduction

Galectins are an evolutionary ancient family of small soluble proteins with affinity for β-D-galactopyranoside-containing glycoconjugates and a conserved amino acid sequence motif [1,2]. By their carbohydrate-binding activity they can cross-link glycoproteins, resulting in a variety of effects, such as regulation of cell adhesion, intracellular glycoprotein traffic, and cell signaling [3–5]. These effects in turn affect cell behavior in inflammation, immunity and cancer, and galectins appear to be rate limiting in some such pathophysiological conditions, e.g., based on effects in null mutant mice and other model systems [6–10]. This has stimulated development of galectin inhibitors as potential drug candidates, but different galectins have a different tissue distribution and function. Although all bind glycoconjugates containing β-galactose residues, each galectin may have a different affinities for larger natural glycans and for artificial small molecule ligands. Hence, there is an important need for selective galectin-inhibitors, that, for example, distinguish between the two most studied galectins in humans, galectin-1 and galectin-3.

The carbohydrate binding site of galectins is a concave groove and long enough to hold about a tetrasaccharide and based on this the carbohydrate binding site of galectins has been described as a combination of four subsites (A–D) together with an additional one less defined fifth subsite E [3]. Within this groove, subsite C is conserved among galectins, made up of the defining amino acid sequence motif and binds β-galactopyranosides by H-bond interaction with 4-OH, 6-OH and the ring

5-O, and CH-π interaction of the α-side of the pyranose ring with a Trp residue. The neighbouring sites, however, vary among galectins, and can be targeted for selective inhibitor development. To do this, previous inhibitor design has derivatized the positions on galactose not engaged by subsite C, namely C1, C2, and C3 [11]. Gulose is a rare saccharide not found in mammals, but can potentially bind galectins because it is structurally similar to galactose with the only difference being the stereoconfiguration at C3. Hence, the C3 is epimeric with the OH axial instead of equatorial in the galectin bound pyranose form. Here, we show that derivatization at C3 in gulose offers a new space for galectin inhibitor design and surprisingly selective inhibitors of galectin-1. In particular, amide-functionalised C3-methyl gulopyranosides are shown to be apparently selective towards human galectin-1.

2. Results and Discussion

2.1. Synthesis of Methyl 3-Deoxy-3-C-(methyl)-β-D-gulopyranosides and galectin inhibition evaluation

The synthesis of the 3-C-methyl-gulo derivatives was initiated by Dess–Martin periodinane oxidation [12] of the known methyl 2,4,6-tri-O-benzyl-β-D-galactopyranoside **11** to afford the corresponding keto derivative **12** in 84% yield (Scheme 1). Methylenation of **12** with Petasis reagent gave the olefin **13** in 79% yield. Next, the olefin **13** was subjected to hydroboration with 9-borabicyclo-[3.3.1]nonane (9-BBN) [12], followed by oxidative cleavage of the carbon-boron bond with alkaline hydrogen peroxide to afford the corresponding gulo and galacto isomers **14a** (36%) and **14b** (24%), which were separated by flash column chromatography at a ratio 3:2. Both the gulo and galacto derivatives **14a** and **14b** were separately subjected to hydrogenation [13] in the presence of Pd(OH)$_2$-C to give the desired methyl 3-deoxy-3-C-(hydroxymethyl)-β-D-gulopyranoside **1a** and methyl 3-deoxy-3-C-(hydroxymethyl)-β-D-galactopyranoside **1b** in yields of 51% and 63%, respectively. Evaluation of **1a** and **1b** as inhibitors of human galectin-1, 2, 3, 4N (N-terminal domain) 4C (C-terminal domain), 7, 8N, 8C, 9N, and 9C in a reported competitive fluorescence anisotropy assay [14,15] revealed that the gulo derivative **1a** was selective for galectin-1 with a dissociation constant of 1300 μM, which is about an order of magnitude better than for the virtually unselective reference compound methyl β-D-galactopyranoside **32** (Figure 1, Table 1).

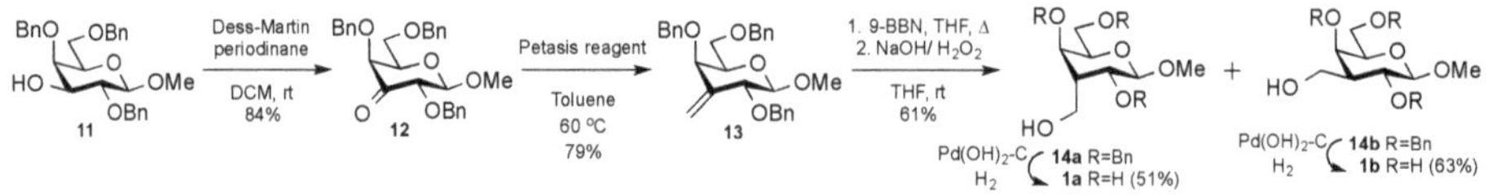

Scheme 1. Synthesis of methyl 3-deoxy-3-C-(hydroxymethyl)-β-D-gulopyranoside **1a**, methyl 3-deoxy-3-C-(hydroxymethyl)-β-D-galactopyranoside **1b**.

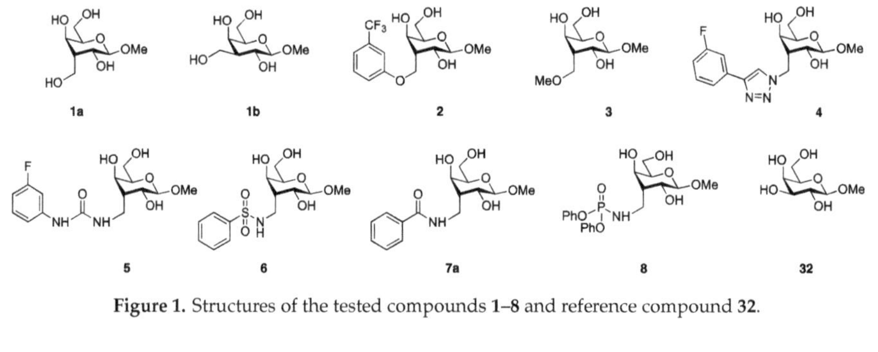

Figure 1. Structures of the tested compounds **1–8** and reference compound **32**.

Table 1. K_d-values (mM)[a] of compounds **1a–1b**, **2–3**, **7a**, **8**, and the reference methyl β-D-galactopyranoside **32** against human galectin-1, 2, 3, 4N, 4C, 7, 8N, 8C, 9N, and 9C in a competitive fluorescence polarization assay [15,16].

Compounds	Galectin									
	1	2	3	4N [b]	4C [c]	7	8N [b]	8C [c]	9N[b]	9C [c]
1a	1.3 ± 0.15	ND [d]	NB [e]	NB	ND	NB	NB	3.7 ± 0.02	NB	NB
1b	NB	NB	NB	>4	NB	NB	NB	NB	NB	NB
2	NB	NB	NB	NB	NB	NB	NB	NB	NB	NB
3	NB	ND	NB	ND	ND	ND	NB	NB	ND	ND
4	NB	NB	NB	NB	NB	NB	NB	NB	NB	NB
5	NB	NB	NB	NB	NB	NB	NB	NB	NB	NB
6	>10	ND	NB	NB	NB	NB	NB	NB	ND	NB
7a	1.8 ± 0.15	NB	NB	NB	NB	NB	NB	NB	ND	NB
8	>10	ND	NB	NB	NB	NB	NB	NB	ND	>5
32	>10 [16]	13 [17]	4.4 [16]	6.6 [17]	10 [17]	4.8 [16]	6.3 [16]	>30 [18]	3.3 [16]	8.6 [19]

[a] The data are average and SEM (standard error of mean) of 4–8 single-triple point measurements. [b] N-terminal domain. [c] C-terminal domain. [d] Not determined. [e] Not binding at the highest concentration tested: 4 mM.

In stark contrast, the galacto derivative **1b** did not bind any galectin tested, except for a weak binding to galectin-4N. This observation encouraged us to further explore the 3-*C*-methyl gulopyranoside scaffold for the discovery of galectin-1-selective inhibitors. Hence, we initiated synthetic efforts toward replacing the hydroxymethyl of **1a** with amide, ether, urea, and triazole functionalities. An aryl ether was synthesized following a recently reported iodonium-salt mediated reaction [20] to give the aryl ether **15**, which after hydrogenolysis [13] of the benzyl protecting groups gave **2** (Scheme 2). The hydroxymethyl **14a** was methylated with methyl iodide to give the methyl ether **16**, which after debenzylation gave the 3-methoxymethyl guloside **3**. Treatment of **14a** with methanesulfonyl chloride furnished the corresponding gulo mesylate, which was then directly treated with NaN$_3$ in dry DMF at 80 °C to provide the gulo azide, **17** in 83% yield. The gulo azide **17** was treated with 1-ethynyl-3-fluorobenzene in the presence of the CuI and DIPEA catalytic system [21] in dry dichloromethane to give the triazole **18** within 48 h in 86% yield. Debenzylation provided the desired triazole-derived methyl guloside **4**. The urea **20** was obtained via reduction of the azide **17** to give the amine **19**, followed by reaction with 3-fluorophenylisocyanate. Debenzylation [13] of **20** afforded the target gulo urea derivative **5** in 66% yield. The amine **19** was treated with benzensulfonyl chloride, benzoyl chloride, and diphenyl phosphoryl chloride in the presence of Et$_3$N to give the protected sulfonamide **21**, amide **22a**, and diphenylphosphonamide **23**, which were subjected for hydrogenolysis [13] in the presence of Pd(OH)$_2$ to get the unprotected amides **6**, **7a**, and **8**.

Evaluation of aryl ether **2**, methyl ether **3**, triazole **4**, urea **5**, sulfonamide **6**, benzamide **7a**, and phosphonamide **8** derived methyl gulosides' affinities for the human galectin-1, 2, 3, 4N, 4C, 7, 8N, 8C, 9N, and 9C showed that most of the gulo derivatives were inactive as ligands for galectins, the benzamide **7a** displayed moderate affinity, similar to that of **1a**, for galectin-1 and with excellent selectivity (Table 1). Particularly noteworthy was that **7a** also had a significantly better affinity for galectin-1 than the simple reference monosaccharide methyl β-D-galactopyranoside **32**. Furthermore, the hydroxylmethyl group of **1a** plays an important role in the interaction with galectin-1, as the corresponding methyl ether **3** binds galectin-1 significantly worse than **1a** does.

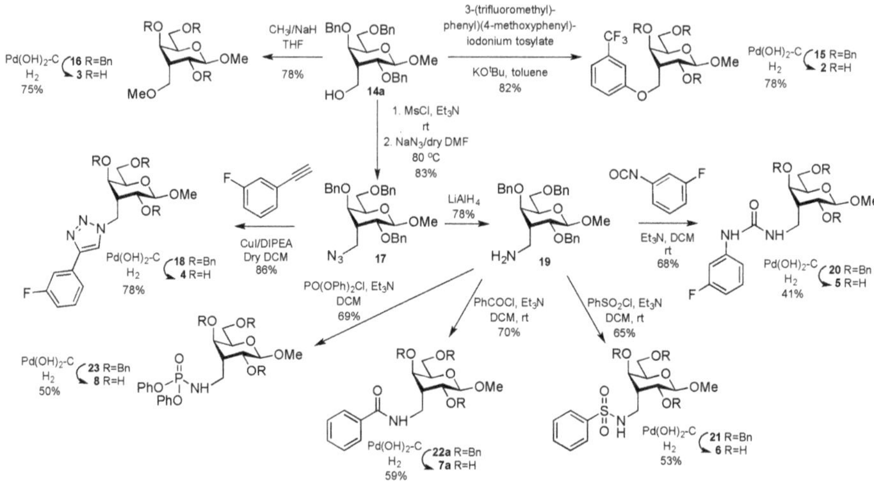

Scheme 2. Synthesis of methyl 3-deoxy-3-*C*-methyl-β-D-gulopyranoside ether **2–3**, triazole **4**, urea **5**, sulfonamide **6**, amide **7a**, and phosphonamide **8** derivatives.

2.2. Synthesis and Optimization of 3-Deoxy-3-C-Amidomethyl-β-D-Gulopyranoside Derivatives as Galectin-1 Inhibitors

The observation that the amide **7a** showed moderate affinity but high selectivity for galectin-1 prompted us to prepare a series **7c–7l** of benzamide analogs carrying selected different substituents at different positions, including four fluorbenzamide expected to possess improved metabolic stability and pharmacokinetic properties, as well as a reference acetamide analog **7b** (Scheme 3). Furthermore, in order to investigate the role of the gulo 3-*C*-methyl substituent, the 3-OH **9** and 3-benzamido **10** gulosides were synthesized (Scheme 3). Hydrolysis of the known 4,6-*O*-benzylidene gulose derivative, **24** [22] with 80% AcOH at 80 °C gave the diol **25** in 91% yield, which upon Zemplen de-*O*-acetylation [23] afforded the target methyl β-D-gulopyranoside **9** in 93% yield. Selective 3-*O*-triflation of methyl 4,6-*O*-benzylidene-β-D-galactopyranoside **26** [24], followed by one-pot benzoylation of 2-*O*-hydroxyl gave **27**. The crude triflate **27** was subsequently converted into the 3-azido-3-deoxy-guloside **28** by treatment with sodium azide in DMF. De-benzylidenation with 80% AcOH at 80 °C and subsequent benzoylation afforded **29** in 43% yield over four steps from **26**. Azide hydrogenation gave **30**, which upon benzoylation and de-*O*-benzoylation gave the benzamide **10**.

An immediate observation upon evaluating the affinities of **7b–7l** and **9–10** (Figure 2, Table 2) was that the acetamide **7b** displays a similar affinity for galectin-1 as the benzamides **7a** and **7c–7k**. Hence, the phenyl moieties of **7a** and **7c–7k** do not contribute to enhancing the affinity for galectin-1. However, the phenyl moieties and substitution patterns of **7a** and **7c–7k** influence the selectivity over other galectins, as six substituted amides (**7a**, **7d**, and **7f–7i**) retained high selectivity for galectin-1 over the other galectins. The pentafluorophenyl **7g** turned out to be the best β-D-gulopyranoside-based monosaccharide inhibitor for human galectin-1 with 14-fold improved affinity over the reference methyl β-D-galactopyranoside **32**. The larger biphenyl **7l** did not bind galectin-1, which suggests that the galectin-1 site accommodating the axial gulo substituent is limited in size. Evaluation of the guloside **9** revealed that while it is similar to the reference galactoside **32** in the affinity for galectin-1, it displays a much higher selectivity in that it is inactive against the other galectins under the evaluation conditions used. Unfortunately, extensive molecular dynamics and docking analyses to explain the selective galectin-1 binding to 3-*C*-methyl-gulosides were inconclusive as such calculations cannot provide reliable relative affinities of bound ligands. Hence, it remains to find a plausible structural explanation for this selectivity. Interestingly, the benzamide **10** showed no binding to galectin-1 under the assay conditions but instead had improved binding to and selectivity for galectin-4N. Hence, while

3-*C*-methyl gulosides represent an interesting structural class for the discovery of selective galectin-1 inhibitors, 3-*C*-amido gulosides may represent a novel structural class for galectin-4 inhibitor discovery.

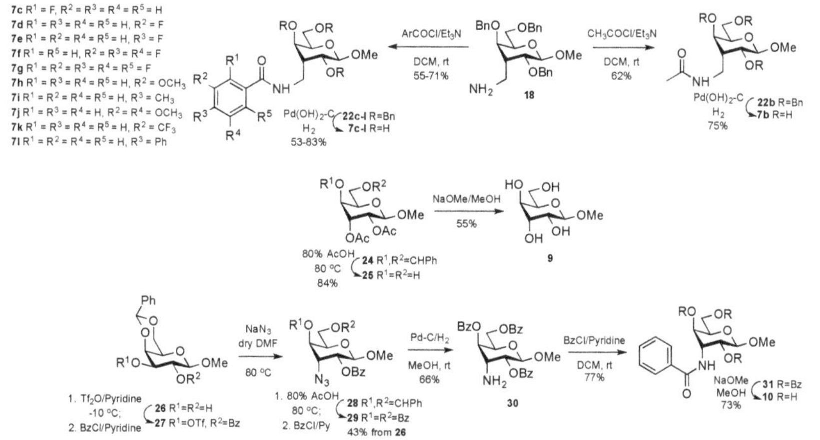

Scheme 3. Synthesis of 3-deoxy-3-*C*-amidomethyl-β-ᴅ-gulo derivatives **7b–7l**, methyl β-ᴅ-gulopyranoside **9**, and methyl 3-deoxy-3-*N*-benzamido-β-ᴅ-gulopyranoside **10**.

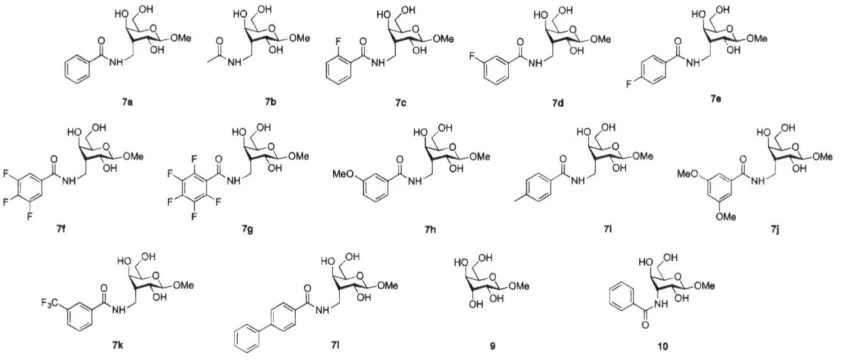

Figure 2. Structures of all tested compounds **7a–7l** and **9–10**.

Table 2. K_d-values (mM)[a] of compounds **7a–7l**, **9**, and **10** against human galectin-1, 2, 3, 4N, 4C, 7, 8N, 8C, 9N, and 9C in a competitive fluorescence polarization assay.

Compounds	Galectin									
	1	2	3	4N [b]	4C [c]	7	8N [b]	8C [c]	9N [b]	9C [c]
7a	1.8 ± 0.15	NB [d]	NB	NB	NB	NB	NB	NB	ND [e]	NB
7b	1.5 ± 0.08	NB	NB	1.9 ± 0.05	NB	NB	NB	2.7 ± 0.5	NB	NB
7c	1.9 ± 0.04	NB	NB	2.2 ± 0.16	NB	NB	NB	NB	NB	NB
7d	1.9 ± 0.4	NB	NB	NB	NB	NB	NB	NB	NB	NB
7e	1.7 ± 0.06	NB	1.6 ± 0.03	1.7 ± 0.07	NB	NB	NB	NB	NB	NB
7f	2.5 ± 0.4	NB	NB	NB	NB	NB	NB	NB	NB	NB
7g	0.7 ± 0.005	NB	NB	NB	NB	NB	NB	NB	NB	NB
7h	3.2 ± 0.5	NB	NB	NB	NB	NB	NB	NB	NB	NB
7i	2.3 ± 0.4	NB	NB	NB	NB	NB	NB	NB	NB	NB
7j	1.8 ± 0.04	NB	NB	2 ± 0.4	NB	NB	NB	2.6 ± 0.6	NB	NB
7k	1.8 ± 0.07	NB	NB	1.9 ± 0.1	NB	NB	NB	NB	NB	NB
7l	NB	NB	NB	NB	NB	NB	NB	NB	NB	NB
9	10 ± 0.25	10 ± 1.5	NB	ND	11 ± 1.2	NB	NB	NB	NB	NB
10	NB	NB	NB	1.3 ± 0.2	NB	ND	NB	NB	NB	NB

[a] The data are average and SEM of 4–8 single-triple point measurements. [b] N-terminal domain. [c] C-terminal domain. [d] Not binding at the highest concentration tested: 4 mM. [e] Not determined.

3. Materials and Methods

3.1. General Methods Experimental Procedures

All reactions were carried out in oven-dried glassware. All solvents and reagents were mainly purchased from Sigma-Aldrich or Fluka and were used without further purification or synthesized via the literature protocol. TLC analysis was performed on pre-coated Merck silica gel 60 F_{254} plates using UV light and charring solution (10 mL conc. H_2SO_4/90 mL EtOH). Flash column chromatography was done on SiO_2 purchased from Aldrich (technical grade, 60 Å pore size, 230–400 mesh, 40–63 μm). All NMR spectra were recorded with the Bruker DRX 400 MHz spectrometer (400 MHz for ^{1}H, 100 MHz for ^{13}C (125 MHz ^{13}C for compound **7k** with the Bruker Avance III 500 MHz spectrometer equipped with a broadband observe SMART probe, Fällanden, Switzerland), 376 MHz for ^{19}F, 162 MHz for ^{31}P, ESI) at ambient temperature using $CDCl_3$ or CD_3OD as solvents. Chemical shifts are given in ppm relative to the residual solvent peak (^{1}H NMR: $CDCl_3$ δ 7.26; CD_3OD δ 3.31; ^{13}C NMR: $CDCl_3$ δ 77.16; CD_3OD δ 49.00) with multiplicity (*b* = broad, *s* = singlet, *d* = doublet, *t* = triplet, *q* = quartet, *quin* = quintet, *sext* = sextet, *hept* = heptet, *m* = multiplet, *td* = triplet of doublets, *dt* = doublet of triplets), coupling constants (in Hz) and integration. Copies of nmr spectra are provided in the supplementary information. High-resolution mass analysis was obtained using the Micromass Q-TOF mass spectrometer. Analytical data is given if the compound is novel or not fully characterized in the literature. Final compounds were further purified via HPLC before evaluation of galectin affinity. All tested compounds were >95% pure according to the analytical HPLC analysis.

3.2. Methyl 2,4,6-Tri-O-Benzyl-β-D-Xylo-Hex-3-Ulopyranoside **12**

Into a solution of alcohol **11** (8.1 g, 17.45 mmol) in dry dichloromethane (250 mL) Dess–Martin periodinane (9.62 g, 22.68 mmol, 1.3 equiv.) was added, under nitrogen atmosphere and the reaction mixture was stirred for 4 h (TLC heptane/EtOAc, 3:1, R_f 0.5). After that, a saturated $NaHCO_3$ solution (400 mL) was added and the mixture was stirred for 30 min. Then, the organic layer was collected and washed successively with the saturated $Na_2S_2O_3$ solution (2 × 250 mL). The organic layer was collected, dried over Na_2SO_4, filtered and concentrated in vacuo. Flash chromatography of the crude material (heptane/EtOAc, 7:2) afforded ketone **12** (6.45 g, 13.955 mmol, yield 80%) as a white solid. $[\alpha]_D^{25}$ −72.3 (c 1.4, $CHCl_3$). ^{1}H NMR ($CDCl_3$, 400 MHz): 7.47–7.21 (m, 15H, Ar*H*), 4.76 (d, 1H, *J* 12.0 Hz, C*H*$_2$Ph), 4.73 (d, 1H, *J* 12.0 Hz, C*H*$_2$Ph), 4.58 (d, 1H, *J* 12.0 Hz, C*H*$_2$Ph), 4.51 (d, 1H, *J* 12.0 Hz, C*H*$_2$Ph), 4.48 (d, 1H, $J_{1,2}$ 7.6 Hz, H-1), 4.44 (d, 1H, $J_{1,2}$ 7.6 Hz, H-2), 4.43 (d, 1H, *J* 11.6 Hz, C*H*$_2$Ph), 4.35 (d, 1H, *J* 11.6 Hz, C*H*$_2$Ph), 3.95 (d, 1H, *J* 1.2 Hz, H-4), 3.83–3.75 (m, 3H, H-5, H-6a, H-6b), 3.61 (s, 3H, OC*H*$_3$). ^{13}C NMR ($CDCl_3$, 100 MHz): 203.8, 137.7, 137.2, 136.4, 128.29, 128.27, 127.26, 128.0, 127.8, 127.6, 127.5, 104.9, 82.1, 80.7, 73.5, 73.4, 72.3, 72.1, 67.5, 57.1. HRMS calcd for $C_{28}H_{30}O_6{}^+NH_4{}^+$ $(M+NH_4)^+$: 480.2386, found: 480.2378.

3.3. Methyl 2,4,6-Tri-O-Benzyl-3-Deoxy-3-C-Methylene-β-D-Xylo-Hex-3-Ulopyranoside **13**

Into a solution of ketone **12** (6.3 g, 13.63 mmol) in dry toluene (100 mL) bis (cyclopentadienyl) dimethyltitanium was added, 5 wt% in toluene (125 mL, 30 mmol, 2.2 equiv.), under nitrogen atmosphere and the reaction mixture was stirred for 48 h at 65 °C in the dark. After that, the reaction mixture (TLC heptane/EtOAc, 4:1, R_f 0.47) was concentrated in vacuo and flash chromatography of the crude material (heptane/EtOAc, 10:1–5:1) afforded methylene derivative **13** (4.6 g, 9.99 mmol, yield 71%) as a light-yellow oil. $[\alpha]_D^{25}$ −40.3 (c 1.1, $CHCl_3$). ^{1}H NMR ($CDCl_3$, 400 MHz): 7.49–7.28 (m, 15H, Ar*H*), 5.61 (t, 1H, $J_{2,H-7a}$ 2.0 Hz, C*H*$_2$), 5.20 (t, 1H, $J_{2,H-7b}$ 2.0 Hz, C*H*$_2$), 5.00 (d, 1H, *J* 12.0 Hz, C*H*$_2$Ph), 4.78 (d, 1H, *J* 12.0 Hz, C*H*$_2$Ph), 4.65 (d, 1H, *J* 12.0 Hz, C*H*$_2$Ph), 4.58 (d, 1H, *J* 12.0 Hz, C*H*$_2$Ph), 4.56 (d, 1H, *J* 12.0 Hz, C*H*$_2$Ph), 4.36 (d, 1H, *J* 7.6 Hz, H-1), 4.28 (d, 1H, *J* 12.0 Hz, C*H*$_2$Ph), 4.14 (dt, 1H, $J_{1,2}$ 7.6 Hz, $J_{2,H-7a}$, $J_{2,H-7b}$ 2.0 Hz, H-2), 4.03 (d, 1H, *J* 0.4 Hz, H-4), 3.91–3.79 (m, 3H, H-5, H-6a, H-6b), 3.65 (s, 3H, OC*H*$_3$). ^{13}C NMR ($CDCl_3$, 100 MHz): 142.2, 138.5, 138.2, 137.9, 128.4, 128.3, 128.2, 128.0, 127.9,

127.7, 127.62, 127.59, 127.5, 113.7, 104.9, 77.7, 77.3, 76.6, 73.7, 73.6, 69.2, 69.0, 56.7. HRMS calcd for $C_{29}H_{32}O_5+NH_4^+$ $(M+NH_4)^+$: 478.2593, found: 478.2607.

3.4. Methyl 2,4,6-Tri-O-Benzyl-3-Deoxy-3-C-Hydroxymethyl-β-D-Gulopyranoside 14a and Methyl 2,4,6-Tri-O-Benzyl-3-Deoxy-3-C-Hydroxymethyl-β-D-Galactopyranoside 14b

A solution of **13** (4.6 g, 9.99 mmol) in dry THF (150 mL) was treated with a 9-BBN solution in THF (0.5 M, 125 mL) and heated to reflux for 24 h. After that, the solution was cooled to 0 °C and a 10% aqueous sodium hydroxide solution (100 mL) and a 30% hydrogen peroxide solution (100 mL) were added simultaneously within 5 min and stirring continued for another 30 min. Then, diethyl ether (200 mL) was added followed by careful addition of a 20% aqueous sodium hydrogen sulfite solution (7 mL). This mixture was stirred further for 60 min and extracted with diethyl ether, and the combined organic layers were dried with Na_2SO_4, filtered, and concentrated in vacuo (TLC heptane/EtOAc, 2:1 (double run), R_f 0.48 for **14a**, R_f 0.4 for **14b**). Flash chromatography (Heptane/EtOAc, 8:1 to 2:1) of the residue afforded a gulo-isomer, **14a** (1.74 g, 3.638 mmol) and galacto-isomer, **14b** (1.16 g, 2.426 mmol) to be ≈3:2 in favor of the guloisomer at an overall yield of 61% (2.9 g, 6.064 mmol). Gulo-isomer **14a**: $[\alpha]_D^{25}$ −25.7 (c 1.3, $CHCl_3$). ^{1}H NMR ($CDCl_3$, 400 MHz): 7.36–7.20 (m, 15H, Ar*H*), 4.82 (d, 1H, *J* 12.0 Hz, C*H*$_2$Ph), 4.65 (d, 1H, *J* 6.4 Hz, H-1), 4.57 (d, 1H, *J* 11.6 Hz, C*H*$_2$Ph), 4.54 (d, 1H, *J* 11.6 Hz, C*H*$_2$Ph), 4.52 (d, 1H, *J* 11.6 Hz, C*H*$_2$Ph), 4.47 (d, 1H, *J* 12.0 Hz, C*H*$_2$Ph), 4.41 (d, 1H, *J* 11.6 Hz, C*H*$_2$Ph), 3.95–3.88 (m, 2H, H-4, H-5), 3.73 (dd, 1H, $J_{1,2}$ 6.4 Hz, $J_{2,3}$ 5.2 Hz, H-2), 3.74–3.57 (m, 4H, H-6a, H-6b, C*H*$_2$OH), 3.54 (s, 3H, OC*H*$_3$), 2.53–2.47 (m, 1H, H-3), 2.35 (bs, 1H, CH$_2$O*H*). ^{13}C NMR ($CDCl_3$, 100 MHz): 138.2, 138.1, 138.0, 128.6, 128.52, 128.47, 128.2, 128.1, 128.03, 127.96, 127.9, 127.8, 101.2, 77.2, 74.8, 73.8, 73.6, 73.4, 71.9, 69.5, 62.0, 56.5, 41.6. HRMS calcd for $C_{29}H_{34}O_6+NH_4^+$ $(M+NH_4)^+$: 496.2699, found: 496.2700. Galacto-isomer **14b**: $[\alpha]_D^{25}$ −13.4 (c 0.9, $CHCl_3$). ^{1}H NMR ($CDCl_3$, 400 MHz): 7.39–7.28 (m, 15H, Ar*H*), 4.92 (d, 1H, *J* 11.2 Hz, C*H*$_2$Ph), 4.65 (d, 1H, *J* 11.2 Hz, C*H*$_2$Ph), 4.60–4.52 (m, 4H, C*H*$_2$Ph), 4.41 (d, 1H, $J_{1,2}$ 7.6 Hz, H-1), 3.90 (d, 1H, $J_{3,4}$ 2.8 Hz, H-4), 3.82 (dd, 1H, *J* 4.8 Hz, *J* 7.2 Hz, C*H*$_2$OH), 3.73–3.55 (m, 8H, H-5, H-6a, H-6b, H-2, C*H*$_2$OH, OC*H*$_3$), 2.04 (bs, 1H, CH$_2$O*H*), 1.87–1.82 (m, 1H, H-3). ^{13}C NMR ($CDCl_3$, 100 MHz): 138.5, 138.1, 137.8, 128.6, 128.53, 128.52, 128.4, 128.3, 128.03, 127.98, 127.8, 106.4, 76.5, 76.2, 74.8, 74.7, 74.6, 73.7, 68.6, 62.2, 56.8, 47.3. HRMS calcd for $C_{29}H_{34}O_6+H^+$ $(M+H)^+$: 479.2434, found: 479.2434.

3.5. Methyl 2,4,6-Tri-O-Benzyl-3-Deoxy-3-C-(3-Trifluoromethylphenoxymethyl)-β-D-Gulopyranoside 15

Compound **14a** (80 mg, 0.17 mmol) was stirred in a 25 mL round-bottom flask in toluene (2 mL) for 3 min. A mixture of 3-(trifluoromethyl)phenyl)(4-methoxyphenyl)iodonium tosylate (140 mg, 0.25 mmol) and potassium tert-butoxide (28.5 mg, 0.25 mmol) were added under air and the mixture turned yellow. The reaction was stirred for 3 h, when the TLC showed almost complete consumption of the starting material (TLC heptane/EtOAc, 3:1, R_f 0.48). The mixture was then diluted with EtOAc (10 mL) and filtered. Then the volatiles were removed under reduced pressure, and the residue was subjected to column chromatography (heptane/EtOAc, 8:1 to 4:1) to provide the purified product **15** (92.6 mg, 0.15 mmol, 89%) as a colorless oil. $[\alpha]_D^{25}$ −70.9 (c 0.8, $CHCl_3$). ^{1}H NMR ($CDCl_3$, 400 MHz): 7.40–7.22 (m, 17H, Ar*H*), 7.08 (bs, 1H, Ar*H*), 7.02 (dd, 1H, *J* 8.0 Hz, *J* 2.4 Hz, Ar*H*), 4.79 (d, 1H, *J* 12.0 Hz, C*H*$_2$Ph), 4.62 (d, 1H, *J* 6.0 Hz, H-1), 4.58 (d, 1H, *J* 12.0 Hz, C*H*$_2$Ph), 4.57 (d, 1H, *J* 11.6 Hz, C*H*$_2$Ph), 4.54 (d, 1H, *J* 12.4 Hz, C*H*$_2$Ph), 4.50 (s, 2H, C*H*$_2$Ph), 4.24 (dd, 1H, *J* 6.0 Hz, *J* 9.6 Hz, H-3a′), 4.19–4.15 (m, 1H, H-5), 4.08 (dd, 1H, *J* 9.6 Hz, *J* 8.0 Hz, H-3b′), 3.87 (dd, 1H, *J* 5.2 Hz, *J* 2.8 Hz, H-4), 3.85 (t, 1H, *J* 6.0 Hz, H-2), 3.80 (dd, 1H, *J* 10.0 Hz, *J* 6.8 Hz, H-6a), 3.72 (dd, 1H, *J* 10.0 Hz, *J* 5.2 Hz, H-6b), 3.57 (s, 1H, OC*H*$_3$), 2.76–2.70 (m, 1H, H-3). ^{13}C NMR ($CDCl_3$, 100 MHz): 158.8, 138.30, 138.27, 137.9, 131.9 (q, *J* 32.1 Hz), 130.0, 128.49, 128.46, 128.3, 128.0, 127.9, 127.83, 127.77, 124.1 (q, *J* 271 Hz), 118.0, 117.6 (q, *J* 3.8 Hz), 111.5 (q, *J* 3.7 Hz), 101.3, 74.7, 73.6, 73.5, 73.3, 72.8, 71.9, 69.8, 64.6, 56.4, 39.6. ^{19}F NMR ($CDCl_3$, 376 MHz): −62.6. HRMS calcd for $C_{36}H_{41}F_3NO_6+NH_4^+$ $(M+NH_4)^+$: 640.2886, found: 640.2895.

3.6. Methyl 2,4,6-Tri-O-Benzyl-3-Deoxy-3-C-Methoxymethyl-β-D-Gulopyranoside **16**

Compound **14a** (57 mg, 0.12 mmol) was stirred in a 5 mL round-bottom flask in dry THF (2 mL) for 5 min at 0 °C. Into the solution, NaH (6 mg, 0.24 mmol) was added and the stirring was continued at 0 °C for 5 min. Then, into the reaction mixture iodomethane dropwise was added and the reaction temperature increased to rt gradually. Stirring continued overnight when the TLC showed almost complete consumption of the starting material (TLC heptane/EtOAc, 3:2, R_f 0.53). Then, NaH was quenched with EtOAc and the volatiles were removed under reduced pressure. The residue was subjected to column chromatography (heptane/EtOAc, 6:1 to 3:1) to provide the purified product **16** (46 mg, 0.09 mmol, 78%). $[\alpha]_D^{25}$ −62.5 (c 1.2, CHCl$_3$). ^{1}H NMR (CDCl$_3$, 400 MHz): 7.35–7.20 (m, 15H, Ar*H*), 4.76 (d, 1H, *J* 12.0 Hz, C*H*$_2$Ph), 4.58 (d, 1H, *J* 12.0 Hz, C*H*$_2$Ph), 4.57 (d, 1H, *J* 11.6 Hz, C*H*$_2$Ph), 4.53 (d, 1H, $J_{1,2}$ 6.4 Hz, H-1), 4.48 (d, 2H, *J* 12.4 Hz, C*H*$_2$Ph), 4.37 (d, 1H, *J* 11.6 Hz, C*H*$_2$Ph), 4.11–4.07 (m, 1H, H-5), 3.75–3.71 (m, 3H, H-2, H-4, H-6a), 3.67–3.61 (m, 2H, H-6b, C*H*$_2$OCH$_3$), 3.56–3.50 (m, 4H, C*H*$_2$OCH$_3$, OC*H*$_3$), 3.29 (s, 3H, CH$_2$OC*H*$_3$), 2.58–2.52 (m, 3H, H-3). ^{13}C NMR (CDCl$_3$, 100 MHz): 138.8, 138.43, 138.39, 128.5, 128.43, 128.36, 128.1, 127.94, 128.90, 127.8, 127.74, 127.69, 127.66, 101.8, 75.1, 74.7, 73.7, 73.3, 73.2, 71.8, 70.2, 69.2, 59.0, 56.4, 40.2. HRMS calcd for C$_{17}$H$_{21}$NO$_6$+H$^+$ (M+H)$^+$: 335.1369, found: 335.1369.

3.7. Methyl 2,4,6-Tri-O-Benzyl-3-Deoxy-3-C-Azidomethyl-β-D-Gulopyranoside **17**

Into a stirred solution of **14a** (1.6 g, 3.35 mmol) in DCM (25 mL) containing Et$_3$N (890 µL, 6.69 mmol) at 0 °C MsCl (390 µL, 5.02 mmol) was added dropwise over 5 min, and the solution was stirred for 4 h at rt (TLC heptane/EtOAc, 1:1, R_f 0.31). The solution was extracted with 1N HCl (2 × 50 mL) followed by sat'd NaHCO$_3$ (2 × 50 mL), and the organic layer was dried (Na$_2$SO$_4$). The solvent was removed by rotary evaporation to give a yellow liquid that was dissolved in dry DMF (10 mL). Sodium azide (1.3 g, 20.08 mmol) was added and the solution was heated at 80 °C for 6 h to give a yellowish-brown mixture. The mixture was cooled at rt, water (50 mL) was added, and the mixture was extracted with EtOAc (2 × 40 mL). The organic layer was washed with brine (50 mL) and dried (Na$_2$SO$_4$). The solvent was removed by rotary evaporation to give a yellow liquid that was then purified by flash chromatography (Heptane/EtOAc 8:1 to 3:1) to give compound **17** (1.4 g, 2.78 mmol, 83% from **14a**) as a colorless liquid. $[\alpha]_D^{25}$ −5.2 (c 0.8, CHCl$_3$). ^{1}H NMR (CDCl$_3$, 400 MHz): 7.38–7.20 (m, 15H, Ar*H*), 4.76 (d, 1H, *J* 12.0 Hz, C*H*$_2$Ph), 4.56 (d, 1H, *J* 12.0 Hz, C*H*$_2$Ph), 4.55 (d, 1H, *J* 11.6 Hz, C*H*$_2$Ph), 4.51 (d, 1H, $J_{1,2}$ 6.4 Hz, H-1), 4.48 (d, 1H, *J* 12.4 Hz, C*H*$_2$Ph), 4.44 (d, 1H, *J* 12.0 Hz, C*H*$_2$Ph), 4.41 (d, 1H, *J* 12.0 Hz, C*H*$_2$Ph), 4.09–4.05 (m, 1H, H-5), 3.77–3.64 (m, 5H, H-2, H-4, H-6a, H-6b, C*H*$_2$N$_3$), 3.50 (s, 3H, OC*H*$_3$), 3.39 (dd, 1H, *J* 12.4 Hz, *J* 8.4 Hz, C*H*$_2$N$_3$), 2.42–2.36 (m, 1H, H-3). ^{13}C NMR (CDCl$_3$, 100 MHz): 138.4, 138.3, 137.9, 128.53, 128.52, 128.50, 128.2, 128.02, 127.95, 127.9, 127.8, 100.8, 75.0, 73.7, 73.6, 73.4, 72.5, 72.0, 69.6, 56.4, 48.6, 39.7. HRMS calcd for C$_{29}$H$_{33}$N$_3$O$_5$+NH$_4$$^+$ (M+NH$_4$)$^+$: 521.2764, found: 521.2775.

3.8. Methyl 2,4,6-Tri-O-Benzyl-3-Deoxy-3-C-[4-(3-Fluorophenyl)-1H-1,2,3-Triazol-1-Yl-Methyl]-β-D-Gulopyranoside **18**

A solution of azide **17** (53 mg, 0.10 mmol) in dichloromethane (2 mL), 1-Ethynyl-3-fluorobenzene (18.1 µL, 0.16 mmol), CuI (10 mol%, 2 mg) and DIPEA (28 µL, 0.16 mmol) were added and the mixture was stirred at room temperature for 48 h (TLC heptane/EtOAc, 2:1, R_f 0.58). The solvent was removed under reduced pressure, and the residue was dissolved in EtOAc (10 mL) and the solution was washed with sat. NH$_4$Cl (10 mL), brine (10 mL), dried over Na$_2$SO$_4$ and concentrated in vacuo. The product was purified by flash column chromatography (heptane/EtOAc, 6:1 to 1:1) to give the corresponding triazole, **18** as white amorphous solid (56.4 mg, 0.09 mmol, 86%). $[\alpha]_D^{25}$ −63 (c 0.6, CHCl$_3$). ^{1}H NMR (CDCl$_3$, 400 MHz): 7.52–7.04 (m, 20H, Ar*H*), 4.80 (d, 1H, *J* 11.6 Hz, C*H*$_2$Ph), 4.63 (dd, 1H, *J* 6.8 Hz, *J* 14.0 Hz, H-3′), 4.60–4.44 (m, 7H, H-1, H-3′, C*H*$_2$Ph), 4.35 (d, 1H, *J* 11.6 Hz, C*H*$_2$Ph), 4.26–4.22 (m, 1H, H-5), 3.79 (dd, 1H, *J* 10.0 Hz, *J* 7.2 Hz, H-6a), 3.74 (dd, 1H, *J* 6.4 Hz, *J* 3.2 Hz, H-4), 3.71 (dd, 1H, *J* 10.0 Hz, *J* 5.2 Hz, H-6b), 3.79 (t, 1H, *J* 4.8 Hz, H-2), 3.51 (s, 1H, OC*H*$_3$), 2.68–2.61 (m, 1H, H-3). ^{13}C NMR (CDCl$_3$, 100 MHz): 163.3 (d, *J* 244 Hz), 146.8, 138.19, 138.16, 137.4, 132.8 (d, *J* 8.3 Hz), 130.5 (d, *J* 8.4 Hz),

128.7, 128.52, 128.51, 128.3, 128.2, 128.0, 127.9, 127.8, 121.3 (d, *J* 2.7 Hz), 120.6, 115.0 (d, *J* 22 Hz), 112.7 (d, *J* 23 Hz), 100.1, 75.2, 73.52, 73.47, 73.1, 72.5, 72.2, 69.4, 56.4, 47.4, 41.1. ^{19}F NMR (CDCl$_3$, 376 MHz): −112.7. HRMS calcd for C$_{37}$H$_{38}$FN$_3$O$_5$+H$^+$ (M+H)$^+$: 624.2874, found: 624.2884.

3.9. Methyl 2,4,6-Tri-O-Benzyl-3-Deoxy-3-C-(Aminomethyl)-β-ᴅ-Gulopyranoside 19

Into a stirred solution of **17** (1.31 g, 2.60 mmol) in dry THF (20 mL) at 0 °C, LiAlH$_4$ (148 mg, 3.9 mmol) was added in portions over 5 min under nitrogen atmosphere, and the solution was stirred for 1 h at rt (TLC DCM/MeOH, 15:1, R$_f$ 0.44). After 30 min, TLC was checked which shows complete conversion of the azide into amine. Then, the reaction was quenched EtOAc and the reaction mixture was filtered through a pad of Celite® (St. Louis, MO, USA). Then, the filtrate was concentrated in vacuo and the crude was purified by column chromatography (DCM:MeOH 25:1) to give compound **19** (969 mg, 2.03 mmol, yield 78%) as a colorless oil. [α]$_D^{25}$ −36.2 (c 1.1, CHCl$_3$). ^{1}H NMR (CDCl$_3$, 400 MHz): 7.35–7.22 (m, 15H, Ar*H*), 4.80 (d, 1H, *J* 12.0 Hz, C*H*$_2$Ph), 4.57–4.54 (m, 3H, H-1, C*H*$_2$Ph), 4.51 (d, 1H, *J* 12.0 Hz, C*H*$_2$Ph), 4.47 (d, 1H, *J* 12.0 Hz, C*H*$_2$Ph), 4.42 (d, 1H, *J* 11.6 Hz, C*H*$_2$Ph), 3.97–3.93 (m, 1H, H-5), 3.74–3.65 (m, 4H, H-2, H-4, H-6a, H-6b), 3.52 (s, 3H, OC*H*$_3$), 3.08 (dd, 1H, *J* 6.4 Hz, *J* 12.8 Hz, C*H*$_2$NH$_2$), 2.68 (dd, 1H, *J* 12.8 Hz, *J* 6.4 Hz, C*H*$_2$NH$_2$), 2.32–2.27 (m, 1H, H-3), 1.99 (bs, 2H, CH$_2$N*H*$_2$). ^{13}C NMR (CDCl$_3$, 100 MHz): 138.6, 138.3, 138.1, 128.53, 128.49, 128.4, 128.2, 128.0, 127.9, 127.83, 127.81, 127.77, 101.3, 76.2, 75.2, 73.6, 73.4, 72.9, 71.7, 69.7, 56.4, 42.7, 39.7. HRMS calcd for C$_{29}$H$_{35}$NO$_5$+H$^+$ (M+H)$^+$: 478.2593, found: 478.2603.

3.10. Methyl 2,4,6-Tri-O-Benzyl-3-Deoxy-3-C-(3-Fluorophenylureidomethyl)-β-ᴅ-Gulopyranoside 20

A solution of amine **19** (61 mg, 0.13 mmol) in dry dichloromethane (2 mL), Et$_3$N (35.6 μL, 0.26 mmol) was added and the mixture was stirred at room temperature for 5 min under N$_2$ atmosphere. Then into the solution phenyl isocyanate (29.2 μL, 0.26 mmol) was added and the solution was stirred at rt for 12 h (TLC heptane/EtOAc, 1:1, R$_f$ 0.32). The solvent was removed under reduced pressure, and the residue was dissolved in EtOAc (10 mL) and the solution was washed with brine (10 mL), dried over Na$_2$SO$_4$ and concentrated in vacuo. The product was purified by flash column chromatography (heptane/EtOAc, 5:1 to 2:1) to give the corresponding semicarbazide **20** as a colorless oil (53.4 mg, 0.09 mmol, yield 68%). [α]$_D^{25}$ −83.1 (c 0.8, CHCl$_3$). ^{1}H NMR (CDCl$_3$, 400 MHz): 7.39–7.12 (m, 17H, Ar*H*), 6.85 (dd, 1H, *J* 1.2 Hz, *J* 8.0 Hz, Ar*H*), 6.71–6.66 (m, 1H, Ar*H*), 6.11 (bs, 1H, NHCON*H*C$_6$H$_4$F), 5.30 (bs, 1H, N*H*CONHC$_6$H$_4$F), 4.79 (d, 1H, *J* 11.6 Hz, C*H*$_2$Ph), 4.59 (d, 1H, *J* 6.0 Hz, H-1), 4.55 (d, 1H, *J* 12.0 Hz, C*H*$_2$Ph), 4.49 (d, 1H, *J* 12.4 Hz, C*H*$_2$Ph), 4.48–4.42 (m, 3H, C*H*$_2$Ph), 4.03–3.99 (m, 1H, H-5), 3.74–3.65 (m, 4H, H-2, H-4, H-6a, H-6b), 3.51 (s, 1H, OC*H*$_3$), 3.42 (dd, 1H, *J* 14.0 Hz, *J* 5.6 Hz, C*H*$_2$NHCONHC$_6$H$_4$F), 3.34 (dd, 1H, *J* 14.0 Hz, *J* 7.6 Hz, C*H*$_2$NHCONHC$_6$H$_4$F), 2.42–2.36 (m, 1H, H-3). ^{13}C NMR (CDCl$_3$, 100 MHz): 163.3 (d, *J* 243 Hz), 154.9, 140.6 (d, *J* 11 Hz), 138.4, 138.2, 137.9, 130.2 (d, *J* 9.5 Hz), 128.8, 128.5, 128.3, 128.2, 128.0, 127.9, 127.8, 114.8 (d, *J* 2.7 Hz), 109.7 (d, *J* 21.2 Hz), 106.9 (d, *J* 26 Hz), 100.7, 75.3, 73.54, 73.49, 73.0, 72.1, 69.4, 56.5, 39.8, 39.1. ^{19}F NMR (CDCl$_3$, 376 MHz): −111.6. HRMS calcd for C$_{36}$H$_{40}$FN$_2$O$_6$+H$^+$ (M+H)$^+$: 615.2886, found: 615.2870.

3.11. General Procedure for the Synthesis of Amides 21, 22a–22l, and 23

To a solution of the amine (1 eq) in dry DCM (2 mL per 0.1 mmol) Et$_3$N (2 eq) was added. Into the solution, acid chloride or anhydride (1.5 eq) was added and the solution was stirred at rt for 8 h. After that, 1(N) HCl solution was added to the reaction mixture and extracted with DCM and washed successively with saturated NaHCO$_3$. After evaporating the solvents in vacuo, the crude material thus obtained was purified by flash chromatography using heptane–EtOAc (5:1 to 1:1) to give pure amides as colorless oils.

3.11.1. Methyl 2,4,6-Tri-O-Benzyl-3-Deoxy-3-C-Phenylsulfonamidomethyl-β-ᴅ-Gulopyranoside 21

Compound **21** (TLC heptane/EtOAc, 2:1, R$_f$ 0.21) was prepared according to the general procedure 3.11 from the amine **19** (55 mg, 0.12 mmol). Obtained as a colorless oil in 65% yield (46.2 mg, 0.07 mmol).

$[\alpha]_D^{25}$ −55.7 (c 0.7, CHCl$_3$). ^{1}H NMR (CDCl$_3$, 400 MHz): 7.73–7.17 (m, 20H, Ar*H*), 5.23 (dd, 1H, *J* 5.2 Hz, *J* 6.8 Hz, CH$_2$N*H*SO), 4.75 (d, 1H, *J* 11.6 Hz, C*H*$_2$Ph), 4.53 (d, 1H, *J* 12.0 Hz, C*H*$_2$Ph), 4.50 (d, 1H, *J* 12.0 Hz, C*H*$_2$Ph), 4.45 (d, 1H, *J* 12.0 Hz, C*H*$_2$Ph), 4.42 (d, 1H, *J* 6.0 Hz, H-1), 4.39 (d, 1H, *J* 11.2 Hz, C*H*$_2$Ph), 4.36 (d, 1H, *J* 11.2 Hz, C*H*$_2$Ph), 3.90–3.87 (m, 1H, H-5), 3.70–3.60 (m, 4H, H-2, H-4, H-6a, H-6b), 3.47 (s, 3H, OC*H*$_3$), 3.17–3.10 (m, 1H, C*H*$_2$NHSO), 3.00–2.93 (m, 1H, C*H*$_2$NHSO), 2.41–2.35 (m, 1H, H-3). ^{13}C NMR (CDCl$_3$, 100 MHz): 139.7, 138.1, 137.9, 137.6, 132.6, 129.1, 128.7, 128.5, 128.2, 128.1, 127.94, 127.91, 127.8, 127.1, 100.5, 76.3, 74.8, 73.54, 73.50, 72.8, 71.8, 69.3, 56.3, 42.2, 39.1. HRMS calcd for C$_{35}$H$_{39}$NO$_7$S+NH$_4{}^+$ (M+NH$_4$)$^+$: 635.2788, found: 635.2791.

3.11.2. Methyl 2,4,6-Tri-*O*-Benzyl-3-Deoxy-3-*C*-(Benzamidomethyl)-β-D-Gulopyranoside **22a**

Compound **22a** (TLC heptane/EtOAc, 2:1, R$_f$ 0.27) was prepared according to the general procedure *3.11* from the amine **19** (43 mg, 0.09 mmol). Obtained as a colorless oil in 70% yield (37 mg, 0.06 mmol). $[\alpha]_D^{25}$ −42.4 (c 0.8, CHCl$_3$). ^{1}H NMR (CDCl$_3$, 400 MHz): 7.43–7.22 (m, 20H, Ar*H*), 7.03 (t, 1H, *J* 5.6 Hz, CH$_2$N*H*CO), 4.89 (d, 1H, *J* 11.2 Hz, C*H*$_2$Ph), 4.69 (d, 1H, *J*$_{1,2}$ 6.0 Hz, H-1), 4.57 (d, 1H, *J* 12.0 Hz, C*H*$_2$Ph), 4.55–4.46 (m, 4H, C*H*$_2$Ph), 4.07–4.03 (m, 1H, H-5), 3.82 (dd, 1H, *J*$_{1,2}$ 6.0 Hz, *J*$_{2,3}$ 4.8 Hz, H-2), 3.79–3.68 (m, 4H, H-4, H-6a, H-6b, C*H*$_2$NHCO), 3.60–3.53 (m, 4H, OC*H*$_3$, C*H*$_2$NHCO), 2.51–2.46 (m, 1H, H-3). ^{13}C NMR (CDCl$_3$, 100 MHz): 166.8, 138.3, 138.2, 137.8, 134.2, 131.2, 128.7, 128.53, 128.51, 128.47, 128.3, 128.2, 128.0, 127.8, 126.9, 100.9, 77.9, 75.9, 74.3, 73.5, 73.1, 72.2, 69.4, 56.5, 39.8, 39.3. HRMS calcd for C$_{36}$H$_{39}$NO$_6$+H$^+$ (M+H)$^+$: 582.2856, found: 582.2851.

3.11.3. Methyl 2,4,6-Tri-*O*-Benzyl-3-Deoxy-3-*C*-(Acetamidomethyl)-β-D-Gulopyranoside **22b**

Compound **22b** (TLC heptane/EtOAc, 1:1, R$_f$ 0.4) was prepared according to the general procedure *3.11* from the amine **19** (49 mg, 0.10 mmol). Obtained as a colorless oil in 62% yield (33 mg, 0.06 mmol). $[\alpha]_D^{25}$ −31.6 (c 0.8, CHCl$_3$). ^{1}H NMR (CDCl$_3$, 400 MHz): 7.39–7.21 (m, 15H, Ar*H*), 6.05 (bs, 1H, N*H*COCH$_3$), 4.83 (d, 1H, *J* 11.6 Hz, C*H*$_2$Ph), 4.59 (d, 1H, *J*$_{1,2}$ 6.4 Hz, H-1), 4.55 (d, 1H, *J* 12.0 Hz, C*H*$_2$Ph), 4.52–4.42 (m, 4H, C*H*$_2$Ph), 3.98 (td, 1H, *J* 6.0 Hz, *J* 2.8 Hz, H-5), 3.74–3.64 (m, 4H, H-2, H-4, H-6a, H-6b), 3.53 (s, 3H, OC*H*$_3$), 3.51–3.45 (m, 1H, C*H*$_2$NHCO), 3.32–3.26 (s, 1H, C*H*$_2$NHCO), 2.36–2.30 (m, 1H, H-3), 1.74 (s, 3H, NHCOC*H*$_3$). ^{13}C NMR (CDCl$_3$, 100 MHz): 169.9, 138.4, 138.3, 137.9, 128.7, 128.5, 128.2, 128.12, 128.09, 127.94, 127.86, 127.7, 100.8, 77.1, 75.4, 73.8, 73.5, 73.0, 72.0, 69.4, 56.5, 39.7, 38.4, 23.2. HRMS calcd for C$_{31}$H$_{37}$NO$_6$+H$^+$ (M+H)$^+$: 520.2699, found: 520.2704.

3.11.4. Methyl 2,4,6-Tri-*O*-Benzyl-3-Deoxy-3-*C*-(2-Fluorobenzamidomethyl)-β-D-Gulopyranoside **22c**

Compound **22c** (TLC heptane/EtOAc, 2:1, R$_f$ 0.19) was prepared according to the general procedure *3.11* from the amine **19** (49 mg, 0.10 mmol). Obtained as a colorless oil in 59% yield (31.5 mg, 0.06 mmol). $[\alpha]_D^{25}$ −41.7 (c 0.7, CHCl$_3$). ^{1}H NMR (CDCl$_3$, 400 MHz): 8.05–8.01 (td, 1H, *J* 8.0 Hz, *J* 1.6 Hz, Ar*H*), 7.48–7.43 (m, 1H, Ar*H*), 7.36–7.16 (m, 17H, Ar*H*), 7.07–7.02 (m, 1H, Ar*H*), 4.84 (d, 1H, *J* 11.6 Hz, C*H*$_2$Ph), 4.65 (d, 1H, *J*$_{1,2}$ 6.8 Hz, H-1), 4.62 (d, 1H, *J* 12.0 Hz, C*H*$_2$Ph), 4.57 (d, 1H, *J* 12.0 Hz, C*H*$_2$Ph), 4.48 (d, 1H, *J* 12.0 Hz, C*H*$_2$Ph), 4.46 (d, 1H, *J* 11.6 Hz, C*H*$_2$Ph), 4.40 (d, 1H, *J* 11.6 Hz, C*H*$_2$Ph), 4.07 (td, 1H, *J* 6.0 Hz, *J* 2.8 Hz, H-5), 3.79–3.58 (m, 6H, H-2, H-4, H-6a, H-6b, C*H*$_2$NHCO), 3.54 (s, 3H, OC*H*$_3$), 2.49–2.43 (m, 1H, H-3). ^{13}C NMR (CDCl$_3$, 100 MHz): 163.2 (d, *J* 3.0 Hz), 160.6 (d, *J* 247 Hz), 138.3, 138.2, 137.8, 133.1 (d, *J* 9.0 Hz), 132.0 (d, *J* 2.0 Hz), 128.46, 128.45, 128.4, 128.3, 128.2, 127.9, 127.8, 127.7, 124.7 (d, *J* 3.1 Hz), 121.4 (d, *J* 12 Hz), 116.1 (d, *J* 24 Hz), 101.0, 76.3, 75.1, 73.6, 73.5, 72.9, 72.0, 69.5, 56.5, 40.0, 38.5. ^{19}F NMR (CDCl$_3$, 376 MHz): -113.4. HRMS calcd for C$_{36}$H$_{38}$FNO$_6$+NH$_4{}^+$ (M+NH$_4$)$^+$: 617.3027, found: 617.3025.

3.11.5. Methyl 2,4,6-Tri-*O*-Benzyl-3-Deoxy-3-*C*-(3-Fluorobenzamidomethyl)-β-D-Gulopyranoside **22d**

Compound **22d** (TLC heptane/EtOAc, 2:1, R$_f$ 0.24) was prepared according to the general procedure *3.11* from the amine **19** (46 mg, 0.10 mmol). Obtained as a colorless oil in 67% yield (48 mg, 0.06 mmol). $[\alpha]_D^{25}$ −61.8 (c 0.7, CHCl$_3$). ^{1}H NMR (CDCl$_3$, 400 MHz): 7.36–6.99 (m, 20H, N*H*CO, Ar*H*), 4.89 (d, 1H, *J* 11.2 Hz, C*H*$_2$Ph), 4.69 (d, 1H, *J*$_{1,2}$ 6.0 Hz, H-1), 4.58 (d, 1H, *J* 12.0 Hz, C*H*$_2$Ph), 4.54–4.47 (m, 4H, C*H*$_2$Ph),

4.07 (td, 1H, *J* 6.4 Hz, *J* 2.8 Hz, H-5), 3.82 (dd, 1H, $J_{1,2}$ 6.0 Hz, $J_{2,3}$ 4.8 Hz, H-2), 3.81–3.66 (m, 4H, H-4, H-6a, H-6b, C*H*$_2$NHCO), 3.60–3.54 (m, 4H, C*H*$_2$NHCO, OC*H*$_3$), 2.50–2.44 (m, 1H, H-3). ^{13}C NMR (CDCl$_3$, 100 MHz): 165.5 (d, *J* 2.2 Hz), 162.7 (d, *J* 246 Hz), 138.2, 138.0, 137.7, 136.6 (d, *J* 6.8 Hz), 130.0 (d, *J* 7.8 Hz), 128.7, 128.53, 128.50, 128.4, 128.29, 128.25, 128.0, 127.9, 127.8, 122.1 (d, *J* 3.0 Hz), 118.2 (d, *J* 22 Hz), 114.4 (d, *J* 23 Hz), 100.8, 78.8, 75.8, 74.3, 73.5, 73.0, 72.2, 69.3, 56.5, 39.6, 39.5. ^{19}F NMR (CDCl$_3$, 376 MHz): −111.9. HRMS calcd for C$_{36}$H$_{38}$FNO$_6$+H$^+$ (M+H)$^+$: 600.2761, found: 600.2772.

3.11.6. Methyl 2,4,6-Tri-*O*-Benzyl-3-Deoxy-3-*C*-(4-Fluorobenzamidomethyl)-β-D-Gulopyranoside 22e

Compound **22e** (TLC heptane/EtOAc, 2:1, R$_f$ 0.2) was prepared according to the general procedure *3.11* from the amine **19** (51 mg, 0.11 mmol). Obtained as a colorless oil in 71% yield (45.4 mg, 0.08 mmol). $[\alpha]_D^{25}$ +51.9 (c 0.6, CHCl$_3$). ^{1}H NMR (CDCl$_3$, 400 MHz): 7.36–7.23 (m, 17H, Ar*H*), 7.05 (t, 1H, *J* 5.6 Hz, N*H*CO), 6.89–6.84 (m, 2H, Ar*H*), 4.90 (d, 1H, *J* 11.2 Hz, C*H*$_2$Ph), 4.70 (d, 1H, $J_{1,2}$ 6.0 Hz, H-1), 4.58 (d, 1H, *J* 12.0 Hz, C*H*$_2$Ph), 4.54–4.48 (m, 4H, C*H*$_2$Ph), 4.05 (td, 1H, *J* 6.0 Hz, *J* 2.8 Hz, H-5), 3.83 (dd, 1H, $J_{1,2}$ 6.0 Hz, $J_{2,3}$ 4.8 Hz, H-2), 3.80–3.67 (m, 4H, H-4, H-6a, H-6b, C*H*$_2$NHCO), 3.59–3.52 (m, 4H, C*H*$_2$NHCO, OC*H*$_3$), 2.52–2.46 (m, 1H, H-3). ^{13}C NMR (CDCl$_3$, 100 MHz): 165.7, 164.5 (d, *J* 250 Hz), 138.2, 138.1, 130.3 (d, *J* 3.0 Hz), 129.1 (d, *J* 8.9 Hz), 128.8, 128.6, 128.53, 128.52, 128.33, 128.28, 128.0, 127.9, 127.8, 115.4 (d, *J* 22 Hz), 100.8, 78.1, 76.0, 74.4, 73.5, 73.1, 72.2, 69.3, 56.5, 39.61, 39.58. ^{19}F NMR (CDCl$_3$, 376 MHz): −108.8. HRMS calcd for C$_{36}$H$_{38}$FNO$_6$+NH$_4$$^+$ (M+NH$_4$)$^+$: 617.3027, found: 617.3038.

3.11.7. Methyl 2,4,6-Tri-*O*-Benzyl-3-Deoxy-3-*C*-(3,4,5-Trifluorobenzamidomethyl)-β-D-Gulopyranoside 22f

Compound **22f** (TLC heptane/EtOAc, 2:1, R$_f$ 0.18) was prepared according to the general procedure *3.11* from the amine **19** (49 mg, 0.10 mmol). Obtained as a colorless oil in 53% yield (34.6 mg, 0.06 mmol). $[\alpha]_D^{25}$ −73.6 (c 0.8, CHCl$_3$). ^{1}H NMR (CDCl$_3$, 400 MHz): 7.36–6.93 (m, 18H, Ar*H*), 4.87 (d, 1H, *J* 11.2 Hz, C*H*$_2$Ph), 4.68 (d, 1H, $J_{1,2}$ 5.6 Hz, H-1), 4.58 (d, 1H, *J* 11.6 Hz, C*H*$_2$Ph), 4.51–4.48 (m, 4H, C*H*$_2$Ph), 4.07 (td, 1H, *J* 6.4 Hz, *J* 3.6 Hz, H-5), 3.82–3.53 (m, 9H, H-2, H-4, H-6a, H-6b, OC*H*$_3$, C*H*$_2$NHCO), 2.46–2.40 (m, 1H, H-3). ^{13}C NMR (CDCl$_3$, 100 MHz): 163.7, 151.0 (ddd, *J* 3.4 Hz, *J* 10.2 Hz, *J* 251 Hz), 141.9 (dt, *J* 15.2 Hz, *J* 255 Hz), 138.2, 137.9, 137.7, 130.4–130.2 (m), 128.8, 128.6, 128.53, 128.47, 128.29, 128.27, 128.2, 127.90, 127.85, 111.5 (dd, *J* 6.1 Hz, *J* 16 Hz), 100.5, 78.1, 75.8, 74.3, 73.6, 73.0, 72.3, 69.3, 56.5, 40.0, 39.3. ^{19}F NMR (CDCl$_3$, 376 MHz): -132.1 (d, *J* 20 Hz), −155.7 (t, *J* 20 Hz). HRMS calcd for C$_{36}$H$_{36}$F$_3$NO$_6$+NH$_4$$^+$ (M+NH$_4$)$^+$: 653.2838, found: 653.2845.

3.11.8. Methyl 2,4,6-Tri-*O*-Benzyl-3-Deoxy-3-*C*-(2,3,4,5,6-Pentafluorobenzamidomethyl)-β-D-Gulopyranoside 22g

Compound **22g** (TLC heptane/EtOAc, 2:1, R$_f$ 0.17) was prepared according to the general procedure *3.11* from the amine **19** (45 mg, 0.09 mmol). Obtained as a colorless oil in 49% yield (31 mg, 0.05 mmol). $[\alpha]_D^{25}$ −75.7 (c 0.5, CHCl$_3$). ^{1}H NMR (CDCl$_3$, 400 MHz): 7.36–7.20 (m, 15H, Ar*H*), 6.81 (t, 1H, *J* 5.6 Hz, C*H*$_2$NHCO), 4.89 (d, 1H, *J* 11.2 Hz, C*H*$_2$Ph), 4.65 (d, 1H, $J_{1,2}$ 6.0 Hz, H-1), 4.57 (d, 1H, *J* 12.0 Hz, C*H*$_2$Ph), 4.51 (d, 1H, *J* 11.2 Hz, C*H*$_2$Ph), 4.48 (d, 1H, *J* 12.0 Hz, C*H*$_2$Ph), 4.47 (d, 1H, *J* 12.0 Hz, C*H*$_2$Ph), 4.42 (d, 1H, *J* 12.0 Hz, C*H*$_2$Ph), 4.06 (td, 1H, *J* 6.8 Hz, *J* 3.6 Hz, H-5), 3.81 (dd, 1H, $J_{1,2}$ 6.0 Hz, $J_{2,3}$ 3.6 Hz, H-2), 3.78–3.69 (m, 3H, H-4, H-6a, H-6b), 3.63–3.58 (m, 2H, C*H*$_2$NHCO), 3.54 (s, 1H, OC*H*$_3$), 2.45–3.39 (m, 1H, H-3). ^{13}C NMR (CDCl$_3$, 100 MHz): 156.9, 145.1–144.9 (m), 142.6–142.4 (m), 140.9–140.6 (m), 138.8–138.5 (m), 138.2, 137.9, 137.7, 136.3–136.0 (m), 128.5, 128.2, 128.1, 128.0, 127.9, 127.8, 111.9–111.5 (m), 106.4, 100.3, 77.8, 75.4, 73.9, 73.5, 72.4, 69.3, 56.5, 39.8, 38.9. ^{19}F NMR (CDCl$_3$, 376 MHz): −140.5 to −140.6 (m, 2F), −151.7 (t, 1F, *J* 21 Hz), −160.1 to −160.3 (m, 2F). HRMS calcd for C$_{36}$H$_{34}$F$_5$NO$_6$+NH$_4$$^+$ (M+NH$_4$)$^+$: 689.2650, found: 689.2656.

3.11.9. Methyl 2,4,6-Tri-*O*-Benzyl-3-Deoxy-3-*C*-(3-Methoxybenzamidomethyl)-β-D-Gulopyranoside 22h

Compound **22h** (TLC heptane/EtOAc, 1:1, R$_f$ 0.45) was prepared according to the general procedure *3.11* from the amine **19** (47 mg, 0.10 mmol). Obtained as a colorless oil in 51% yield (30.7 mg, 0.05 mmol).

$[\alpha]_D^{25}$ −43.2 (c 0.5, CHCl$_3$). ^{1}H NMR (CDCl$_3$, 400 MHz): 7.35–7.21 (m, 17H, Ar*H*), 7.08 (t, 1H, *J* 8.0 Hz, Ar*H*), 7.00 (m, 2H, CH$_2$N*H*CO, Ar*H*), 6.75–6.72 (m, 1H, Ar*H*), 4.87 (d, 1H, *J* 11.6 Hz, C*H*$_2$Ph), 4.68 (d, 1H, *J*$_{1,2}$ 6.4 Hz, H-1), 4.57 (d, 1H, *J* 11.6 Hz, C*H*$_2$Ph), 4.54 (d, 1H, *J* 11.6 Hz, C*H*$_2$Ph), 4.51 (d, 1H, *J* 12.8 Hz, C*H*$_2$Ph), 4.48 (d, 1H, *J* 11.2 Hz, C*H*$_2$Ph), 4.45 (d, 1H, *J* 11.2 Hz, C*H*$_2$Ph), 4.05 (td, 1H, *J* 6.4 Hz, *J* 2.8 Hz, H-5), 3.83 (dd, 1H, *J*$_{1,2}$ 6.4 Hz, *J*$_{2,3}$ 5.2 Hz, H-2), 3.79 (s, 3H, C$_6$H$_4$OC*H*$_3$), 3.78–3.67 (m, 5H, H-2, H-4, H-6a, H-6b, C*H*$_2$NHCO), 3.59–3.53 (m, 4H, CH$_2$N*H*CO, OC*H*$_3$), 2.50–2.45 (m, 1H, H-3). ^{13}C NMR (CDCl$_3$, 100 MHz): 166.8, 159.9, 138.3, 138.2, 137.8, 129.5, 128.7, 128.5, 128.4, 128.3, 128.1, 128.0, 127.9, 127.8, 118.4, 117.7, 112.3, 100.9, 77.6, 75.8, 74.2, 73.5, 73.1, 72.1, 69.4, 56.5, 55.5, 39.8, 39.2. HRMS calcd for C$_{37}$H$_{41}$NO$_7$+H$^+$ (M+H)$^+$: 612.2961, found: 612.2972.

3.11.10. Methyl 2,4,6-Tri-*O*-Benzyl-3-Deoxy-3-*C*-(p-Toluamidomethyl)-β-D-Gulopyranoside **22i**

Compound **22i** (TLC heptane/EtOAc, 2:1, R$_f$ 0.24) was prepared according to the general procedure *3.11* from the amine **19** (51 mg, 0.11 mmol). Obtained as a colorless oil in 61% yield (38.8 mg, 0.07 mmol). $[\alpha]_D^{25}$ −56.2 (c 0.5, CHCl$_3$). ^{1}H NMR (CDCl$_3$, 400 MHz): 7.35–7.21 (m, 17H, Ar*H*), 7.04 (d, 1H, *J* 7.6 Hz, Ar*H*), 6.96 (t, 1H, *J* 6.0 Hz, CH$_2$N*H*CO), 4.88 (d, 1H, *J* 11.2 Hz, C*H*$_2$Ph), 4.68 (d, 1H, *J*$_{1,2}$ 6.4 Hz, H-1), 4.57 (d, 1H, *J* 12.0 Hz, C*H*$_2$Ph), 4.53 (d, 1H, *J* 11.2 Hz, C*H*$_2$Ph), 4.51 (d, 1H, *J* 12.0 Hz, C*H*$_2$Ph), 4.48 (d, 1H, *J* 12.0 Hz, C*H*$_2$Ph), 4.45 (d, 1H, *J* 11.6 Hz, C*H*$_2$Ph), 4.04 (td, 1H, *J* 6.4 Hz, *J* 2.8 Hz, H-5), 3.82 (dd, 1H, *J*$_{1,2}$ 6.4 Hz, *J*$_{2,3}$ 5.2 Hz, H-2), 3.78–3.67 (m, 3H, H-4, H-6a, H-6b, C*H*$_2$NHCO), 3.58–3.52 (m, 4H, CH$_2$N*H*CO, OC*H*$_3$), 2.51–2.45 (m, 1H, H-3), 2.36 (s, 3H, C*H*$_3$). ^{13}C NMR (CDCl$_3$, 100 MHz): 166.8, 141.6, 138.3, 138.2, 137.9, 131.4, 130.3, 129.24, 129.16, 128.7, 128.52, 128.51, 128.4, 128.3, 128.1, 128.0, 127.9, 127.8, 126.9, 101.0, 77.8, 75.9, 74.2, 73.5, 73.1, 72.1, 69.4, 56.5, 39.9, 39.2, 21.5. HRMS calcd for C$_{37}$H$_{41}$NO$_6$+H$^+$ (M+H)$^+$: 596.3012, found: 596.3019.

3.11.11. Methyl 2,4,6-Tri-*O*-Benzyl-3-Deoxy-3-*C*-(3,5-Dimethoxybenzamidomethyl)-β-D-Gulopyranoside **22j**

Compound **22j** (TLC heptane/EtOAc, 2:1, R$_f$ 0.22) was prepared according to the general procedure *3.11* from the amine **19** (53 mg, 0.11 mmol). Obtained as a colorless oil in 67% yield (48 mg, 0.07 mmol). $[\alpha]_D^{25}$ −39.5 (c 0.8, CHCl$_3$). ^{1}H NMR (CDCl$_3$, 400 MHz): 7.35–7.18 (m, 15H, Ar*H*), 6.82 (t, 1H, *J* 6.0 Hz, N*H*CO), 6.72 (d, 2H, *J* 2.4 Hz, Ar*H*), 6.54 (t, 1H, *J* 2.4 Hz, Ar*H*), 4.84 (d, 1H, *J* 11.6 Hz, C*H*$_2$Ph), 4.65 (d, 1H, *J*$_{1,2}$ 6.4 Hz, H-1), 4.57 (d, 1H, *J* 12.0 Hz, C*H*$_2$Ph), 4.56 (d, 1H, *J* 12.0 Hz, C*H*$_2$Ph), 4.48 (d, 1H, *J* 11.6 Hz, C*H*$_2$Ph), 4.47 (d, 1H, *J* 11.6 Hz, C*H*$_2$Ph), 4.43 (d, 1H, *J* 11.6 Hz, C*H*$_2$Ph), 4.03 (td, 1H, *J* 6.4 Hz, *J* 2.8 Hz, H-5), 3.78 (dd, 1H, *J*$_{1,2}$ 6.0 Hz, *J*$_{2,3}$ 4.8 Hz, H-2), 3.75–3.65 (m, 10H, H-4, H-6a, H-6b, C*H*$_2$NHCO, 2 × OC*H*$_3$), 3.57–3.52 (m, 4H, CH$_2$N*H*CO, OC*H*$_3$), 2.47–2.41 (m, 1H, H-3). ^{13}C NMR (CDCl$_3$, 100 MHz): 167.1, 160.9, 163.3, 138.3, 138.2, 137.8, 136.8, 128.7, 128.5, 128.2, 128.1, 128.0, 127.93, 127.88, 127.8, 104.9, 103.6, 100.9, 77.2, 75.6, 73.9, 73.5, 73.1, 72.1, 69.5, 56.5, 55.6, 39.9, 39.0. HRMS calcd for C$_{38}$H$_{43}$NO$_8$+H$^+$ (M+H)$^+$: 642.3066, found: 642.3067.

3.11.12. Methyl 2,4,6-tri-*O*-Benzyl-3-Deoxy-3-*C*-(3-Trifluoromethylbenzamidomethyl)-β-D-Gulopyranoside **22k**

Compound **22k** (TLC heptane/EtOAc, 2:1, R$_f$ 0.25) was prepared according to the general procedure *3.11* from the amine **19** (43 mg, 0.09 mmol). Obtained as a colorless oil in 55% yield (32.2 mg, 0.05 mmol). $[\alpha]_D^{25}$ −38.7 (c 0.8, CHCl$_3$). ^{1}H NMR (CDCl$_3$, 400 MHz): 8.00 (s, 1H, Ar*H*), 7.69–7.66 (m, 1H, Ar*H*), 7.34–7.21 (m, 17H, Ar*H*), 7.09 (t, 1H, *J* 6.0 Hz, CH$_2$N*H*CO), 4.88 (d, 1H, *J* 11.6 Hz, C*H*$_2$Ph), 4.69 (d, 1H, *J*$_{1,2}$ 6.0 Hz, H-1), 4.58 (d, 1H, *J* 12.0 Hz, C*H*$_2$Ph), 4.53 (d, 1H, *J* 11.2 Hz, C*H*$_2$Ph), 4.51–4.47 (m, 3H, C*H*$_2$Ph), 4.07 (s td, 1H, *J* 6.4 Hz, *J* 3.2 Hz, H-5), 3.83 (dd, 1H, *J* 6.0 Hz, *J* 4.8 Hz, H-2), 3.81–3.68 (m, 4H, H-2, H-4, H-6a, H-6b, C*H*$_2$NHCO), 3.62–3.59 (m, 1H, C*H*$_2$NHCO), 3.56 (s, 3H, OC*H*$_3$), 2.50–2.45 (m, 1H, H-3). ^{13}C NMR (CDCl$_3$, 100 MHz): 165.5, 138.2, 138.1, 137.8, 135.1, 133.4 131.2 (q, *J* 32 Hz), 129.5, 129.2, 129.1, 128.7, 128.54, 128.52, 128.4, 128.31, 128.25, 128.0, 127.9, 127.8, 125.2, 124.6 (q, *J* 3.7 Hz), 123.7 (q, *J* 271 Hz), 100.8, 77.8, 75.8, 74.3, 73.6, 73.1, 72.2, 69.3, 56.5, 39.62, 39.61. ^{19}F NMR (CDCl$_3$, 376 MHz): −62.7. HRMS calcd for C$_{37}$H$_{38}$F$_3$NO$_6$+H$^+$ (M+H)$^+$: 650.2729, found: 650.2727.

3.11.13. Methyl 2,4,6-Tri-*O*-Benzyl-3-Deoxy-3-*C*-(4-Phenylbenzamidomethyl)-β-D-Gulopyranoside **221**

Compound **221** (TLC heptane/EtOAc, 2:1, R_f 0.32) was prepared according to the general procedure *3.11* from the amine **19** (60 mg, 0.13 mmol). Obtained as a colorless oil in 55% yield (45.5 mg, 0.07 mmol). $[\alpha]_D^{25}$ −47.8 (c 0.9, CHCl$_3$). ^{1}H NMR (CDCl$_3$, 400 MHz): 7.61–7.24 (m, 24H, Ar*H*), 7.10 (t, 1H, *J* 6.0 Hz, C*H*$_2$N*H*CO), 4.92 (d, 1H, *J* 11.2 Hz, C*H*$_2$Ph), 4.72 (d, 1H, *J*$_{1,2}$ 6.4 Hz, H-1), 4.59 (d, 1H, *J* 12.0 Hz, C*H*$_2$Ph), 4.57–4.48 (m, 4H, C*H*$_2$Ph), 4.08 (td, 1H, *J* 6.0 Hz, *J* 2.8 Hz, H-5), 3.85 (dd, 1H, *J*$_{1,2}$ 6.0 Hz, *J*$_{2,3}$ 4.8 Hz, H-2), 3.82–3.70 (m, 4H, H-4, H-6a, H-6b, C*H*$_2$NHCO), 3.64–3.59 (m, 1H, C*H*$_2$NHCO), 3.58 (s, 3H, OC*H*$_3$), 2.55–2.50 (m, 1H, H-3). ^{13}C NMR (CDCl$_3$, 100 MHz): 166.5, 143.9, 140.3, 138.3, 138.2, 137.9, 132.9, 129.0, 128.7, 128.53, 128,49, 128.3, 128.2, 128.02, 127.98, 127.9, 127.8, 127.4, 127.24, 127.15, 100.9, 77.9, 75.9, 74.3, 73.5, 73.1, 72.2, 69.4, 56.5, 39.8, 39.4. HRMS calcd for C$_{42}$H$_{43}$NO$_6$+NH$_4^+$ (M+NH$_4$)$^+$: 675.3434, found: 675.3433.

3.11.14. Methyl 2,4,6-tri-*O*-Benzyl-3-Deoxy-3-*C*-(Diphenylphosphonamidomethyl)-β-D-Gulopyranoside **23**

Compound **23** (TLC heptane/EtOAc, 2:1, R_f 0.26) was prepared according to the general procedure *3.11* from the amine **19** (52 mg, 0.11 mmol). Obtained as a colorless oil in 69% yield (53.3 mg, 0.08 mmol). $[\alpha]_D^{25}$ −77.8 (c 0.7, CHCl$_3$). ^{1}H NMR (CDCl$_3$, 400 MHz): 7.36–7.13 (m, 20H, Ar*H*), 4.76 (d, 1H, *J* 12.0 Hz, C*H*$_2$Ph), 4.55–49 (m, 3H, H-1, C*H*$_2$Ph), 4.45 (d, 1H, *J* 12.0 Hz, C*H*$_2$Ph), 4.34 (d, 1H, *J* 11.6 Hz, C*H*$_2$Ph), 4.28 (d, 1H, *J* 11.6 Hz, C*H*$_2$Ph), 3.95–92 (m, 1H, H-5), 3.71–3.58 (m, 5H, H-2, H-4, H-6a, H-6b, N*H*PO(OPh)$_2$), 3.47 (s, 3H, OC*H*$_3$), 3.40–3.31 (m, 1H, C*H*$_2$NHSO), 3.15–3.09 (m, 1H, C*H*$_2$NHSO), 2.33–2.27 (m, 1H, H-3). ^{13}C NMR (CDCl$_3$, 100 MHz): 150.9 (dd, *J* 6.7 Hz, *J* 2.4 Hz), 138.22, 138.21, 137.9, 128.6, 128.51, 128.45, 128.1, 128.0, 127.93, 127.87, 127.8, 125.0 (d, *J* 4.3 Hz), 120.3 (d, *J* 5.0 Hz, *J* 8.0 Hz), 100.8, 76.1, 74.9, 73.6, 73.4, 72.8, 71.8, 69.5, 56.5, 42.1 (d, *J* 1.8 Hz), 40.2. ^{31}P NMR (CDCl$_3$, 162 MHz): -1.01. HRMS calcd for C$_{41}$H$_{44}$PNO$_8$+H$^+$ (M+H)$^+$: 710.2883, found: 710.2889.

3.12. *General Procedure the Synthesis of* **1a**, **1b**, **2–6**, **7a–7l**, *and* **8**

A solution of crude in EtOAc/isopropanol (1:3) was stirred with Pd(OH)$_2$/C (10% wt., 1 mg per 4 mg of crude) under hydrogen atmosphere at room temperature for 12 h. All the hydrogenation reactions were carried out in an EtOAc-isopropanol mixture (1:3, 4 mL). After the completion of the reaction (as indicated by TLC), the reaction mixture was filtered through a Celite bed and washed with methanol. The filtrate was concentrated under reduced pressure and purified through the flash column (DCM:MeOH) to get the desired compounds as white amorphous solids or colorless oils.

3.12.1. Methyl 3-Deoxy-3-*C*-Hydroxymethyl-β-D-Gulopyranoside **1a**

Compound **1a** (TLC, DCM/MeOH, 5:1, R_f 0.41) was prepared according to the general procedure *3.12* from the alcohol **14a** (63 mg, 0.13 mmol). Obtained as a white amorphous solid in 51% yield (14 mg, 0.07 mmol) from flash column chromatography (DCM:MeOH 12:1–5:1). $[\alpha]_D^{25}$ −50.7 (c 0.6, CH$_3$OH). ^{1}H NMR (CD$_3$OD, 400 MHz): 4.39 (d, 1H, *J* 7.6 Hz, H-1), 3.96 (dd, 1H, *J*$_{3,4}$ 4.0 Hz, *J*$_{4,5}$ 2.0 Hz, H-4), 3.92 (dd, 1H, *J* 11.2 Hz, *J* 5.6 Hz, C*H*$_2$OH), 3.86 (dd, 1H, *J*$_{1,2}$ 7.6 Hz, *J*$_{2,3}$ 6.0 Hz H-2), 3.84–3.74 (m, 3H, H-5, H-6a, H-6b), 3.67 (dd, 1H, *J* 11.2 Hz, *J* 8.4 Hz, C*H*$_2$OH), 3.51 (s, 3H, OC*H*$_3$), 2.32–2.26 (m, 1H, H-3). ^{13}C NMR (CD$_3$OD, 125 MHz): 104.0, 76.1, 69.0, 68.3, 63.1, 59.6, 56.8, 49.0. HRMS calcd for C$_8$H$_{16}$O$_6$-H$^+$ (M-H)$^+$: 207.0869, found: 207.0865.

3.12.2. Methyl 3-Deoxy-3-*C*-Hydroxymethyl-β-D-Galactopyranoside **1b**

Compound **1b** (TLC, DCM/MeOH, 5:1, R_f 0.40) was prepared according to the general procedure *3.12* from the alcohol **14b** (46 mg, 0.10 mmol). Obtained as a white amorphous solid in 63% yield (12.6 mg, 0.06 mmol) from flash column chromatography (DCM:MeOH 12:1–6:1). $[\alpha]_D^{25}$ −32.1 (c 0.5, CH$_3$OH). ^{1}H NMR (CD$_3$OD, 500 MHz): 4.16 (d, 1H, *J* 7.6 Hz, H-1), 3.97 (d, 1H, *J*$_{3,4}$ 2.4 Hz, H-4), 3.90 (dd, 1H, *J* 10.4 Hz, *J* 4.4 Hz, C*H*$_2$OH), 3.78 (dd, 1H, *J* 10.8 Hz, *J* 8.4 Hz, H-2), 3.72 (dd, 1H, *J* 5.6 Hz,

H-6a, H-6b), 3.55–3.51 (m, 4H, H-5, OCH_3), 3.44 (dd, 1H, J 10.8 Hz, J 7.6 Hz, CH_2OH), 1.73–1.66 (m, 1H, H-3). ^{13}C NMR (CD$_3$OD, 125 MHz): 107.6, 79.8, 68.8, 67.1, 62.7, 61.2, 57.1, 49.0. HRMS calcd for C$_8$H$_{16}$O$_6$+Na$^+$ (M+Na)$^+$: 231.0845, found: 231.0840.

3.12.3. Methyl 3-Deoxy-3-C-(3-Trifluoromethylphenoxymethyl)-β-D-Galactopyranoside 2

Compound **2** (TLC, DCM/MeOH, 10:1, R$_f$ 0.5) was prepared according to the general procedure *3.12* from the ether **15** (53 mg, 0.09 mmol). Obtained as a colorless oil in 75% yield (22.5 mg, 0.06 mmol) from flash column chromatography (DCM:MeOH 20:1–12:1). $[\alpha]_D^{25}$ −12.8 (c 0.7, CH$_3$OH). ^{1}H NMR (CD$_3$OD, 400 MHz): 7.46 (t, 1H, J 8.0 Hz, ArH), 7.23–7.21 (m, 3H, ArH), 4.50 (d, 1H, J 7.2 Hz, H-1), 4.35 (dd, 1H, J 4.8 Hz, J 10.0 Hz, CH_2OH), 4.22 (t, 1H, J 8.8 Hz, CH_2OH), 4.08 (bs, 1H, H-4), 3.97–3.94 (m, 2H, H-2, H-5), 3.78 (d, 2H, J 6.0 Hz, H-6a, H-6b), 3.54 (s, 3H, OCH_3), 2.64–2.58 (m, 1H, H-3). ^{13}C NMR (CD$_3$OD, 100 MHz): 160.6, 132.8 (q*, J 32 Hz), 131.4, 125.5 (q*, J 270 Hz), 119.3, 118.3 (br q, J 3.7 Hz), 112.4 (br q, J 3.8 Hz), 104.1, 76.1, 68.2, 68.1, 65.9, 62.9, 56.8, 46.1. ^{19}F NMR (CD$_3$OD, 376 MHz): −64.2. HRMS calcd for C$_{15}$H$_{20}$F$_3$O$_6$+H$^+$ (M+H)$^+$: 353.1212, found: 353.1208.

*Only two peaks from the q are observed: See Supplementary information page S43)

3.12.4. Methyl 3-Deoxy-3-C-Methoxymethyl-β-D-Galactopyranoside 3

Compound **3** (TLC, DCM/MeOH, 10:1, R$_f$ 0.5) was prepared according to the general procedure *3.12* from **16** (36 mg, 0.07 mmol). Obtained as a colorless oil in 64% yield (10.4 mg, 0.05 mmol). $[\alpha]_D^{25}$ −33.4 (c 0.5, CH$_3$OH) from flash column chromatography (DCM:MeOH 15:1–9:1). ^{1}H NMR (CD$_3$OD, 400 MHz): 4.41 (d, 1H, J 7.6 Hz, H-1), 3.93 (dd, 1H, $J_{4,5}$ 3.2 Hz, $J_{3,4}$ 2.0 Hz, H-4), 3.88–3.82 (m, 2H, H-2, H-5), 3.73 (d, 2H, J 6.0 Hz, H-6a, H-6b), 3.65 (dd, 1H, J 10.0 Hz, J 4.8 Hz, CH_2OCH$_3$), 3.57 (dd, 1H, J 10.0 Hz, J 4.8 Hz, CH_2OCH$_3$), 3.50 (s, 3H, OCH_3), 3.33 (s, 3H, OCH_3), 2.38–2.34 (m, 1H, H-3). ^{13}C NMR (CD$_3$OD, 100 MHz): 104.1, 76.3, 70.2, 68.7, 68.5, 63.1, 59.1, 56.8, 46.6. HRMS calcd for C$_9$H$_{18}$O$_6$+Na$^+$ (M+Na)$^+$: 245.1001, found: 245.1004.

3.12.5. Methyl 3-Deoxy-3-C-[4-(3-Fluorophenyl)-1H-1,2,3-Triazol-1-Ylmethyl]-β-D-Galactopyranoside 4

Compound **4** (TLC, DCM/MeOH, 10:1, R$_f$ 0.43) was prepared according to the general procedure *3.12* from triazole **18** (55 mg, 0.09 mmol). Obtained as a colorless oil in 78% yield (24.3 mg, 0.07 mmol) from flash column chromatography (DCM:MeOH 20:1–9:1). $[\alpha]_D^{25}$ −20.5 (c 0.7, CH$_3$OH). ^{1}H NMR (CD$_3$OD, 400 MHz): 8.40 (s, 1H, ArH), 7.65–7.05 (m, 4H, ArH), 4.80 (dd, 1H, J 14.4 Hz, J 6.0 Hz, CH_2N$_3$C$_8$H$_5$F), 4.63 (dd, 1H, J 14.4 Hz, J 9.2 Hz, CH_2OH), 4.48 (d, 1H, J 6.4 Hz, H-1), 4.02–3.99 (m, 1H, H-5), 3.82–3.78 (m, 4H, H-2, H-4, H-6a, H-6b), 3.53 (s, 3H, OCH_3), 2.72–2.66 (m, 1H, H-3). ^{13}C NMR (CD$_3$OD, 100 MHz): 164.3 (d, J 243 Hz), 147.7 (d, J 3.1 Hz), 134.0 (d, J 8.3 Hz), 131.8 (d, J 8.4 Hz), 123.4, 122.4 (d, J 2.5 Hz), 115.9 (d, J 21 Hz), 113.2 (d, J 23 Hz), 103.6, 75.8, 68.0, 67.4, 62.8, 56.8, 48.2, 46.8. HRMS calcd for C$_{16}$H$_{20}$FN$_3$O$_5$+H$^+$ (M+H)$^+$: 354.1465, found: 354.1462.

3.12.6. Methyl 3-Deoxy-3-C-(3-Fluorophenylureido)Methyl-β-D-Galactopyranoside 5

Compound **5** (TLC, DCM/MeOH, 10:1, R$_f$ 0.44) was prepared according to the general procedure *3.12* from **20** (50 mg, 0.0814 mmol). Obtained as a colorless oil in 41% yield (11.5 mg, 0.03 mmol) from flash column chromatography (DCM:MeOH 12:1–5:1). $[\alpha]_D^{25}$ −17.3 (c 0.6, CH$_3$OH). ^{1}H NMR (CD$_3$OD, 400 MHz): 7.34 (dt, 1H, J 12.0 Hz, J 2.0 Hz, ArH), 7.24–6.64 (m, 3H, ArH), 4.44 (d, 1H, J 7.2 Hz, H-1), 3.90–3.83 (m, 3H, H-2, H-4, H-5), 4.22 (t, 1H, J 8.8 Hz, CH_2OH), 4.08 (bs, 1H, H-4), 3.97–3.94 (m, 2H, H-2, H-5), 3.78–3.76 (d, 2H, H-6a, H-6b), 3.52 (s, 3H, OCH_3), 3.47 (dd, 1H, J 14.4 Hz, J 6.4 Hz, CH_2NHCONH), 3.41 (dd, 1H, J 14.4 Hz, J 7.6 Hz, CH_2NHCONH), 2.27–2.21 (m, 1H, H-3). ^{13}C NMR (CD$_3$OD, 100 MHz): 164.5 (d, J 240 Hz), 158.0, 143.0 (d, J 11 Hz), 131.0 (d, J 9.8 Hz), 115.2 (d, J 3.2 Hz), 109.4 (d, J 22 Hz), 106.7 (d, J 22 Hz), 103.8, 76.1, 69.2, 69.0, 63.0, 56.9, 46.8, 38.0. HRMS calcd for C$_{15}$H$_{21}$FN$_2$O$_6$+H$^+$ (M+H)$^+$: 354.1462, found: 345.1459.

3.12.7. Methyl 3-Deoxy-3-*C*-(Phenylsufonamido)Methyl-β-D-Galactopyranoside **6**

Compound **6** (TLC, DCM/MeOH, 10:1, R_f 0.45) was prepared according to the general procedure *3.12* from amide **21** (39 mg, 0.06 mmol). Obtained as a colorless oil in 53% yield (11.6 mg, 0.03 mmol) from flash column chromatography (DCM:MeOH 20:1–10:1). $[\alpha]_D^{25}$ −21.4 (c 0.6, CH₃OH). ¹H NMR (CD₃OD, 400 MHz): 7.88–7.56 (m, 5H, Ar*H*), 4.26 (d, 1H, *J* 7.2 Hz, H-1), 3.93 (d, 1H, *J* 3.6 Hz, H-4), 3.80 (dd, 1H, *J*₁,₂ 7.2 Hz, *J*₂,₃ 6.0 Hz, H-2), 3.76–3.70 (m, 3H, H-5, H-6a, H-6b), 3.20 (dd, 1H, *J* 5.2 Hz, *J* 11.2 Hz, C*H*₂NH), 3.20 (dd, 1H, *J* 11.2 Hz, *J* 10.0 Hz, C*H*₂NH), 2.26–2.21 (m, 1H, H-3). ¹³C NMR (CD₃OD, 100 MHz): 141.7, 133.7, 130.3, 128.0, 103.5, 75.7, 68.4, 67.9, 63.2, 56.8, 46.3, 40.6. HRMS calcd for C₁₄H₂₁NO₇S+H⁺ (M+H)⁺: 348.1117, found: 348.1115.

3.12.8. Methyl 3-Deoxy-3-*C*-Benzamidomethyl-β-D-Gulopyranoside **7a**

Compound **7a** (TLC, DCM/MeOH, 10:1, R_f 0.41) was prepared according to the general procedure *3.12* from the amide **22a** (35 mg, 0.05 mmol). Obtained as a colorless oil in 59% yield (11 mg, 0.04 mmol) from flash column chromatography (DCM:MeOH 20:1–10:1). $[\alpha]_D^{25}$ −6.5 (c 0.6, CH₃OH). ¹H NMR (CD₃OD, 400 MHz): 7.82–7.80 (m, 2H, Ar*H*), 7.56–7.52 (m, 1H, Ar*H*), 7.49–7.44 (m, 2H, Ar*H*), 4.48 (d, 1H, *J*₁,₂ 6.8 Hz, H-1), 3.95 (td, 1H, *J* 6.0 Hz, *J* 2.4 Hz, H-5), 3.89–3.86 (m, 2H, H-2, H-4), 3.79 (d, 1H, *J*₆ₐ,₆ᵦ 12.4 Hz, *J*₅,₆ₐ 5.6 Hz, H-6a), 3.75 (dd, 1H, *J*₆ₐ,₆ᵦ 12.4 Hz, *J*₅,₆ᵦ 5.6 Hz, H-6b), 3.68 (dd, 1H, *J* 14.0 Hz, *J* 6.4 Hz, C*H*₂NH), 3.61 (dd, 1H, *J* 11.2 Hz, *J* 6.4 Hz, C*H*₂NH), 3.54 (s, 3H, OC*H*₃), 2.42–2.35 (m, 1H, H-3). ¹³C NMR (CD₃OD, 100 MHz): 170.5, 135.6, 132.7, 129.6, 128.2, 103.7, 76.0, 69.0, 68.7, 63.1, 56.8, 46.4, 37.8. HRMS calcd for C₁₄H₁₉NO₆+H⁺ (M+H)⁺: 289.1291, found: 298.1289.

3.12.9. Methyl 3-Deoxy-3-*C*-Acetamidomethyl-β-D-Gulopyranoside **7b**

Compound **7b** (TLC, DCM/MeOH, 10:1, R_f 0.42) was prepared according to the general procedure *3.12* from the amide **22b** (27 mg, 0.05 mmol). Obtained as a colorless oil in 75% yield (9.7 mg, 0.04 mmol) from flash column chromatography (DCM:MeOH 15:1–7:1). $[\alpha]_D^{25}$ −1.5 (c 0.6, CH₃OH). ¹H NMR (CD₃OD, 400 MHz): 4.40 (d, 1H, *J*₁,₂ 6.8 Hz, H-1), 3.88–3.84 (m, 1H, H-5), 3.82–371 (m, 4H, H-2, H-4, H-6a, H-6b), 3.51 (s, 3H, OC*H*₃), 3.44 (dd, 1H, *J* 14.0 Hz, *J* 6.0 Hz, C*H*₂NH), 3.38 (dd, 1H, *J* 14.0 Hz, *J* 8.8 Hz, C*H*₂NH), 2.24–2.18 (m, 1H, H-3), 1.95 (s, 3H, NHCOC*H*₃). ¹³C NMR (CD₃OD, 100 MHz): 173.6, 103.6, 75.9, 68.7, 68.5, 63.1, 56.8, 46.3, 37.1, 22.6. HRMS calcd for C₁₀H₁₉NO₆+H⁺ (M+H)⁺: 250.1291, found: 250.1291.

3.12.10. Methyl 3-Deoxy-3-*C*-(2-Fluorobenzamidomethyl)-β-D-Galactopyranoside **7c**

Compound **7c** (TLC, DCM/MeOH, 10:1, R_f 0.4) was prepared according to the general procedure 3.12 from the amide **22c** (33 mg, 0.06 mmol). Obtained as a colorless oil in 69% yield (12.5 mg, 0.04 mmol) from flash column chromatography (DCM:MeOH 20:1–9:1). $[\alpha]_D^{25}$ −9.3 (c 0.5, CH₃OH). ¹H NMR (CD₃OD, 400 MHz): 7.76 (td, 1H, *J* 7.6 Hz, *J* 2.0 Hz, Ar*H*), 7.56–7.50 (m, 1H, Ar*H*), 7.28 (td, 1H, *J* 7.6 Hz, *J* 0.8 Hz, Ar*H*), 7.20 (ddd, 1H, *J* 11.2 Hz, *J* 8.4 Hz, *J* 0.8 Hz, Ar*H*), 4.48 (d, 1H, *J*₁,₂ 7.2 Hz, H-1), 3.93–3.86 (m, 3H, H-2, H-4, H-5), 3.77 (d, 2H, *J*₅,₆ₐ, *J*₅,₆ᵦ 5.6 Hz, H-6a, H-6b), 3.66 (d, 2H, *J* 7.2 C*H*₂NH), 3.53 (s, 3H, OC*H*₃), 2.42–2.34 (m, 1H, H-3). ¹³C NMR (CD₃OD, 100 MHz): 166.7, 161.4 (d, *J* 248 Hz), 134.2 (d, *J* 8.8 Hz), 131.6 (d, *J* 2.3 Hz), 125.7 (d, *J* 3.4 Hz), 123.9 (d, *J* 14 Hz), 131.6 (d, *J* 23 Hz), 103.7, 76.0, 69.0, 68.8, 63.1, 56.9, 46.1, 38.1. ¹⁹F NMR (CD₃OD, 376 MHz): −116.0. HRMS calcd for C₁₅H₂₀FNO₆+H⁺ (M+H)⁺: 330.1353, found: 330.1352.

3.12.11. Methyl 3-Deoxy-3-*C*-(3-Fluorobenzamidomethyl)-β-D-Galactopyranoside **7d**

Compound **7d** (TLC, DCM/MeOH, 10:1, R_f 0.45) was prepared according to the general procedure *3.12* from the amide **22d** (35 mg, 0.06 mmol). Obtained as a colorless oil in 59% yield (11.3 mg, 0.03 mmol) from flash column chromatography (DCM:MeOH 20:1–10:1). $[\alpha]_D^{25}$ −14.6 (c 0.6, CH₃OH). ¹H NMR (CD₃OD, 400 MHz): 7.65–7.26 (m, 4H, Ar*H*), 4.47 (d, 1H, *J* 7.2 Hz, H-1), 3.94 (td, 1H, *J* 6.0 Hz, *J* 2.0 Hz, H-5), 3.88–3.85 (m, 2H, H-2, H-4), 3.77 (d, 2H, *J* 5.6 Hz, H-6a, H-6b), 3.67 (dd, 1H, *J* 14.0 Hz, *J*

6.4 Hz, C*H*$_2$OH), 3.67 (dd, 1H, *J* 14.0 Hz, *J* 9.2 Hz, C*H*$_2$OH), 3.53 (s, 3H, OC*H*$_3$), 2.41–2.35 (m, 1H, H-3). ^{13}C NMR (CD$_3$OD, 100 MHz): 168.9 (d, *J* 2.7 Hz), 164.1 (d, *J* 244 Hz), 138.0 (d, *J* 6.8 Hz), 131.6 (d, *J* 7.9 Hz), 124.0 (d, *J* 2.9 Hz), 119.4 (d, *J* 22 Hz), 115.2 (d, *J* 23 Hz), 103.6, 75.9, 68.9, 68.6, 63.1, 56.8, 46.4, 37.8. HRMS calcd for C$_{15}$H$_{20}$FNO$_6$+H$^+$ (M+H)$^+$: 330.1353, found: 330.1354.

3.12.12. Methyl 3-Deoxy-3-*C*-(4-Fluorobenzamidomethyl)-β-D-Galactopyranoside **7e**

Compound **7e** (TLC, DCM/MeOH, 10:1, R$_f$ 0.44) was prepared according to the general procedure *3.12* from the amide **22e** (40 mg, 0.07 mmol). Obtained as a colorless oil in 53% yield (11.6 mg, 0.04 mmol) from flash column chromatography (DCM:MeOH 20:1–9:1). $[\alpha]_D^{25}$ −16.4 (c 0.5, CH$_3$OH). ^{1}H NMR (CD$_3$OD, 400 MHz): 7.89–7.84 (m, 2H, Ar*H*), 7.22–7.16 (m, 2H, Ar*H*), 4.47 (d, 1H, *J*$_{1,2}$ 7.2 Hz, H-1), 3.94 (td, 1H, *J* 6.0 Hz, *J* 3.0 Hz, H-5), 3.88–3.85 (m, 2H, H-2, H-4), 3.78 (d, 1H, *J*$_{6a,6b}$ 11.6 Hz, *J*$_{5,6a}$ 5.6 Hz, H-6a), 3.75 (d, 1H, *J*$_{6a,6b}$ 11.6 Hz, *J*$_{5,6b}$ 5.6 Hz, H-6b), 3.66 (d, 1H, *J* 14.0 Hz, *J* 6.4 Hz, C*H*$_2$NH), 3.61 (d, 1H, *J* 14.0 Hz, *J* 6.4 Hz, C*H*$_2$NH), 3.53 (s, 3H, OC*H*$_3$), 2.40–2.34 (m, 1H, H-3). ^{13}C NMR (CD$_3$OD, 100 MHz): 169.3, 166.2 (d, *J* 249 Hz), 131.9 (d, *J* 3.0 Hz), 130.8 (d, *J* 8.9 Hz), 116.4 (d, *J* 22 Hz), 103.7, 75.9, 69.0, 68.6, 63.1, 56.8, 46.4, 37.8. ^{19}F NMR (CD$_3$OD, 376 MHz): −110.7. HRMS calcd for C$_{15}$H$_{20}$FNO$_6$+H$^+$ (M+H)$^+$: 330.1353, found: 330.1354.

3.12.13. Methyl 3-Deoxy-3-*C*-(3,4,5-Trifluorobenzamidomethyl)-β-D-Galactopyranoside **7f**

Compound **7f** (TLC, DCM/MeOH, 10:1, R$_f$ 0.47) was prepared according to the general procedure *3.12* from the amide **22f** (31 mg, 0.05 mmol). Obtained as a colorless oil in 70% yield (12.5 mg, 0.03 mmol) from flash column chromatography (DCM:MeOH 20:1–9:1). $[\alpha]_D^{25}$ −13.5 (c 0.6, CH$_3$OH). ^{1}H NMR (CD$_3$OD, 400 MHz): 7.66–7.59 (m, 2H, Ar*H*), 4.45 (d, 1H, *J* 6.8 Hz, H-1), 4.00 (td, 1H, *J* 6.0 Hz, *J* 2.0 Hz, H-5), 3.87–3.84 (m, 2H, H-2, H-4), 3.76 (d, 2H, *J* 5.6 Hz, H-6a, H-6b), 3.66 (dd, 1H, *J* 14.0 Hz, *J* 6.0 Hz, C*H*$_2$NH), 3.59 (dd, 1H, *J* 14.0 Hz, *J* 9.2 Hz, C*H*$_2$NH), 3.53 (s, 3H, OC*H*$_3$), 2.40–2.40–2.34 (m, 1H, H-3). ^{13}C NMR (CD$_3$OD, 100 MHz): 166.7, 152.3 (ddd, 248.3 Hz, *J* 9.8 Hz, *J* 3.8 Hz), 143.0 (dt, *J* 254 Hz, *J* 16 Hz,), 132.2–132.0 (m), 113.1 (dd, *J* 17 Hz, *J* 6.1 Hz), 103.7, 75.9, 68.8, 68.5, 63.1, 56.8, 46.3, 38.0. ^{19}F NMR (CD$_3$OD, 376 MHz): −135.7 (d, *J* 20 Hz). −159.1 (t, *J* 20 Hz). HRMS calcd for C$_{15}$H$_{18}$F$_3$NO$_6$+H$^+$ (M+H)$^+$: 366.1164, found: 366.1161.

3.12.14. Methyl 3-Deoxy-3-*C*-(2,3,4,5,6-Pentafluorobenzamidomethyl)-β-D-Galactopyranoside **7g**

Compound **7g** (TLC, DCM/MeOH, 10:1, R$_f$ 0.38) was prepared according to the general procedure *3.12* from the amide **22g** (30 mg, 0.04 mmol). Obtained as a colorless oil in 83% yield (14.9 mg, 0.04 mmol) from flash column chromatography (DCM:MeOH 20:1–8:1). $[\alpha]_D^{25}$ −18.9 (c 0.6, CH$_3$OH). ^{1}H NMR (CD$_3$OD, 400 MHz): 4.46 (d, 1H, *J* 6.4 Hz, H-1), 3.94–3.85 (m, 4H, H-2, H-4, H-5), 3.79 (dd, 1H, *J* 6.0 Hz, *J* 11.2 Hz, H-6a), 3.79 (dd, 1H, *J* 5.2 Hz, *J* 11.2 Hz, H-6a), 3.68 (dd, 1H, *J* 6.4 Hz, *J* 14.4 Hz, C*H*$_2$NHCO), 3.62 (dd, 1H, *J* 14.4 Hz, *J* 8.8 Hz, C*H*$_2$NHCO), 3.53 (s, 3H, OC*H*$_3$), 2.38–2.32 (m, 1H, H-3). ^{13}C NMR (CD$_3$OD, 100 MHz): 159.9, 146.4, 144.0, 140.2, 137.6, 103.6, 75.8, 68.5, 68.4, 63.2, 56.9, 46.3, 37.7. ^{19}F NMR (CD$_3$OD, 376 MHz): −143.8 to −143.2 (m), −155.2 to −155.3 (m), −163.7 to −163.9 (m). HRMS calcd for C$_{15}$H$_{16}$F$_5$NO$_6$+H$^+$ (M+H)$^+$: 402.0976, found: 402.0974.

3.12.15. Methyl 3-Deoxy-3-*C*-(3-Methoxybenzamidomethyl)-β-D-Galactopyranoside **7h**

Compound **7h** (TLC, DCM/MeOH, 10:1, R$_f$ 0.42) was prepared according to the general procedure *3.12* from the amide **22h** (28 mg, 0.03 mmol). Obtained as a colorless oil in 66% yield (10.3 mg, 0.03 mmol) from flash column chromatography (DCM:MeOH 20:1–10:1). $[\alpha]_D^{25}$ −9.5 (c 0.5, CH$_3$OH). ^{1}H NMR (CD$_3$OD, 400 MHz): 8.47 (t, *J* 5.2 Hz, CON*H*), 7.38–7.33 (m, 3H, Ar*H*), 7.08 (m, 1H, Ar*H*), 4.47 (d, 1H, *J* 7.2 Hz, H-1), 3.94 (td, 1H, *J* 5.6 Hz, *J* 2.0 Hz, H-5), 3.88–3.85 (m, 2H, H-2, H-4), 3.84 (s, 3H, OC*H*$_3$), 3.78 (d, 2H, *J* 5.6 Hz, H-6a, H-6b), 3.66–3.62 (m, 2H, C*H*$_2$NN), 3.53 (s, 3H, OC*H*$_3$), 2.41–2.35 (m, 1H, H-3). ^{13}C NMR (CD$_3$OD, 100 MHz): 170.3, 161.3, 137.0, 130.7, 120.3, 118.5, 113.6, 103.7, 75.9, 70.0, 68.6, 63.1, 56.9, 55.9, 46.4, 37.8. HRMS calcd for C$_{16}$H$_{23}$NO$_7$+H$^+$ (M+H)$^+$: 342.1553, found: 342.1555.

3.12.16. Methyl 3-Deoxy-3-*C*-(p-Toluamidomethyl)-β-D-Galactopyranoside **7i**

Compound **7i** (TLC, DCM/MeOH, 10:1, R_f 0.48) was prepared according to the general procedure *3.12* from the amide **22i** (32 mg, 0.05 mmol). Obtained as a colorless oil in 53% yield (11.9 mg, 0.04 mmol) from flash column chromatography (DCM:MeOH 20:1–10:1). $[\alpha]_D^{25}$ −4.8 (c 0.5, CH_3OH). ^1H NMR (CD_3OD, 400 MHz): 7.72–7.69 (m, 2H, ArH), 7.27 (d, 2H, J 8.0 Hz, ArH), 4.47 (d, 1H, J 7.2 Hz, H-1), 3.94 (td, 1H, J 5.6 Hz, J 2.0 Hz, H-5), 3.88–3.85 (m, 2H, H-2, H-4), 3.77 (d, 2H, J 5.6 Hz, H-6a, H-6b), 3.66–3.62 (m, 2H, CH_2NN), 3.53 (s, 3H, OCH_3), 2.39–2.34 (m, 1H, H-3, CH_3). ^{13}C NMR (CD_3OD, 100 MHz): 170.4, 143.4, 132.7, 130.2, 128.2, 103.7, 76.0, 69.0, 68.7, 63.2, 56.9, 46.4, 37.7, 21.4. HRMS calcd for $C_{16}H_{23}NO_6$+H^+ (M+H)$^+$: 326.1604, found: 326.1603.

3.12.17. Methyl 3-Deoxy-3-*C*-(3,5-Dimethoxybenzamidomethyl)-β-D-Galactopyranoside **7j**

Compound **7j** (TLC, DCM/MeOH, 10:1, R_f 0.43) was prepared according to the general procedure *3.12* from the amide **22j** (24 mg, 0.04 mmol). Obtained as a colorless oil in 62% yield (10 mg, 0.03 mmol) from flash column chromatography (DCM:MeOH 20:1–9:1). $[\alpha]_D^{25}$ −25.7 (c 0.5, CH_3OH). ^1H NMR (CD_3OD, 400 MHz): 6.96 (d, 1H, J 2.0 Hz, ArH), 6.63 (t, 1H, J 2.0 Hz, ArH), 4.47 (d, 1H, $J_{1,2}$ 7.2 Hz, H-1), 3.94 (td, 1H, J 5.6 Hz, J 2.0 Hz, H-5), 3.88–3.85 (m, 2H, H-2, H-4), 3.82 (s, 6H, 2×OCH_3), 3.77 (d, 2H, J 6.0 Hz), 3.67–3.57 (m, 2H, CH_2NH), 3.53 (s, 3H, OCH_3), 2.40–2.34 (m, 1H, H-3). ^{13}C NMR (CD_3OD, 100 MHz): 170.2, 162.4, 137.6, 106.1, 104.5, 103.7, 75.9, 69.0, 68.6, 63.1, 56.8, 56.0, 46.4, 37.8. HRMS calcd for $C_{17}H_{25}NO_8$+H^+ (M+H)$^+$: 372.1658, found: 372.1663.

3.12.18. Methyl 3-Deoxy-3-*C*-(3-Trifluoromethylbenzamidomethyl)-β-D-Galactopyranoside **7k**

Compound **7k** (TLC, DCM/MeOH, 10:1, R_f 0.51) was prepared according to the general procedure *3.12* from the amide **22k** (25 mg, 0.04 mmol). Obtained as a colorless oil in 66% yield (11.2 mg, 0.03 mmol) from flash column chromatography (DCM:MeOH 20:1–10:1). $[\alpha]_D^{25}$ −3.9 (c 0.6, CH_3OH). ^1H NMR (CD_3OD, 400 MHz): 8.13 (s, 1H, ArH), 8.08 (t, 1H, J 8.0 Hz, ArH), 7.85 (d, 1H, J 8.0 Hz, ArH), 7.68 (t, 1H, J 8.0 Hz, ArH), 4.47 (d, 1H, $J_{1,2}$ 7.2 Hz, H-1), 3.95 (td, 1H, J 6.0 Hz, J 2.0 Hz, H-5), 3.89–3.86 (m, 2H, H-2, H-4), 3.79 (dd, 1H, $J_{6a,6b}$ 12.0 Hz, $J_{5,6a}$ 6.0 Hz, H-6a), 3.76 (dd, 1H, $J_{6a,6b}$ 12.0 Hz, $J_{5,6b}$ 6.0 Hz, H-6b), 3.69 (dd, 1H, J 13.6 Hz, J 5.6 Hz, CH_2NH), 3.63 (dd, 1H, J 13.6 Hz, J 9.2 Hz, CH_2NH), 3.54 (s, 3H, OCH_3), 2.43–2.37 (m, 1H, H-3). ^{13}C NMR (CD_3OD, 125 MHz): 168.7, 136.7, 132.0 (q, J 32 Hz), 131.9, 130.6, 129.2 (q J 3.6 Hz), 125.4 (q, J 270 Hz), 125.1 (q, J 4.0 Hz), 103.7, 75.9, 68.9, 68.6, 63.1, 56.9, 46.5, 37.8. ^{19}F NMR (CD_3OD, 376 MHz): −64.2. HRMS calcd for $C_{16}H_{20}F_3NO_6$+H^+ (M+H)$^+$: 380.1321, found: 380.1321.

3.12.19. Methyl 3-Deoxy-3-*C*-(4-Phenylbenzamidomethyl)-β-D-Galactopyranoside **7l**

Compound **7l** (TLC, DCM/MeOH, 10:1, R_f 0.54) was prepared according to the general procedure *3.12* from the amide **22l** (39 mg, 0.06 mmol). Obtained as a colorless oil in 67% yield (15.4 mg, 0.04 mmol) from flash column chromatography (DCM:MeOH 20:1–12:1). $[\alpha]_D^{25}$ −21.4 (c 0.7, CH_3OH). ^1H NMR (CD_3OD, 400 MHz): 7.70 (dd, 2H, J 6.8 Hz, J 2.0 Hz, ArH), 7.71 (dd, 2H, J 6.8 Hz, J 2.0 Hz, ArH), 7.65 (m, 2H, ArH), 7.48–7.44 (m, 2H, ArH), 7.40–7.35 (m, 1H, ArH), 4.49 (d, 1H, $J_{1,2}$ 7.2 Hz, H-1), 3.96 (td, 1H, J 6.0 Hz, J 2.4 Hz, H-5), 3.82–3.75 (m, 2H, H-6a, H-6b), 3.73–3.62 (m, 2H, CH_2NH), 3.54 (s, 3H, OCH_3), 2.44–2.38 (m, 1H, H-3). ^{13}C NMR (CD_3OD, 100 MHz): 170.1, 145.7, 141.2, 134.2, 130.0, 129.1, 128.8, 128.11, 128.07, 103.7, 76.0, 69.0, 68.7, 63.2, 56.9, 46.4, 37.8. HRMS calcd for $C_{21}H_{25}NO_6$+H^+ (M+H)$^+$: 388.1760, found: 388.1761.

3.12.20. Methyl 3-Deoxy-3-*C*-(Diphenylphosphonamidomethyl)-β-D-Galactopyranoside **8**

Compound **8** (TLC, DCM/MeOH, 10:1, R_f 0.45) was prepared according to the general procedure 3.12 from the amide **23** (43 mg, 0.06 mmol). Obtained as a colorless oil in 50% yield (13.3 mg, 0.03 mmol) from flash column chromatography (DCM:MeOH 15:1–9:1). $[\alpha]_D^{25}$ −18.6 (c 0.7, CH_3OH). ^1H NMR (CD_3OD, 400 MHz): 7.40 (t, 4H, J 8.0 Hz, ArH), 7.29–7.21 (m, 6H, ArH), 4.36 (d, 1H, J 7.2 Hz, H-1), 3.90

(dd, 1H, *J* 4.0 Hz, *J* 2.0 Hz, H-4), 3.83 (dd, 1H, *J* 7.2 Hz, *J* 5.6 Hz, H-2), 3.79–3.76 (m, 1H, H-5), 3.71 (dd, 1H, $J_{5,6a}$ 10.4 Hz, $J_{6a,6b}$ 4.4 Hz, H-6a), 3.71 (dd, 1H, $J_{5,6b}$ 10.8 Hz, $J_{6a,6b}$ 4.4 Hz, H-6b), 3.51 (s, 3H, OCH_3), 3.50–3.44 (m, 1H, CH_2NH), 3.17–3.08 (m, 1H, CH_2NH), 2.28–2.22 (m, 1H, H-3). ^{13}C NMR (CD$_3$OD, 100 MHz): 152.2 (dd, *J* 6.2 Hz, *J* 2.7 Hz), 130.9, 126.3 (d, *J* 3.1 Hz), 121.4 (dd, *J* 4.6 Hz, *J* 11.1 Hz), 103.6, 75.7, 68.7, 68.1, 63.2, 56.8, 47.4 (d, 5.7 Hz), 39.2. ^{31}P NMR (CD$_3$OD, 162 MHz): −1.0. HRMS calcd for C$_{20}$H$_{26}$PNO$_8$+H$^+$ (M+H)$^+$: 440.1474, found: 440.1470.

3.13. Methyl 2,3-Di-O-Acetyl-β-D-Gulopyranoside **25**

Compound **24** (300 mg, 0.82 mmol) was dissolved in 80% aqueous AcOH (5 mL) and the solution was stirred at 80 °C for 2 h. When the TLC (TLC, heptane/EtOAc, 1:2, R$_f$ 0.39) showed complete consumption of the starting material, the solvents were evaporated under reduced pressure and co-evaporated twice with toluene (10 mL). Then, the crude was purified via flash chromatography (Heptane/EtOAc, 3:1–1:2) to obtain pure compound **25** (191 mg, 0.69 mmol, 84%) as a white foam. $[\alpha]_D^{25}$ −30.4 (c 0.8, CHCl$_3$). ^{1}H NMR (CDCl$_3$, 400 MHz): 5.35 (t, 1H, $J_{2,3}$ 3.6 Hz, H-3), 5.10 (dd, 1H, $J_{1,2}$ 8.0 Hz, $J_{2,3}$ 3.6 Hz, H-2), 4.68 (d, 1H, *J* 8.0 Hz, H-1), 3.93–3.89 (m, 4H, H4, H-5, H-6a, H-6b), 3.52 (s, 3H, OCH_3), 2.11 (s, 3H, COCH_3), 2.03 (s, 3H, COCH_3). ^{13}C NMR (CDCl$_3$, 100 MHz): 170.0, 169.9, 99.9, 73.3, 70.6, 69.0, 68.4, 62.8, 56.9, 20.98, 20.95. HRMS calcd for C$_{11}$H$_{18}$O$_8$+Na$^+$ (M+Na)$^+$: 301.0899, found: 301.0898.

3.14. Methyl β-D-Gulopyranoside **9**

Compound **25** (120 mg, 0.43 mmol) was dissolved in MeOH (3 mL). NaOMe (1.0 mL, 0.5 M in MeOH) was added and the solution was stirred at room temperature for 2 h (TLC, DCM/MeOH, 5:1, R$_f$ 0.3). The solution was neutralized with DOWEX 50 W H$^+$ resin, filtered and the solvents were evaporated under reduced pressure and the crude was purified via flash chromatography (DCM/MeOH, 7:1–3:1) to obtain pure compound **9** (46 mg, 0.23 mmol, 55%) as a colorless oil. $[\alpha]_D^{25}$ −19.2 (c 0.9, CH$_3$OH). ^{1}H NMR (D$_2$O, 400 MHz): 4.60 (d, 1H, $J_{1,2}$ 8.4 Hz, H-1), 4.05 (t, 1H, $J_{3,4}$ 3.6 Hz, H-4), 4.00–3.96 (m, 1H, H-5), 3.80 (dd, 1H, $J_{3,4}$ 3.6 Hz, $J_{4,5}$ 1.2 Hz, H-4), 3.75 (dd, 1H, $J_{6a,6b}$ 12.0 Hz, $J_{5,6a}$ 6.4 Hz, H-6a), 3.76 (dd, 1H, $J_{6a,6b}$ 12.0 Hz, $J_{5,6a}$ 4.8 Hz, H-6b), 3.76 (dd, 1H, $J_{1,2}$ 8.4 Hz, $J_{2,3}$ 3.6 Hz, H-2), 3.56 (s, 3H, OCH_3). ^{13}C NMR (D$_2$O, 100 MHz): 101.5, 73.8, 71.1, 69.3, 68.0, 61.0, 56.9. HRMS calcd for C$_7$H$_{14}$O$_6$+Na$^+$ (M+Na)$^+$: 217.0688, found: 217.0687.

3.15. Methyl 3-Azido-2,4,6-Tri-O-Benzoyl-3-Deoxy-β-D-Gulopyranoside **29**

Triflic anhydride (235 μL, 1.4 mmol) was added dropwise to a stirred solution of **26** (400 mg, 1.4 mmol) in DCM (10 mL) and pyridine (451 μL, 5.6 mmol) at −30 °C and under N$_2$ atmosphere after which the reaction was allowed to reach rt under 2 h. BzCl (179 μL, 1.54 mmol) was added and the reaction was stirred for another 2 h before the reaction was diluted with DCM (25 mL) and washed with saturated NaHCO$_3$ (2 × 25 mL). The combined aqueous phases were extracted with DCM (40 mL). The pooled organic phases were dried over MgSO$_4$ and concentrated to give crude **27**. Sodium azide (637 mg, 9.8 mmol) was added to the crude **27** (≤1.4 mmol) in DMF (15 mL) and the reaction was stirred overnight at 70 °C under N$_2$ atmosphere. The reaction was concentrated to give crude **28**, which was dissolved in 90% AcOH (20 mL) and heated at 80 °C for 3 h. The solvent was evaporated in vacuo and co-evaporated with toluene to remove the residual AcOH. The residue was dissolved in pyridine (15 mL), into the solution catalytic amount of DMAP and benzoyl chloride (488 μL, 4.2 mmol) was added subsequently. The solution was stirred at room temperature for 4 h when TLC (heptane/EtOAc, 4:1, R$_f$ 0.48) showed complete conversion of the starting material to a faster moving spot. The solvents were evaporated in vacuo and co-evaporated with toluene to remove residual pyridine. The solid residue thus obtained was dissolved in EtOAc (50 mL) and washed with 1 N HCl (50 mL), followed by saturated NaHCO$_3$ and brine (50 mL). The organic layer was collected, dried (Na$_2$SO$_4$), filtered and evaporated in vacuo. The crude was purified by flash chromatography using heptane/EtOAc (6:1 to 5:2) as the eluent to afford pure compound **29** (324 mg, 0.61 mmol, 43% over four steps) as a white amorphous mass. $[\alpha]_D^{25}$ −45.3 (c 0.7, CHCl$_3$). ^{1}H NMR (CDCl$_3$, 400 MHz): 8.14–7.39 (m, 15H, Ar*H*),

5.51 (dd, 1H, $J_{1,2}$ 7.6 Hz, $J_{2,3}$ 4.0 Hz, H-2), 5.41 (dd, 1H, $J_{3,4}$ 4.0 Hz, $J_{3,4}$ 0.8 Hz, H-4), 5.00 (d, 1H, J 7.6 Hz, H-1), 4.66–4.61 (m, 1H, H-5), 4.52–4.45 (m, 3H, H-3, H-6a, H-6b), 3.58 (s, 3H, OCH_3). ^{13}C NMR (CDCl$_3$, 100 MHz): 166.0, 165.3, 165.2, 133.8, 133.6, 133.2, 130.02, 129.98, 129.7, 129.5, 129.0, 128.7, 128.6, 128.5, 128.4, 99.6, 70.3, 70.1, 69.5, 62.4, 60.1, 57.0. HRMS calcd for $C_{28}H_{25}N_3O_8 + NH_4^+$ $(M+NH_4)^+$: 549.1985, found: 549.1989.

3.16. Methyl 3-Amino-2,4,6-Tri-O-Benzoyl-3-Deoxy-β-D-Gulopyranoside 30

A solution of **29** (201 mg, 0.3784 mmol) in MeOH (7 mL) was stirred with Pd(OH)$_2$/C (10% wt., 1 mg per 5 mg of crude, 40 mg) under hydrogen atmosphere at room temperature for 2 h. After the completion of the reaction (as indicated by TLC, heptane/EtOAc, 1:1, R$_f$ 0.26), the reaction mixture was filtered through a Celite bed and washed with methanol. The filtrate was concentrated under reduced pressure and purified through flash column (heptane/EtOAc, 4:1–1:1) to get the desired compound as a white amorphous solid. Yield: 126 mg (0.2494 mmol, 66%). $[\alpha]_D^{25}$ −39.9 (c 0.8, CHCl$_3$). ^{1}H NMR (CDCl$_3$, 400 MHz): 8.13–7.38 (m, 15H, ArH), 5.38 (dd, 1H, $J_{1,2}$ 7.2 Hz, $J_{2,3}$ 4.0 Hz, H-2), 5.29 (dd, 1H, $J_{3,4}$ 4.4 Hz, $J_{4,5}$ 2.4 Hz, H-4), 5.11 (d, 1H, J 7.2 Hz, H-1), 4.83–4.79 (m, 1H, H-5), 4.64 (dd, 1H, dd, 1H, $J_{6a,6b}$ 11.2 Hz, $J_{5,6a}$ 6.8 Hz, H-6a), 4.51 (dd, 1H, dd, 1H, $J_{6a,6b}$ 11.2 Hz, $J_{5,6b}$ 6.0 Hz, H-6b), 3.90 (t, 1H, $J_{3,4}$, $J_{2,3}$ 4.0 Hz, H-3), 3.57 (s, 3H, OCH_3), 1.97 (bs, 1H, NH_2). ^{13}C NMR (CDCl$_3$, 100 MHz): 166.3, 165.9, 165.3, 133.7, 133.5, 133.2, 130.1, 129.9, 129.8, 128.69, 128.67, 128.5, 99.3, 72.1, 71.7, 70.2, 63.3, 57.1, 50.6. HRMS calcd for $C_{28}H_{27}NO_8 + H^+$ $(M+H)^+$: 506.1815, found: 506.1817.

3.17. Methyl 3-Benzamido-2,4,6-Tri-O-Benzoyl-3-Deoxy-β-D-Gulopyranoside 31

Compound **30** was dissolved in pyridine (3 mL), into the solution catalytic amount of DMAP and benzoyl chloride (29 μL, 0.2464 mmol) was added subsequently. The solution was stirred at room temperature for 3 h when TLC (heptane/EtOAc, 1:1, R$_f$ 4.8) showed complete conversion of the starting material to a faster moving spot. The solvents were evaporated in vacuo and co-evaporated with toluene to remove residual pyridine. The solid residue thus obtained was dissolved in EtOAc (7 mL) and washed with 1 (N) HCl (5 mL), followed by saturated NaHCO$_3$ and brine (5 mL). The organic layer was collected, dried over Na$_2$SO$_4$, filtered and evaporated in vacuo. The crude was purified by flash chromatography using heptane/EtOAc (7:1 to 3:1) as the eluent to afford pure compound **31** (77 mg, 0.1265 mmol, 77%) as a white amorphous solid. $[\alpha]_D^{25}$ −48.8 (c 0.6, CHCl$_3$). ^{1}H NMR (CDCl$_3$, 400 MHz): 8.11–7.29 (m, 20H, ArH), 6.60 (d, 1H, $J_{3,NHCOPh}$ 8.4 Hz, NHCOPh), 5.96 (dd, 1H, J 10.8 Hz, J 6.0 Hz, H-4), 5.55 (t, 1H, J 2.8 Hz, H-2), 5.34–5.29 (m, 1H, H-3), 4.99 (d, 1H, $J_{1,2}$ 2.8 Hz, H-1), 4.92 (dd, 1H, $J_{6a,6b}$ 11.6 Hz, $J_{5,6a}$ 5.6 Hz, H-6a), 4.86 (dd, 1H, $J_{6a,6b}$ 11.6 Hz, $J_{5,6b}$ 6.4 Hz, H-6a), 4.77 (dd, 1H, $J_{6a,6b}$ 12.4 Hz, $J_{4,5}$ 6.0 Hz, H-5), 3.61 (s, 3H, OCH_3). ^{13}C NMR (CDCl$_3$, 100 MHz): 167.4, 166.8, 166.2, 165.5, 133.9, 133.8, 133.7, 133.0, 131.7, 130.0, 129.9, 129.7, 129.6, 129.1, 128.7, 128.59, 128.57, 128.3, 127.0, 99.5, 72.4, 71.8, 68.4, 64.5, 60.4, 56.8, 46.3. HRMS calcd for $C_{35}H_{31}NO_9 + H^+$ $(M+H)^+$: 610.2077, found: 610.2081.

3.18. Methyl 3-Benzamido-3-Deoxy-β-D-Gulopyranoside 10

Compound **31** (54 mg, 0.0886 mmol) was dissolved in MeOH (2 mL). NaOMe (0.5 mL, 0.5 M in MeOH) was added and the solution was stirred at room temperature for 12 h (TLC, DCM/MeOH, 10:1, R$_f$ 0.4). The solution was neutralized with DOWEX 50 W H+ resin, filtered and the solvents were evaporated under reduced pressure and the residue was purified by a short flash column using DCM–MeOH (9:1) to afford the compound **10** (19.2 mg, 0.0646 mmol, 73%). $[\alpha]_D^{25}$ −18.3 (c 0.6, CH$_3$OH). ^{1}H NMR (CD$_3$OD, 400 MHz): 7.83 (d, 2H, J 7.6 Hz, ArH), 7.57–7.44 (m, 3H, ArH), 4.70 (d, 1H, $J_{1,2}$ 8.4 Hz, H-1), 4.48–4.44 (m, 1H, H-3), 4.01 (dd, 1H, $J_{3,4}$ 3.6 Hz, $J_{4,5}$ 1.2 Hz, H-4), 3.94 (dd, 1H, $J_{1,2}$ 7.6 Hz, $J_{2,3}$ 5.2 Hz, H-2), 3.84 (td, 1H, $J_{5,6a}$, $J_{5,6a}$ 6.0 Hz, $J_{4,5}$ 1.6 Hz, H-5), 3.77 (dd, 1H, $J_{6a,6b}$ 11.2 Hz, $J_{5,6a}$ 6.0 Hz, H-6a), 3.74 (dd, 1H, $J_{6a,6b}$ 11.2 Hz, $J_{5,6a}$ 6.0 Hz, H-6b), 3.57 (s, 3H, OCH_3). ^{13}C NMR (CD$_3$OD, 100 MHz): 171.4, 164.6, 135.9, 132.7, 129.5, 128.6, 103.4, 75.7, 68.8, 67.8, 62.6, 56.9, 56.0. HRMS calcd for $C_{14}H_{19}NO_6 + H^+$ $(M+H)^+$: 298.1291, found: 298.1289.

3.19. Expression Constructs, Expression, and Purification of Recombinant Galectins

Human galectin-1 [25], galectin-2 [26], galectin-3 [27], galectin-4N [19], galectin-4C [19], galectin-8N [28], galectin-8C [28], galectin-9N [29], and galectin-9C [30], were expressed and purified as described earlier. Human galectin-7 was expressed using a pET3c plasmid in *E. coli* BL21-star. The plasmid containing expression optimized DNA encoding the full human galectin-7 sequence (NCBI Reference Sequence: NP_002298.1) was obtained from GenScript (Piscataway, NJ, USA). Bacterial culture and induction, and galectin purification was essential as described for galectin-3 expressed with the same vector [27]; a typical yield was 1.5–2 mg/L culture. Lactose was removed by chromatography on a PD-10 column (Amersham Biosciences) with repeated ultrafiltration with Centriprep (Amicon).

3.20. Fluorescence Polarization Assay

Fluorescence polarization experiments were carried out either with a POLARStar plate reader and FLUOstar Galaxy software or with a PheraStarFS plate reader and PHERAstar Mars version 2.10 R3 software (BMG, Offenburg, Germany). The dissociation constant (K_d) values were determined in PBS as described earlier [18,19]. Specific conditions for galectin-1, 2, 3, 4N, 4C, 8N, 8C, 9N, and 9C were kept as reported [29]. Experiments were performed at room temperature with human galectin-7 at 5 μM and the fluorescent probe β-D-galactopyranosyl-(1–4)-2-acetamido-2-deoxy-β-d-glucopyranosyl-(1–3)-β-d-galactopyranosyl-(1–4)-(*N*1-fluorescein-5-yl-carbonylaminomethylcarbonyl)-β-D-glucopyranosylamine [29] at 0.02 μM. All the compounds in Table 1 except 32 were dissolved in a neat DMSO at 100 mM and diluted in PBS to three to six different concentrations to be tested in duplicate. Average K_d values and SEMs were calculated from 2–8 single-triple point measurements showing between 30%–70% inhibition.

4. Conclusions

In summary, we report the synthesis and discovery of 3-*C*-methyl-guloside derivatives as highly selective galectin-1 inhibitors with 3-*C*-benzamidomethyl-3-deoxy-gulosides being the most selective structural class. The reason for the exceptional galectin-1-selectivites discovered remains to be elucidated as molecular modelling failed to provide insight into this and experimental structural studies by X-ray diffraction or nmr spectroscopy are likely necessary. Although the galectin-1 affinities are in the high-μM to low mM range, they are significantly higher affinity than that of simple galactosides, such as methyl β-D-galactopyranoside, and thus points towards a novel structural class and synthetic route towards the discovery of galectin-1 inhibitors with high selectivity. This is important in light of the roles of galectin-1 in tumor progression and immune regulation [31,32].

Supplementary Materials: Supplementary materials can be found at http://www.mdpi.com/1422-0067/20/15/3786/s1.

Author Contributions: K.B.P. and M.M. contributed equally to the synthesis and characterization of all compounds. K.B.P wrote the major part of the manuscript. H.L. supervised and analyzed the result of the fluorescence polarization assay. U.J.N. conceived the study, analyzed the data and co-wrote the paper. The manuscript was written with contributions from all authors. All authors have given consent to the final version of the manuscript.

Funding: We thank the Swedish Research Council (Grant No. 621-2016-03667), a project grant awarded by the Knut and Alice Wallenberg Foundation (KAW 2013.0022), and Galecto Biotech AB, Sweden for financial support.

Acknowledgments: We thank Barbro Kahl-Knutson for excellent assistance with determining affinities by fluorescence polarization and Sofia Essén for excellent assistance with hrms and analytical hplc experiments.

Conflicts of Interest: H.L. and U.J.N. are shareholders in Galecto Biotech AB, Sweden, a company developing galectin inhibitors.

Abbreviations

Ac	Acetyl
Bn	Benzyl
DCM	Dichloromethane
THF	Tetrahydrofuran
DMF	Dimethylformamide
DIPEA	Diisopropylethylamine
AcOH	AcOH
EtOAc	EtOAc
TLC	Thin layer chromatography
HPLC	High-performance liquid chromatography
HRMS	High resolution mass spectrometry
DMSO	Dimethylsulfoxide
μM	Micromolar
mM	Milimolar
9-BBN	9-Borabicyclo[3.3.1]nonane
DMAP	4-Dimethylaminopyridine

References

1. Lella, S.D.; Sundblad, V.; Cerliani, J.P.; Guardia, C.M.; Estrin, D.A.; Vasta, G.R.; Rabinovich, G.A. When Galectins Recognize Glycans: From Biochemistry to Physiology and Back Again. *Biochemistry* **2011**, *50*, 7842–7857. [CrossRef] [PubMed]
2. Yang, R.-Y.; Rabinovich, G.A.; Liu, F.-T. Galectins: Structure, function and therapeutic potential. *Expert Rev. Mol. Med.* **2008**, *10*, e17. [CrossRef] [PubMed]
3. Johannes, L.; Jacob, R.; Leffler, H. Galectins at a glance. *J. Cell Sci.* **2018**, *131*, 1–9. [CrossRef] [PubMed]
4. Rabinovich, G.A.; Croci, D.O. Regulatory Circuits Mediated by Lectin-Glycan Interactions in Autoimmunity and Cancer. *Immunity* **2012**, *36*, 322–335. [CrossRef]
5. Rabinovich, G.A.; Toscano, M.A. Turning "sweet" on immunity: Galectin–glycan interactions in immune tolerance and inflammation. *Nat. Rev. Immunol.* **2009**, *9*, 338–352. [CrossRef]
6. Costa, A.F.; Gomes, A.M.; Kozlowski, E.O.; Stelling, M.P.; Pavão, M.S.G. Extracellular galectin-3 in tumor progression and metastasis. *Front Oncol.* **2014**, *4*, 1–9.
7. Cousin, J.M.; Cloninger, M.J. The Role of Galectin-1 in Cancer Progression, and Synthetic Multivalent Systems for the Study of Galectin-1. *Int. J. Mol. Sci.* **2016**, *17*, 1566. [CrossRef]
8. Chang, W.-A.; Tsai, M.-J.; Kuo, P.-L.; Hung, J.-Y. Role of galectins in lung cancer. *Oncol Lett.* **2017**, *14*, 5077–5084.
9. Diao, B.; Liu, Y.; Zhang, Y.; Xie, J.; Gong, J. The role of galectin-3 in the tumorigenesis and progression of pituitary tumors. *Oncol. Lett.* **2018**, *15*, 4919–4925. [CrossRef]
10. Califice, S.; Castronovo, V.; Brule, F.V.D. Galectin-3 and cancer. *Int. J. Oncol.* **2004**, *25*, 983–992.
11. Laaf, D.; Bojarová, P.; Elling, L.; Kren, V. Galectin-Carbohydrate Interactions in Biomedicine and Biotechnology. *Trends Biotechnol.* **2019**, *37*, 402–415. [CrossRef] [PubMed]
12. Notz, W.; Hartel, C.; Waldscheck, B.; Schmidt, R.R. De Novo Synthesis of a Methylene-Bridged Neu5Ac-r-(2,3)-Gal *C*-Disaccharide. *J. Org. Chem.* **2001**, *66*, 4250–4260. [CrossRef] [PubMed]
13. Manickam, G.; Ghoshal, A.; Subramaniam, P.; Li, Y. More Efficient Palladium Catalyst for Hydrogenolysis of Benzyl Groups. *Synthetic Commun.* **2006**, *36*, 925–928.
14. Sörme, P.; Kahl-Knutson, B.; Wellmar, U.; Nilsson, U.J.; Leffler, H. Fluorescence Polarization to Study Galectin–Ligand Interactions. *Methods Enzymol.* **2003**, *362*, 504–512. [PubMed]
15. Sörme, P.; Kahl-Knutson, B.; Huflejt, M.U.; Nilsson, U.J.; Leffler, H. Fluorescence polarization as an analytical tool to evaluate galectin–ligand interactions. *Anal. Biochem.* **2004**, *334*, 36–47.
16. Cumpstey, I.; Carlsson, S.; Leffler, H.; Nilsson, U.J. Synthesis of a phenyl thio-β-ᴅ-galactopyranoside library from 1,5-difluoro-2,4-dinitrobenzene: Discovery of efficient and selective monosaccharide inhibitors of galectin-7. *Org. Biomol. Chem.* **2005**, *3*, 1922–1932. [CrossRef] [PubMed]

17. Öberg, C.-T.; Blanchard, H.; Leffler, H.; Nilsson, U.J. Protein subtype-targeting through ligand epimerization: Talose-selectivity of galectin-4 and galectin-8. *Bioorg. Med. Chem. Lett.* **2008**, *18*, 3691–3694. [CrossRef]

18. Pal, K.B.; Mahanti, M.; Huang, X.; Persson, S.; Sundin, A.P.; Zetterberg, F.R.; Oredsson, S.; Leffler, H.; Nilsson, U.J. Quinoline–galactose hybrids bind selectively with high affinity to a galectin-8 N-terminal domain. *Org. Biomol. Chem.* **2018**, *16*, 6295–6305. [CrossRef] [PubMed]

19. Mandal, S.; Rajput, V.K.; Sundin, A.P.; Leffler, H.; Mukhopadhyay, B.; Nilsson, U.J. Galactose-amidine derivatives as selective antagonists of galectin-9. *Can. J. Chem.* **2016**, *94*, 936–939. [CrossRef]

20. Tolnai, G.L.; Nilsson, U.J.; Olofsson, B. Efficient *O*-Functionalization of Carbohydrates with Electrophilic Reagents. *Angew. Chem. Int. Ed. Engl.* **2016**, *55*, 11226–11230. [CrossRef]

21. Peterson, K.; Kumar, R.; Stenström, O.; Verma, P.; Verma, P.R.; Håkansson, M.; Kahl-Knutsson, B.; Zetterberg, F.; Leffler, H.; Akke, M.; et al. Systematic Tuning of Fluoro-galectin-3 Interactions Provides Thiogalactoside Derivatives with Single-Digit nM Affinity and High Selectivity. *J. Med. Chem.* **2018**, *61*, 1164–1175. [CrossRef]

22. Öberg, C.T.; Noresson, A.-L.; Delaine, T.; Larumbe, A.; Tejler, J.; von Wachenfeldt, H.; Nilsson, U.J. Synthesis of 3-azido-3-deoxy-β-D-galactopyranosides. *Carbohydr. Res.* **2009**, *344*, 1282–1284.

23. Zemplén, G. Abbau der reduzierenden Biosen, I. Direkte Konstitutionsermitt-lung der cellobiose. *Ber. Dtsch. Chem. Ges.* **1926**, *59*, 1254.

24. Wong, C.-H.; Maris-Varas, F.; Hung, S.-C.; Marron, T.G.; Lin, C.-C.; Gong, K.W.; Weitz-Schimidt, G. Small Molecules as Structural and Functional Mimics of Sialyl Lewis X Tetrasaccharide in Selectin Inhibition: A Remarkable Enhancement of Inhibition by Additional Negative Charge and/or Hydrophobic Group. *J. Am. Chem. Soc.* **1997**, *119*, 8152–8158. [CrossRef]

25. Salomonsson, E.; Larumbe, A.; Tejler, J.; Tullberg, E.; Rydberg, H.; Sundin, A.; Khabut, T.; Frejd, T.; Lobsanov, Y.D.; Rini, J.M.; et al. Monovalent Interactions of Galectin-1. *Biochemistry* **2010**, *49*, 9518–9532. [CrossRef]

26. Gitt, M.A.; Massa, S.M.; Leffler, H.; Barondes, S.H. Isolation and expression of a gene encoding L-14-II, a new human soluble lactose-binding lectin. *J. Biol. Chem.* **1992**, *267*, 10601–10606.

27. Massa, S.M.; Cooper, D.N.; Leffler, H.; Barondes, S.H. L-29, an endogenous lectin, binds to glycoconjugate ligands with positive cooperativity. *Biochemistry* **1993**, *32*, 260–267. [CrossRef]

28. Carlsson, S.; Öberg, C.T.; Carlsson, M.C.; Sundin, A.; Nilsson, U.J.; Smith, D.; Cummings, R.D.; Almkvist, J.; Karlsson, A.; Leffler, H. Affinity of galectin-8 and its carbohydrate recognition domains for ligands in solution and at the cell surface. *Glycobiology* **2007**, *17*, 663–676. [CrossRef]

29. Öberg, C.T.; Carlsson, S.; Fillion, E.; Leffler, H.; Nilsson, U.J. Efficient and Expedient Two-Step Pyranose-Retaining Fluorescein Conjugation of Complex Reducing Oligosaccharides: Galectin Oligosaccharide Specificity Studies in a Fluorescence Polarization Assay. *Bioconj. Chem.* **2003**, *14*, 1289–1297. [CrossRef]

30. Delaine, T.; Collins, P.; MacKinnon, A.; Sharma, G.; Stegmayr, J.; Rajput, V.K.; Mandal, S.; Cumpstey, I.; Larumbe, A.; Salameh, B.A.; et al. Galectin 3-Binding Glycomimetics that Strongly Reduce Bleomycin-Induced Lung Fibrosis and Modulate Intercellular Glycan Recognition. *ChemBioChem* **2016**, *17*, 1759–1770. [CrossRef]

31. Méndez-Huergo, S.P.; Blidner, A.G.; Rabinovich, G.A. Galectins: Emerging regulatory checkpoints linking tumor immunity and angiogenesis. *Curr. Opin. Immun.* **2017**, *45*, 8–15. [CrossRef]

32. Kaltner, H.; Toegel, S.; Caballero, G.G.; Manning, J.C.; Ledeen, R.W.; Gabius, H.-J. Galectins: Their network and roles in immunity/tumor growth control. *Histochem. Cell. Biol.* **2016**, *147*, 239–256. [CrossRef]

International Journal of
Molecular Sciences

Article

Analysis of Procollagen C-Proteinase Enhancer-1/Glycosaminoglycan Binding Sites and of the Potential Role of Calcium Ions in the Interaction

Jan Potthoff [1,2,†], Krzysztof K. Bojarski [1,†] , Gergely Kohut [3,4] , Agnieszka G. Lipska [1], Adam Liwo [1], Efrat Kessler [5], Sylvie Ricard-Blum [6] and Sergey A. Samsonov [1,*]

1 Faculty of Chemistry, University of Gdańsk, ul. Wita Stwosza 63, 80-308 Gdańsk, Poland; potthoff.jan@googlemail.com (J.P.); krzysztof.bojarski@ug.edu.pl (K.K.B.); lypstyk@gmail.com (A.G.L.); adam.liwo@ug.edu.pl (A.L.)
2 Institute of Chemistry and Biochemistry, Free University of Berlin, Takustr. 3, 14195 Berlin, Germany
3 Institute of Materials and Environmental Chemistry, Research Centre for Natural Sciences, Hungarian Academy of Sciences, Magyar tudósok körútja 2, 1117 Budapest, Hungary; kohut.gergely@ttk.mta.hu
4 MTA-ELTE Research Group of Peptide Chemistry, Hungarian Academy of Sciences, Eötvös Loránd University, Budapest 112, P.O. Box 32, 1518 Budapest, Hungary
5 Maurice and Gabriela Goldschleger Eye Research Institute, Tel-Aviv University Sackler Faculty of Medicine, Sheba Medical Center, Tel-Hashomer, 52621 Ramat Gan, Israel; ekessler@post.tau.ac.il
6 Institute of Molecular and Supramolecular Chemistry and Biochemistry, UMR 5246, CNRS, INSA Lyon, CPE, University Claude Bernard Lyon 1, Univ Lyon, 69622 Villeurbanne CEDEX, France; sylvie.ricard-blum@univ-lyon1.fr
* Correspondence: sergey.samsonov@ug.edu.pl; Tel.: +48-58-523-5166
† These authors contributed equally to this work.

Received: 13 September 2019; Accepted: 9 October 2019; Published: 10 October 2019

Abstract: In this study, we characterize the interactions between the extracellular matrix protein, procollagen C-proteinase enhancer-1 (PCPE-1), and glycosaminoglycans (GAGs), which are linear anionic periodic polysaccharides. We applied molecular modeling approaches to build a structural model of full-length PCPE-1, which is not experimentally available, to predict GAG binding poses for various GAG lengths, types and sulfation patterns, and to determine the effect of calcium ions on the binding. The computational data are analyzed and discussed in the context of the experimental results previously obtained using surface plasmon resonance binding assays. We also provide experimental data on PCPE-1/GAG interactions obtained using inhibition assays with GAG oligosaccharides ranging from disaccharides to octadecasaccharides. Our results predict the localization of GAG-binding sites at the amino acid residue level onto PCPE-1 and is the first attempt to describe the effects of ions on protein-GAG binding using modeling approaches. In addition, this study allows us to get deeper insights into the in silico methodology challenges and limitations when applied to GAG-protein interactions.

Keywords: procollagen C-proteinase enhancer-1; glycosaminoglycans; computational analysis of protein-glycosaminoglycan interactions; calcium ions; fragment-based docking

---

1. Introduction

Glycosaminoglycans (GAGs) are anionic periodic linear polysaccharides, which are composed of periodic disaccharide units [1] and play a key role in many biologically relevant processes by interacting with their numerous and diverse protein targets such as cytokines and growth factors in the extracellular matrix [2–5]. However, the molecular mechanisms underlying GAG-mediated interactions are not fully understood, and experimental techniques alone are not sufficient for gaining insights into them [6]. Molecular modeling approaches are not only complementary to experiments, but also provide additional and crucial details, which are experimentally inaccessible. In our previous work, we successfully applied molecular docking and molecular dynamics methodologies in order to model protein-GAG interactions. In particular, we have modeled the effects of GAG binding on chemokines [7,8], growth factors [9,10] and other proteins [11,12], which allowed us to investigate the fundamental questions related to these interactions such as their specificity, the role of multipose character of GAG binding and polarity of binding poses of these periodic molecules.

In this work, we model interactions of GAGs with procollagen C-proteinase enhancer-1 (PCPE-1, encoded by gene *PCOLCE*), a glycoprotein which plays an important role in the assembly of the extracellular matrix [13,14]. Lacking proteolytic activity on its own, PCPE-1 enhances C-terminal procollagen processing, mediated by tolloid-like proteinases such as bone morphogenetic protein 1 (BMP-1) and mammalian tolloid (mTLD) designated BMP-1/tolloid-like proteinases (BTPs) [14–17]. PCPE-1 expression is upregulated in fibrosis [18,19]. PCPE-1 comprises two complement, sea urchin protein Uegf, BMP-1 (CUB) domains [20] and a netrin-like (NTR) domain [21]. Although neither an X-ray nor an NMR structure is available for full-length PCPE-1, X-ray structure of CUB1-CUB2 domains (PDB ID: 6FZV, 2.7 Å) in a complex with C-propeptide of procollagen [22] and NMR structure of the NTR domain (PDB ID: 1UAP) are available [23]. In the structure of the active CUB1-CUB2 fragment of PCPE-1 bound to the C-propeptide trimer of procollagen III (CPIII), two Ca^{2+} ions participate in the formation of the interface between the CUB1-CUB2 domains and the procollagen III molecule [22]. Often, CUB domains bind Ca^{2+}, and Ca^{2+} coordination involves acidic amino acid residues (i.e., Tyr-Glu-Asp-Asp motif) [24]. A conserved calcium binding site has indeed been identified in the CUB1 domain of PCPE-1, and mutational analysis of this site confirmed that PCPE-1 stimulating activity requires a calcium binding motif in the CUB1 domain, which is highly conserved among CUB-containing proteins [25]. A low-resolution structure of the full-length PCPE-1 protein was proposed based on small angle X-ray scattering (SAXS), analytical ultracentrifugation and transmission electron microscopy, showing that PCPE-1 is a rod-like molecule, with a length of 150 Å [26]. PCPE-1 binds to heparin (HP) as shown using affinity chromatography [27] and surface plasmon resonance (SPR) binding assays [28], and the binding is mediated by the NTR domain. Heparan sulfate (HS) and dermatan sulfate (DS) but not chondroitin sulfate (CS) inhibit PCPE-1-HP interactions. HP also binds to BMP-1 [29]. HS could thus potentially act as a scaffold to assemble BMP-1, PCPE-1 and procollagen together at the cell surface [28]. Therefore, the characterization of PCPE-1/GAG interactions at the atom level is important for the detailed understanding of PCPE-1 functions.

The aim of this work is to get deeper insights into PCPE-1/GAG interactions using both SPR inhibition assays and in silico techniques to complement the experimental data obtained in the previous [28] and present work. Modeling approaches were used to build structural models of full-length PCPE-1 and to determine GAG specific binding to PCPE-1 and its domains. We analyzed the binding of PCPE-1 to GAGs of different types, lengths and sulfation patterns, which were rationally and systematically chosen to match those used in experiments. We also investigate the potential role of Ca^{2+} in these interactions [28] and evaluate the challenges of in silico methodology to study protein-GAG interactions [30]. The results reported here contribute to the understanding of the biologically relevant PCPE-1/GAG interaction and, for the first time, systematically predict the structural positions and the effects of Ca^{2+} ions on protein-GAG complexes.

2. Results and Discussion

2.1. Experimental Results

We have previously shown that DS, HS and HP but not CS inhibited the binding of soluble PCPE-1 to immobilized HP [28]. Here, we investigated the effect of HP oligosaccharides of various length as inhibitors of PCPE-1 binding to HP in order to determine the optimal size of HP required to bind to PCPE-1. There was a trend towards an increase in inhibition of PCPE-1-HP interaction with the length of HP oligosaccharides from dp2 to dp8, and then from dp14 to full-length HP chains (Figure 1). HP decasaccharides and dodecasaccharides (dp10 and dp12, respectively) inhibited the binding of PCPE-1 to HP to a lesser extent than the HP octasaccharide (dp8). The oligosaccharides used for inhibition experiments were separated according to their size and not to their sulfation pattern and/or charges. They thus contain a mixture of oligosaccharides of the same size displaying a different number of sulfate groups in different positions of their sequences resulting in different binding motifs with likely different inhibitory efficiencies. This heterogeneity might be more pronounced in dp10 and dp12, leading to a lower global inhibition by these oligosaccharides than by the octasaccharide.

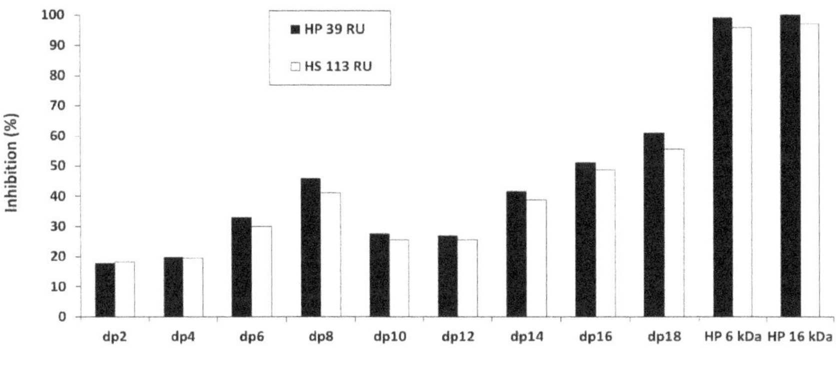

Figure 1. Surface plasmon resonance (SPR) inhibition assays. Inhibition of the binding of recombinant human procollagen C-proteinase enhancer-1 (PCPE-1) to biotinylated heparin (HP) and heparan sulfate (HS) captured on a streptavidin sensor chip (39 and 113 resonance units (RU) respectively) by HP oligosaccharides of different degrees of polymerization (dp2-dp18) and by HP (6 and 16 kDa) at a concentration of 5 µg/mL.

Then we applied the in silico approaches we have previously developed to analyze the binding of PCPE-1 to GAGs at the atomic level and to determine if these interactions were exclusively electrostatic-driven or if other factors modulate the binding strength.

2.2. Modeling the Full Structure of PCPE-1

We created two ensembles of full-length PCPE-1 structures using the UNRES (from UNited RESidue) coarse-grained (CG) approach to determine the structure of the linker located between the CUB1-CUB2 and the NTR domains. In the first one, the structures of the linkers were optimized, and the domain structures were restrained, while in the second one, SAXS derived restraints were used additionally in order to reproduce the experimental data [26] (see Section 3.4 for more details). Five most probable structural models were obtained for both ensembles. For HP binding analysis we used the first three models obtained without SAXS restraints and one model obtained with SAXS restraints (SAXS Model) (Table 1). The radii of gyration of the models obtained without SAXS restraints were significantly lower than those of the elongated structures restrained using SAXS data. As expected, the SAXS Model had a radius of gyration in agreement with the experimental value calculated using SAXS (41 ± 3 Å versus 43 ± 1 Å [26]. The obtained model was also consistent with the length of the protein determined experimentally (150 Å). Poisson-Boltzmann surface area (PBSA) calculations applied to these 4 models suggest that potential binding regions for HP were located in the NTR domain for Model 3, at the interface of the linker and the NTR domain for Model 2 and SAXS Model, and at the common interface of all domains (CUB1-CUB2, linker and NTR) in Model 1 (Figure 2, Supplementary Figure S1).

Table 1. Models of the full-length PCPE-1 obtained using UNRES coarse-grained (CG) simulations.

Model	Restraints	Probability	Radius of Gyration (Å)
1		34	22.2
2		32	24.8
3	CUB1-CUB2, NTR domains	18	22.6
4		8	22.8
5		8	22.6
1		39	43.5
2		21	44.5
3	CUB1-CUB2, NTR domains + SAXS-based	17	43.5
4		14	43.3
5		9	43.9

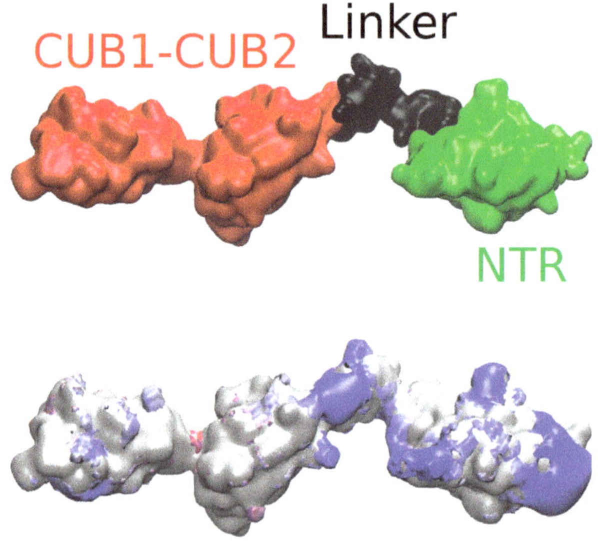

Figure 2. Small angle X-ray scattering (SAXS) Model (upper panel). Netrin-like (NTR) domain: green; CUB1-CUB2: red; the interdomain linker between the CUB2 and NTR domains: black. Positive electrostatic potential isosurfaces (2.0 kcal/mol · e^{-1}) in the absence of Ca^{2+} ions obtained by Poisson-Boltzmann surface area (PBSA) calculations (bottom panel).

2.3. PCPE-1 Interactions with Glycosaminoglycans

We modeled and analyzed the binding of PCPE-1 and its domains, NTR and CUB1-CUB2, with the following GAGs: chondroitin sulfate-6 (CS6) made of two GalNAc6S-GlcA or three disaccharide units (dp4 and dp6, respectively), dermatan sulfate comprised of three GalNAc6S-IdoA disaccharide units (dp6), and heparin (HP) made of one, two and three GlcNS6S-IdoA2S disaccharide units (dp2, dp4, and dp6 respectively). These GAGs were selected for the following reasons: to compare the in-silico data with the experimental ones previously obtained with these GAGs [28] and to investigate the effects of epimerization, length and sulfation pattern of GAGs on binding. Conventional docking approaches are severely limited in terms of the size of GAGs and can be effectively used only for the GAGs with a length up to dp6 [31]. Therefore, we used HP oligosaccharides of different lengths, from dp2 to dp6, to determine the effect of the GAG length on the binding to PCPE-1. Since HP is the strongest binder, the results obtained with HP oligosaccharides of different lengths should be the most representative. Furthermore, the GAGs studied here were selected in order to systematically evaluate the changes in binding to PCPE-1 according to the GAG length (dp4-dp6 for CS6 and dp2-dp6 for HP), the epimerization of glucuronic acid (CS6 dp6 and DS dp6), the increase in the number and position of sulfated groups (i.e., the sulfation pattern) and the net charge of the oligosaccharides (CS6, DS and HP).

Several clusters of docking solutions were obtained for each GAG tested. The polarity of the binding poses was analyzed because the orientation of the GAG chain was shown to be non-random for the IL-8 chemokine [7] and determinant for the binding specificity of the C-X-C motif chemokine ligand 14 [8], suggesting an important functional role of GAG polarity in their interactions with proteins. Then, for the most diverse binding poses within these clusters, molecular dynamics (MD) simulations were performed with binding free energy post-processing calculations and per residue binding free energy decomposition. We would like to emphasize that choosing a proper procedure of pose selection for such analysis is very challenging, since it is unclear how many clusters and solutions within each cluster should be representative, which part of the trajectory should be analyzed in terms of the free energy, if only the best scored pose from a cluster or all the poses should be taken into account for the further calculations, and how to weight their contributions in the latter case. The answers to these questions are dependent on the molecular systems and on the particular goal of the modeling study. These methodology-oriented aspects of protein-GAG modeling will be further discussed below.

2.3.1. The NTR-Domain

Among the found clusters of docking solutions, for CS6 dp4 and HP dp2, dp4 and dp6, one major cluster was observed, while there was more uniform distribution of the solutions between several clusters for CS6 dp6 and DS dp6 (Table 2). This suggests that for those molecules, especially for CS6 where the clusters are especially diverse, multipose binding might be quite probable. GAG multipose binding was previously identified both experimentally and computationally for TIMP-3, which is homologous with the NTR domain [11]. Most solutions were localized near the C-terminal α-helix of the NTR domain except for CS6 dp6 (Figures 3 and 4). The size of the clusters obtained by molecular docking was not correlated with their corresponding free binding energies calculated from the MD simulation. This means that molecular docking alone was not able to properly score the solutions, although the Autodock 3 (AD3) scoring function is one of the most successful scoring schemes when applied to GAG complexes [10]. Similarities of the binding regions for the docking solutions post-processed by MD-based binding free energy decomposition per residue are reflected in Tables 3 and 4 for the obtained clusters and for each GAG ligand respectively. According to the binding free energy values obtained for the NTR domain bound to GAGs compared to the experimental complexes from the PDB [31] and given that no dissociation of these complexes was observed, we assume that the binding of the analyzed GAGs to NTR is stable. The binding strength, evaluated by the calculation of free binding energy, of CS6 dp4 and CS6 dp6 did not significantly differ, but the cluster location of CS6 dp6 differed from those of CS dp4, DS dp6 and HP dp2, dp4 and dp6. Only the third biggest cluster for CS6 dp6 was located in the region overlapping with those of other analyzed GAGs. CS6 dp6 and longer CS6 oligosaccharides might thus bind NTR differently from CS6 dp4 and other GAGs. Therefore, although the binding strength was similar for CS6 and DS, their preferred binding sites were distinct for these two GAGs, which differ only in the epimerization of glucuronic acid. This could potentially explain the results from surface plasmon resonance binding assays, which showed that CS6 did not inhibit PCPE-1 binding to HP whereas DS did [28]. Whereas DS competes with HP for the same binding site on PCPE-1, CS6 binds to a different region, which would allow HP oligosaccharides to remain bound to the NTR domain. Similar computational approaches were successfully applied to demonstrate the experimentally proven differences in binding strength between DS and CS6 interacting via the same binding pose to IL-8 [7,32] In contrast, the binding differences for those GAGs were related to certain differences in the binding pose for CXCL14 [8]. This suggests that for protein-GAG complexes, the predictive power of the computational methods is dramatically dependent on the protein involved and the distribution of the clusters on its surface, which is, in turn, also sensitive to a particular clustering procedure. HP binds the NTR domain stronger than CS6 and DS, while its increase in length stabilizes the interaction suggesting a key role of electrostatic interactions, although few hydrophobic amino acid residues (leucine and valine) and polar, uncharged, amino acid residues (asparagine and glutamine) were predicted to interact with the analyzed GAGs (Figure 4). Most clusters revealed a bias towards specific polarity of GAG binding poses, although this trend was less pronounced for HP dp6 and DS dp6 (Table 2). This suggests that our docking approach is able to distinguish GAG polarity, which is an important methodological finding and will allow us to investigate one of the potential parameters underlying the specificity of protein-GAG interactions [8].

Table 2. Molecular docking molecular dynamics (MD)-based analysis summary for NTR-GAG interaction.

GAG	[1] m, ε	[2] #	[3] Size	[4] ΔG (kcal/mol)	[5] Polarity
CS6, dp4	3, 2	1	19	-42.0 ± 6.6; -48.3 ± 7.7; -41.3 ± 6.6	17/2
		2	6	-30.1 ± 16.0; -63.3 ± 7.1	6/0
		3	4	-34.4 ± 9.6; -38.4 ± 8.6	2/2
		4	3	-46.7 ± 10.5	3/0
CS6, dp6	3, 2	1	3	-56.6 ± 9.0	3/0
		2	3	-33.9 ± 9.2	3/0
		3	3	-36.8 ± 7.1; -64.2 ± 11.8	3/0
DS, dp6	3, 2	1	6	-35.5 ± 6.3; -41.5 ± 6.8	5/1
		2	4	-36.7 ± 6.6	4/0
		3	3	-63.7 ± 8.3	3/0
		4	3	-37.8 ± 8.2	2/1
HP, dp2	3, 2	1	25	-44.9 ± 9.3; -41.1 ± 7.3; -23.1 ± 7.6	25/0
		2	12	-27.9 ± 9.2	12/0
		3	9	-42.0 ± 9.0	9/0
		4	3	-27.7 ± 8.9; -28.7 ± 5.6	3/0
HP, dp4	3, 2	1	32	-39.0 ± 7.2; -29.4 ± 10.4	21/11
		2	3	-53.9 ± 7.2	3/0
		3	3	-50.6 ± 11.5; -57.4 ± 8.6	2/1
HP, dp6	3, 2	1	15	-69.5 ± 7.8; -56.7 ± 7.4; -43.5 ± 9.7; -54.0 ± 14.8; -80.5 ± 10.7	9/6
		2	7	-68.3 ± 11.0; -44.7 ± 7.5; -57.1 ± 10.6; -55.4 ± 9.1	4/3
		3	6	-50.1 ± 10.0; -44.6 ± 9.5; -65.7 ± 11.5; -61.8 ± 13.6	4/2

[1] DBSCAN parameters m, the minimal neighborhood size, and ε, neighborhood search radius [33]; [2] cluster number; [3] cluster size (number of solutions); [4] free energy of binding obtained by MM-GBSA; [5] the polarity of a GAG binding pose was defined as its preferred orientation in relation to the reducing and non-reducing end.

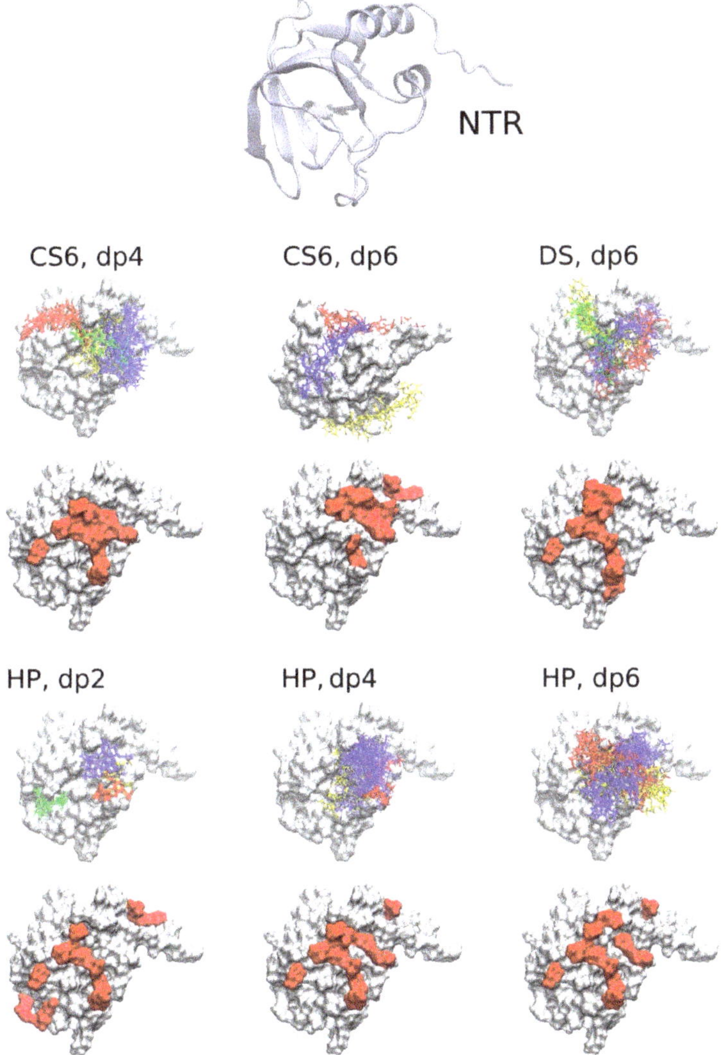

Figure 3. Molecular docking and molecular mechanics-generalized born surface area (MM-GBSA) for NTR-glycosaminoglycan (GAG) complexes. The structure of the NTR domain is shown in cartoon representation at the top. For each GAG, the analyzed clusters of docking solutions are shown in blue, red, yellow and green (from the most to the less populated cluster); the top 10 residues binding to GAGs according to MM-GBSA calculations averaged per GAG are highlighted in red surface. Note that the clusters for CS6 dp6 are shown for a different protein spatial orientation to allow for a better visualization. In addition, averaging the per-residue energy for very different clusters could be misleading as shown for CS6 dp6: the residues shown in red do not overlap with the surface patches where the most representative clusters of solutions are located.

```
318     C P K Q C R R T G T L Q S N F C A S S L V V T A T V K S M V R E P G E G L A V T   357
CS6 dp4           * *   * *
CS6 dp6             *   * * *
DS dp6        *   * *
HP dp2            * *                                               *
HP dp4            * *   * *
HP dp6            * *   * *

358     V S L I G A Y K T G G L D L P S P P T G A S L K F Y V P C K Q C P P M K K G V S   397
CS6 dp4                                                                       *
CS6 dp6
DS dp6                                                                        *
HP dp2                                                            *           *
HP dp4                                                                        *
HP dp6                                                                        *

398     Y L L M G Q V E E N R G P V L P P E S F V V L H R P N Q D Q I L T N L S K R K C   437
CS6 dp4                                       * *   *   *     *
CS6 dp6                                                 *   * *     *             * *
DS dp6                                          *   *   * * * *
HP dp2                                          *   *   *                         * *
HP dp4                                          *   *   *     *
HP dp6                                          *   *   *   *                     *
```

Figure 4. NTR amino acid residues identified in the top 10 for binding GAGs according to MM-GBSA calculations per cluster are labeled as an asterisk.

Table 3. Similarity of GAG binding poses for the NTR domain as of common amino acid residues identified in the top 10 for binding according to MM-GBSA calculations per cluster.

GAG	CS6, dp4				CS6, dp6				DS, dp6				HP, dp2			HP, dp4			HP, dp6		
CS6, dp4	10	7	7	6	6	4	7	5	6	6	4	7	5	5	5	6	6	6	7	7	7
	7	10	6	9	3	2	6	6	6	7	5	6	7	6	6	5	8	7	8	7	9
	7	6	10	5	4	3	5	6	5	7	4	7	4	6	5	4	5	7	7	7	6
	6	9	5	10	3	2	5	6	6	6	4	5	7	5	5	6	9	6	7	6	8
CS6, dp6	6	3	4	3	10	7	5	2	5	3	1	4	3	1	1	5	3	2	5	5	4
	4	2	3	2	7	10	5	2	4	2	1	5	2	1	1	4	2	1	4	3	3
	7	6	5	5	5	5	10	5	5	4	5	6	5	4	4	7	5	4	7	6	7
DS, dp6	5	6	6	6	2	2	5	10	5	6	6	6	6	7	6	5	6	8	7	7	7
	6	6	5	6	5	4	5	5	10	5	5	4	6	4	4	6	7	5	7	6	6
	6	7	7	6	3	2	4	6	5	10	6	6	6	6	6	3	5	7	7	8	6
	4	5	4	4	1	1	5	6	5	6	10	4	4	5	5	3	4	6	5	5	5
HP, dp2	7	6	7	5	4	5	6	6	4	6	4	10	5	5	5	5	5	6	7	7	7
	5	7	4	7	3	2	5	6	6	6	4	5	10	5	5	7	7	5	7	7	7
	5	6	6	5	1	1	4	7	4	6	5	5	5	10	6	3	5	7	6	6	6
	5	6	5	5	1	1	4	6	4	6	5	5	5	6	10	3	5	6	6	6	6
HP, dp4	6	5	4	6	5	4	7	5	6	3	3	5	7	3	3	10	7	3	6	5	6
	6	8	5	9	3	2	5	6	7	5	4	5	7	5	5	7	10	6	7	6	8
	6	7	7	6	2	1	4	8	5	7	6	6	5	7	6	3	6	10	6	7	7
HP, dp6	7	8	7	7	5	4	7	7	7	7	5	7	7	6	6	6	7	6	10	9	9
	7	7	7	6	5	3	6	7	6	8	5	7	7	6	6	5	6	7	9	10	8
	7	9	6	8	4	3	7	7	6	6	5	7	7	6	6	6	8	7	9	8	10

Each line/column in front/below each GAG reflects a separate cluster, for which average values were taken into account.

Table 4. Similarity of GAG binding poses for the NTR domain as of the number of common amino acid residues identified in the top 10 for binding according to MM-GBSA calculations per GAG.

GAG	CS6, dp4	CS6, dp6	DS, dp6	HP, dp2	HP, dp4	HP, dp6
CS6, dp4	10	5	7	6	9	8
CS6, dp6	5	10	4	4	6	6
DS, dp6	7	4	10	6	7	7
HP, dp2	6	4	6	10	7	7
HP, dp4	9	6	7	7	10	9
HP, dp6	8	6	7	7	9	10

2.3.2. CUB1-CUB2 Domains

Although there is no experimental evidence suggesting that CUB1-CUB2 domains of PCPE-1 directly interact with GAGs, the differences in binding of NTR and full-length PCPE-1 to HP and HS [28] indicate that CUB1-CUB2 domains could affect GAG binding to the full-length PCPE-1 protein. Therefore, we analyzed the potential binding of these domains to GAGs using the same procedure as above.

All the predicted binding poses were either located in the cleft region between the CUB domains or bridged both CUB domains (Supplementary Figures S2 and S3). In both cases, such potential binding would lead to restricted movements of the CUB domains relative to each other, which, in turn, would affect the overall flexibility of PCPE-1 and its ability to recognize and to bind its partners. Calculated GAG free binding energies were essentially less favorable than those calculated for the NTR domain (Supplementary Table S1), which is consistent with NTR being responsible for GAG binding in PCPE-1. No binding poses of the analyzed GAGs or the structures that can be obtained from them by GAG chain elongation were found to be in close proximity to the Ca^{2+} binding sites or at the interface with procollagen peptides [23]. In a number of cases, the binding poses predicted by molecular docking were unstable (ΔG higher than -15 kcal/mol), and the GAG dissociated from the protein. Such behavior was typically observed for HP oligosaccharides and is explained by the repulsion of these highly charged molecules by the negatively charged residues of the CUB1-CUB2 domain. Poses corresponding to the binding of CS6 and DS, which are less negatively charged than HP, were globally more stable. However, some binding poses were very stable and comparable with those found in the NTR domain (e.g., cluster 2, solution 2 for CS6 dp4). In such cases, bound GAGs protruded deeply into the cleft between the CUB1 and CUB2 domains forming strong van der Waals interactions in addition to the electrostatic interactions, which are believed to be the driving force in the formation of protein-GAG complexes [31,34]. As reported for the NTR domain, highly significant differences in free energy were found for GAGs within and beyond the same clusters. One major cluster was found for CS6 of various length in contrast to what was observed for other GAG analyzed. No correlation was found between the size of clusters and their free binding energies. The comparison between the observed clusters and the data averaged for different GAGs in terms of the most important protein binding residues showed high similarities for all GAGs, suggesting weak and rather unspecific binding to CUB1-CUB2 domains (Supplementary Figure S2, Supplementary Tables S2 and S3). Interestingly, for CS6 dp4 the differences between the clusters were more prominent than the differences of these clusters with those obtained for other GAGs. The increase in length of HP from dp2 to dp6 did not modify the potential interaction pattern with the CUB1-CUB2 domains. All clusters revealed strong polarity preferences except for the DS clusters.

2.3.3. Full PCPE-1

GAG binding was characterized with full-length PCPE-1 models obtained using UNRES CG simulations and HP dp6 as a ligand. Binding to Model 1, which was the most probable model among the ones obtained without the SAXS-based restraints, was significantly stronger than to Models 2, 3 and the SAXS Model (Table 5, Supplementary Table S4), as well as to the NTR domain (t-test, *p*-value < 0.05). Binding to Model 3 was also significantly stronger than to Model 2 and to the NTR domain. All clusters of HP dp6 solutions obtained for Model 1 were located in the region formed by the same residues of the NTR domain, the linker and the CUB1-CUB2 domains. For Model 2, the first cluster was located differently from the second and the third clusters. All the clusters correspond to the residues belonging predominantly to the NTR domain and the linker, but also partially to the CUB1-CUB2 domains. For Model 3, only NTR residues contributed to the binding of HP dp6. Cluster 1 was the most representative for Model 3 according to the molecular docking results, although not the most favorable according to MM-GBSA calculations, which again points to the essential differences in molecular docking and MD-based scoring. In the SAXS Model, GAG binding occurred at the NTR/linker interface. In all cases, the clusters were located in PCPE-1 patches corresponding to the positive electrostatic potential shown in Figure 2 and Figure S1.

Table 5. Molecular docking MD-based analysis summary for PCPE-1 SAXS Model/HP dp6 interaction.

[1] m, ε	[2] #	[3] Size	[4] ΔG, kcal/mol	[5] Top$_{\text{MM-GBSA}}$ 10 Residues for GAG Binding	[6] Polarity
2, 2.64	1	4	$-62.4.8 \pm 19.0$; -54.9 ± 9.1 -49.6 ± 18.6	R435, K436, R275, R288, K279, K299, K365, K434, N331, K295	4/0
	2	3	-50.1 ± 9.7; -79.0 ± 17.0; -38.1 ± 9.4	K436, R435, K365, K299, K434, K271, K295, R288, K165, K279	3/0
	3	3	-30.8 ± 10.7; -36.0 ± 7.8; -42.3 ± 10.6	K299, K436, K279, K365, K271, K434, K295, K165, Q282, R435	2/1

[1] DBSCAN parameters m, the minimal neighborhood size, and ε, neighborhood search radius [33]; [2] cluster number; [3] cluster size; [4] free energy of binding obtained by MM-GBSA; [5] residues identified in the top 10 for binding according to MM-GBSA calculations per cluster ordered by the impact (starting from the most favorable one). [6] The polarity of a GAG binding pose was defined as its preferred orientation in relation to the reducing and non-reducing end.

2.4. The Potential Role of Ca^{2+} in PCPE-1 Interactions with Glycosaminoglycans

2.4.1. Prediction of Ca^{2+} Binding Sites

According to the experimental data, the interactions between both full-length PCPE-1 and the NTR domain with HP and HS are cation-dependent [28]. Therefore, we attempted to analyze the impact of Ca^{2+} ions on HP binding in silico, which allowed us to evaluate the available computational tools in terms of sensitivity and prediction power to account for divalent ions in such calculations. As a first step, we applied three different approaches (see Section 3.7 for details) to annexin V protein, which has 9 experimentally identified occupied Ca^{2+} binding sites, some of which are occupied upon HP binding [35]. The IonCom server predicted correctly eight out of nine experimentally known binding sites, while FoldX and MD approaches correctly predicted six binding sites (Table 6).

Table 6. Ca^{2+} predictions for annexin V and PCPE-1 domains: number of the binding sites predicted are provided.

Protein	PDB ID	Experimental Structure	Method		
			FoldX	IonCom	[1] MD
Annexin V	1G5N	9	6	8	6
NTR	1UAP	0	0	0	2
					3
					1
					1
					1
CUB1-CUB2	6FZV	2	2	2	2
					2
					1
					2
					1

[1] Five repetitions of the MD simulations were performed for PCPE-1 domains.

Furthermore, we performed MM-GBSA calculations to estimate if the strength of the Ca^{2+} binding in these experimentally known binding sites correlated with the predictions (Table 7). As shown in the table, the total energies of interactions were positive despite the fact that all the ions were stable during the entire MD simulation performed in explicit solvent. This reflects the fact that the implicit continuous solvent model in MM-GBSA fails to properly account for the strength of binding for these divalent ions in terms of the full binding free energy. At the same time, in vacuo electrostatic energy was highly negative and could be meaningful for comparing binding sites since the studied interactions were electrostatically driven. A t-test performed for the in vacuo electrostatic energy values did not point out any statistical differences between the sites, which were properly predicted and the ones which the MD-based approach failed to predict.

Table 7. MM-GBSA free energy calculations (per Ca^{2+} ion) for the experimentally known Ca^{2+} binding sites in annexin V.

Ca^{2+} Number (X-Ray)	[1] ΔG, kcal/mol	[2] ΔG_{ele}, kcal/mol	[3] FoldX	[3] IonCom	[3] MD
319	57.2 ± 4.7	−310.4 ± 10.3	+	+	+
320	47.5 ± 4.8	−264.8 ± 15.7	+	+	−
321	36.5 ± 3.5	−296.0 ± 10.9	−	+	+
322	59.7 ± 4.9	−380.5 ± 9.5	+	+	−
323	36.4 ± 3.5	−332.4 ± 7.9	−	−	+
324	62.4 ± 4.4	−376.6 ± 8.1	+	+	+
325	47.5 ± 6.1	−413.2 ± 13.0	+	+	+
326	39.3 ± 3.7	−312.2 ± 9.2	−	+	+
327	59.2 ± 4.7	−302.3 ± 8.6	+	+	−

[1] and [2]: ΔG and ΔG_{ele} stand for the total and in vacuo electrostatic MM-GBSA free energies, respectively. [3] Plus and minus reflect whether the method was capable of predicting the corresponding experimentally detected binding site correctly.

We applied three ion-binding site prediction methods to the NTR and the CUB1-CUB2 domains of PCPE-1 (Table 6). Neither FoldX nor IonCom found any Ca^{2+} binding site for the NTR domain, while the MD approach identified from one to three binding sites, one of which being consistent through all five repetitions of MD simulations. The fact that these methods did not agree with MD simulations could be due to conformational changes of negatively charged amino acid side chains during the MD simulation, allowing them to come close to each other and to coordinate calcium ions. FoldX and IonCom used static structures, which prevents the dynamics required for the coordination of Ca^{2+}. For CUB1-CUB2 domains, all methods were consistent and predicted two Ca^{2+} binding sites identical to those found in the CUB1-CUB2 domain complexed with procollagen (PDB ID: 6FZV). This means that CUB1-CUB2 domains in PCPE-1 could be already prebound to Ca^{2+} ions when the interaction with the procollagen is established. Furthermore, we compared the predicted Ca^{2+} binding sites for PCPE-1 domains in terms of electrostatic energies obtained from MM-GBSA calculations with the corresponding energies for annexin V in order to estimate their strength (Table 8). CUB1-CUB2 binding sites were energetically comparable with those of annexin V. The Ca^{2+} binding site in CUB2 was stronger than in CUB1 as suggested by more favorable electrostatic energies and by the fact that the Ca^{2+} binding site in CUB2 was identified by MD in all five MD replicas, while the Ca^{2+} binding site of CUB1 was correctly identified in three MD simulations. The Ca^{2+} binding sites predicted for the NTR domain were significantly weaker, and only one of them was found in all MD replicas. This suggests that this site may be unoccupied when the NTR domain is in solution and not bound to a GAG. The experimental evidence that the NTR domain binding to HP is dependent on divalent cations [28] leads to the hypothesis that Ca^{2+} ions could potentially bind within the interface of the NTR-HP complex.

Table 8. MM-GBSA free energy calculations (per Ca^{2+} ion) for the predicted Ca^{2+} binding sites in PCPE-1 domains and corresponding Ca^{2+} binding site occupancy in 100 ns MD simulation.

PCPE-1 Domain	Ca^{2+} Site	^{2} ΔG_{ele}, kcal/mol	Site Occupancy, ns
NTR, MD1	E405, E406, N407	-116.5 ± 20.4	65
	G367, D370	-58.3 ± 15.7	40
NTR, MD2	E405, E406, N407	-125.8 ± 14.7	85
	D314/N-terminus of NTR	-38.6 ± 19.0	35
	G367, D370	-49.6 ± 14.2	90
NTR, MD3	E405, E406, N407	-120.3 ± 15.9	75
NTR, MD4	E405, E406, N407	-123.4 ± 12.2	45
NTR, MD5	E405, E406, N407 ^{1} E405, E406, N407/G367, D370	-51.7 ± 13.7; -156.2 ± 33.6	65 25
CUB1-CUB2 (X-ray, PDB ID: 6FZV)	E85, Y92, D93, D134	-363.2 ± 10.8	100
	Y180, E208, D216, D258	-466.5 ± 12.5	100
CUB1-CUB2, MD1	E85, Y92, D93, D134	-389.9 ± 19.3	25
	Y180, E208, D216, D258	-302.9 ± 24.3	90
CUB1-CUB2, MD2	E85, Y92, D93, D134	-368.3 ± 18.7	85
	Y180, E208, D216, D258	-371.9 ± 11.8	90
CUB1-CUB2, MD3	Y180, E208, D216, D258	-389.2 ± 18.6	85
CUB1-CUB2, MD4	E85, Y92, D93, D134	-293.7 ± 9.0	75
	Y180, E208, D216, D258	-374.0 ± 16.1	95
CUB1-CUB2, MD5	Y180, E208, D216, D258	-521.4 ± 10.8	95

^{1} In the course of this simulation, G367 and D370 moved towards E405, E406 and N407 to coordinate Ca^{2+}. (MD: molecular dynamics, 1–5: replicas). ^{2} ΔG_{ele} stands for the in vacuo electrostatic MM-GBSA free energy.

We calculated electrostatic potential isosurfaces for the NTR and CUB1-CUB2 domains in the presence and in the absence of Ca^{2+} ions using the PBSA approach to predict how the electrostatic properties of the protein were affected by Ca^{2+} ions binding, which, in turn, could have an impact on GAG binding. For this, we used two Ca^{2+} binding sites corresponding to the X-ray structure (PDB

ID: 6FZV) of the CUB1-CUB2 domain, and the weak Ca^{2+} binding site predicted in the NTR domain (Figure 5). Major differences in both positive and negative electrostatic potential shape were observed for the NTR domain. This is not only explained by the direct effect of Ca^{2+} positive charge but also by the fact that E405, E406 and N407 were moved closer to each other to coordinate the cation, which also affects the topology of the isosurface. For CUB1-CUB2 domains, the largest positively charged patch of the potential isosurface was not noticeably affected by the presence Ca^{2+} ions. To sum up, the predicted 3 Ca^{2+} binding sites for both PCPE-1 domains, when occupied, could potentially affect GAG binding. This potential effect is analyzed below.

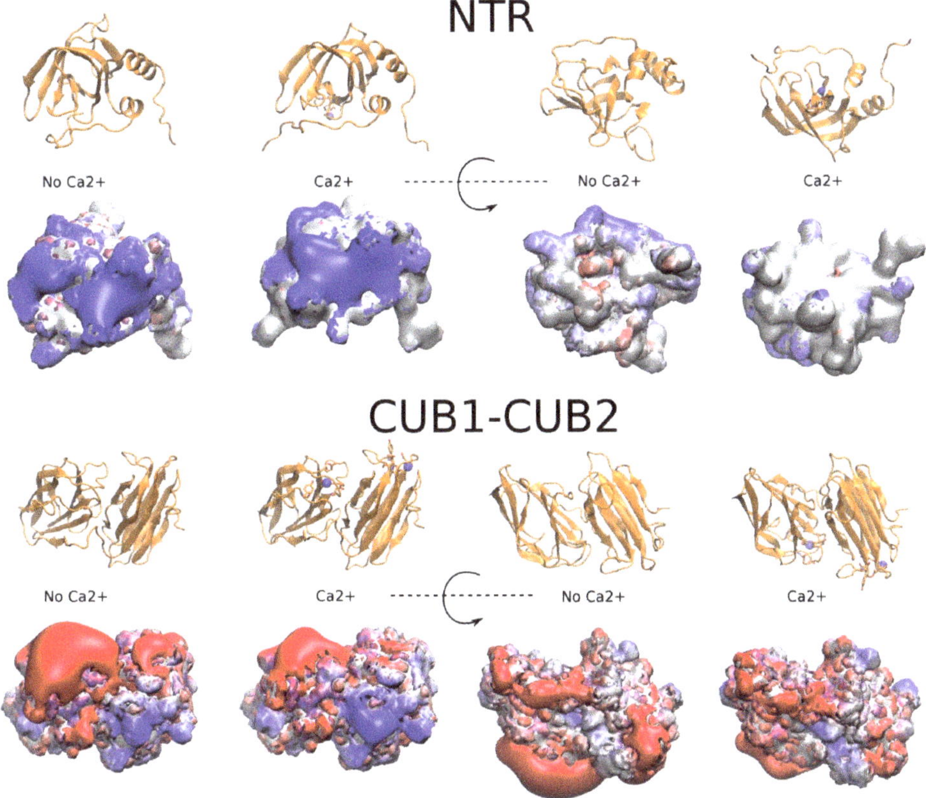

Figure 5. Electrostatic potential isosurfaces (blue, positive; red, negative) of NTR (-2.5 kcal/mol·e^{-1} and 1.0 kcal/mol·e^{-1}) and CUB1-CUB2 (-3 kcal/mol·e^{-1} and 3 kcal/mol·e^{-1}) domains in the presence and in the absence of Ca^{2+} ions obtained by PBSA calculations. Protein domains are shown in cartoon with the residues coordinating Ca^{2+} ions in licorice representation; Ca^{2+} ions: blue spheres.

2.4.2. PCPE-1 Interactions with Glycosaminoglycans in the Presence of Ca^{2+} Ions

In order to analyze the potential impact of Ca^{2+} ions on PCPE-1-GAG interactions, HP dp2, dp4 and dp6 were docked onto NTR, CUB1-CUB2 and the full-length protein in the presence of Ca^{2+} ions, followed by MD-based analysis. Two Ca^{2+} ions were prebound to CUB1-CUB2 and one to the NTR domain. Despite the differences in electrostatic properties of these domains described above, no significant changes in docking results or binding free energies were observed when compared with the domains without prebound Ca^{2+} (Supplementary Figures S4 and S5 and Supplementary Tables S5 and S6). In both cases, the binding occurred in the regions distant from the Ca^{2+} ion binding sites.

Only one cluster (HP dp6, cluster 3 in the NTR domain) was found to be close to the Ca^{2+} ion. In the corresponding binding poses the ion was coordinated by a sulfate group from the terminal GlcNS(6S) residue. However, the binding poses from this cluster were significantly less favorable than those located distantly from the Ca^{2+} ion.

We also analyzed the impact of Ca^{2+} ions on the GAG binding to the full-length PCPE-1 (Table 9, Supplementary Table S7, Figure 6, Supplementary Figure S6). For Models 1 and 2 and the SAXS Model, we did not observe any effect of Ca^{2+} ions on the location of the structural clusters nor the direct participation of the Ca^{2+} ions in binding HP dp6. For Model 3, there was a significant difference between the clusters observed in the absence and in the presence of Ca^{2+} ions (e.g., cluster 2) related to the essential changes of the electrostatic potential on the protein surface in the presence of Ca^{2+}. In terms of the free energies of binding, the presence of Ca^{2+} ions did not significantly affect binding to Models 1 or 2 or the SAXS Model. For cluster 2 of Model 3, which was relocated in the presence of Ca^{2+} ions, the free energy of binding became less favorable than in the absence of Ca^{2+} ions.

Table 9. Molecular docking MD-based analysis summary for PCPE-1 SAXS Model/Ca^{2+}/HP dp6 interaction.

[1] m, ε	[2] #	[3] Size	[4] ΔG, kcal/mol	[5] Top$_{MM\text{-}GBSA}$ 10 Residues for GAG Binding	[6] Polarity
2, 2.8	1	6	−58.8 ± 12.2; −56.0 ± 19.7; −58.6 ± 13.7	R435, K436, K434, K365, K299, K279, K295, R288, P438, K271	5/1
	2	4	−79.5 ± 15.6; −32.6 ± 11.0; −42.8 ± 10.0	K436, R435, K279, R288, K365, K299, K434, Q282, G281, K287	3/1
	3	3	−30.8 ± 10.7; −70.7 ± 13.2; −56.6 ± 18.4	R435, K299, K436, K365, K434, R275, K279, K295, K271, K305	3/0

[1] DBSCAN parameters *m*, the minimal neighborhood size. and *ε*, neighborhood search radius [33]; [2] cluster number; [3] cluster size; [4] free energy of binding obtained by MM-GBSA; [5] residues identified in the top 10 for binding according to MM-GBSA calculations per cluster ordered by the impact (starting from the most favorable one). [6] The polarity of a GAG binding pose was defined as its preferred orientation in relation to the reducing and non-reducing end.

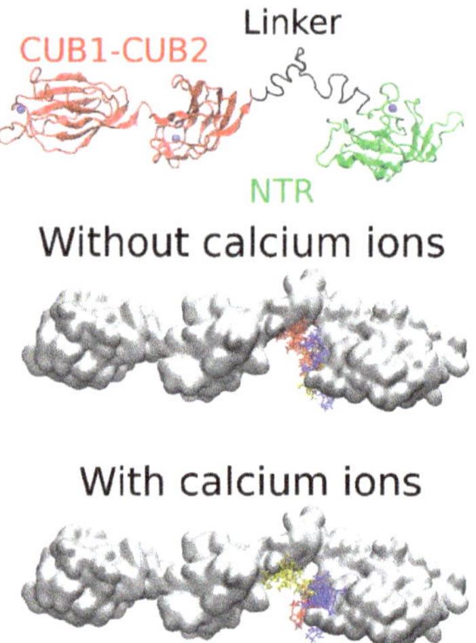

Figure 6. Molecular docking results for the models of the full-length PCPE-1 SAXS Model in the absence and presence of Ca^{2+} ions and HP dp6. The clusters of docking solutions are shown in blue, red and yellow (from the most to the least populated clusters). NTR domain: green; CUB1-CUB2: red; the interdomain linker between the CUB2 and NTR domains: black.

To summarize our attempts to determine the impact of Ca^{2+} on GAG binding to PCPE-1, our approach to consider Ca^{2+} ions as a part of the protein did not detect any favorable effect of these divalent ions on GAG binding in contrast to the experimental data [28]. Therefore, we hypothesize that Ca^{2+} ions might bind to GAGs rather than to PCPE-1 prior to the complex formation. The binding of divalent ions to GAGs has been experimentally reported for Zn^{2+}, Mn^{2+}, Cu^{2+}, Ca^{2+}, Co^{2+}, Na^+, Mg^{2+}, Fe^{3+}, Ni^{2+}, Al^{3+} and Sr^{2+} ions [36,37]. The crucial role of cations in protein-GAG interactions was experimentally shown for amyloid precursor protein [38], HP cofactor II [39], endostatin [40,41], FGF1 and IL-7 [42]. Experimental data obtained using mass spectrometry [43], NMR [44], gel-filtration chromatography [45,46] and infrared spectroscopy [47] indicate that GAGs interact with divalent ions, and that these interactions affect GAG structure and conformational properties. Divalent ions will be integrated in GAG structure in our future work on protein-GAG complexes. Another potential role of Ca^{2+} ions could be to stabilize PCPE-1 structure, which would affect its interactions with GAGs.

2.4.3. Predicting Longer GAG Binding Poses Using the Fragment-Based Approach

We calculated the binding poses of long (dp11) HP chains on full-length PCPE-1 in the absence and the presence of three Ca^{2+} ions, two bound to CUB1-CUB2 and one to the NTR domain (Table 10, Table S8). HP dp11 was the longest GAG that we managed to assemble in these docking experiments, and it was used to model the scenario when the GAGs longer than dp6 are bound to the protein. The increase in length of the HP chain from dp6 to dp11 significantly stabilized (p-value < 0.05) the interactions with PCPE-1 Models 1 and 2. The addition of Ca^{2+} strengthened the interactions of HP dp11 with PCPE-1 for Models 1, 2 and the SAXS Model but not for Model 3. Similarly to what was observed for HP dp6, Model 1 was the strongest in terms of HP dp11 binding. Ca^{2+} ions did not affect the binding sites of HP dp6 on the surface of Models 1, 2 or the SAXS Model but changed binding to Model 3, as described for HP dp6 (Figure 7, Supplementary Figure S7). In all the cases, docked HP dp6 structures overlapped very well with those of HP dp11. It could be concluded that docking HP dp6 defines the core binding unit of a HP chain. The increase in binding affinity with the increase of HP length agrees with the experimental trend observed in this study (Figure 1).

Table 10. Fragment-based molecular docking MD analysis summary for PCPE-1 SAXS Model/Ca^{2+}/HP dp11 interaction.

[1] #	[2] Ca^{2+}	[3] ΔG, kcal/mol	[4] Top$_{MM\text{-}GBSA}$ 10 Residues for GAG Binding
1	–	-65.6 ± 12.3	K299, R288, R435, K436, K295, K293, K305, K287, K365, P298
2	–	-64.2 ± 11.2	K436, K434, R275, K279, K295, R435, K365, R288, K299, K287
3	–	-94.7 ± 12.4	K295, K436, R435, K365, K434, K293, K299, R288, K305, V294
1	+	-93.3 ± 12.3	R435, K279, K295, K436, K305, K299, K434, N331, K271, R324
2	+	-73.8 ± 11.9	R435, K434, K436, K299, K295, K279, P298, K293, K287, K305
3	+	-102.5 ± 14.7	K436, K299, K295, R435, R275, K293, K279, K434, K305, P441

[1] Pose number; [2] Ca^{2+} presence; [3] free energy of binding obtained by MM-GBSA; [4] residues identified in the top 10 for binding according to MM-GBSA calculations per cluster ordered by the impact (starting from the most favorable one).

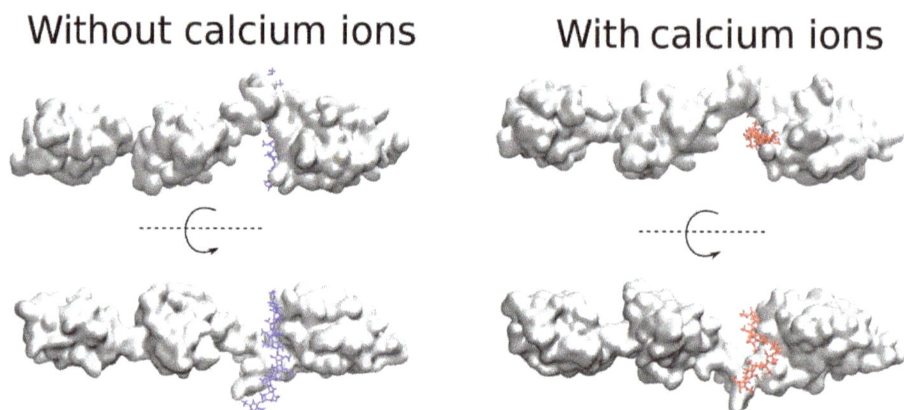

Figure 7. Molecular docking results for the models of the full PCPE-1 SAXS Model in the absence (in blue) and the presence (in red) of Ca^{2+} ions and HP dp11 corresponding to the most favorable free binding energies.

3. Materials and Methods

3.1. Surface Plasmon Resonance (SPR) Binding Assays

The SPR measurements were performed on a BIAcore 3000 instruments (GE Healthcare, Uppsala, Sweden), and the data were analtzed with the BIAevaluation 3.1 Software (GE Healthcare, Uppsala, Sweden) as previously described [28]. Inhibition assays of PCPE-1 binding to HP and HS by HP oligosaccharides (from dp2 up to dp8, generous gift of Rabia Sadir and Hugues Lortat-Jacob, Institut de Biologie Structurale, Grenoble, France) were carried out as previously described [28]. Briefly, HP (Sigma, St Quentin Fallavier, France) and HS (Celsus Laboratories Inc, Cincinnati, OH, USA) from porcine intestinal mucosa were biotinylated and captured on streptavidin previously immobilized on a CM4 sensor chip (GE Healthcare, Uppsala, Sweden). Human recombinant PCPE-1 (1 μM) [28] was incubated with HP oligosaccharides (5 μg/mL) in 10 mM Hepes pH 7.5 + NaCl 0.15 M + P20 0.005% (HBS) + 5 mM $CaCl_2$ for one hour before injection over immobilized HP (39 RUs) and HS (113 RUs) at a flow rate of 30 μL/min for 4 min. The percentages of inhibition were calculated relative to PCPE-1 binding level incubated in the same conditions with HBS + 5 mM $CaCl_2$. The running buffer was HBS and the temperature was set at 25 °C.

3.2. Structures

3.2.1. Protein Structures

The structure of the N-terminal CUB1-CUB2 domains was obtained from the X-ray structure of CUB1-CUB2 fragment of PCPE-1 bound to the C-propeptide trimer of procollagen III (PDB ID: 6FVZ, 2.7 Å). In this structure, the residues 33–275 are resolved (here and further, the numeration of the sequence corresponds to the UniProtKB ID: Q15113). However, the structure of the linker (151–157) between two CUB domains was not determined due to its flexibility [22]. The structure of the NTR domain (313–442) was also obtained from the PDB (PDB ID: 1UAP, 1st NMR model) [23]. The structures of the linker between two CUB domains (151–157) as well as the interdomain linker (276–312) were built in xLeap module of AMBER16, refined using an MD approach and then used for modeling the full-length structure of PCPE-1 with a CG MD approach as described in the Section 3.4.

3.2.2. Glycosaminoglycan Structures

The following GAG structures were used for molecular docking: CS6 dp4, CS6 dp6, DS dp6, HP dp2, dp4, dp3, dp6 with (IdoA(2S) ring in $^1C_4$ conformation, as this conformation is clearly predominant in HP [9,48]). These structures were built in our previous work [49].

3.3. Electrostatic Potential Calculations

The PBSA approach as implemented in AmberTools within AMBER16 package [50] was used to calculate electrostatic potential isosurfaces corresponding to the analyzed proteins. This method proved to be successful for GAG binding site prediction on an extensive protein-GAG dataset from the PDB [31]. The default value for the grid spacing of 1.0 Å and ff99SB force field parameters were used. The electrostatic potential isosurfaces were analyzed in VMD [51]. For visualization, such values of positive and negative electrostatic potential were chosen for each molecular system so that the data were as informative as possible for GAG binding site propensities.

3.4. Coarse-Grained MD Simulations

In order to calculate the conformations of the PCPE-1 linkers (151–157 and 276–312 sequences) to obtain a model of the full-length protein, we applied a CG multiplexed replica exchange molecular dynamics (MREMD) [52,53] approach as implemented in UNRES (United Residues) [54,55], as previously described. The protocol was similar to that used in our previous work [12]. Distance restraints were imposed on the domains during the MREMD simulations. Additionally, in one of the simulations, SAXS-derived restraints [22] were imposed [56]. Each MREMD simulation consisted of 20 trajectories run at temperatures from 265 K to 370 K. Each trajectory consisted of 3.5×10^7 MD steps with 4.89 fs length for simulations with the restraints on the domains only and 7.1×10^4 steps for simulations with additional SAXS-derived restraints. The lower number of steps for simulations with information from the SAXS experiment was used because the radius of gyration of maintained structures was obtained already after that time. Only conformations from the last quarter of the simulation were taken into further analysis with the use of the weighted histogram analysis method (WHAM) [56]. The next step was minimum variance cluster analysis [57] of the conformational ensemble at T = 300 K, which enabled us to obtain five clusters, ranked according to summary probabilities of the ensembles and containing the most probable structures to the cluster with the least probable structures. For each cluster, one representative structure, closest to the cluster centroid, was selected as the representative conformation. The last step was the conversion of CG structures into all-atom ones using the PULCHRA [58] and SCWRL [59] algorithms.

3.5. Molecular Docking

3.5.1. Autodock 3

The docking simulations of GAG ligands to the PCPE-1 were performed with Autodock 3 (AD3) [60], which was previously shown to yield the best performance among docking programs for GAG ligands [10,31]. The blind docking procedure was used: the whole protein surface was available for the ligand when sampling a potential binding site. For all proteins, we used $127 \times 127 \times 127$ grid points for AD3 runs. However, because of the differences in protein size, the grid step was different for different proteins to contain the whole protein molecule within a single grid box: the default value of 0.375 Å was used for the NTR domain, while 0.5 Å was used for CUB1-CUB2 domains, and the grid step of 0.6–1.0 Å was used for different models of the full-length PCPE-1 protein due to their essentially bigger sizes. All GAG ligands were docked to the protein with the use of the Lamarckian genetic algorithm. The initial population size was 300, 10^5 generations, $9995 \cdot 10^5$ energy evaluations and 1000 independent runs with up to 33 torsional angle degrees of freedom were carried out. 1000 docked structures for each molecular system (protein-GAG pair) were obtained and further analyzed; the 50 top-scored ones (according to AD3 scoring function) were chosen for further clustering with the DBSCAN algorithm [33]. The clustering parameters, neighborhood search radii and minimal numbers of cluster members were manually selected for each system individually in order to yield 2-4 representative clusters. The distance metric used for clustering, which is defined as the root-mean-square of atomic distances for the nearest atoms of the same type, takes into account the periodic nature of GAGs, which is more appropriate for those ligands than the classical root-mean-square deviation (RMSD) [61]. Within each cluster, those poses which were different from one another to be categorized into subgroups were selected for further analysis. Such a procedure was used to account for the multiple pose binding previously observed for GAG ligands [11,62]. The GAG glycosidic linkages from the obtained docking poses were visually filtered in order to avoid incorrect geometries that could be produced by AD3 [63].

3.5.2. Fragment-Based Approach

In order to dock longer HP to the full-length PCPE-1 protein, which is unfeasible for the AD3 protocol described above due to the limitation of the available number of the degrees of freedom and, in general, because of the computationally very expensive conformational sampling required for simulations with such ligands, a fragment-based docking approach we recently developed was applied [64]. In brief, first, HP dp3 of both types, IdoA(2S)-GlcNS(6S)-IdoA(2S) and GlcNS(6S)-IdoA(2S)-GlcNS(6S), were docked using the AD3 docking procedure described in Section 3.5.1. Then, all 1000 solutions for each HP dp3 fragments were used to assemble ~10^5 dp11 GAG chains applying the approach standard parameters [64]. This was followed by the refinement and all-atom conversion of the chains with a slight modification in the original scripts to avoid the RMSD-based selection of the best fitting structures compared to the experimental ones due to the lack of proper atomistic experimental PCPE-1/GAG complex structures. In particular, instead of the previously described way of selecting structures, a simple RMSD-based clustering (cutoff 4 Å) was performed to filter out the duplicates and find the most relevant structures. From the resulted ~5-40 atomistic structures the ones with significantly different docking poses were selected and refined together with the full protein using MD simulations applying the same procedure as described in Section 3.6, except for the minimization procedure where at the first step of 10^4–10^5 steepest descent minimization cycles were applied before the conjugate gradient minimization step.

3.6. All-Atom MD Simulations and MM-GBSA Free Energy Calculations

All-atom molecular dynamics (MD) simulations of the PCPE-1, PCPE-1/Ca^{2+} and PCPE-1/Ca^{2+}/GAG complexes obtained by molecular docking were performed with the use of the AMBER16 MD package [50]. Periodic boundary conditions in a truncated octahedron TIP3P water box with at least 8 Å distance from the solute to the periodic box border were used. Na^+ and Cl^- monovalent counterions were used to neutralize the system. ff99SB force field parameters for protein [65] and the GLYCAM06 [66] for GAGs were used, respectively. Prior to MD production runs, two energy-minimization steps were performed: first, 500 steepest descent cycles and 1000 conjugate-gradient cycles with harmonic force restraints on solute (10 kcal/mol/$Å^2$), then, 3000 steepest-descent cycles and 3000 conjugate-gradient cycles without restraints. After the minimization, the system was heated up to 300 K for 10 ps with harmonic force restraints on solute (10 kcal/mol/$Å^2$), equilibrated for 100 ps at 300 K and 10^5 Pa in isothermal isobaric ensemble (NTP). This was followed by a 100 ns MD production run in the same NTP ensemble. The SHAKE algorithm, 2 fs time integration, 8 Å cutoff for non-bonded interactions, and the particle mesh Ewald method were used. The trajectories were analyzed using the cpptraj module of AMBER Tools [46]. Free-energy calculations and per-residue energy decomposition were done using molecular mechanics-generalized born surface area (MM-GBSA) model igb = 2 [67] for protein-GAG and protein-Ca^{2+} complexes for the parts of the trajectory where convergence in terms of RMSD was obtained. The obtained energy values account explicitly for the enthalpy and implicitly for the solvent entropy. For this reason, the reported energies should not be strictly interpreted as full free energy of binding: the entropic contribution to binding was not taken into account explicitly. Entropy calculations were shown to dramatically increase the overall noise in the free binding energies when used within MM-GBSA free energy calculation schemes in general [68] and for GAG containing systems particularly [69].

3.7. Ca²⁺ Ion Position Prediction

We applied and compared several approaches to predict the Ca^{2+} binding sites on the surface of PCPE-1 and its domains. Prior to applying these approaches to PCPE-1, we analyzed their performance for the complex between an annexin V and HP, where Ca^{2+} ions are known to be stable and to contribute to GAG binding (PDB ID: 1G5N, 2.7 Å) [35]. In this structure, 9 Ca^{2+} were resolved.

3.7.1. Molecular Dynamics Approach

We used the MD approach with the protocols described in Section 3.6 to predict the binding sites of Ca^{2+} on the surface of protein. The length of each MD simulation run was 100 ns. In these calculations, Ca^{2+} ions were placed randomly in the periodic box. For annexin V, 9 Ca^{2+} were used corresponding to the number of Ca^{2+} ions observed in the experimental structure. For CUB1-CUB2, NTR domains, three ions were used. The number of ions was chosen in order to sample effectively the protein surface within a reasonable simulation time. For the CUB1-CUB2 and the NTR domains, the simulations were repeated 5 times. The trajectories were analyzed, and the for the frames where the Ca^{2+} ions were stably bound in terms of RMSD convergence for the coordination complex, MM-GBSA free energy calculations were performed. The obtained values for the predicted Ca^{2+} binding sites were compared with the corresponding energies obtained from the simulations of annexin V-Ca^{2+} and CUB1-CUB2-Ca^{2+} crystal structures. The latter was extracted from the structure of the complex of CUB1-CUB2 with procollagen peptide (PDB ID: 6FVZ, 2.7 Å).

3.7.2. FoldX and IonCom

We used the scripts of FoldX available at http://foldx.embl.de and online ion ligand binding site prediction tool IonCom at https://zhanglab.ccmb.med.umich.edu/IonCom for Ca^{2+} binding site predictions. FoldX represents a tool with an implemented empirical force field developed for effective evaluation of the contribution of mutations on the stability, folding and dynamics of proteins and nucleic acids [70,71]. As an output FoldX yields the positions of predicted ions on the surface of the protein. In contrast, IonCom utilizes ab initio training and template-based information to output a list of protein residues potentially involved in ion binding [72]. Both programs were used with the default parameters.

3.8. Visualization and Data Analysis

VMD [51], Chimera [73] and Pymol [74] were used for structural analysis visualization, MD trajectory analysis, as well as for the graphics production. R package was used for data analysis [75].

4. Conclusions

We report here the computational analysis of the interactions of full-length PCPE-1 and its domains with GAGs of various lengths and sequences. This model of full-length PCPE-1 based on SAXS restraints is in agreement with the experimental values of its radius of gyration and length. The full-length protein binds GAGs through the NTR domain and the interdomain linker, while the binding to CUB1-CUB2 domains is weaker, likely non-specific, and less energetically favorable. GAG preferential binding to the NTR domain is mostly electrostatically-driven. CS6 is predicted to bind to a different site of the NTR domain than the other GAGs, which may account for the experimental differences previously observed between CS6 and DS/HP [28]. Fragment-based docking of longer GAG oligosaccharides results in overlap with the docking poses obtained for shorter GAGs and corresponds to more favorable interactions than those established by shorter oligosaccharides in agreement with the strongest inhibition of PCPE-1-HP interactions by longer HP oligosaccharides reported in this study. Our results suggest that calcium ions may bind to GAGs before they interact with PCPE-1 or may stabilize the structure and conformation of full-length PCPE-1. Although we have used several computational approaches to predict Ca^{2+} binding sites on the protein surface, considering calcium ions as a part of the protein receptor for docking is not an approach applicable to all systems.

From a methodological point of view, we have shown that the size of the clusters identified by molecular docking is not correlated with their free binding energies obtained in the MD simulation. We have successfully applied, for the first time, fragment-based docking to dp11 oligosaccharides, which will be useful for the computational characterization of protein interactions with long GAGs, which are challenging to study using conventional docking approaches.

Supplementary Materials: Supplementary Materials can be found at http://www.mdpi.com/1422-0067/20/20/5021/s1.

Author Contributions: J.P., K.K.B. obtained and analyzed the data. G.K., A.G.L, A.L. adapted the software to the system, obtained and analyzed the data and wrote the manuscript. S.R.-B. and S.A.S. are responsible for the conceptualization of the work and wrote the manuscript. S.R.-B. and E.K. performed and interpreted SPR binding assays. S.A.S supervised J.P., G.K., K.K.B., A.G.L. and acquired the funding.

Funding: This research was funded by National Science Centre of Poland, grant numbers UMO-2018/30/E/ST4/00037 (S.A.S.), UMO-2017/25/B/ST4/01026 (A.L.), UMO-2018/31/N/ST4/01677 (K.K.B.).

Acknowledgments: Computational resources were provided by the Polish Grid Infrastructure (PL-GRID, grants protgags, gagstr, kchtkat), ZIH at TU Dresden (grant p_gag) and our cluster at the Faculty of Chemistry, University of Gdansk. We thank Rabia Sadir (Institut de Biologie Structurale, UMR 5075, Grenoble, France) for her generous gift of heparin oligosaccharides.

Conflicts of Interest: The authors declare no conflicts of interest.

Abbreviations

AD3	Autodock 3
BMP-1	Bone morphogenetic protein-1
CG	Coarse-grained
CS	Chondroitin sulfate
CUB	Complement, sea urchin protein Uegf, BMP-1
DS	Dermatan sulfate
GAG	Glycosaminoglycan
HP	Heparin
HS	Heparan sulfate
MD	Molecular dynamics
MM-GBSA	Molecular mechanics-generalized born surface area
mTLD	Mammalian tolloid
NTR	Netrin-like domain
PBSA	Poisson-Boltzmann surface area
PCPE-1	Procollagen C-proteinase enhancer-1
SAXS	Small angle X-ray scattering
UNRES	United residue force field

References

1. Esko, J.D.; Kimata, K.; Lindahl, U. Proteoglycans and Sulfated Glycosaminoglycans. In *Essentials of Glycobiology*, 2nd ed.; Varki, A., Cummings, R.D., Esko, J.D., Freeze, H.H., Stanley, P., Bertozzi, C.R., Hart, G.W., Etzler, M.E., Eds.; Cold Spring Harbor Laboratory Press: Cold Spring Harbor, NY, USA, 2009; pp. 1–784.
2. Pomin, V.H.; Mulloy, B. Glycosaminoglycans and Proteoglycans. *Pharmaceuticals* **2018**, *11*, 27. [CrossRef]
3. Proudfoot, A.E. Chemokines and Glycosaminoglycans. *Front. Immunol.* **2015**, *6*, 246. [CrossRef]
4. Shute, J. Glycosaminoglycan and chemokine/growth factor interactions. *Handb. Exp. Pharmacol.* **2012**, *207*, 307–324.

5. Iozzo, R.V.; Zoellerm, J.J.; Nyströmm, A. Basement membrane proteoglycans: Modulators Par Excellence of cancer growth and angiogenesis. *Mol. Cells* **2009**, *27*, 503–513. [CrossRef] [PubMed]

6. Almond, A. Multiscale modeling of glycosaminoglycan structure and dynamics: Current methods and challenges. *Curr. Opin. Struct. Biol.* **2018**, *50*, 58–64. [CrossRef] [PubMed]

7. Pichert, A.; Samsonov, S.A.; Theisgen, S.; Thomas, L.; Baumann, L.; Schiller, J.; Beck-Sickinger, A.G.; Huster, D.; Pisabarro, M.T. Characterization of the interaction of interleukin-8 with hyaluronan, chondroitin sulfate, dermatan sulfate and their sulfated derivatives by spectroscopy and molecular modeling. *Glycobiology* **2012**, *22*, 134–145. [CrossRef] [PubMed]

8. Penk, A.; Baumann, L.; Huster, D.; Samsonov, S.A. NMR and Molecular Modeling Reveal Specificity of the Interactions between CXCL14 and Glycosaminoglycans. *Glycobiology* **2019**. [CrossRef] [PubMed]

9. Bojarski, K.K.; Sieradzan, A.K.; Samsonov, S.A. Molecular Dynamics Insights into Protein-Glycosaminoglycan Systems from Microsecond-Scale Simulations. *Biopolymers* **2019**. [CrossRef]

10. Uciechowska-Kaczmarzyk, U.; Chauvot de Beauchene, I.; Samsonov, S.A. Docking software performance in protein-glycosaminoglycan systems. *J. Mol. Graph. Mod.* **2019**, *90*, 42–50. [CrossRef]

11. Rother, S.; Samsonov, S.A.; Hofmann, T.; Blaszkiewicz, J.; Köhling, S.; Schnabelrauch, M.; Möller, S.; Rademann, J.; Kalkhof, S.; von Bergen, M.; et al. Structural and functional insights into the interaction of sulfated glycosaminoglycans with tissue inhibitor of metalloproteinase-3—A possible regulatory role on extracellular matrix homeostasis. *Acta Biomater.* **2016**, *45*, 143–154. [CrossRef]

12. Vallet, S.D.; Miele, A.E.; Uciechowska-Kaczmarzyk, U.; Liwo, A.; Duclos, D.; Samsonov, S.A.; Ricard-Blum, S. Insights into the structure and dynamics of lysyl oxidase propeptide, a flexible protein with numerous partners. *Sci. Rep.* **2018**, *8*, 11768. [CrossRef] [PubMed]

13. Ricard-Blum, S.; Ruggiero, F.; van der Rest, M. The collagen superfamily. *Top. Curr. Chem.* **2005**, *247*, 35–84.

14. Vadon-Le Goff, S.; Hulmes, D.J.S.; Moali, C. BMP-1/tolloid-like proteinases synchronize matrix assembly with growth factor activation to promote morphogenesis and tissue remodeling. *Matrix Biol.* **2015**, *44–46*, 14–23. [CrossRef] [PubMed]

15. Takahara, K.; Lyons, G.E.; Greenspan, D.S. Bone morphogenetic protein-1 and a mammalian tolloid homologue (mTld) are encoded by alternatively spliced transcripts which are differentially expressed in some tissues. *J. Biol. Chem.* **1994**, *269*, 32572–32578.

16. Kessler, E.; Takahara, K.; Biniaminov, L.; Brusel, M.; Greenspan, D.S. Bone Morphogenetic Protein-1: The Type I Procollagen C-Proteinase. *Science* **1996**, *271*, 360–362. [CrossRef]

17. Li, S.W.; Sieron, A.L.; Fertala, A.; Hojima, Y.; Arnold, W.V.; Prockop, D.J. The C-proteinase that processes procollagens to fibrillar collagens is identical to the protein previously identified as bone morphogenic protein-1. *Proc. Natl. Acad. Sci. USA* **1996**, *93*, 5127–5130. [CrossRef]

18. Kessler, E.; Hassoun, E. Procollagen C-Proteinase Enhancer 1 (PCPE-1) in Liver Fibrosis. *Methods Mol. Biol.* **2019**, *1944*, 189–201.

19. Hassoun, E.; Safrin, M.; Ziv, H.; Pri-Chen, S.; Kessler, E. Procollagen C-proteinase enhancer 1 (PCPE-1) as a plasma marker of muscle and liverf in mice. *PLoS ONE* **2016**, *11*, e0159606.

20. Takahara, K.; Kessler, E.; Biniaminov, L.; Brusel, M.; Eddy, R.L.; Jani-Saitfl, S.; Shows, T.B.; Greenspan, D.S. Type I procollagen COOH-terminal proteinase enhancer protein: Identification, primary structure, and chromosomal localization of the cognate human gene (PCOLCE). *J. Biol. Chem.* **1994**, *269*, 26280–26285.

21. Bányai, L.; Patthy, L. The NTR module: Domains of netrins, secreted frizzled related proteins, and type I procollagen C-proteinase enhancer protein are homologous with tissue inhibitors of metalloproteases. *Protein Sci.* **1999**, *8*, 1636–1642. [CrossRef]

22. Liepinsh, E.; Banyai, L.; Pintacuda, G.; Trexler, M.; Patthy, L.; Otting, G. NMR structure of the netrin-like domain (NTR) of human type I procollagen C-proteinase enhancer defines structural consensus of NTR domains and assesses potential proteinase inhibitory activity and ligand binding. *J. Biol. Chem.* **2003**, *278*, 25982–25989. [CrossRef] [PubMed]

23. Pulido, D.; Sharma, U.; Vadon-Le Goff, S.; Hussain, S.A.; Cordes, S.; Mariano, N.; Bettler, E.; Moali, C.; Aghajari, N.; Hohenester, E.; et al. Structural Basis for the Acceleration of Procollagen Processing by Procollagen C-Proteinase Enhancer-1. *Structure* **2018**, *26*, 1384–1392. [CrossRef] [PubMed]

24. Gaboriaud, C.; Gregory-Pauron, L.; Teillet, F.; Thielens, N.M.; Bally, I.; Arlaud, G.J. Structure and properties of the Ca^{2+}-binding CUB domain, a widespread ligand-recognition unit involved in major biological functions. *Biochem. J.* **2011**, *439*, 185–193. [CrossRef] [PubMed]

25. Blanc, G.; Font, B.; Eichenberger, D.; Moreau, C.; Ricard-Blum, S.; Hulmes, D.J.S.; Moali, C. Insights into how CUB domains can exert specific functions while sharing a common fold. Conserved and specific features of the CUB1 domain contribute to the molecular basis of procollagen C-proteinase enhancer-1 activity. *J. Biol. Chem.* **2007**, *282*, 16924–16933. [CrossRef]

26. Bernocco, S.; Steiglitz, B.M.; Svergun, D.I.; Petoukhov, M.V.; Ruggiero, F.; Ricard-Blum, S.; Ebel, C.; Geourjon, C.; Deleage, G.; Font, B.; et al. Low resolution structure determination shows procollagen C-proteinase enhancer to be an elongated multidomain glycoprotein. *J. Biol. Chem.* **2003**, *278*, 7199–7205. [CrossRef]

27. Moschcovich, L.; Bernocco, S.; Font, B.; Rivkin, H.; Eichenberger, D.; Chejanovsky, N.; Hulmes, D.J.; Kessler, E. Folding and activity of recombinant human procollagen C-proteinase enhancer. *Eur. J. Biochem.* **2001**, *268*, 2991–2996. [CrossRef]

28. Weiss, T.; Ricard-Blum, S.; Moschcovich, L.; Wineman, E.; Mesilaty, S.; Kessler, E. Binding of procollagen C-proteinase enhancer-1 (PCPE-1) to heparin/heparan sulfate: Properties and role in PCPE-1 interaction with cells. *J. Biol. Chem.* **2010**, *285*, 33867–33874. [CrossRef]

29. Bekhouche, M.; Kronenberg, D.; Vadon-Le Goff, S.; Bijakowski, C.; Lim, N.H.; Font, B.; Kessler, E.; Colige, A.; Nagase, H.; Murphy, G.; et al. Role of the netrin-like domain of procollagen C-proteinase enhancer-1 in the control of metalloproteinase activity. *J. Biol. Chem.* **2010**, *285*, 15950–15959. [CrossRef]

30. Sankaranarayanan, N.V.; Nagarajan, B.; Desai, U.R. So you think computational approaches to understanding glycosaminoglycan-protein interactions are too dry and too rigid? Think again! *Curr. Opin. Struct. Biol.* **2018**, *50*, 91–100. [CrossRef]

31. Samsonov, S.A.; Pisabarro, M.T. Computational analysis of interactions in structurally available protein-glycosaminoglycan complexes. *Glycobiology* **2016**, *26*, 850–861. [CrossRef]

32. Möbius, K.; Nordsieck, K.; Pichert, A.; Samsonov, S.A.; Thomas, L.; Schiller, J.; Kalkhof, S.; Pisabarro, M.T.; Beck-Sickinger, A.G.; Huster, D. Investigation of lysine side chain interactions of Interleukin-8 with Heparin and other glycosaminoglycans studied by a methylation-NMR approach. *Glycobiology* **2013**, *23*, 1260–1269. [CrossRef] [PubMed]

33. Ester, M.; Kriegel, H.; Sander, J.; Xu, X. A density-based algorithm for discovering clusters in large spatial databases with noise. In Proceedings of the International Conference on Knowledge Discovery and Data Mining, Portland, OR, USA, 2–4 August 1996; pp. 226–231.

34. Imberty, A.; Lortat-Jacob, H.; Pérez, S. Structural view of glycosaminoglycan-protein interactions. *Carbohydr. Res.* **2007**, *342*, 430–439. [CrossRef] [PubMed]

35. Capila, I.; Hernáiz, M.J.; Mo, Y.D.; Mealy, T.R.; Campos, B.; Dedman, J.R.; Linhardt, R.J.; Seaton, B.A. Annexin V—Heparin oligosaccharide complex suggests heparan sulfate—Mediated assembly on cell surfaces. *Structure* **2001**, *9*, 57–64. [CrossRef]

36. Woodhead, N.E.; Long, W.F.; Williamson, F.B. Binding of zinc ions to heparin. Analysis by equilibrium dialysis suggests the occurrence of two, entropy-driven, processes. *Biochem J.* **1986**, *237*, 281–284. [CrossRef] [PubMed]

37. Stevic, I.; Parmar, N.; Paredes, N.; Berry, L.R.; Chan, A.K.C. Binding of heparin to metals. *Cell Biochem. Biophys.* **2011**, *59*, 171–178. [CrossRef] [PubMed]

38. Multhaup, G.; Bush, A.I.; Pollwein, P.; Masters, C.L. Binding of heparin to metals. *FEBS Lett.* **1994**, *335*, 151–154. [CrossRef]

39. Eckert, R.; Ragg, H. Zinc ions promote the interaction between heparin and heparin cofactor II. *FEBS Lett.* **2003**, *541*, 121–125. [CrossRef]

40. Ricard-Blum, S.; Féraud, O.; Lortat-Jacob, H.; Rencurosi, A.; Fukai, N.; Dkhissi, F.; Vittet, D.; Imberty, A.; Olsen, B.R.; van der Rest, M. Characterization of endostatin binding to heparin and heparan sulfate by surface plasmon resonance and molecular modeling: Role of divalent cations. *J. Biol. Chem.* **2004**, *279*, 2927–2936. [CrossRef]

41. Han, Q.; Fu, Y.; Zhou, H.; He, Y.; Luo, Y. Contributions of Zn(II)-binding to the structural stability of endostatin. *FEBS Lett.* **2007**, *581*, 3027–3032. [CrossRef]

42. Zhang, F.; Liang, X.; Beaudet, J.M.; Lee, Y.; Linhardt, R.J. The Effects of Metal Ions on Heparin/Heparin Sulfate-Protein Interactions. *J. Biomed. Technol. Res.* **2014**, *1*, 10. [CrossRef]

43. Seo, Y.; Schenauer, M.R.; Leary, J.A. Biologically Relevant Metal-Cation Binding Induces Conformational Changes in Heparin Oligosaccharides as Measured by Ion Mobility Mass Spectrometry. *Int. J. Mass Spectrom.* **2011**, *303*, 191–198. [CrossRef] [PubMed]

44. Lerner, L.; Torchia, D.A. A multinuclear NMR study of the interactions of cations with proteoglycans, heparin, and Ficoll. *J. Biol. Chem.* **1986**, *261*, 12706–12714. [PubMed]

45. Parrish, R.F.; Fair, W.R. Selective binding of zinc ions to heparin rather than to other glycosaminoglycans. *Biochem. J.* **1981**, *193*, 407–410. [CrossRef] [PubMed]

46. Grushka, E.; Cohen, A.S. The Binding of Cu(II) and Zn(II) Ions by Heparin. *Anal. Lett.* **1982**, *15*, 1277–1288. [CrossRef]

47. Grant, D.; Long, W.F.; Williamson, F.B. Infrared spectroscopy of heparin-cation complexes. *Biochem. J.* **1987**, *244*, 143–149. [CrossRef]

48. Sattelle, B.; Shakeri, J.; Almond, A. Does microsecond sugar ring flexing encode 3D-shape and bioactivity in the heparanome? *Biomacromolecules* **2013**, *14*, 1149–1159. [CrossRef]

49. Samsonov, S.; Bichmann, L.; Pisabarro, M.T. Coarse-grained model of glycosaminglycans. *J. Chem. Inf. Mod.* **2015**, *55*, 114–124. [CrossRef]

50. Case, D.A.; Ben-Shalom, I.Y.; Brozell, S.R.; Cerutti, D.S.; Cheatham, T.E., III; Cruzeiro, V.W.D.; Darden, T.A.; Duke, R.E.; Ghoreishi, D.; Gilson, M.K.; et al. *AMBER16*; University of California: San Francisco, CA, USA, 2018.

51. Humphrey, W.; Dalke, A.; Schulten, K. VMD: Visual Molecular Dynamics. *J. Mol. Graph.* **1996**, *14*, 33–38. [CrossRef]

52. Rhee, Y.M.; Pande, V.S. Multiplexed-replica exchange molecular dynamics method for protein folding simulation. *Biophys. J.* **2003**, *84*, 775–786. [CrossRef]

53. Hansmann, U.H.E.; Okamoto, Y. Comparative study of multicanonical and simulated annealing algorithms in the protein folding problem. *Physica A* **1994**, *212*, 415–437. [CrossRef]

54. Liwo, A.; Baranowski, M.; Czaplewski, C.; Gołaś, E.; He, Y.; Jagieła, D.; Krupa, P.; Maciejczyk, M.; Makowski, M.; Mozolewska, M.A.; et al. A unified coarse-grained model of biological macromolecules based on mean-field multipole-multipole interactions. *J. Mol. Model.* **2014**, *20*, 2306. [CrossRef] [PubMed]

55. Liwo, A.; Czaplewski, C.; Ołdziej, S.; Scheraga, H.A. Computational techniques for efficient conformational sampling of proteins. *Curr. Opin. Struct. Biol.* **2008**, *18*, 134–139. [CrossRef] [PubMed]

56. Karczyńska, A.S.; Mozolewska, M.A.; Krupa, P.; Giełdoń, A.; Liwo, A.; Czaplewski, C. Prediction of protein structure with the coarse-grained UNRES force field assisted by small X-ray scattering data and knowledge-based information. *Proteins* **2018**, *86*, 228–239. [CrossRef] [PubMed]

57. Kumar, S.; Bouzida, D.; Swendsen, R.H.; Kollman, P.A.; Rosenberg, J.M. The weighted histogram analysis method for free-energy calculations on biomolecules. I. The method. *J. Comput. Chem.* **1992**, *13*, 1011–1021. [CrossRef]

58. Murtagh, F.; Heck, A. Multivariate Data Analysis. In *Astrophysics and Space Science Library*; Kluwer Academic Publishers: Strasbourg, France, 1987; pp. 55–109.

59. Rotkiewicz, P.; Skolnick, J. Fast procedure for reconstruction of full-atom protein models from reduced representations. *J. Comput. Chem.* **2008**, *29*, 1460–1465. [CrossRef]

60. Morris, G.M.; Goodsell, D.S.; Halliday, R.S.; Huey, R.; Hart, W.E.; Belew, R.K.; Olson, A.J. Automated docking using a Lamarcklan algorithm an empirical binding free energy function. *J. Comput. Chem.* **1998**, *19*, 1639–1662. [CrossRef]

61. Samsonov, S.; Gehrcke, J.-P.; Pisabarro, M.T. Flexibility and explicit solvent in molecular dynamics-based docking of protein-glycosaminoglycan systems. *J. Chem. Inf. Mod.* **2014**, *54*, 582–592. [CrossRef]

62. Joseph, R.R.; Mosier, P.D.; Desai, U.R.; Rajarathnam, K. Solution NMR characterization of chemokine CXCL8/IL-8 monomer and dimer binding to glycosaminoglycans: Structural plasticity mediates differential binding interactions. *Biochem. J.* **2015**, *472*, 121–133. [CrossRef]

63. Nivedha, A.K.; Makeneni, S.; Foley, B.L.; Tessier, M.B.; Woods, R.J. Importance of ligand conformational energies in carbohydrate docking: Sorting the wheat from the chaff. *J. Comput. Chem.* **2014**, *35*, 526–539. [CrossRef]

64. Samsonov, S.A.; Zacharias, M.; Chauvot de Beauchene, I. Modeling large protein-glycosaminoglycan complexes using a fragment-based approach. *J. Comput. Chem.* **2019**, *40*, 1429–1439. [CrossRef]

65. Hornak, V.; Abel, R.; Okur, A.; Strockbine, B.; Roitberg, A.; Simmerling, C. Comparison of multiple Amber force fields and development of improved protein backbone parameters. *Proteins* **2006**, *65*, 712–725. [CrossRef] [PubMed]

66. Kirschner, K.N.; Yongye, A.B.; Tschampel, S.M.; González-Outeiriño, J.; Daniels, C.R.; Foley, B.L.; Woods, R.J. GLYCAM06: A generalizable biomolecular force field. Carbohydrates. *J. Comput. Chem.* **2008**, *4*, 622–655. [CrossRef] [PubMed]

67. Onufriev, A.; Case, D.A.; Bashford, D. Effective Born radii in the generalized Born approximation: The importance of being perfect. *J. Comput. Chem.* **2002**, *23*, 1297–1304. [CrossRef] [PubMed]

68. Homeyer, N.; Gohlke, H. Free Energy Calculations by the Molecular Mechanics Poisson-Boltzmann Surface Area Method. *Mol. Inf.* **2012**, *31*, 114–122. [CrossRef] [PubMed]

69. Gandhi, N.S.; Mancera, R.L. Free energy calculations of glycosaminoglycan-protein interactions. *Glycobiology* **2009**, *19*, 1103–1115. [CrossRef]

70. Schymkowitz, J.; Borg, J.; Stricher, F.; Nys, R.; Rousseau, F.; Serrano, L. The FoldX web server: An online force field. *Nucleic Acids Res.* **2005**, *33*, W382–W388. [CrossRef]

71. Schymkowitz, J.W.; Rousseau, F.; Martins, I.C.; Ferkinghoff-Borg, J.; Stricher, F.; Serrano, L. Prediction of water and metal binding sites and their affinities by using the Fold-X force field. *Proc. Natl. Acad. Sci. USA* **2005**, *102*, 10147–10152. [CrossRef]

72. Hu, X.; Dong, Q.; Yang, J.; Zhang, Y. Recognizing metal and acid radical ion binding sites by integrating ab initio modeling with template-based transferals. *Bioinformatics* **2016**, *32*, 3260–3269. [CrossRef]

73. Pettersen, E.F.; Goddard, T.D.; Huang, C.C.; Couch, G.S.; Greenblatt, D.M.; Meng, E.C.; Ferrin, T.E. UCSF Chimera—A visualization system for exploratory research and analysis. *J. Comput. Chem.* **2004**, *25*, 1605–1612. [CrossRef]

74. *The PyMOL Molecular Graphics System*, Version 1.2r3pre; Schrödinger, LLC: New York, NY, USA, 2002.

75. R Core Team. *R: A Language and Environment for Statistical Computing*; R Foundation for Statistical Computing: Vienna, Austria, 2013; ISBN 3-900051-07-0. Available online: http://www.R-project.org/ (accessed on 1 May 2017).

International Journal of
Molecular Sciences

Article

In Silico Study of the Resistance to Organophosphorus Pesticides Associated with Point Mutations in Acetylcholinesterase of Lepidoptera: *B. mandarina, B. mori, C. auricilius, C. suppressalis, C. pomonella, H. armígera, P. xylostella, S. frugiperda,* and *S. litura*

Francisco Reyes-Espinosa [1], Domingo Méndez-Álvarez [1], Miguel A. Pérez-Rodríguez [2], Verónica Herrera-Mayorga [1,3], Alfredo Juárez-Saldivar [1], María A. Cruz-Hernández [1] and Gildardo Rivera [1,*]

[1]   Laboratorio de Biotecnología Farmacéutica, Centro de Biotecnología Genómica, Instituto Politécnico Nacional, Reynosa 88710, Mexico; frelibi@hotmail.com (F.R.-E.); doomadv@hotmail.com (D.M.-Á.); veronica_qfb@hotmail.com (V.H.-M.); ajuarezs1500@gmail.com (A.J.-S.); tonitacruz@gmail.com (M.A.C.-H.)
[2]   Departamento de Botánica, Universidad Autónoma Agraria Antonio Narro, Saltillo 25315, Mexico; miguel_cbg@hotmail.com
[3]   Departamento de Ingeniería Bioquímica, Unidad Académica Multidisciplinaria Mante, Universidad Autónoma de Tamaulipas, Mante 89840, Mexico
*   Correspondence: gildardors@hotmail.com; Tel.: +52-899-9243627 (ext. 87758)

Received: 22 March 2019; Accepted: 13 May 2019; Published: 15 May 2019

Abstract: An in silico analysis of the interaction between the complex-ligands of nine acetylcholinesterase (AChE) structures of Lepidopteran organisms and 43 organophosphorus (OPs) pesticides with previous resistance reports was carried out. To predict the potential resistance by structural modifications in Lepidoptera insects, due to proposed point mutations in AChE, a broad analysis was performed using computational tools, such as homology modeling and molecular docking. Two relevant findings were revealed: (1) Docking results give a configuration of the most probable spatial orientation of two interacting molecules (AChE enzyme and OP pesticide) and (2) a predicted ΔG_b. The mutations evaluated in the form 1 acetylcholinesterase (AChE-1) and form 2 acetylcholinesterase (AChE-2) structures of enzymes do not affect in any way (there is no regularity of change or significant deviations) the values of the binding energy (ΔG_b) recorded in the AChE–OPs complexes. However, the mutations analyzed in AChE are associated with a structural modification that causes an inadequate interaction to complete the phosphorylation of the enzyme.

Keywords: acetylcholinesterase; resistance; organophosphorus; pesticides; molecular modeling; lepidopterous; insects

---

1. Introduction

A great diversity of organisms belonging to the order Lepidoptera is of great economic interest. As a study model, nine structures of Lepidoptera acetylcholinesterase (AChE) were treated: *Bombyx mandarina, Bombyx mori, Chilo auricilius, Chilo suppressalis, Cydia pomonella, Helicoverpa armígera, Plutella xylostella, Spodoptera frugiperda,* and *Spodoptera litura.* The mulberry silkworm, *B. mori,* is a Lepidopteran insect of great economic importance because of its use in natural silk fiber production and because it is a valuable insect model that has greatly enhanced our understanding of the biology of insects, including many agricultural pests. This insect is also often used for the production of recombinant eukaryotic proteins or as a model organism for pest control studies. The life cycle of the mulberry silkworm is well described; its genome was sequenced in 2004 [1]. *B. mandarina Moore* (Lepidoptera: Bombycidae) is an endangered wild Indian mulberry silkworm species [2]. The striped rice stem borer, *C. suppressalis*, is one of the most important rice pests in East Asia, India, and Indonesia. The main host plant of *C. suppressalis* is rice, maize, and many wild hosts [3]. The cotton bollworm, *H. armigera*, causes serious losses, in particular to cotton, tomatoes, and maize. The most important crop hosts, in which *H. armigera* is a major pest, are cotton, pigeonpea, chickpea, tomato, sorghum, and cowpea; other hosts include groundnut, okra, peas, field beans (*Lablab* spp.), soybeans, lucerne, *Phaseolus* spp., other Leguminosae, tobacco, potatoes, maize, flax, a number of fruits (Prunus, Citrus), forest trees, and a range of vegetable crops [4]. The resistance to pyrethroids in *H. armigera* can be conferred through three separate mechanisms: Detoxification by mixed function oxidases (metabolic resistance), nerve insensitivity, and delayed penetration [5]. The diamondback moth (DBM) in Mexico, *P. xylostella*, is one of the most studied insect pests in the world, yet it is among the 'leaders' of the most difficult pests to control. The DBM is a highly invasive species and it has shown resistance to almost every insecticide. The natural host plant range of the DBM is limited to Brassicaceae (also called Cruciferae), which is characterized by having glucosinolates, which are sulfur-containing secondary plant compounds. Cruciferous vegetables (such as cauliflower, cabbage, garden cress, bok choy, broccoli, and similar green leaf vegetables) are crop species that are cultivated for food production, and their weeds serve as alternate hosts. Some populations of DBM have also been found to infest non-cruciferous plants [6]. The tobacco caterpillar, *S. litura*, is one of the most important insect pests of agricultural crops in the Asian and African tropics. Among the main crop species attacked by *S. litura* in the tropics are *Colocasia esculenta*, cotton, flax, groundnuts, jute, lucerne, maize, rice, soybeans, tea, tobacco, and vegetables (aubergines, *Brassica*, *Capsicum*, cucurbit vegetables, *Phaseolus*, potatoes, sweet potatoes, and species of *Vigna*). Other hosts include ornamentals, wild plants, weeds, and shade trees [7]. *S. litura* have developed resistance to many commercially available pesticides, such as profenofos [8].

In order to have sustainable agriculture and improve public health, effective and appropriate pesticide management is necessary. Every year significant economic losses are reported, mainly because of damage to agricultural, forestry, and livestock production, caused by the persistence of insect pests; this fact makes adequate control of pests necessary. In this context, the scientific community continues a joint multidisciplinary effort to elucidate the mechanisms of resistance developed by pest organisms. One relevant contribution by Guo et al. (2017) is the development of a computational pipeline that uses AChE to detect resistance mutations of AChE in insect RNA-Sequencing data that facilitates the full use of large-scale genetic data obtained by next-generation sequencing [9]. A recent study by Brevik et al. (2018) reported that the median duration between the introduction of an insecticide and the first report of resistance was 66 generations (95% CI 60–78 generations) [10].

The prevalence of resistant insects is influenced by different factors that can be grouped into three categories: (1) Biological factors, such as generation time, number of offspring per generation, and migration; (2) genetic factors that include the frequency and dominance of the resistance gene, fitness of the resistance genotype, and the number of different resistance alleles; and (3) operational factors in which man intervenes, such as treatment, persistence, and insecticide chemistry, allusive to timing and dosage of insecticide application [11].

The term resistance to insecticides refers to a hereditary change in the sensitivity of a pest population that is reflected in recurrent failure to perform its insecticidal action, generating inadequate pest control [11–13]. There are several ways insects can become resistant: Behavioral resistance (resistant insects may detect or recognize a danger and avoid the toxin; they simply stop feeding), penetration resistance (resistant insects may absorb the toxin more slowly than susceptible insects), metabolic resistance (resistant insects may detoxify or destroy the toxin faster than susceptible insects), and altered target-site resistance (the toxin binding site becomes modified to reduce the insecticide's effects); often, more than one of these mechanisms occurs at the same time [11].

According to the mechanisms of action of pesticides, more than 25 types of resistance have been identified and at least 55 types of chemical species [12,13]. One main group is acetylcholinesterase (AChE) inhibitors, which are divided into two subgroups, cataloged as carbamates and organophosphates; both affect the nervous system. AChE is a hydrolytic enzyme that acts on acetylcholine (ACh)—its natural substrate, a neurotransmitter—generating choline and acetic acid. In the presence of organophosphorus (OPs) pesticides, AChE is phosphorylated and, as a consequence, is inhibited [14,15].

The inhibition process involves several stages, outlined in Figure 1. The first is the affinity of OPs for AChE, which determines the reversible inhibition of the enzyme (described by the affinity constant, $K_a = k_{+1}/k_{-1}$) [14]. The second is the phosphorylation constant (known as k_{+2} or k_p), which governs the rate of formation of the stable covalent bond, causing permanent inhibition of the enzyme and the release of a leaving group (BH, Figure 1) [16]. The third is a hydrolysis reaction considered in homologous kinetic systems and widely studied in lines of research concerning self-reactivating cholinesterases [14,15]. The fourth process is known as the aging of phosphorylation [17,18]. It consists of the hydrolysis of one of the alkyl residues of the phosphate group bound to the active site, giving rise to a very stable complex, characterized by an acid group in the phosphoric center. The irreversibility of the reaction is established by the magnitude of the constant k_{+4} [18,19]. Regarding the ideal characteristics that a good OP pesticide must meet, at least four kinetic criteria are established in order to provide an optimal pesticide activity as well as safety for mammals: (1) K_a (pest) < K_a (mammals), (2) K_{+2} (pest) > K_{+2} (mammals), (3) K_{+3} (pest) < K_{+3} (mammals), and (4) K_{+4} (pest) > K_{+4} (mammals) [15].

$$\text{AChEH} + \text{AB} \underset{k_{-1}}{\overset{k_{+1}}{\rightleftharpoons}} \text{AChEHAB} \xrightarrow{k_{+2}} \text{BH} + \text{AChEA} \xrightarrow{k_{+3}} \text{AChEH} + \text{AOH}$$
$$\text{AChEA} \xleftarrow{} \quad k_{+4} \searrow \text{AChEA}'$$

Figure 1. General scheme of the inhibition of acetylcholinesterase (AChE) by an organophosphorus (OP) inhibitor. Where AChEH is the free enzyme; AB, the OP molecule; AChEA', the dealkylated form of the phosphorylated enzyme (AChEH) [14,15].

The main interest of the research work is performed in an in silico study of the resistance to organophosphorus (OPs) pesticides associated with point mutations in AChE of Lepidoptera, using computational tools to elucidate the structural basis of the mechanism of resistance. The study considers the two forms of AChE present in Lepidoptera, known as form 1 acetylcholinesterase (AChE-1) and form 2 acetylcholinesterase (AChE-2), which present about 40% sequence identity. The OPs molecules evaluated include compounds identified in cases of resistance of Lepidoptera to OP insecticide, consulted in the Arthropod Pesticide Resistance Database (APRD) [20], as well as the database of the Insecticide Resistance Action Committee (IRAC) [21].

The study also includes two non-OPs molecules, acetylcholine (ACh)—a natural substrate of AChE—and a non-OP pesticide, psoralen. Psoralen is a naturally occurring furocoumarin, found in *Psoralea*, which is a genus in the legume family (Fabaceae). *Psoralea corylifolia* is an important medicinal plant that is used and widely studied in several traditional medicines to cure various diseases (e.g., anti-carcinogenic activity, anti-depressant activity, skin related problem/leukoderma, Alzheimer's disease) [22,23].

2. Results

2.1. Computational Protocol

The implemented computational protocol consisted of three stages: (a) Modeling by homology using the SWISS-MODEL program [24,25], (b) point mutations performed with the FoldX program [26,27], and (c) molecular docking using Autodock 4.2 [28]. The AChE molecular models, as well as the respective modified models, were constructed satisfactorily (see details about results for Homology Modeling in Table A1, Appendix A). The quality of the models built by homology modeling was analyzed using the multiple 3D alignment option provided by the PDBeFold service (summary showed in Table 1) [29]. In this 3D alignment analysis, the structures that were used as templates for constructed models were used as a reference, PDB 5YDJ (AChE of *Anopheles gambiae*) and 1DX4 (AChE of *Drosophila melanogaster*) for AChE-1 and AChE-2 enzymes, respectively.

Table 1. Alignment results (performed with PDBeFOLD web server [29]). N^{res}, number of aligned residues; $\%^{SI}$, % Sequence Identity; root-mean-square deviation (RMSD); quality of alignment (Q^{score}), with 1 being the highest score.

Model	AChE-1				AChE-2			
	N^{res}	$\%^{SI}$	RMSD	Q^{score}	N^{res}	$\%^{SI}$	RMSD	Q^{score}
B. mandarina	539	71.8	0.4837	0.935	551	62.9	0.3079	0.808
B. mori	539	72.0	0.4837	0.935	551	62.9	0.3079	0.808
C. auricilius	540	71.0	0.3956	0.941	548	63.1	0.3553	0.810
C. suppressalis	540	71.0	0.4593	0.936	493	62.7	0.5486	0.883
C. pomonella	551	44.1	1.3909	0.772	540	40.4	1.4337	0.678
H. armígera	540	72.0	0.3577	0.944	551	63.1	0.3514	0.806
P. xylostella	541	71.6	0.3695	0.942	551	63.3	0.3626	0.805
S. litura	540	72.0	0.3354	0.946	551	63.3	0.3784	0.804

2.2. Molecular Docking of OPs on AChE Enzymes

All the predicted ΔG_b from the docking of 43 OPs molecules (in addition to acetylcholine and psoralen molecules) on AChE enzymes of Lepidoptera (*B. mandarina*, *B. mori*, *C. auricilius*, *C. suppressalis*, *C. pomonella*, *H. armígera*, *P. xylostella*, and *S. litura*) are presented in three sections: (1) AChE-1 and AChE-2, both wild type, (2) modified AChE-1, and (3) modified AChE-2. In order to carry out an analysis of results, all the predicted ΔG_b are presented graphically (Figures 2–4, respectively); the molecules evaluated are plotted against the estimated free binding energy (predicted ΔG_b) obtained. In each section, the order of the molecules is presented according to the predicted ΔG_b recorded in *B. mori* (e.g., Figures 2a, 3a and 4a).

2.3. Docking of OPs Molecules on AChE-1 and AChE-2 (Wild Type)

Three global assessments of AChE wild type docking results are as follows: (1) The natural substrate molecule, acetylcholine (ACh) records values of $-5.3 < \Delta G_b < -4.30$ Kcal/mol in complex with AChE-1, and in complex with AChE-2 $-5.85 < \Delta G_b < -4.43$ Kcal/mol. (2) Psoralen (non-OP molecule) records a narrow range of ΔG_b values $-6.94 < \Delta G_b < -6.23$ Kcal/mol, in complex with both AChE enzymes. (3) In the evaluation of the AChE enzymes (AChE-1 and AChE-2 wild type) by docking of OP (shown in Figure 2a–h), 33 OPs were identified that registered $\Delta G_b^{acetylcholine} > \Delta G_b^{OPs}$, lower ΔG_b values with respect to the ΔG_b value recorded for ACh. These molecules were leptophos (32c), cyanofenphos (28c), tetrachlorvinphos (39c), phosalone (14c), chlorfenvinphos (42c), profenofos (37c), dialifos (29c), bromophos-ethyl (27c), isoxathion (31c), prothiofos (36c), O-ethyl O-(4-nitrophenyl) phenylphosphonothioate known as EPN (21c), azinphos-methyl (3c), phoxim 43c), sulprofos (38c), diazinon (5c), chlorpyrifos (4c), pirimiphos-methyl (34c), bromophos (23c), quinalphos (30c), phosmet (17c), triazophos (35c), ronnel (16c), carbophenothion-methyl (20c), phenthoate (22c), chlorpyrifos-methyl (26c), mephosfolan (18c), cyanophos (25c), parathion (1c), ethion (9c), fenitrothion (33c), dioxabenzofos (24c), methidathion (19c), and malathion (10c). Then 10 OPs were identified that register $\Delta G_b{}^{acetylcholine} \leq \Delta G_b{}^{OPs}$, that is, to register a binding energy with a magnitude equal to or greater than the natural substrate: These OPs were parathion-methyl (12c), disulfoton (8c), dicrotophos (41c), naled (13c), dimethoate (7c), trichlorfon (15c), mevinphos (40c), methamidophos (11c), dichlorvos (6c), and acephate (2c). In Figure 3, only the predicted ΔG_b results of ACh, psoralen, and the 10 OPs (molecules: 2c, 6c, 7c, 8c, 11c, 12c, 13c, 15c, 40c, and 41c) with the highest recorded predicted ΔG_b are shown.

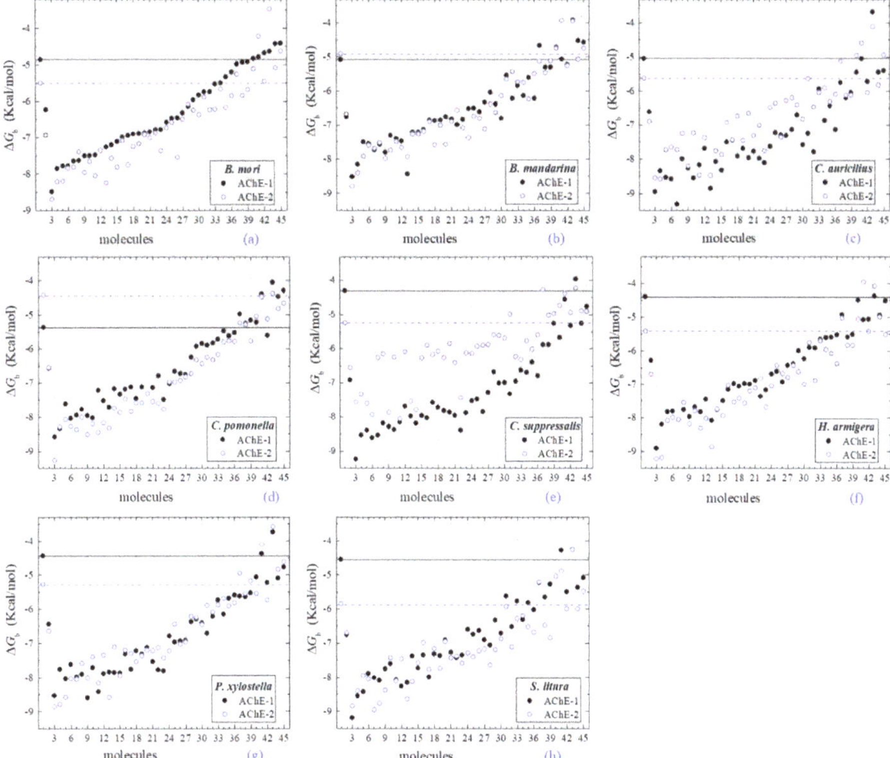

Figure 2. Molecular docking results of AChEs (AChE-1 and AChE-2, both wild type) in complex with molecules. AChE of the following organisms: *B. mori* (**a**), *B. mandarina* (**b**), *C. auricilius* (**c**), *C. pomonella* (**d**), *C. suppressalis* (**e**), *H. armígera* (**f**), *P. xylostella* (**g**), and *S. litura* (**h**). The molecules evaluated are the following: 1, acetylcholine; 2, psoralen; the OPs molecules: 3, leptophos (32c); 4, cyanofenphos (28c); 5, tetrachlorvinphos (39c); 6, phosalone (14c); 7, chlorfenvinphos (42c); 8, profenofos (37c); 9, dialifos (29c); 10, bromophos-ethyl (27c); 11, isoxathion (31c); 12, prothiofos (36c); 13, EPN (21c); 14, azinphos-methyl (3c); 15, phoxim (43c); 16, sulprofos (38c); 17, diazinon (5c); 18, chlorpyrifos (4c); 19, pirimiphos-methyl (34c); 20, bromophos (23c); 21, quinalphos (30c); 22, phosmet (17c); 23, triazophos (35c); 24, ronnel (16c); 25, carbophenothion-methyl (20c); 26, phenthoate (22c); 27, chlorpyrifos-methyl (26c); 28, mephosfolan (18c); 29, cyanophos (25c); 30, parathion (1c); 31, ethion (9c); 32, fenitrothion (33c); 33, dioxabenzofos (24c); 34, methidathion (19c); 35, malathion (10c); 36, parathion-methyl (12c); 37, disulfoton (8c); 38, dicrotophos (41c); 39, naled (13c); 40, dimethoate (7c); 41, trichlorfon (15c); 42, mevinphos (40c); 43, methamidophos (11c); 44, dichlorvos (6c); and 45, acephate (2c). ΔG_b, the estimated energy of binding. Continuous line shows the energy threshold recorded in AChE-1$^{wild\ type}$-ACh complex. Dotted line shows the energy threshold in AChE-2$^{wild\ type}$-ACh complex. Docking was performed using the AutoDock4.2 program.

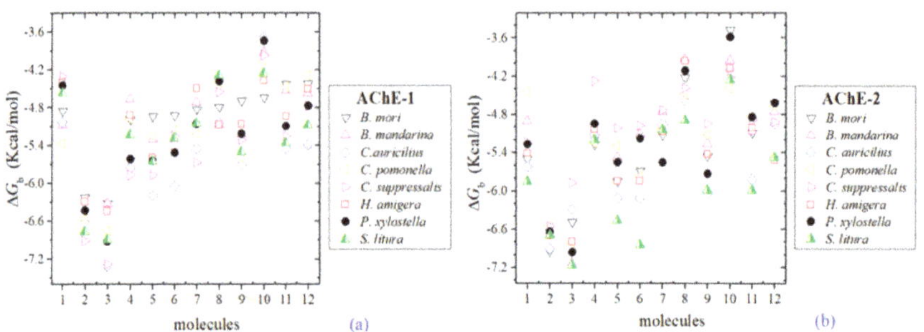

Figure 3. Predicted ΔG_b results of the low inhibition score of OPs molecules in AChE-1 (**a**) and AChE-2 (**b**), both wild type. ΔG_b, estimated free energy of binding; molecules: 1, acetylcholine; 2, psoralen; 3, chlorpyrifos-methyl; 4, disulfoton; 5, dicrotophos; 6, naled; 7, dimethoate; 8, trichlorfon; 9, mevinphos; 10, methamidophos; 11, dichlorvos; and 12, acephate.

2.4. Docking of OPs Molecules on Modified AChE-1

Predicted ΔG_b results of the AChE-1 wild type and modified AChE-1 enzymes are shown in Figure 4a–g. A global assessment is that 33 OPs in complex with modified AChE-1 (molecules: 1c, 3c, 4c, 5c, 9c, 10c, 14c, 16c, 17c, 18c, 19c, 20c, 21c, 22c, 23c, 24c, 25c, 26c, 27c, 28c, 29c, 30c, 31c, 32c, 33c, 34c, 35c, 36c, 37c, 38c, 39c, 42c, and 43c) register ΔG_b values below the threshold registered for AChE-1 in complex with acetylcholine (i.e., ΔG_b of AChE-1 wild type-ACh complex > ΔG_b of modified AChE-1-OP complex) this is $\Delta G_b{}^{acetylcholine} > \Delta G_b{}^{OPs}$. In Figure 4h, only the results of ACh, psoralen, and the 10 OPs (molecules: 2c, 6c, 7c, 8c, 11c, 12c, 13c, 15c, 40c, and 41c) with the highest recorded predicted ΔG_b are shown, which is a lower inhibition score recorded ($\Delta G_b{}^{acetylcholine} \leq \Delta G_b{}^{OPs}$). Docking results of the 11c molecule, methamidophos, recorded a $\Delta G_b{}^{ACh} < \Delta G_b{}^{11c}$ in the AChE-1 (both wild type and modified) on the seven evaluated organisms. In the AChE-1 (both wild type and modified) of *P. liture* and *P. xylostella*, nine OPs molecules (molecules: 2c, 6c, 7c, 8c, 12c, 13c, 15c, 40c, and 41c) presented a similar trend $\Delta G_b{}^{ACh} > \Delta G_b{}^{OPs}$. But in AChE-1 of *B. mandarina* and *C. auricilius*, these nine OPs molecules presented $\Delta G_b{}^{ACh} < \Delta G_b{}^{OPs}$. To identify the non-covalent interactions between the AChE-1 enzymes of *P. xylostella* and their ligands, the respective molecular models (presented in Figure 4) were analyzed with PLIP (Protein-Ligand Interaction Profiler) [30]; data shown in Figure 5.

2.5. Docking of OPs Molecules on Modified AChE-2

Docking results of the molecules evaluated on AChE-2 (wild type and modified) showed some appreciable variations in the magnitude of ΔG_b registered among the enzymes of the 8 evaluated organisms (see Figure 6). The global assessment is (1) that 29 OPs were identified that register $\Delta G_b{}^{ach} < \Delta G_b{}^{OPs}$ (molecules: 1c, 3c, 4c, 5c, 14c, 16c, 17c, 20c, 21c, 22c, 23c, 24c, 25c, 26c, 27c, 28c, 29c, 30c, 31c, 32c, 33c, 34c, 35c, 36c, 37c, 38c, 39c, 42c, and 43c) (see Figure 6a–h), (2) that six OPs were identified that register $\Delta G_b{}^{ACh} > \Delta G_b{}^{OPs}$ (2c, acephate, 6c, dichlorvos, 7c, dimethoate, 8c, disulfoton, 11c, methamidophos, and 15c, trichlorfon, except for AChE-2 that were only 2c, 11c, and 15c), and (3) the another 8 compounds (9c, ethion, 10c, malathion, 12c, parathion-methyl, 13c, naled, 18c, mephosfolan,, 19c, methidathion, 40c, mevinphos, and 41c, dicrotophos) registered a different magnitude in the AChE enzymes evaluated. Predicted ΔG_b results of the AChE-2 wild type and modified AChE-2 enzymes are shown in Figure 6a–h.

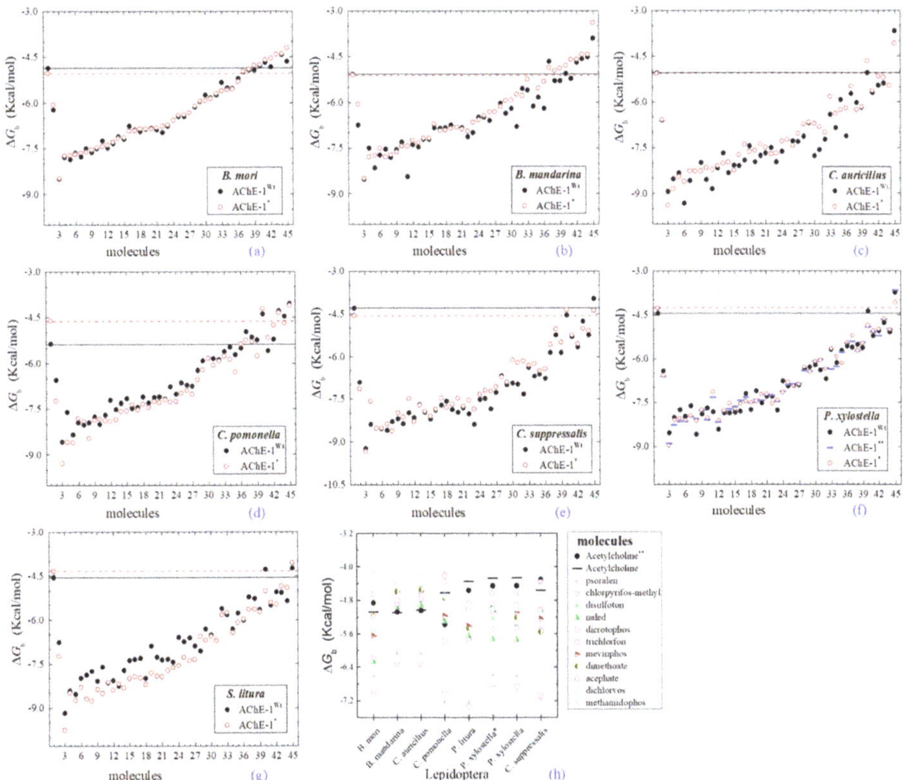

Figure 4. Predicted ΔG_b results of OPs molecules in AChE-1 (^{Wt}, wild type, full circles in **a–g**, and, *, modified, empty circles in **a–g**). ΔG_b, estimated free energy of binding; point mutations on AChE-1: A201S, G227A, and L452S to *B. mori* (**a**) and to *B. mandarina* (**b**); A201S to *C. auricilius* (**c**) and to *C. suppressalis* (**e**); F290V to *C. pomonella* (**d**); A201S, G227A, and F290A to *P. xylostella* (**f**) and *S. litura* (**g**); A201S and G227A to *P. xylostella* * (**f**); **, evaluated in AChE-1 wild type. The molecules evaluated are the following: 1, acetylcholine; 2, psoralen; the OPs molecules: 3, 32c; 4, 39c; 5, 28c; 6, 42c; 7, 14c; 8, 29c; 9, 37c; 10, 27c; 11, 21c; 12, 31c; 13, 36c; 14, 43c; 15, 3c; 16, 35c; 17, 4c; 18, 5c; 19, 23c; 20, 30c; 21, 34c; 22, 38c; 23, 17c; 24, 16c; 25, 20c; 26, 22c; 27, 26c; 28, 18c; 29, 25c; 30, 33c; 31, 1c; 32, 9c; 33, 10c; 34, 19c; 35, 24c; 36, 12c, parathion-methyl; 37, 8c, disulfoton; 38, 13c, naled; 39, 41c, dicrotophos; 40, 15c, trichlorfon; 41, 40c, mevinphos; 42, 7c, dimethoate; 43, 2c, acephate; 44, 6c, dichlorvos; and 45, 11c, methamidophos. In **a–g**, a continuous line shows the energy threshold recorded in the AChE-1$^{wild\ type}$-ACh complex and a dotted line shows the energy threshold in the AChE-1^{modified}-ACh complex. (**h**) Results of the ten OPs with lower inhibition score recorded ($\Delta G_b^{acetylcholine} < \Delta G_b^{OPs}$) in **a–g**: These OPs molecules are 2c, 6c, 7c, 8c, 11c, 12c, 13c, 15c, 40c, and 41c. ΔG_b, the estimated energy of binding. Docking was performed using the AutoDock4.2 program.

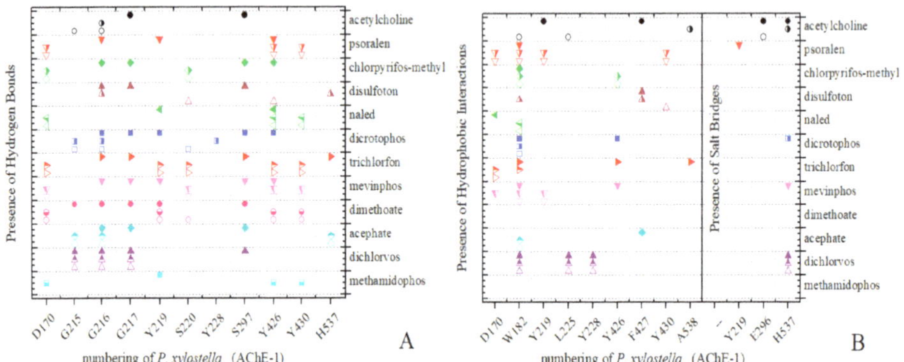

Figure 5. PLIP (Protein-Ligand Interaction Profiler) results in AChE-1 of *P. xylostella*. Identification of noncovalent interactions; (**A**) hydrogen bonds, (**B**) hydrophobic interactions and salt bridges. Full symbol, AChE-1 wild type; empty symbol, modified AChE-1 by A201S and G227A; half full symbol, modified AChE-1 by A201S, G227A, and F290A.

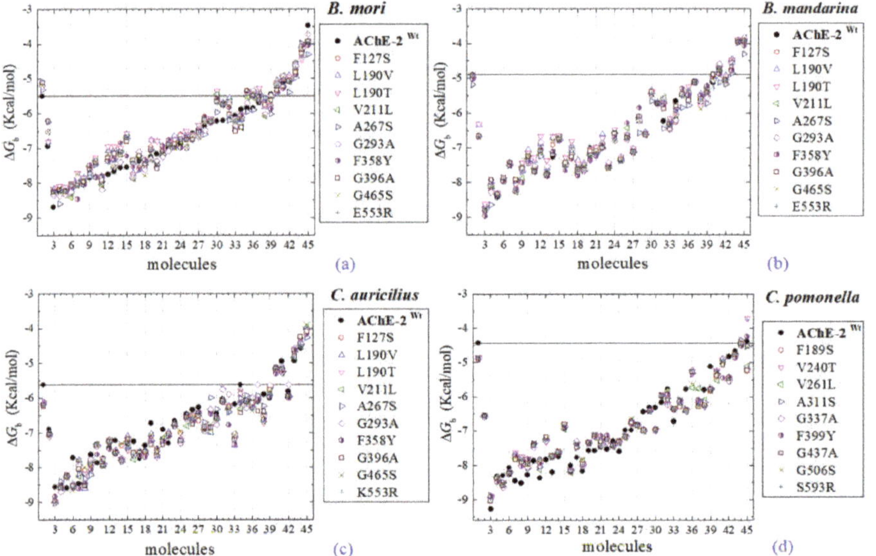

Figure 6. *Cont.*

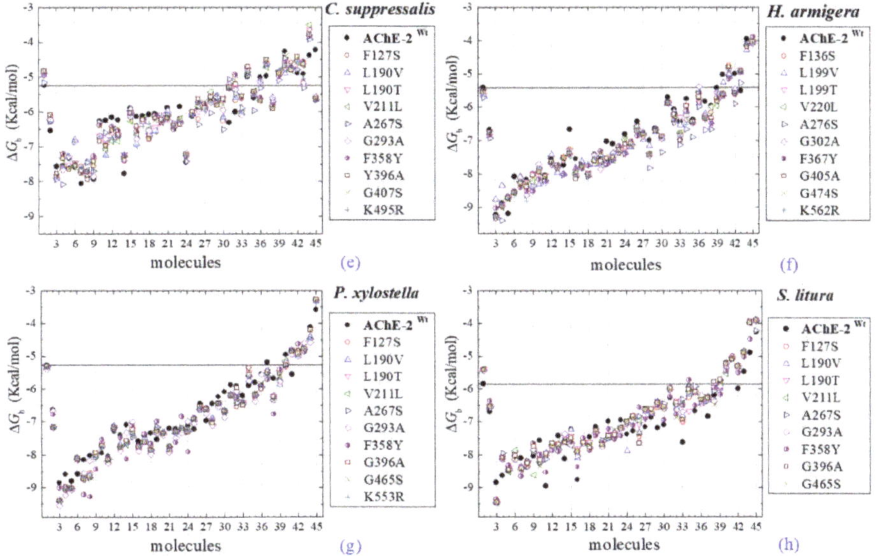

Figure 6. Molecular docking results of AChEs (AChE-2 wild type and AChE-2 modified) in complex with molecules. AChE-2 ([wt], wild type enzyme) of the following organisms: *B. mori* (**a**), *B. mandarina* (**b**), *C. auricilius* (**c**), *C. pomonella* (**d**), *C. suppressalis* (**e**), *H. armígera* (**f**), *P. xylostella* (**g**), and *S. litura* (**h**). The point mutations on AChE-2 are indicated in box, respectively. The molecules evaluated are the following: 1, acetylcholine; 2, psoralen; the OPs: 3, 32c; 4, 21c; 5, 28c; 6, 39c; 7, 31c; 8, 29c; 9, 14c; 10, 43c; 11, 42c; 12, 5c; 13, 27c; 14, 3c; 15, 22c; 16, 37c; 17, 35c; 18, 36c; 19, 4c; 20, 34c; 21, 38c; 22, 30c; 23, 23c; 24, 17c; 25, 16c; 26, 20c; 27, 26c; 28, 1c; 29, 25c; 30, 33c; 31, 24c; 32, 10c, malathion; 33, 18c, mephosfolan; 34, 9c, ethion; 35, 12c, parathion-methyl; 36, 41c, dicrotophos; 37, 13c, naled; 38, 19c, methidathion; 39, 40c, mevinphos; 40, 8c, disulfoton; 41, 7c, dimethoate; 42, 6c, dichlorvos; 43, 2c, acephate; 44, 15c, trichlorfon; and 45, 11c, methamidophos. ΔG_b, the estimated energy of binding. A continuous line shows the energy threshold recorded in AChE-2$^{\text{wild type}}$-ACh complex. Docking was performed using the AutoDock4.2 program.

3. Discussion

3.1. Construction of Molecular Models

The molecular models of the respective AChE enzymes are reliable. In the evaluation of their quality, they register acceptable scores both in their construction (Table A1, Appendix A) and in the 3D alignment, with respect to the reference crystallographic structure (Table 1). The certainty of the structural predictions is based on the quality of the molecular models (e.g., root-mean-square deviation (RMSD) and quality of alignment (Q^{score})). In our study, the constructed models are reliable and reproduce the experimental reference evidence.

3.2. Docking Results of Acetylcholine Evaluations on AChE

The docking results of the AChE–ACh complex, obtained in the evaluation of AChE-1, form 1 of AChE, and AChE-2, form 2 of AChE (both wild type and modified enzymes), are highlighted below.

(A) AChE-1 and AChE-2 (both wild type enzymes), the energy threshold registered for ACh shows a difference in magnitude between $\Delta G_b{}^{\text{AChE-1 wild type}}$ and $\Delta G_b{}^{\text{AChE-2 wild type}}$ (see Figure 6), the exception is AChE of *B. mandarina* (Figure 2b) (the energies recorded to AChE-2-ACh complexes were equal in magnitude, $\Delta G_b{}^{\text{AChE-1 wild type}} = \Delta G_b{}^{\text{AChE-2 wild type}}$). The result obtained in AChE of *B. mandarina* suggests that there is no difference in the affinity of AChEs ($\Delta G_b{}^{\text{AChE-2 wild type}}$ and $\Delta G_b{}^{\text{AChE-2 wild type}}$) for the substrate. Whereas, in the other systems, a difference in affinity between AChE and ACh is predicted, which could be confirmed by catalytic activity studies of the enzymes.

(B) AChE-1$^{\text{wild type}}$ and AChE-1^{modified}, as seen in Figure 4, six studied systems (*B. mori*—Figure 4a, *B. mandarina*—Figure 4b, *C. auricilius*—Figure 4c, *C. suppressalis*—Figure 4e, *P. xylostella*—Figure 4f, and *S. litura*—Figure 4g) showed that no difference or significant deviations was observed between the energy threshold registered for ACh in $\Delta G_b{}^{\text{AChE-1 wild type}}$ and $\Delta G_b{}^{\text{AChE-1 modified}}$. Only AChE-1 of *C. pomonella* (Figure 4d) in complex with ACh recorded an absolute relative energy value near to 0.7 Kcal/mol. This is a $\Delta G_b{}^{\text{AChE-1 modified}}$ greater than the $\Delta G_b{}^{\text{AChE-1 wild type}}$. This result suggests a decrease in affinity for the substrate *ace*. For the other AChE-1–ACh complexes analyzed, there was no variation between the two enzymes (AChE-1$^{\text{wild type}}$ and AChE-1^{modified}). This result suggests that the mutations evaluated in AChE-1 do not affect the affinity of *ace*.

(C) AChE-2$^{\text{wild type}}$ and AChE-2^{modified}, all evaluations of ACh in AChE-2 did not record a variation between $\Delta G_b{}^{\text{AChE wild type}}$ and $\Delta G_b{}^{\text{AChE modified}}$; the energy thresholds of ACh in AChE-2 evaluations do not register a variation (Figure 6).

3.3. Docking Results of Psoralen Evaluations on AChE

The docking results of the AChE–psoralen complex recorded a score -6.8 Kcal/mol $< \Delta G_b <$ -6.0 Kcal/mol. This result was consistent in all the docking evaluations on AChE (AChE-1 and AChE-2, both wild type and modified with slight variations (see Figure 2a–h, Figures 4a–g and 6a–h)). The results obtained suggest that the specific modifications made to AChE do not affect the recorded energy. Results presented in Figure 5 supports this assertion; the profile of interactions recorded in the AChE-1 of *P. xylostella* in complex with psoralen shows the presence of hydrogen bonds (recorded in residue Y426) and hydrophobic interactions (recorded in residue W182), present in both AChE-1$^{\text{wild type}}$ and AChE-1^{modified}. The results of the energy predictions of the enzyme AChE of Lepidoptera in the study of psoralen are promising for an in vitro study, mainly for two reasons: The first is its characteristic of being a natural molecule (no-OP) with inhibitory activity to AChE, and the second, is the scores psoralen obtained in the binding to AChE register a $\Delta G_b < -6.0$ Kcal/mol in complex with both AChE enzymes (AChE-1 and AChE-2) of the Lepidoptera organisms studies here.

3.4. Resistance-Associate Mutation on AChE

A review of the cases of registered incidences of resistance of Lepidoptera exclusively to the mechanism of action (MoA) of AChE inhibitors by OP (source: www.pesticideresistance.org, consulted in 2018-August) resulted in the top nine resistant Lepidoptera species shown in Table 2. There is a great consistency between the results presented in Figures 4 and 5 and the information contained in Table 2. Eight OPs (chlorpyrifos-methyl (26c), disulfoton (8c), dicrotophos (41c), trichlorfon (15c), mevinphos (40c), dimethoate (7c), acephate (2c), and dichlorvos (6c), in complex with AChE-1 wild type of *P. xylostella*, presented a hydrogen bond between S297 and O3 from ligand (OP insecticide). This interaction was very important in the effect of irreversible inhibition (Figure 1). In the respective modified enzymes, this interaction is absent—AChE-1 was modified by A201S and G227A and by A201S, G227A, and F290A. Predicted ΔG_b results of AChE enzymes in complex with OPs molecules recorded an consistent energy values in the majority of the evaluations, reflected in the profile of the graphics presented in Figure 2a–h, Figure 4a–g and Figure 6a–h, according to the chemical composition of the OP molecule. However, the energy predictions between wild and modified enzymes (AChE-1 and AchE-2) do not show a relevant effect on the recorded magnitude; that is, the proposed mutations do not give rise to an enzyme insensitivity. Molecular models of modified AChE (with point mutations) provide information on predicted ΔG_b, in addition to the possible conformation of the interaction mode. Docking results also showed a different location of the binding site of OPs on the enzyme, as a consequence of a change in the electronic atmosphere caused by the point mutation (e.g., acephate (2c) in complex with AChE-1 of *P. xylostella*, Figure 7).

Table 2. Top nine Lepidoptera organisms involved in resistance to OPs. [£] Studied in this work * Period of 2010–2016 [20].

Genus Species	# Cases:		OPs Insecticide
Plutella xylostella [£]	862	279 *	1c, 2c*, 4c*, 5c, 6c, 7c, 10c, 11c, 12c, 13c, 15c, 18c, 19c, 21c, 22c, 24c, 26c, 28c, 28c, 29c, 30c, 31c, 32c, 33c, 34c, 35c, 36c, 37c, 40c*, and 43c
Helicoverpa armígera [£]	856	129 *	3c, 4c*, 10c, 12c, 30c, 31c, 37c*, and 43c
Spodoptera litura [£]	644	251 *	4c*, 5c, 6c, 10c, 15c, 30c, 35c, 37c*, 42c, and 43c*
Spodoptera exigua	525	303 *	4c*, 12c, 30c, and 37c*
Cydia pomonella [£]	193	46 *	1c, 3c*, 4c*, 12c*, and 17c
Chilo suppressalis [£]	148	79 *	1c, 4c, 5c, 11c, 15c, 21c, 33c, and 35c
Earias vittella	128	32 *	4c, 35c, and 37c
Spodoptera frugiperda [£]	125	57 *	2c*, 4c, 5c, 6c, 10c, 12c,15c, and 38c
Heliothis virescens	120	—	1c, 12c, 16c, 17c, 27c, and 38c

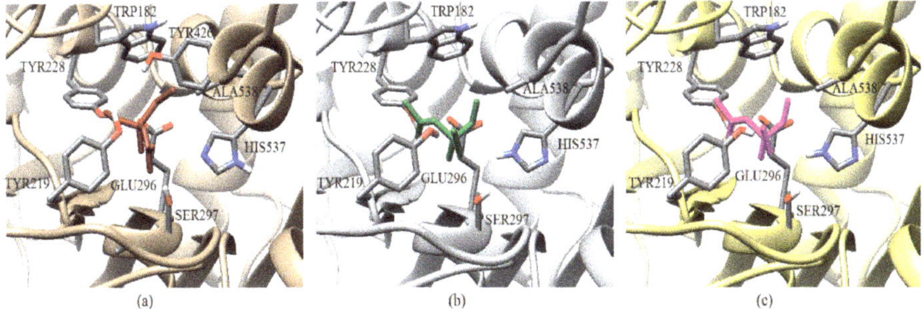

Figure 7. Graphical representation of the acephate binding on AChE-1 of *P. xylostella*. Spatial localization of acephate structure on the AChE-1 enzyme as a result of docking is shown in three molecular models: Conformation acquired on AChE-1 wild type, acephate is chain in red (**a**), conformation acquired on modified AChE-1 (A201S and G227A, and -A201S, G227A, and F290A, respectively), the acephate is chain in green and magenta (**b** and **c**, respectively). The main interactions are shown in Figure 5. The location of SER297 is shown, which is an important residue for the effect of enzyme phosphorylation. HIS537 in AChE-1 of *P. xilostella* is stabilized by GLU296 and GLU423. The pKa values of ionizable residues could be predicted with PROPKA [31].

The contributions of the present study about modified AChE-2 of Lepidoptera in complex with OP are mainly two: (1) Docking results give a configuration of the most probable spatial orientation of two interacting molecules (AChE enzyme and OP pesticide), and (2) a predicted ΔG_b. Regarding the predicted ΔG_b results obtained from the evaluations in wild type and modified AChE-2, the results are not conclusive, regarding the effect of the proposed mutations in the direction of generating insensitivity of the enzyme. Only small variations were obtained in the magnitude of the predicted energy, which could be due to the variation of the estimate.

4. Materials and Methods

4.1. Homology Modeling

The AChE sequences (form 1 known as AChE-1 and form 2 known as AChE-2) of the nine lepidopterous organisms were retrieved from GenBank of the National Centre for Biotechnology Information [32]. The accession numbers of AChE-1 and AChE-2 were the following: EU262633.2 and EU262632.2 of *B. mandarina*; NM_001043915.1 and NM_001114641.1 of *B. mori*; KF574430.1 and KF574431.1 of *C. auricilius*; EF453724.1 and EF470245.1 of *C. suppressalis*; DQ267977.1 and DQ267976.1 of *C. pomonella*; JF894118.1 and JF894119.1 of *H. armigera*; JQ085429.1 and AY061975.1 of *P. xylostella*; AQQ79918.1 and AQQ79919.1 of *S. litura*; and AGK44160.1 (AChE-1) of *S. frugiperda*. Therefore, the molecular models of AChEs were constructed by homology using the SWISS-MODEL program, operating in the search-templates mode, followed by a user-template mode [22,25]. The PDB structures of AChE that recorded the best score (Appendix A) were selected to be used as templates.

4.2. Point Mutations on the AChE Structures

Once the 3D models of the wild type AChE enzyme were constructed, we proceeded to perform point mutations, previously described as possibly responsible for diverse resistance to carbamates and organophosphorus compounds in many insect species [33]. The AChE-2 structures of *B. mandarina*, *B. mori*, *C. auricilius*, *C. suppressalis*, *C. pomonella*, *H. armígera*, *P. xylostella*, and *S. litura* were modified in eight residues considered the most important mutations in Diptera organisms [9,34,35] (F78S, L(V)129V/T, V150L, A201S, G227A, F290Y, G328A, and G396S, A(K, S)484R corresponding AChE numbering of *T. californica*) with the FoldX program, using the FoldX Tool BuildModel [26,27]. In addition, molecular models of AChE-1 were generated to construct the modified enzyme by resistance-associated mutations in lepidopterous organisms [9]: A201S, G227A and L452S to *B. mandarina* and to *B. mori*; A201S to *C. auricilius*; A201S to *C. suppressalis*; F290V to *C. pomonella*; A201S, G227A and F290A to *P. xylostella* and *S. litura*, corresponding to the AChE numbering of *T. californica*; in Figure 8, a 3D representation is presented.

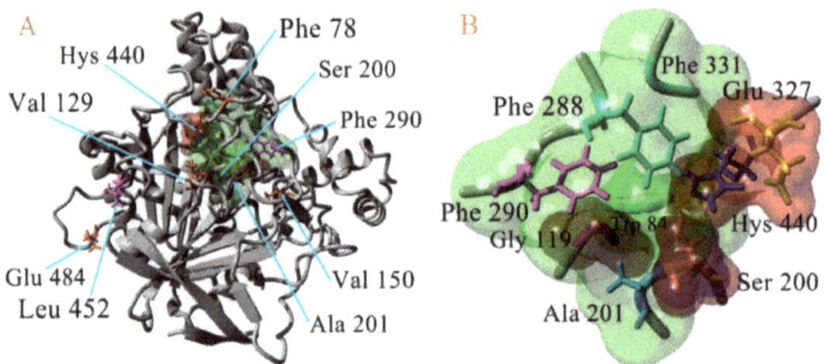

Figure 8. 3D atomic model of AChE, numbering of *T. californica*. (**A**). Location of waste susceptible to proposed point mutations. (**B**) Proximity of catalytic triad (Ser^{200}, Hys^{440}, and Glu^{327}).

4.3. Ligands Used in Docking

All OPs molecules were obtained from the PubChem database [36,37]. The identification codes (CID) of the compounds were the following: CID991 (parathion, 1c), CID1982 (acephate, 2c), CID2268 (azinphos-methyl, 3c), CID2730 (chlorpyrifos, 4c), CID3017 (diazinon, 5c), CID3039 (dichlorvos, 6c), CID3082 (dimethoate, 7c), CID3118 (disulfoton, 8c), CID3286 (ethion, 9c), CID4004 (malathion, 10c), CID4096 (methamidophos, 11c), CID4130 (parathion-methyl, 12c), CID4420 (naled, 13c), CID4793 (phosalone, 14c), CID5853 (trichlorfon, 15c), CID9298 (ronnel, 16c), CID12901 (phosmet, 17c), CID13707 (mephosfolan, 18c), CID13709 (methidathion, 19c), CID13721 (carbophenothion-methyl,

20c), CID16421 (EPN, 21c), CID17435 (phenthoate, 22c), CID16422 (bromophos, 23c), CID19657 (dioxabenzofos, 24c), CID17522 (cyanophos, 25c), CID21803 (chlorpyrifos-methyl, 26c), CID20965 (bromophos-ethyl, 27c), CID25669 (cyanofenphos, 28c), CID25146 (dialifos, 29c), CID26124 (quinalphos, 30c), CID29307 (isoxathion, 31c), CID30709 (leptophos, 32c), CID31200 (fenitrothion, 33c), CID34526 (pirimiphos-methyl, 34c), CID32184 (triazophos, 35c), CID36870 (prothiofos, 36c), CID38779 (profenofos, 37c), CID37125 (sulprofos, 38c), CID5284462 (tetrachlorvinphos, 39c), CID5355863 (mevinphos, 40c), CID5371560 (dicrotophos, 41c), CID5377784 (chlorfenvinphos, 42c), and CID9570290 (phoxim, 43c). In addition, the natural substrate molecule CID187 (acetylcholine (ACh)) and a non-OPs pesticide (CID6199, psoralen) were included as a control in docking evaluations.

4.4. Verification of Computational Methodology

The computational methodology was verified by comparing our results of the binding energy (ΔG_b) predicted with the docking results reported previously (see Appendix B) by Somani et al. (2015) to AChE of humans in complex with psoralen (PDB 1EVE was used) [22] and Ranjan et al. (2018) to AChE of humans in complex with several ligands (PDB 3LII was used) [38]; phosalone (CID6199), dimefox (CID8264), dichlorvos (CID5371560), phoxim ethyl phosphonate (CID6507160), heptenophos (CID62773), and methamidophos (CID4096). These molecules were obtained from the PubChem database [36,37]. The docking evaluation was carried out as described in the molecular docking section.

4.5. Molecular Docking of Ligands Targeting AChE Enzymes

In order to perform the molecular docking, all the molecular models of the AChE (wild type and modified by point mutations) were prepared in the UCSF Chimera program using the tool dockPrep with standard protocol [39] and saved in PDBQT format, and all ligands were saved in the mol2 format, according to the standard protocol. For each protein structure (AChE-1 and AChE-2) a GridBox was built with a dimension of X = 18.8 Å, Y = 18.8 Å and Z = 18.8 Å, using the MGTool program [28]. The coordinates of the center of the GridBox used were $X_1 = -59.62$, $Y_1 = 58.834$, and $Z_1 = 16.476$. The docking was performed using the AutoDock4.2 program [28], which was installed on a Linux Mint 17.3 operating system implemented with a 3.4 GHz Intel Core i7 processor and 23.5 GB RAM. All molecular models were analyzed with PLIP (Protein-Ligand Interaction Profiler) [30].

5. Conclusions

The Lepidoptera family has an important economic impact in the world. However, resistance to insecticides is a very common problem in different countries. This resistance is mainly associated with punctual mutations; therefore, a model that helps us know how this change affects the activity of the insecticides is necessary to obtain or apply better control strategies. Based on the results in this study, the proposed mutations are not associated with the presence of insensitivity to the enzyme. However, the mutations evaluated in the AChE-1 and AChE-2 structures of enzymes do not affect in any way (there is no regularity of change or significant deviations) the values of the binding energy (ΔG_b) recorded in the AChE-OPs complexes. Therefore, it is assumed that the proposed mutations confer resistance due to an inadequate steric interaction that prevents a phosphorylation reaction of the enzyme by the OP molecule and, therefore, is an irreversible inhibition.

Author Contributions: Acquisition of funds, revision, and edition, G.R. and M.A.C.-H.; methodology and Software, D.M.-A., A.J.-S., and V.H.-M.; validation, D.M.-A. and F.R.-E.; data curation, F.R.-E.; formal analysis, M.A.R.-P.; drafting of the initial draft, F.R.-E.

Acknowledgments: This work was supported by the National Council of Science and Technology (CONACyT, Mexico), by grant number Postdoc, CVU 204984 to F.R.E. We wish to express our gratitude to the Instituto Politecnico Nacional for the support granted.

Conflicts of Interest: The authors declare no conflict of interest.

Abbreviations

AChE	acetylcholinesterase
AChE-1[wild type]	form 1 acetylcholinesterase, wild type, without mutations
AChE-1[modified]	form 1 acetylcholinesterase, modified, with puntual mutations
AChE-2 [wild type]	form 2 acetylcholinesterase, wild type, without mutations
AChE-2 [modified]	form 2 acetylcholinesterase, modified, with puntual mutations
OP	organophosphorus
ΔG_b	binding energy
RMSD	root-mean-square deviation
Q[score]	quality of alignment

Appendix A

Results for the Homology Modeling and Multiple Alignment of AChE Enzymes

Once the sequences of AChE enzymes of Lepidoptera indicated in Section 4 were retrieved from The National Center for Biotechnology Information [32], we proceeded to construct the molecular models by homology using the SWISS-MODEL program (operating in search-templates mode, followed by user-template mode) [24,25]. The best recorded score was 5YDJ (AChE of *Anopheles gambiae*) and 1DX4 (AChE of *Drosophila melanogaster*) for AChE-1 and AChE-2, respectively. Data sheets of the SWISS-MODEL homology modelling report is presented in Table A1.

Table A1. Model building report of AChE (AChE-1 and AChE-2, both wild type) of Lepidoptera organisms [24,25].

AChE-1 Model	Template	Seq[I]	GMQE	QMEAN[Zs]	Seq[S]	Range/Covererage
B. mandarina	5YDJ	68.77	0.73	−1.64	0.52	103-642/0.83
B. mori	5YDJ	70.62	0.74	−1.60	0.53	103-642/0.80
C. auricilius	5YDJ	66.39	0.72	−1.70	0.51	114-653/0.88
C. suppressalis	5YDJ	67.46	0.71	−2.27	0.51	87-653/0.85
C. pomonella	5YDJ	66.99	0.72	−1.43	0.51	111-650/0.88
H. armigera	5YDJ	67.27	0.72	−1.45	0.51	115-654/0.87
P. xylostella	5YDJ	67.49	0.74	−1.47	0.51	98-638/0.90
S. litura	5YDJ	67.27	0.73	−1.42	0.51	114-653/0.88
AChE-2 Model	**Template**	**Seq[I]**	**GMQE**	**QMEAN**	**Seq[S]**	**Range/Covererage**
B. mandarina	1DX4	59.10	0.72	−2.38	0.49	53-603/0.87
B. mori	1DX4	60.18	0.74	−2.52	0.49	53-603/0.87
C. auricilius	1DX4	60.81	0.74	−2.51	0.50	53-600/0.86
C. suppressalis	1DX4	58.08	0.71	−3.00	0.48	53-542/0.84
C. pomonella	1DX4	60.72	0.74	−2.46	0.49	53-603/0.87
H. armigera	1DX4	60.36	0.73	−2.90	0.49	62-612/0.86
P. xylostella	1DX4	60.54	0.74	−2.67	0.49	53-603/0.87
S. litura	1DX4	60.54	0.74	−2.23	0.49	53-603/0.87

Sequence Identity (Seq[I]); Global Model Quality Estimation (GMQE); a comprehensive scoring function for model quality assessment (QMEAN[Zs]); Sequence Similarity (Seq[S]).

The parameters recorded in the evaluation of built models refer to the structural aspects linked to the template used. The Global Model Quality Estimation (GMQE) score is expressed in values of 0 to 1. Higher numbers indicate higher reliability, reflecting the expected accuracy of a model built with that alignment and template, as well as the coverage of the target. The comprehensive scoring function for model quality assessment (QMEAN[Zs]) score provides an estimate of the model on a global scale. Values recorded around zero indicate good agreement between the model structure and experimental structures of similar size, and score values of −4.0 or below indicate low quality models. The alignment of the sequences of the molecular models constructed are presented in Figure A1 (AChE-1) and Figure A2 (AChE-2). Alignments were executed with the Crustal Omega tool [40,41].

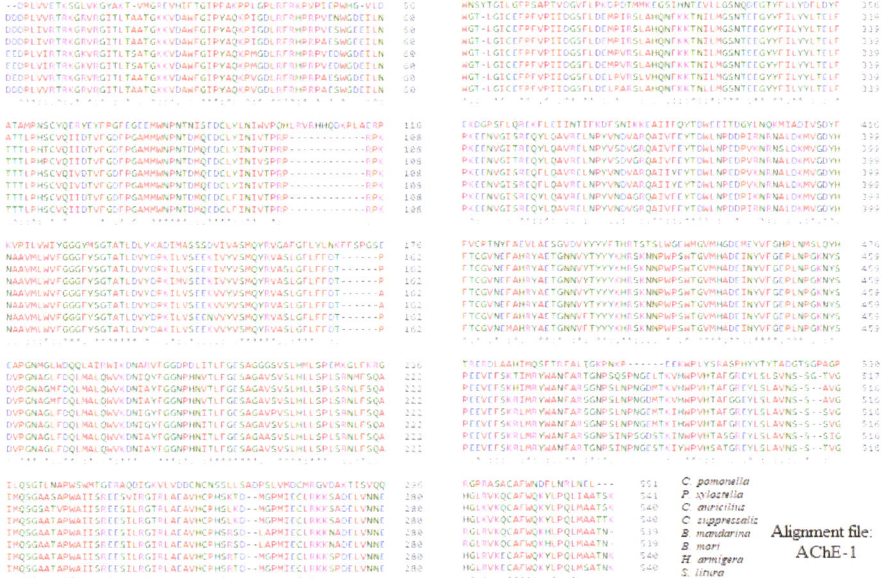

Figure A1. Alignment of the sequences of the constructed molecular models of AChE-1 (form 1 acetylcholinesterase).

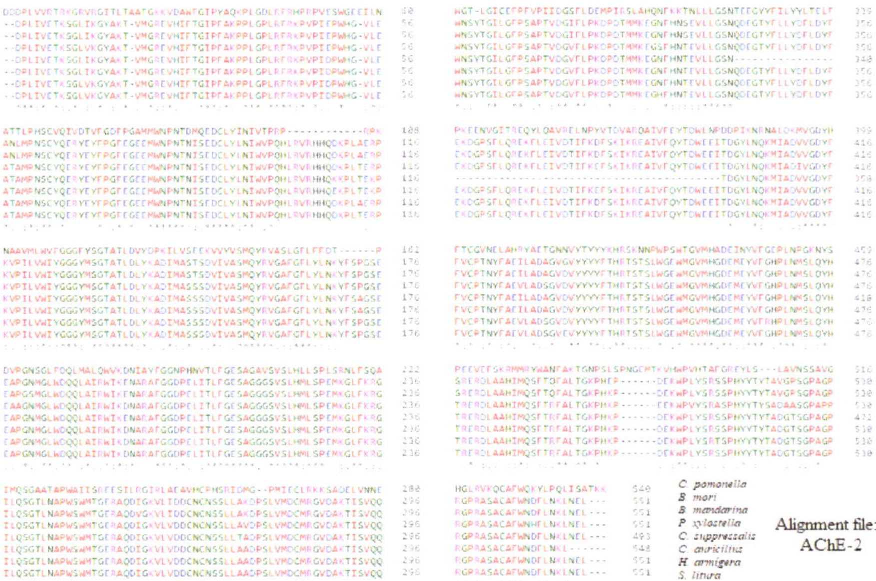

Figure A2. Alignment of the sequences of the constructed molecular models of AChE-2 (form 2 acetylcholinesterase).

Appendix B

Verification of Computational Methodology

The verification of computational methodology consisted of reproducing results of ΔG_b, previously reported by Ranjan et al. 2018 [38], Somani et al. 2015 [22], Chaudhry et al. 2013 [42], and Sharma et al. 2011 [43] approaches the interaction of OP with human AChE, where these authors use crystallographic structures to perform their analyses. In order to detect the variability in reproducibility in energy prediction, we evaluated 28 OPs molecules using our methodology (results not shown). We detected (performing 10 runs, in accordance with the docking protocol for the Autodoc 4.2) program variations in the determination of ΔG_b of 0.5 Kcal/mol; this parameter was obtained by analyzing the variation in the results of the evaluations of the 28 OPs.

The ΔG_b predictions obtained by docking with the purpose of comparing the computational protocol with other computational methodology reporters, in which six molecules on human AChE (PDB: 1EVE) were evaluated, provided results according to those previously reported [22,38] by other research groups (see Table A2). The docking study performed by Somani et al. (2015) and Ranjan et al. (2018) was carried out using the standard glide molecular docking protocol implemented within the Maestro molecular modeling suite by Schrodinger [22,38]. The results register a relative binding energy ($\Delta\Delta G_b$) with a value up to -1.42 Kcal/mol. Considering our variation in the determination of ΔG_b (of 0.5 Kcal/mol), we suggest that the results are comparable.

Table A2. Data sheet of comparison of computational methodologies.

Ligands	ΔG_b *	ΔG_b	$\Delta\Delta G_b$
Psoralen	-6.84 [a]	-6.97	0.13
Dimefox	-4.14 [b]	-4.39	0.25
Dichlorvos	-4.47 [b]	-5.52	1.05
Phoxim ethyl phosphonate	-6.25 [b]	-7.33	1.08
Heptenophos	-5.66 [b]	-6.83	1.17
Methamidophos	-5.87 [b]	-4.45	-1.42

[a], Somani et al., 2015 [22]; [b], Ranjan et al. (2018) [38]; ΔG_b, predicted binding energy in Kcal/mol; *, previously reported; $\Delta\Delta G_b = \Delta G_b{}^* - \Delta G_b$, relative binding energy.

References

1. Hanfu, H.; O'Brochta, D.A. Advanced technologies for genetically manipulating the silkworm *Bombyx mori*, a model Lepidopteran insect. *Proc. Biol. Sci.* **2015**, *282*, 20150487.

2. Roychoudhury, N.; Mishra, R. *Bombyx mandarina Moore* (Lepidoptera: Bombycidae): An Endangered Wild Indian Mulberry Silkworm Species. Tropical Forest Research Institute, Jabalpur, MP, India. *Van Sangyan* **2015**, *2*, 48–50.

3. Invasive Species Compendium, CAB International. *C. suppressalis* (Striped Rice Stem Borer). 2018. Available online: https://www.cabi.org/ISC/datasheet/12855 (accessed on 1 June 2018).

4. Invasive Species Compendium, CAB International. *H. armigera* (Cotton Bollworm). 2018. Available online: https://www.cabi.org/ISC/datasheet/26757 (accessed on 1 June 2018).

5. Forrester, N.; Cahill, M.; Bird, L.; Layland, J. Management of pyrethroid and endosulfan resistance in *Helicoverpa armigera* (Lepidoptera: Noctuidae) in Australia. Pyrethroid resistance: Field resistance mechanisms. *Bull. Entomol. Res.* **1993**, *1*, 132.

6. Invasive Species Compendium, CAB International. *P. xylostella* (Diamondback Moth). 2018. Available online: https://www.cabi.org/ISC/datasheet/42318 (accessed on 1 June 2018).

7. Invasive Species Compendium, CAB International. *S. litura* (Taro Caterpillar). 2018. Available online: https://www.cabi.org/ISC/datasheet/44520 (accessed on 1 June 2018).

8. Abbas, N.; Shad, S.; Razaq, M.; Waheed, A.; Aslam, M. Resistance of *Spodoptera litura* (Lepidoptera: Noctuidae) to profenofos: Relative fitness and cross-resistance. *Crop Prot.* **2014**, *58*, 49–54. [CrossRef]

9. Guo, D.; Luo, J.; Zhou, Y.; Xiao, H.; He, K.; Yin, C.; Xu, J.; Li, F. ACE: An efficient and sensitive tool to detect insecticide resistance-associated mutations in insect acetylcholinesterase from RNA-Seq data. *BMC Bioinform.* **2017**, *18*, 1–9. [CrossRef] [PubMed]

10. Brevik, K.; Schoville, S.; Mota-Sanchez, D.; Chena, Y. Pesticide durability and the evolution of resistance: A novel application of survival analysis. *Pest Manag. Sci.* **2018**, *74*, 1953–1963. [CrossRef]

11. Sakine, U. Insecticides and Advances in Integrated Pest Management. In *Insecticide Resistance*; Perveen, F., Ed.; InTech Press: London, UK, 2012; pp. 469–478.

12. *IRAC Insecticide Resistance Action Committee Mode of Action Classification Scheme*; Prepared by: IRAC International MoA Working Group; Approved by: IRAC Executive Issued; IRAC: Baghdad, Iraq, 2018; pp. 1–26.

13. Sparks, T.; Nauen, R. IRAC: Mode of action classification and insecticide resistance management. *Pestic. Biochem. Physiol.* **2015**, *121*, 122–128. [CrossRef] [PubMed]

14. Aharoni, A.; O'Brien, R. Inhibition of acetylcholinesterases by anionic organophosphorous compounds. *Biochemistry* **1968**, *7*, 1538–1545. [CrossRef]

15. Aldridge, W. The nature of the reaction of organophosphorous compounds and carbamates with esterases. *Bull. World Health Organ.* **1971**, *44*, 25–30. [PubMed]

16. Langel, U.; Jarv, J. Leaving group effects in butyrylcholinesterase reaction with organophosphorous inhibitors. *Biochim. Biophys. Acta* **1978**, *525*, 122–133. [CrossRef]

17. Lushchekina, S.; Patrick, M. Catalytic bioscavengers against organophosphorus agents: Mechanistic issues of self-reactivating cholinesterases. *Toxicology* **2018**, *409*, 91–102. [CrossRef]

18. Lushchekina, S.; Schopfer, L.; Grigorenko, B.; Nemukhin, A.; Varfolomeev, S.; Lockridge, O.; Masson, P. Optimization of cholinesterase-based catalytic bioscavengers against organophosphorus agents. *Front. Pharmacol.* **2018**, *9*, 1–13. [CrossRef]

19. Ordentlich, A.; Kronman, C.; Barak, D.; Stein, D.; Ariel, N.; Marcus, D.; Velan, B.; Shafferman, A. Engineering resistance to 'aging' of phosphorylated human acetylcholinesterase. Role of hydrogen bond network in the active center. *FEBS Lett.* **1993**, *344*, 215–220.

20. Michigan State University Board of Trustees. Arthropod Pesticide Resistance Database (APRD). 2000. Available online: https://www.pesticideresistance.org (accessed on 1 June 2018).

21. Insecticide Resistance Action Committee (IRAC) Resistance Management for Sustainable Agriculture and Improved Public Health. Available online: http://www.irac-online.org (accessed on 1 June 2018).

22. Somani, G.; Kulkarni, C.; Shinde, P.; Shelke, R.; Laddha, K.; Sathaye, S. In vitro acetylcholinesterase inhibition by psoralen using molecular docking and enzymatic studies. *J. Pharm. Bioallied. Sci.* **2015**, *7*, 32–36. [CrossRef]

23. Sadia-Chishty, M. A review on medicinal importance of babchi (*Psoralea corylifolia*). *Int. J. Recent Scien. Res.* **2016**, *7*, 11504–11512.

24. Waterhouse, A.; Bertoni, M.; Bienert, S.; Studer, G.; Tauriello, G.; Gumienny, R.; Heer, F.T.; de Beer, T.A.P.; Rempfer, C.; Bordoli, L.; et al. SWISS-MODEL: Homology modelling of protein structures and complexes. *Nucleic Acids Res.* **2018**, *46*, W296–W303. [CrossRef] [PubMed]

25. Protein Structure Homology-Modelling Server. Available online: https://swissmodel.expasy.org (accessed on 1 June 2018).

26. Van Durme, J.; Delgado, J.; Stricher, F.; Serrano, L.; Schymkowitz, J.; Rousseau, F. A graphical interface for the FoldX forcefield. *Bioinformatics* **2011**, *27*, 1711–1712. [CrossRef]

27. Schymkowitz, J.; Borg, J.; Stricher, F.; Ny, R.; Rousseau, F.; Serrano, L. The FoldX web server: An online force field. *Nucleic Acids Res.* **2005**, *33*, W382–W388. [CrossRef]

28. Morris, G.M.; Huey, R.; Lindstrom, W.; Sanner, M.F.; Belew, R.K.; Goodsell, D.S.; Olson, A J. Autodock4 and AutoDockTools4: Automated docking with selective receptor flexiblity. *J. Comput. Chem.* **2009**, *16*, 2785–2791. [CrossRef]

29. Krissinel, E.; Henrick, K. Secondary-structure matching (SSM), a new tool for fast protein structure alignment in three dimensions. *Acta Crystallogr. D Biol. Crystallogr.* **2014**, *60*, 2256–2268. [CrossRef]

30. Salentin, S.; Schreiber, S.; Haupt, V.; Adasme, M.; Schroeder, M. PLIP: Fully automated protein-ligand interaction profiler. *Nucleic Acids Res.* **2015**, *43*, W443–W447. [CrossRef]

31. Dolinsky, T.J.; Nielsen, J.E.; McCammon, J.A.; Baker, N.A. PDB2PQR: An automated pipeline for the setup, execution, and analysis of Poisson-Boltzmann electrostatics calculations. *Nucleic Acids Res.* **2004**, *32*, W665–W667. [CrossRef]

32. The National Center for Biotechnology Information (NCBI). Available online: http://www.ncbi.nlm.nih.gov (accessed on 1 June 2018).

33. Menozzi, P.; Shi, M.; Lougarre, A.; Tang, Z.; Fournier, D. Mutations of acetylcholinesterase which confer insecticide resistance in *Drosophila melanogaster* populations. *BMC Evol. Biol.* **2004**, *4*, 1–7. [CrossRef] [PubMed]

34. Carvalho, R.; Omoto, C.; Field, L.; Williamson, M.; Bass, C. Investigating the molecular mechanisms of organophosphate and pyrethroid resistance in the fall armyworm *Spodoptera frugiperda*. *PLoS ONE* **2013**, *8*, e62268. [CrossRef] [PubMed]

35. Cassanelli, S.; Reyes, M.; Rault, M.; Manicardi, G.; Sauphanor, B. Acetylcholinesterase mutation in an insecticide-resistant population of the codling moth *Cydia pomonella* (L.). *Insect Biochem. Mol. Biol.* **2006**, *36*, 642–653. [CrossRef]

36. Kim, S.; Chen, J.; Cheng, T.; Gindulyte, A.; He, J.; He, S.; Li, Q.; Shoemaker, B.; Thiessen, P.; Yu, B.; et al. PubChem 2019 update: Improved access to chemical data. *Nucleic Acids Res.* **2019**, *47*, D1102–D1109. [CrossRef]

37. PubChem. Available online: https://pubchem.ncbi.nlm.nih.gov (accessed on 1 July 2018).

38. Ranjan, A.; Chauhan, A.; Jindal, T. In silico and in vitro evaluation of human acetylcholinesterase inhibition by organophosphates. *Environ. Toxicol. Pharmacol.* **2018**, *57*, 131–140. [CrossRef]

39. Pettersen, E.F.; Goddard, T.D.; Huang, C.C.; Couch, G.S.; Greenblatt, D.M.; Meng, E.C.; Ferrin, T.E. UCSF Chimera—A Visualization System for Exploratory Research and Analysis. *J. Comput. Chem.* **2004**, *25*, 1605–1612. [CrossRef]

40. Sievers, F.; Higgins, D.G. Clustal omega. *Curr. Prot. Bioinform.* **2014**, *48*, 3.13.1–3.13.16. [CrossRef]

41. Multiple Sequence Alignment. Clustal Omega. Available online: https://www.ebi.ac.uk/Tools/msa/clustalo/ (accessed on 1 July 2018).

42. Chaudhry, M.; DASS, J.F.; Selvakumar, D.; Kumar, N. In-silico study of acetylcholinesterase inhibition by organophosphate pesticides. *Int. J. Pharm. Bio Sci.* **2013**, *4*, B788–B802.

43. Sharma, A.K.; Gaur, K.; Tiwari, R.K.; Gaur, M.S. Computational interaction analysis of organophosphorus pesticides with different metabolic proteins in humans. *J. Biomed. Res.* **2011**, *25*, 335–347. [CrossRef]

International Journal of
Molecular Sciences

Investigation of Phospholipase Cγ1 Interaction with SLP76 Using Molecular Modeling Methods for Identifying Novel Inhibitors

Neha Tripathi [1], Iyanar Vetrivel [1], Stéphane Téletchéa [2], Mickaël Jean [3],
Patrick Legembre [3,4] and Adèle D. Laurent [1,*]

1   CEISAM UMR CNRS 6230, UFR Sciences et Techniques, Université de Nantes, 44322 Nantes CEDEX 3,
    France; neha.tripathi@univ-nantes.fr (N.T.); iyanar.vetrivel@univ-nantes.fr (I.V.)
2   UFIP UMR CNRS 6286, UFR Sciences et Techniques, Université de Nantes, 44322 Nantes CEDEX 3, France;
    stephane.teletchea@univ-nantes.fr
3   CLCC Eugène Marquis, Equipe Ligue Contre Le Cancer, 35042 Rennes, France;
    mickael.jean@univ-rennes1.fr (M.J.); patrick.legembre@inserm.fr (P.L.)
4   COSS INSERM UMR1242, Université Rennes 1, 35042 Rennes, France
*   Correspondence: Adele.Laurent@univ-nantes.fr; Tel.: +33-(0)251-125-743

Received: 27 August 2019; Accepted: 19 September 2019; Published: 23 September 2019

Abstract: The enzyme phospholipase C gamma 1 (PLCγ1) has been identified as a potential drug target of interest for various pathological conditions such as immune disorders, systemic lupus erythematosus, and cancers. Targeting its SH3 domain has been recognized as an efficient pharmacological approach for drug discovery against PLCγ1. Therefore, for the first time, a combination of various biophysical methods has been employed to shed light on the atomistic interactions between PLCγ1 and its known binding partners. Indeed, molecular modeling of PLCγ1 with SLP76 peptide and with previously reported inhibitors (ritonavir, anethole, daunorubicin, diflunisal, and rosiglitazone) facilitated the identification of the common critical residues (Gln805, Arg806, Asp808, Glu809, Asp825, Gly827, and Trp828) as well as the quantification of their interaction through binding energies calculations. These features are in agreement with previous experimental data. Such an in depth biophysical analysis of each complex provides an opportunity to identify new inhibitors through pharmacophore mapping, molecular docking and MD simulations. From such a systematic procedure, a total of seven compounds emerged as promising inhibitors, all characterized by a strong binding with PLCγ1 and a comparable or higher binding affinity to ritonavir ($\Delta G_{bind} < -25$ kcal/mol), one of the most potent inhibitor reported till now.

Keywords: phospholipase C gamma 1; SLP76; virtual screening; pharmacophore mapping; molecular docking; molecular dynamics

---

1. Introduction

Apoptosis [1] is genetically encoded to provide mechanisms related to organ formation, or to eliminate damaged cells. The mechanism is mediated by regulated interactions between various cellular components [2]. Under pathological conditions, such as cancers, autoimmune disorders and viral infections, dysregulated components might cause the decrease in cellular apoptosis [3]. Therefore, targeting the enzymes and receptors involved in regulation of cellular apoptosis has been established as an important therapeutic strategy for such pathological conditions [4–12]. Amongst the numerous regulators of apoptosis, the multifunctional phospholipase C (PLC) enzymes interact with target proteins to modulate the cellular apoptosis [13–16]. Indeed, PLCs are essential for regulation of several cellular processes as they catalyze the hydrolysis of phosphatidylinositol 4,5-bisphosphate (PIP2) into inositol 1,4,5-triphosphate (IP3) and diacylglycerol (DAG) using Ca^{2+} as cofactor [17]. In mammals, PLCγ isozyme is, particularly, involved in cell growth regulation [17–20], and is constituted of two isoforms (i.e., PLCγ1 and PLCγ2) [17]. The PLCγ1 isoform is constitutively expressed in all cells, whereas PLCγ2 is mainly expressed in immune cells [17]. Particularly, PLCγ1 has been identified to play an important role in the regulation of cell growth and cellular differentiation [17,19], by interacting with various macromolecular targets [17] such as epidermal growth factor, fibroblast growth factor, platelet derived growth factor, vascular endothelial growth factor and cluster of differentiation 95 (CD95) [16,21]. PLCγ1 is also known to be involved and to play an important role in cell invasion, metastasis and progression in cancers [13,15,22].

Structurally, PLCγ1 is a multidomain protein [14,21] (Figure S1), for which the catalytic site is present in a TIM barrel [19]. The catalytic activity of PLCγ1 is controlled by a conformational change in relative orientation of its various domains which, in turn, is governed by the phosphorylation of the Tyr783 residue [17]. The structural complexity of PLCγ1 contributes to the multitude of its biological targets. Particularly, the SH3 domain of PLCγ1 (PLCγ1-SH3) has been reported to contain the binding site for several target proteins, enriched in proline (PXXP motifs) [18]. Experimental studies ascertain the importance of the PLCγ1-SH3 in interactions with several proteins including autoimmune poly-endocrinopathy candidiasisectodermal dystrophy protein (AIRE), colonic and hepatic tumor overexpressed protein (CHTOG) and the gliomatumor suppressor candidate region gene 1 protein [23]. It has also been identified to be essential for mitogenic activity of PLCγ1 [24], as the SH3 domain, in combination with SH2 domains, induces mitogenesis in quiescent fibroblast, indicating its importance for cellular growth [25]. The interaction site for dynamin (a membrane-remodeling GTPase) is also located within PLCγ1-SH3 [26]. Pharmacological involvement of such interactions in various pathological conditions gives rise to the opportunity to identify therapeutic agents, specifically targeting the PLCγ1-SH3 and thus preventing the interaction between PLCγ1 and its cellular targets.

The lymphocyte cytosolic protein 2, also known as SLP76 is a T-cell adaptor protein, which has been structurally characterized to interact with the PLCγ1-SH3 (PDB ID: 1YWO, Figure S1) [18], and thus offers the representative interactions between the PLCγ1-SH3 and its substrates. The PLCγ1-SH3 binds to the XPXXPXR motif of SLP76, which is more specific than a usual PXXP motif. Using this information, Poissonnier et al. have reported the design of an original peptidomimetic inhibitor, by employing molecular modeling studies [16]. In their work, protein-fragment complementation assay (PCA) and in vitro screening of 1280 molecules (Prestwick library) has been performed, identifying the inhibitors of PLCγ1–CD95 interactions which include ritonavir (HIV protease inhibitor), anethole (flavoring agent), daunorubicin (topoisomerase inhibitor), diflunisal (nonsteroidal anti-inflammatory drug), and rosiglitazone (antidiabetic agent which interacts with peroxisome proliferator-activated receptor) [16] (Figure 1). Additionally, a peptidomimetic (named DB550) (Figure 1) was designed on the basis of structural features extracted from the calcium inducing domain (CID) of CD95. These inhibitors were demonstrated to specifically inhibit the interactions of PLCγ1 and CD95. Administration of both ritonavir and DB550 showed therapeutic effects in lupus-prone mice [16]. Overall these findings indicate that targeting the PLCγ1-SH3 is of prime importance for the management of various pathological conditions involving a plethora of immunological conditions and cancers. Availability of three dimensional (3D) complex between various drug targets and their modulators become crucial for facilitating the rational drug design. Although, the experimental techniques, such as X-ray crystallography and NMR, are powerful tools in determining these structures, these is time-consuming and expensive, and not feasible for several proteins. The molecular modeling techniques such as homology modeling, molecular docking, and molecular dynamics simulations offer an appropriate solution for the prediction of intermolecular recognition interactions [9–11,27–29]. Several examples of successful application of molecular modeling techniques for the identification of potential therapeutic agents are available in literature [29–36]. In view of this, the computational methods were utilized in the current study. To the best of our knowledge, combining molecular docking and molecular dynamics (MD) simulations have never been performed on the PLCγ1-SH3.

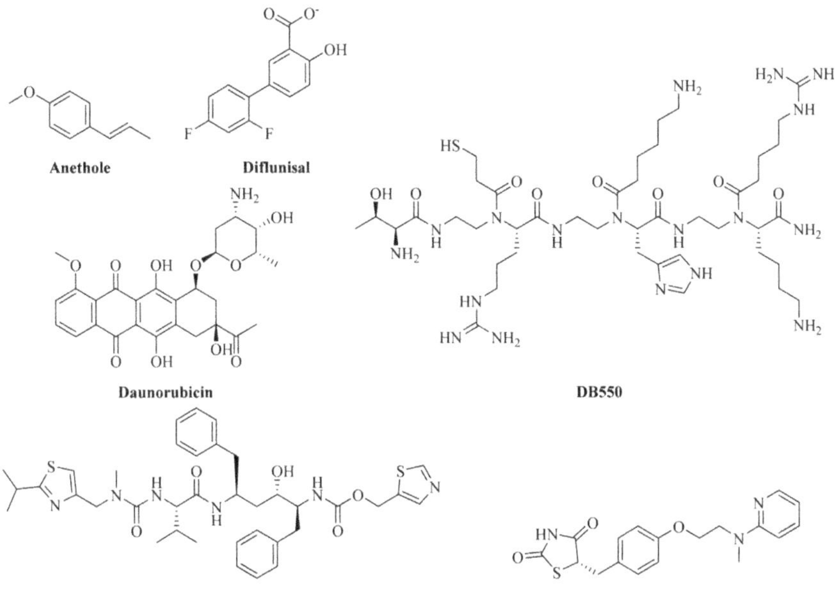

Figure 1. Structural formula of the reported inhibitors of phospholipase C gamma 1–cluster of differentiation 95 (PLCγ1–CD95) interactions [16].

In the present study, the PLCγ1-SLP76 complex (Figure S1) is therefore exploited for identifying novel inhibitors targeting PLCγ1 through a structure-based pharmacophore map to identify key structural features involved at the PLCγ1-SLP76 interface. As a first step, the molecular docking and molecular dynamics simulations of reference compounds, shown in Figure 1, is performed to characterize key residues as well as their binding affinity, so as to obtain reference values. Thereafter, a virtual library of compounds (constituted of 227,228 molecules) was subjected to a systematic virtual screening protocol, from which the top sixteen molecules were considered for an extended work using MD simulations to ensure their stable binding with PLCγ1. After a careful analysis of the MD results, it was found that out of the sixteen molecules, seven were highly promising candidates for inhibiting the interaction between the PLCγ1-SH3 and its target proteins. To the best of our knowledge, this is the first attempt to employ SLP76-based features for drug design against PLCγ1 as well as to screen such a large library of 227,228 compounds.

2. Results and Discussion

2.1. Molecular Recognition of SLP76 and Known Inhibitors by PLCγ1

The PLCγ1-SLP76 crystal structure (PDB ID: 1YWO) [18] offers the opportunity to employ a structure-based drug design strategy for the identification of novel PLCγ1 inhibitors. Deng et al. employed isothermal titration calorimetry to identify a proline rich motif (^{186}PPVPPQRP^{193}) in SLP76 [18]. In the crystal structure, SLP76 forms four H-bonds with PLCγ1 via Asp808, Glu809, Trp828 and Asn844 (Figure 2A) and hydrophobic interactions at the protein–peptide interface through the proline enriched motif (XPXXPXR). Globally, the binding of SLP76 with PLCγ1 is governed by both, structural and electrostatic complementarity (Figure 2A and Figure S2A,B) [18]. Indeed, PLCγ1 possesses an arginine binding site which is characterized by a highly electronegative surface potential due to the presence of acidic Asp808 and Glu809 (Figure S2A). The latter are complementary to the highly electropositive Arg192 of SLP76 (Figure S2B) forming a salt bridge interaction. The ΔG_{bind} value for co-crystallized conformation of SLP76 with PLCγ1 was estimated through MM/GBSA calculation to a high magnitude value, i.e., −85.42 kcal/mol (Figure 2B).

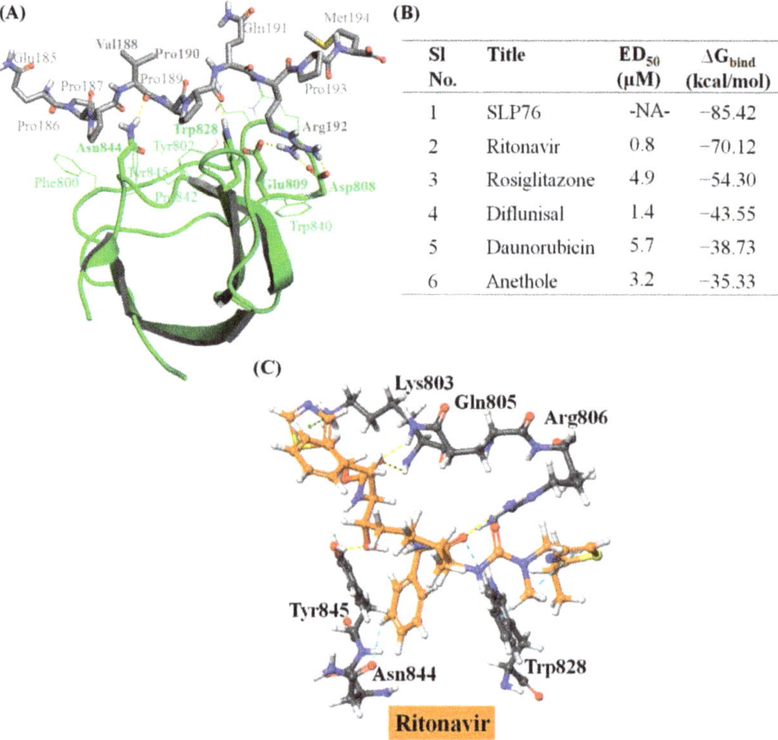

(B)

Sl No.	Title	ED$_{50}$ (μM)	ΔG$_{bind}$ (kcal/mol)
1	SLP76	-NA-	−85.42
2	Ritonavir	0.8	−70.12
3	Rosiglitazone	4.9	−54.30
4	Diflunisal	1.4	−43.55
5	Daunorubicin	5.7	−38.73
6	Anethole	3.2	−35.33

Figure 2. Interactions of SLP76 and reported inhibitors [16] with PLCγ1. (**A**) Main interactions between SLP76 and PLCγ1 in the X-ray crystal structure [18]. Residues, involved in H-bonds are shown in bold and stick. (**B**) Experimental ED$_{50}$ values [16] and calculated MM/GBSA binding energy (ΔG$_{bind}$) for SLP76 (in the crystal structure) and reported inhibitors (after IFD) with PLCγ1. (**C**) Key interactions of ritonavir with PLCγ1. Legend for interactions: H-bonds in yellow; π⋯cation interactions in green; π⋯π stacking interactions in blue; aromatic H-bonds in cyan; salt bridges in magenta. -NA-: Not Applicable.

Molecular docking of the reported PLCγ1 inhibitors (Figure 1) [16] was performed firstly to compare their binding values with SPL76 and secondly to establish selection criteria for the following step, that is the virtual screening. Structural superimposition of the predicted docked poses of these inhibitors in PLCγ1-SH3 reveals that all inhibitors overlap with the SLP76 peptide (Figure S3), especially at the C-terminal of SLP76. We do note, however, that ritonavir and rosiglitazone are slightly less aligned onto the SLP76 *N*-terminal side. ΔG$_{bind}$ values were computed for each reported inhibitor so as to evaluate its correlation with their reported effective dose (ED$_{50}$) [16] (Figure 2B). A direct correlation is rather difficult to establish between the ED$_{50}$ and calculated ΔG$_{bind}$ values, but it is clear that SLP76 and ritonavir have the highest binding affinity following the ED$_{50}$ trend. Among the reported inhibitors, ritonavir is effectively the most potent inhibitor (ED$_{50}$ of 0.8 μM and ΔG$_{bind}$ of −70.12 kcal/mol) of the PLCγ1 and CD95 interaction. Table S1 enlists all the relevant non-covalent interactions between the reported inhibitors [16] and PLCγ1 the molecular docking (IFD) (see supporting information for details).

2.2. Pharmacophore Mapping and Molecular Docking Based Identification of Promising Hits

Reported interactions between SLP76 and PLCγ1 (Figure 2A) can be considered as the important pharmacophoric features of PLCγ1 interacting agents, as also confirmed by the inhibitor binding (Table S1, Figure 2C and Figure S4). A structure-based pharmacophore map was created accordingly followed by a virtual screening helping to firstly identify compounds possessing similar SPL76 specific binding features. Out of a total of thirty-four structural, hydrophobic and electrostatic features present in SLP76 peptide, five of them were kept (Figure 3A) based on the interactions reported in literature [18], and observed in the available crystal structure (Figure 2A). The pharmacophore is thus built in order to selected compounds which contain two H-bond acceptors mimicking Val188 and Pro190, two hydrophobic groups aligning on Pro189 and Pro190 and one positive feature as Arg192 in SPL76 which was also treated as H-bond donor features. From the initial virtual library of compounds (227,228 molecules) used for the virtual screening, 2734 molecules simultaneously exhibit the five selected pharmacophore features (alignment of top 15 molecules with the generated pharmacophore is shown in Figure S5). Analysis of the topological diameter (range 10 to 25), molecular weight (>350 D) and molecular volume (>1000 Å^3) of the 2734 molecules showed that they exhibit a large size (Figure S6A–C), which helps in occupying the ligand binding site as for the SPL76 peptide in the PLCγ1-SH3 and, possibly, could enhance the specificity of the molecule towards PLCγ1. The octanol/water partition coefficient for most of the selected molecules was in the range of 2 to 5 (Figure S6D) indicating their possible ability to permeate through membranes. Thereafter, a systematic molecular docking protocol (Figure S7) was employed to realize the interaction of the 2734 molecules with the PLCγ1-SH3.

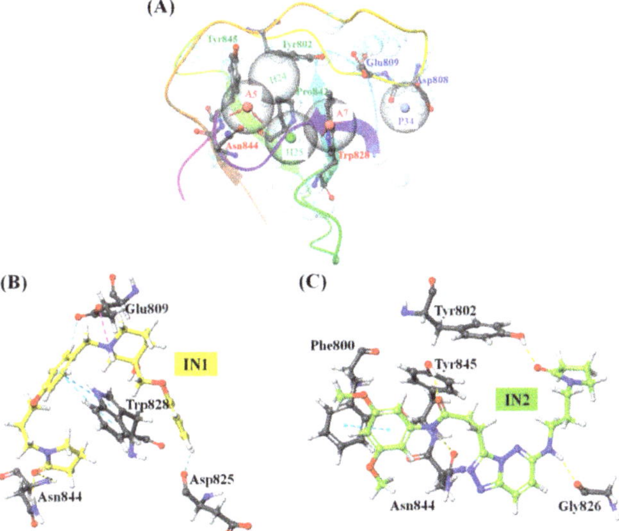

Figure 3. Results from the virtual screening performed for the identification of promising PLCγ1 inhibitors (**A**) Generated pharmacophore hypothesis from PDB ID: 1YWO [18]. Selected five pharmacophore features are shown as large grey spheres, where A: H-bond acceptor. The red arrows indicate the direction of H-bond formation ("A" being the H-bond acceptor); H: hydrophobic group; P: Positive functional group, which is treated equivalent to H-bond donor. Cyan spheres represent excluded receptor volume shell. (**B**) and (**C**) Non-covalent interactions of IN1 and IN2, respectively, with PLCγ1 (see Figure 2C for color legend).

To narrow down the number of compounds a HTVS has been realized filtering compounds specifically interacting with the PLCγ1 arginine binding site (Asp808 and Glu809) and with the

XPXXPXR proline enriched motif (Trp828 and Asn844) of PLCγ1. With the motive of blocking PLCγ1 activity, the presence of interactions within the arginine binding site and at least two H-bonds between screened molecules and PLCγ1 was considered as the selection criterion. Of these two H-bonds, one should be present with the arginine binding site (Asp808 or Glu809) and the other within the XPXXPXR motif recognition site (Trp828 or Asn844). The HTVS helped to filter this set of molecules to 705 compounds (with Glide gscore ≤ −3.5) which were subsequently subjected to molecular docking with higher precision. Evaluation of molecular docking results on the basis of glide gscore, reproducibility of docked conformation and most importantly, structural overlap with cocrystallized ligand SLP76, helped us to identify potential candidates for PLCγ1 inhibition. Final selection after each docking step was based on the calculated ΔG_{bind} value. SP mode and XP mode molecular docking (Figure S7) helped to bring the number of promising hits to 33 molecules characterized by ΔG_{bind} value below −40 kcal/mol, while IFD further narrowed down this number to 16 molecules (Table S2 and Figures S8 and S9). All molecules occupied the similar interaction site as one of the *N*-terminal domain of SLP76 (Figure S10). Molecular interactions between each ligand and PLCγ1 after the last molecular docking step (IFD) are presented in Table 1.

Table 1. Molecular recognition interactions between IN1-IN16 molecules and PLCγ1 after induced fit molecular docking. Residues in bold are also involved in similar interactions with SLP76.

Title	H-Bond	NH⋯π/π⋯π Stacking Interactions	Hydrophobic Interactions	Other Residues within 5Å
IN1	**Glu809, Asn844**	**Trp828**	Tyr802, Gly826, Gly827, **Trp840**, Phe841, **Pro842, Tyr845**	**Gln805, Arg806**, Asp808, Gln824, Asp825, Ser843
IN2	Phe800, Tyr802, Gly826, **Asn844**	**Tyr845**	Leu799, Gly827, Trp828, **Trp840**, Phe841, **Pro842**	**Gln805, Arg806**, Asp808, Glu809
IN3	Arg806, **Asp808**, Gly826, **Asn844**	**Trp828**, Trp840	Tyr802, Gly825, Phe841, **Pro842, Tyr845**	**Gln805**, Glu809, Ser843
IN4	Gln805, Arg806, **Asp808, Glu809**, Tyr845		Tyr802, Trp828, **Trp840, Pro842, Tyr845**,	Lys803, Glu807, Gln824, Asp825, Asn844
IN5	**Asp808, Glu809 Trp828**	Trp840	Tyr802, Gly827, Gly826, Trp829, Phe841, **Pro842, Tyr845**	**Gln805, Arg806**, Gln824, Asp825, Arg830, Ser843, Asn844
IN6	Gln805, Arg806, **Asp808**, Trp840		Tyr802, Trp828, Gly826, Gly827	Gln824, Asp825, Arg830
IN7	**Glu809**, Gly826	**Trp828**, Trp840	**Pro842, Tyr845**, Tyr802, Gly827	**Gln805, Arg806**, Asp808, Asp825, Ser843, Asn844
IN8	Arg806, **Asp808**, Asn844	**Trp828**, Trp840	Tyr802, Gly826, Gly827, Phe841, **Pro842, Tyr845**	**Gln805**, Glu809, Gln824, Asp825, Ser843
IN9	Asp808, **Glu809, Trp828**, Trp840	Arg806, Trp840	Tyr802, **Pro842, Tyr845**	**Gln805**
IN10	**Asp808**, Asn844	Trp840, **Trp828**	**Tyr845**, Tyr802, Gly826, Gly827, Pro842	**Arg806**, Glu809, Gln824, Asp825, Ser843
IN11	Gln805, **Trp828**, Asn844,		Tyr802, Gly826, Gly827, **Trp840**, **Pro842, Tyr845**	Lys803, **Arg806**, Asp808, Glu809, Arg830, Ser843,
IN12	Gln805, Arg806, **Glu809**	**Trp828**	Tyr802, Gly826, **Trp840, Pro842**, Tyr845	Asp801, Lys803, Asp808, Gln824, Asp825, Asn844
IN13	**Asp808**, Gly826, **Trp828**	**Trp828**, Trp840	Tyr802, Gly827, **Pro842, Tyr845**,	**Arg806**, Glu809, Asp825, Ser843, Asn844
IN14	**Asp808, Glu809**, Gly826	**Trp828**	Tyr802, Gly827, **Trp840**	**Arg806**, Gln824, Asp825
IN15	**Asp808**	**Trp828**, Trp840	Tyr802, Gly826, Phe841, **Pro842**	**Gln805, Arg806**, Glu809, Asp825
IN16	**Asp808**, Gly826, Asn844	**Trp828**	Gly827, **Trp840, Pro842, Tyr845**	**Gln805, Arg806**, Glu809, Asp825, Ser843

The 3D molecular recognition interactions for top scoring hits, i.e., IN1 and IN2 are shown in Figure 3B,C, whereas interactions for the other 14 molecules (IN3-IN16) are shown in Figure S11. PLCγ1 residues which participated in H-bonds or salt bridge interactions with all the 16 selected ligands are Gln805, Asp808, Glu809, Trp828, Asn844, and Tyr845 (Table 1). Additional complex stabilization was observed pertaining to NH$\cdots$π/π$\cdots$π interactions with Arg806, Trp828, Trp840, and Tyr845. Hydrophobic interactions were mainly observed with Tyr802, Gly826, Gly827, Trp840, Phe841, Pro842, and Tyr845, for each ligand. As discussed earlier, a careful attention was paid throughout the molecular docking steps to keep two key intermolecular interactions, i.e., at arginine binding site and proline (from XPXXPXR motif) binding site. The ΔG_{bind} values calculated after IFD (Table S3) were comparable to that of ritonavir and SLP76 (Figure 2B), ranging from −78.07 to −56.34 kcal/mol, thus further supporting their candidature as PLCγ1 inhibitors. Interestingly, all selected compounds possessed a basic nature (predicted *pKa* value > 13), facilitating their interaction with the negatively charged arginine binding site of PLCγ1. As shown in Figure 3B,C and Figure S11, a positively charged nitrogen center in these molecules occupied the arginine binding site by interacting with Asp808 or Glu809. These generated complexes were thus taken further for the MD simulations.

2.3. MD Simulations

In order to evaluate the stability of the identified interactions under dynamical conditions and ensure strong binding of ligands with the target, MD simulation is a method of key choice. The generated 16 complexes were submitted to MD simulations for 50 ns to study the system relaxation. Additionally, PLCγ1-SLP76 and PLCγ1-ritonavir complexes were also subjected to MD simulations, as they are considered as reference systems. To ensure reproducibility of the results, each system was simulated in three replicates. Combined cluster analysis (Figure S12) revealed that three replicates behave similarly (keeping 70% as cut-off) for the complexes formed by SLP76, IN1, IN6, IN11, and IN15 with PLCγ1. Indeed, the majority of the three simulation coordinates belongs to one unique cluster. For the systems containing ritonavir, IN2, IN3, IN4, IN5, IN7, IN9, IN12, and IN13, the cluster population was spanned over two clusters, while for PLCγ1 bound to IN8, IN10, IN14, and IN16, at least one of the simulations indicated a wider distribution of the cluster population over the period of simulation run. The RMSD analysis between the clusters in the various systems showed that the inter-cluster distance was <2 Å (Table S4) and the average distance from the centroid for various clusters was <1.5 Å for all the systems. Thus, it can be concluded that the triplicate simulations successfully produced comparable results.

In order to evaluate the stabilized binding of each ligand to PLCγ1 in the generated complexes over the period of simulation, the distribution of each cluster population with time was analyzed (Figure S13). For SLP76, ritonavir, IN1 to IN5, IN8, IN11, and IN13 to IN16, >70% of the frames remained in a single cluster over the last 25 ns of simulation in the three replicate runs. For IN6, IN7, IN10, and IN12, at least two replicates showed an equilibrated trajectory over the entire simulation run. The whole protein RMSD analysis showed that PLCγ1 structure was stabilized during the simulation and showed minimum difference (RMSD < 2.0 Å) in the various complexes as compared to their initial coordinates (Figure S14).

After the global evaluation of the simulation trajectories, we decided to analyze the local behaviors of the molecules at the binding sites. The structural overlap of the ligand position after molecular docking, after the system equilibration and after the production run (for one representative replicate) is shown in Figure S15. For SLP76, the structure overlap was performed between the cocrystallized conformation, equilibrated conformation and the structure after MD simulations. The three structures indicate a clear overlap between each SLP76 conformation (Figure S15A) which indicates the ability of the adopted protocol to maintain the cocrystallized conformation. Compared to the complex generated after molecular docking, the position of the ligand did not change much after system equilibration for majority of the 16 molecules, except for ritonavir and IN11. After the MD simulations, IN1, IN4-IN6, IN8, IN10, and IN13-IN16, were maintained close to the docked pose. The final structure after the MD simulation revealed a significant movement from the docked position for ritonavir, IN3, IN7, IN11, and IN12, while for others no major change in their relative position was observed (Figure S15). For compound IN3, this displacement was mainly observed in the position of pyrazolo (3, 4-*d*) pyrimidinyl ring from the arginine binding site to Proline motif binding site. For IN7, IN11 and IN12 an obvious unbinding of the compound from PLCγ1 was observed during each replica of the molecular dynamics simulations. For a detailed investigation of binding behavior of the identified hits, the RMSD and distance from crucial residues were evaluated (Figures S16 and S17).

2.4. Stable Binding of Identified Molecules to PLCγ1

In order to select the molecules which showed reproducible and stable binding to PLCγ1, RMSD along the simulations and time-dependent distance between the center of mass of the bound ligand and Asn844 of PLCγ1 (COM_dist) were plotted to rapidly identify the unbinding of some ligands (Figures S16 and S17). The cocrystallized peptide SLP76 shows a very stable complexation with PLCγ1 throughout all the simulation runs, as indicated by the stable RMSD value for the entire complex, by the COM_dist and by the ligand RMSD (Figure S16A). Structural overlap of the final coordinates for the three replicates of PLCγ1-SLP76 complex indicates a similar orientation of SLP76 in the binding site, except for its terminal amino acids (Gln185 and Met194). The SLP76 position after MD simulation is highly similar to the one from cocrystallized conformation (RMSD values ranging from 2 to 4 Å). Additionally, the calculated per-nanosecond ΔG_{bind} value for the PLCγ1-SLP76 system indicates a stabilized affinity along the simulation run (Figure S18) with an average ΔG_{bind} value (over last 5 ns) of −50.14 ± 3.96 kcal/mol (Table S5). It can be observed that compared to the ΔG_{bind} value from MM/GBSA calculations (Figure 2B), the value after MD simulations (Table S5) is numerically increased significantly from −85 kcal/mol to −50.14 ± 3.96 kcal/mol, respectively, signifying lowered affinity. From the component analysis of ΔG_{bind} (vdW, electrostatic, etc.), the complexation is dominated by the electrostatic component (−119.92 kcal/mol, Table S5) which is attributed to the strong interaction at the arginine binding site.

The PLCγ1-ritonavir complex generated from the IFD (Figure 2C) was also submitted to MD simulations, which resulted into the stabilized PLCγ1-ritonavir complex (RMSD and COM_dist in Figure S16B), and here again, the calculated ΔG_{bind} value for three replicates (Figure S18) were numerically increased (−28.42 ± 3.31 kcal/mol in Table S5) compared to the one after IFD (−70.12 kcal/mol in Figure 2B), which is much higher (lower affinity) than the ΔG_{bind} of SLP76 (difference of 21.72 kcal/mol). Such a variation can be attributed to the difference in molecular dimension and surface electrostatics between both partners. As we know, SLP76 is an intracellularly present binding partner for PLCγ1, while ritonavir is required to cross the membrane barrier for interacting with its macromolecular drug targets and its smaller molecular size favorably contributes to penetrate the cell membrane. Therefore, normalized ΔG_{bind} value (based on molecular weight or molecular volume) (Table S6) were used to obtain a ΔG_{bind} values accounting for such bias. ΔG_{bind} per unit weight ($\Delta G_{bind\text{-}MW}$) and per unit volume ($\Delta G_{bind\text{-}MV}$) for SLP76 are calculated to be −0.044 kcal/mol and −0.015 kcal/mol, respectively, while for ritonavir $\Delta G_{bind\text{-}MW}$ and $\Delta G_{bind\text{-}MV}$ are −0.040 kcal/mol and −0.013 kcal/mol, respectively. Such normalized ΔG_{bind} values show, as expected, that the binding of ritonavir is highly comparable to SLP76. As ritonavir has been already reported to inhibit the interaction of PLCγ1 with the CD95 death domain by binding at PLCγ1-SH3 [16], we consider the ΔG_{bind} value of ritonavir as a cutoff for the selection of potent inhibitors.

Based on the RMSD and COM_dist values (Figures S16 and S17), molecules showing stable binding to PLCγ1 can be rapidly identified. Three ligands out of the selected 16 compounds, IN7, IN11 and IN12, were released from the binding site of PLCγ1 in at least one of the replicate simulation runs (Figures S16I and S17D,E). Other ligands remained bound at the PLCγ1-SH3. Analysis of ΔG_{bind} over the last 10 ns simulation trajectory (Figure S18) helped to identify molecules with similar binding behavior with PLCγ1 as that of ritonavir. Molecules for which ΔG_{bind} was numerically lower than −25 kcal/mol in all replica simulations (Figure S18) are IN1, IN2, IN3, IN5, IN6, IN8, and IN10 and can be considered equivalent in terms of ΔG_{bind} to ritonavir (Figure 4A and Table S5). Interestingly, $\Delta G_{bind\text{-}MW}$ and $\Delta G_{bind\text{-}MV}$ for all the selected molecules (except for IN7, IN11, and IN12, which were released from the binding site) were lower than SLP76 and ritonavir (Table S5) indicating even a stronger binding towards PLCγ1. The per-residue atomic fluctuation for each system also indicates that the molecules which showed stable binding also present lower fluctuation in PLCγ1 (Figure S19). Contrarily, IN7, IN11, and IN12, which are released from the binding site, induced a higher degree of structural fluctuation within the protein structure. Hereafter, IN7, IN11, and IN12 were not considered for the rest of the analysis.

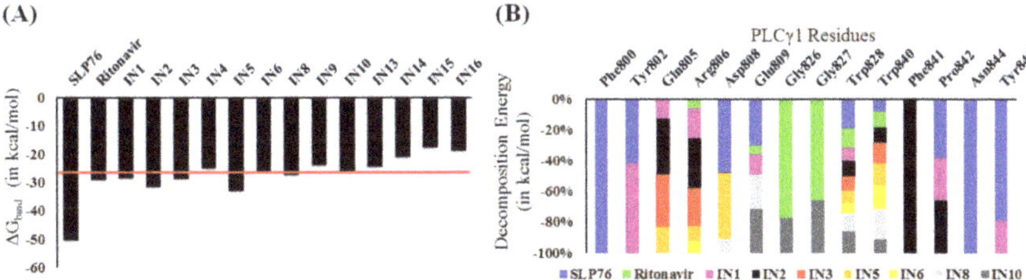

Figure 4. Interaction analysis for the molecular dynamics of the sixteen complexes. (**A**) Average ΔG_{bind} value calculated over the last 5 ns. Red line indicates the cutoff used for final selection of compounds and (**B**) Per-residue decomposition energy analysis for selected potential PLGγ1-inhibtior complexes during molecular dynamics (MD) simulation.

2.5. Molecular Recognition of the Selected Molecules to PLCγ1 Considering MD Simulations

Molecular recognition interactions play a crucial role in ensuring stability of the complex and binding affinity of the molecules to their target. An analysis of per-residue total decomposition energy allowed the identification of key amino acids involved in favorable interactions with ligands. Residues with a significant contribution to the binding energy (cutoff −0.5 kcal/mol) involving the ligands and PLCγ1 are presented as stacked bar plot in Figure 4B. The critical residues for SLP76 binding to PLCγ1-SH3 were Phe800, Tyr802, Asp808, Glu809, Trp828, Trp840, Pro842, Asn844, and Tyr845. Interestingly, these results were in good correlation with the reported important residues (Asp808, Trp840, and Tyr845) (detected from NMR data) [16] for interaction of SLP76 with PLCγ1-SH3. Residues which were identified to be involved in interaction with inhibitors are Gln805, Arg806, Asp808, Glu809, Trp828, and Trp840. Such conclusion was based on their involvement in the complex formation for multiple systems (indicated by presence of multiple colors in the stacked bars). Interestingly, these residues also exhibited lower atomic fluctuation in the presence of bound ligands (Figure S20), and were also reported to be crucial for SLP76 binding to PLCγ1 [18]. Thus, their involvement in interaction with identified compounds, increases the confidence in PLCγ1 inhibiting ability of the selected molecules.

Residues which were involved in H-bond interactions with the selected molecules were identified by analyzing the last 10 ns MD simulation trajectory for the various complexes. Average number of H-bonds (Figure S21) were found to be higher than 3 for IN1, IN3, IN5, IN6, IN10, and IN13 (Table S7). Cumulative H-bond occupancy analysis (Figure 5) helped to identify the residues involved in H-bond interactions. Stacked bars with a high degree of color variation (indicating the presence of H-bond in several PLCγ1-inhibitor complexes) represent the residues important for stabilization of PLCγ1-inhibitor complex, which include Arg806, Asp808, Glu809, and Trp828. These residues were also found to interact with SLP76 during MD simulations via H-bonds. Residues Gln805, Asp825, and Gly827 were involved in H-bond interactions with the inhibitors, but not with SLP76. Their importance in inhibitor binding can be evaluated in vitro; however, this is not covered in the scope of current study.

Based on the extensive MD simulation analysis, performed herein, IN1, IN2, IN3, IN5, IN6, IN8, and IN10 (Figure 6) were proposed as the most potent candidates for PLCγ1 inhibition (Table S2). At the moment, none of these molecules have been evaluated for any kind of biological activity according to the ChEMBL database [37], making those highly interesting compounds for further developments. In vitro evaluation of their binding to PLCγ1 and subsequent, interference in the interaction of PLCγ1 with its cellular targets would be of great therapeutic relevance.

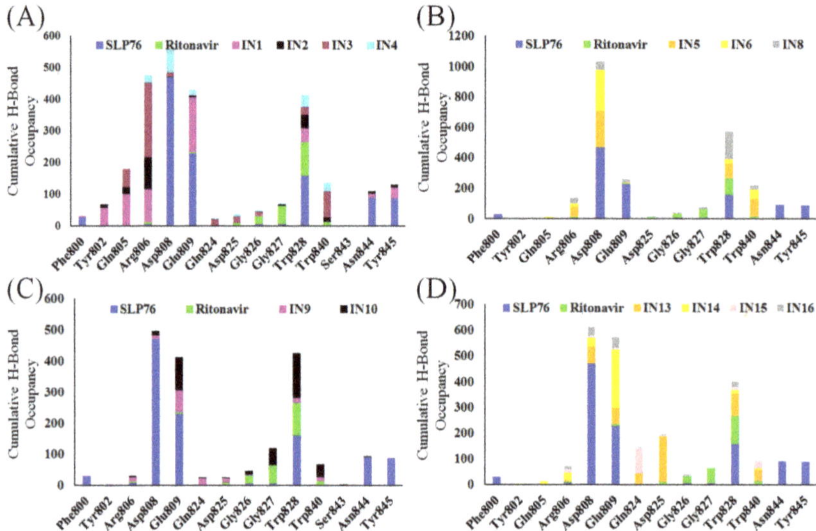

Figure 5. H-bond occupancy analysis for the PLCγ1 residues in various systems over the last 10 ns in various complexes after MD simulations.

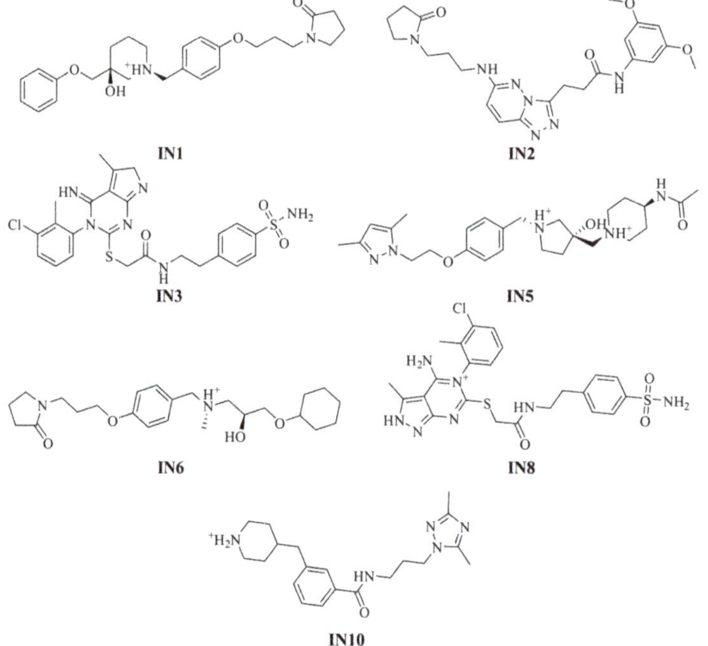

Figure 6. Selected potential inhibitors PLCγ1 after the three replicate MD simulations.

3. Materials and Methods

3.1. Pharmacophore Modeling

Starting from the available crystallographic structure of PLCγ1-SLP76 complex (PDB ID: 1YWO) [18], a pharmacophore model was defined based on all potential pharmacophore features of the ligand complementary to the substrate-binding site using the PHASE module [38,39] of the Schrödinger software package, version 2018-2 [40]. Structural features of SLP76 which facilitate its binding to the PLCγ1-SH3 were identified and five of them were selected from a collection of fourteen features. An excluded volume shell based on van der Waals radii was also taken into consideration to mimic the receptor binding site while generating the pharmacophore model. Eventually, from the 227,228 compounds arising from various libraries (Table S8), 2734 molecules exhibit simultaneously the five features defined by the pharmacophore model. These were subjected to molecular docking.

3.2. Molecular Docking-Based Virtual Screening

The crystal structure of the complex between the PLCγ1-SH3 (from *Rattus norvegicus*) and SLP76 (from *Homo sapiens*) (PDB ID: 1YWO) [18] was considered for molecular docking. High sequence identity (95%) and sequence similarity (98%) between the PLCγ1-SH3 from *Rattus norvegicus* (UniProt ID: P10686) and from *Homo sapiens* (UniProt ID: P19174) (Figure S22A) permits the use of such structure for the molecular modeling studies. Indeed, the three non-identical residues of PLCγ1 (Ile813, Glu825 and Ile846 in *Homo sapiens* and Thr813, Asp825 and Val846 in *R. norvegicus*) are not involved in the interaction with SLP76 (Figure S22B,C). Using the Protein Preparation Wizard module [41] of Schrödinger software package, version 2018-2 [42], pre-processing of the macromolecular structure was performed, i.e., addition of missing hydrogens, removal of water molecules beyond 5 Å and assignment of the right bond order. The *protassign* utility of the Protein Preparation Wizard module was employed for optimization of ionization state using PROPKA, for predicting p*Ka* values in proteins (pH 7.0 ± 2.0) and orientations of side chain functional groups (e.g., hydroxy group in Ser, Thr and Tyr; amino group in Asn and Gln). A restrained minimization of the complex was then performed (cutoff root mean square deviation (RMSD) 0.3 Å) with the help of *impref* utility, so as to avoid any steric clashes.

The 2734 molecules obtained after the pharmacophore filtering were prepared using the LigPrep module of Schrödinger software package, version 2018-2 [41,43]. For the high-throughput virtual screening (HTVS) step, the ionization states of these molecules were not considered, whereas for subsequent steps, these molecules were subjected to preparation in LigPrep, generating their ionization states (using Epik ionizer [44–46], pH 7.0 ± 2.0). For a comparative analysis, reported PLCγ1 inhibitors [16], i.e., anethole, daunorubicin, diflunisal, ritonavir and rosiglitazone (Figure 1) were also considered and submitted to molecular docking at the SLP76 binding site in PLCγ1-SH3.

The interaction grid for molecular docking was generated with the Receptor Grid Generation module of Schrödinger software package at the centroid of bound ligand in the prepared PLCγ1-SLP76 complex (grid center: 19.29, 2.63, 25.99). The size of the interaction grid was extended up to 10 Å as inner box and additional 20 Å as outer box. Molecular docking was performed using the Glide module of Schrödinger software package [47–50] in four steps (Figure S7), i.e., (i) high-throughput virtual screening (HTVS), (ii) Standard Precision (SP) mode docking, (iii) eXtra Precision (XP) mode docking and (iv) Induced Fit Docking (IFD) [51–53]. For HTVS, only one pose was considered, whereas for subsequent steps, 20 poses were generated for each molecule (with all parameters at their default values and by employing the OPLS_2005 force field) [54]. After each step, results were subjected to a pose filtering for the presence of crucial hydrogen bond (H-bond) interactions with PLCγ1 (via Asp808/Glu809 and Trp828/Asn844), evaluation for structural overlap with cocrystallized ligand, reproducibility of the docked conformation and glide docking score. Molecular Mechanics-Generalized Born Surface Area (MM/GBSA) based binding free energy (ΔG_{bind}) were computed for the complexation of selected molecules with PLCγ1, using Prime module [55]. Molecules with a ΔG_{bind} value lower than −40 kcal/mol were taken for next steps. For the hit selection after IFD, the ΔG_{bind} cutoff was kept to −55 kcal/mol. The sixteen molecules, named hereafter as INX (where X = 1 to 16) were further considered for the MD simulations. Previously reported PLCγ1 inhibitors [16] (Figure 1) were also submitted to IFD and, subsequent MM/GBSA ΔG_{bind} calculations.

3.3. Molecular Dynamics Simulations

In order to evaluate the stability of sixteen complexes generated from molecular docking, PLCγ1-ritonavir complex and PLCγ1-SLP76 complex, MD simulations were carried out using the AMBER18 package [56]. The General Amber Force Field (GAFF) [57] and Amber ff99SB force field [58] were employed for ligands and protein preparation, respectively. The AM1-bcc method (semi-empirical with bond charge correction) [59] of the antechamber module from Amber tools 18 [56] was utilized for deriving charges on the ligands. TIP3P water model [60] was used for solvation (cubic box; 15 Å × 15 Å × 15 Å). Each system was neutralized by adding counter ions and an ionic concentration of 0.15 M was maintained by adding additional Na^+ and Cl^- ions. All systems were subjected to minimization and gradual heating (from 0 to 300 K, under NVT ensemble). Thereafter, density equilibration (under NPT ensemble) and equilibration (1 ns under NPT ensemble) were performed sequentially at 310 K and 1 atm pressure (pressure relaxation time of 2.0 ps and time step of 2 fs). Finally, three replica of the production run for 50 ns were performed under NPT ensemble for each system using a cutoff distance of 12 Å for non-bonded interactions. Long-range electrostatic interactions were treated with the Particle-Mesh Ewald (PME) method [61]. Bulk effect was simulated by enabling periodic boundary conditions. All covalent bonds containing hydrogen atoms were constrained using the SHAKE algorithm [62]. Ptraj module [63] of Amber tools [56] and Visual Molecular Dynamics software (VMD) [64] were used for trajectory analysis. Combined clustering analysis was performed, for the three replicate MD simulations, Ptraj module [63] to evaluate the reproducibility of the results and ligand binding during the simulation. A hierarchical agglomerative (bottom-up) approach was employed as clustering algorithm (number of clusters: 5) and the best-fit coordinate RMSD between all the heavy atoms was considered as the parameter for clustering. ΔG_{bind} values were also calculated using MM/GBSA method [65] over the last 10 ns of MD simulations trajectory.

4. Conclusions

Involvement of PLCγ1 in a number of cellular processes makes it an important drug target for a number of pathological and disease conditions, including immunological disorders and cancers. The PLCγ1-SH3 is known to be involved in interaction with several proteins, regulating a number of cellular processes. It has been proposed as an important target domain for the design of anti-PLCγ1 agents. The occupied binding site of PLCγ1-SH3 prevents the interaction of PLCγ1 with the target adaptor proteins, thus leading to the modification of cellular responses including cell proliferation, differentiation of cell death. Therefore, identification of compounds which can efficiently and stably bind to PLCγ1-SH3 was undertaken through computer aided drug design (CADD) study.

A systematic virtual screening was performed by employing a pharmacophore mapping based on the SLP76 peptide, molecular docking and molecular dynamics (MD) simulations. In this process, a large collection of 227,228 compounds was evaluated against the pharmacophore filtering which helped to identify 2734 compounds with potential features to bind at the PLCγ1-SH3. These molecules were then submitted to molecular docking in an increasing degree of precision, shortlisting sixteen compounds. Under static conditions, they exhibited a significant degree of binding affinity and important molecular recognitions with the PLCγ1. To evaluate the binding of the identified hits to PLCγ1 under dynamical conditions, MD simulations in triplicate were undertaken for each of the 16 complexes. System stability and binding energy analyses helped to identify compounds IN1, IN2, IN3, IN5, IN6, IN8, and IN10 (Figure 6) as promising candidates for inhibiting the interaction of PLCγ1 with its target proteins as they exhibit a stable binding at PLCγ1-SH3. Additionally, identified important molecular recognitions can help to streamline drug discovery against PLCγ1. Residues which participated in the stable binding of inhibitors to the protein are Gln805, Arg806, Asp808, Glu809, Asp825, Gly827, and Trp828. These results are in agreement with the reported experimental data [16]. To the best of our knowledge, this work is the first report of a systematic application of CADD for identification of inhibitors against PLCγ1. These molecules can be taken up further for in vitro evaluation of their PLCγ1 inhibiting effect.

Supplementary Materials: Supplementary materials can be found at http://www.mdpi.com/1422-0067/20/19/4721/s1.

Author Contributions: A.D.L. and P.L. conceived the project. N.T., I.V., and A.D.L. performed the formal analysis and wrote the first draft. S.T., M.J., P.L., and A.D.L. review and edit the manuscript. All authors discussed the results and progress in all stages and given approval to the final version of the manuscript.

Funding: A.D.L., I.V. and N.T. thank the Institut National Contre le Cancer (INCA) and the Région Pays de la Loire, for the financial support during their post-doctoral research. I.V. and N.T thank Centre national de la recherche scientifique (CNRS) for their post-doctoral research fellowship. This study has also been supported by ANR-17-CE15-0027, INCa PLBIO-2018, Ligue contre le cancer équipe labellisée 2016-2019, Fondation ARC 2017-2019 (PL).

Acknowledgments: We do warmly thank D. Vercauteren for the careful reading and fruitful comments. Authors thank CCIPL (Centre de Calcul Intensif des Pays de Loire) and Aymeric Blondel for the technical support.

Conflicts of Interest: The authors declare no conflict of interest.

Abbreviations

3D	Three Dimensional
ΔG_{bind}	Binding energy
AMBER	Assisted Model Building with Energy Refinement
CADD	Computer Aided Drug Design
CD95	Cluster of differentiation 95
CID	Calcium inducing domain
COM_dist	Center of Mass of the Bound Ligand and AsnA844 of PLCγ1
GAFF	General Amber Force Field
H-bond	Hydrogen-bond
HTVS:	High-throughput virtual screening
IFD	Induced fit docking
IP3	Inositol 1,4,5-triphosphate
MD	Molecular dynamics
MM/GBSA	Molecular Mechanics-Generalized Born Surface Area
PDB	Protein Data bank
PIP2	Phosphatidylinositol 4,5-bisphosphate
PLCγ1	Phospholipase C gamma 1
PLCγ1-SH3	SH3 domain of PLCγ1
PME	Particle-Mesh Ewald
RMSD	Root mean square deviation
SP	Standard Precision
TIP3P	Three-site Transferrable Intermolecular Potential
VMD	Visual Molecular Dynamics software
XP	eXtra Precision

References

1. Perl, A.; Gergely, P.J.; Puskas, F.; Banki, K. Metabolic switches of T-cell activation and apoptosis. *Antioxidants Redox Signal.* **2002**, *4*, 427–443. [CrossRef] [PubMed]
2. Solary, E.; Dubrez, L.; Eymin, B. The role of apoptosis in the pathogenesis and treatment of diseases. *Eur. Respir. J.* **1996**, *9*, 1293–1305. [CrossRef] [PubMed]
3. Thompson, C.B. Apoptosis in the pathogenesis and treatment of disease. *Science* **1995**, *267*, 1456–1462. [CrossRef] [PubMed]

4. Sun, Y. E3 ubiquitin ligases as cancer targets and biomarkers. *Neoplasia* **2006**, *8*, 645–654. [CrossRef] [PubMed]

5. Hunter, A.M.; LaCasse, E.C.; Korneluk, R.G. The inhibitors of apoptosis (IAPs) as cancer targets. *Apoptosis* **2007**, *12*, 1543–1568. [CrossRef] [PubMed]

6. Faustman, D.; Davis, M. TNF receptor 2 pathway: Drug target for autoimmune diseases. *Nat. Rev. Drug Discov.* **2010**, *9*, 482–493. [CrossRef] [PubMed]

7. Paplomata, E.; O'regan, R. The PI3K/AKT/mTOR pathway in breast cancer: Targets, trials and biomarkers. *Ther. Adv. Med. Oncol.* **2014**, *6*, 154–166. [CrossRef]

8. Isono, F.; Fujita-Sato, S.; Ito, S. Inhibiting RORγt/Th17 axis for autoimmune disorders. *Drug Discov. Today* **2014**, *19*, 1205–1211. [CrossRef]

9. Nayak, A.; Satapathy, S.R.; Das, D.; Siddharth, S.; Tripathi, N.; Bharatam, P.V.; Kundu, C.N. Nanoquinacrine induced apoptosis in cervical cancer stem cells through the inhibition of hedgehog-GLI1 cascade: Role of GLI-1. *Sci. Rep.* **2016**, *6*, 20600. [CrossRef]

10. Das, S.; Tripathi, N.; Preet, R.; Siddharth, S.; Nayak, A.; Bharatam, P.V.; Kundu, C.N. Quinacrine induces apoptosis in cancer cells by forming a functional bridge between TRAIL-DR5 complex and modulating the mitochondrial intrinsic cascade. *Oncotarget* **2017**, *8*, 248–267. [CrossRef]

11. Das, S.; Tripathi, N.; Siddharth, S.; Nayak, A.; Nayak, D.; Sethy, C.; Bharatam, P.V.; Kundu, C.N. Etoposide and doxorubicin enhance the sensitivity of triple negative breast cancers through modulation of TRAIL-DR5 axis. *Apoptosis* **2017**, *22*, 1205–1224. [CrossRef]

12. Fasching, P.; Stradner, M.; Graninger, W.; Dejaco, C.; Fessler, J. Therapeutic potential of targeting the Th17/Treg axis in autoimmune disorders. *Molecules* **2017**, *22*, E134. [CrossRef]

13. Sala, G.; Dituri, F.; Raimondi, C.; Previdi, S.; Maffucci, T.; Mazzoletti, M.; Rossi, C.; Iezzi, M.; Lattanzio, R.; Piantelli, M.; et al. Phospholipase Cγ1 is required for metastasis development and progression. *Cancer Res.* **2008**, *68*, 10187–10196. [CrossRef]

14. Yang, Y.R.; Choi, J.H.; Chang, J.-S.; Kwon, H.M.; Jang, H.-J.; Ryu, S.H.; Suh, P.-G. Diverse cellular and physiological roles of phospholipase C-γ1. *Adv. Enzyme Regul.* **2012**, *52*, 138–151. [CrossRef] [PubMed]

15. Lattanzio, R.; Piantelli, M.; Falasca, M. Role of phospholipase C in cell invasion and metastasis. *Adv. Biol. Regul.* **2013**, *53*, 309–318. [CrossRef]

16. Poissonnier, A.; Guégan, J.P.; Nguyen, H.T.; Best, D.; Levoin, N.; Kozlov, G.; Gehring, K.; Pineau, R.; Jouan, F.; Morere, L.; et al. Disrupting the CD95–PLCγ1 interaction prevents Th17-driven inflammation. *Nat. Chem. Biol.* **2018**, *14*, 1079–1089. [CrossRef]

17. Kadamur, G.; Ross, E.M. Mammalian Phospholipase, C. *Annu. Rev. Physiol.* **2012**, *75*, 127–154. [CrossRef]

18. Deng, L.; Velikovsky, C.A.; Swaminathan, C.P.; Cho, S.; Mariuzza, R.A.; Huber, R. Structural basis for recognition of the T Cell adaptor protein SLP-76 by the SH3 domain of phospholipase Cγ1. *J. Mol. Biol.* **2005**, *352*, 1–10. [CrossRef]

19. Gierschik, P.; Buehler, A.; Walliser, C. Activated PLCγ breaking loose. *Structure* **2012**, *20*, 1989–1990. [CrossRef]

20. Bunney, T.D.; Esposito, D.; Mas-Droux, C.; Lamber, E.; Baxendale, R.W.; Martins, M.; Cole, A.; Svergun, D.; Driscoll, P.C.; Katan, M. Structural and functional integration of the PLCγ interaction domains critical for regulatory mechanisms and signaling deregulation. *Structure* **2012**, *20*, 2062–2075. [CrossRef]

21. Koss, H.; Bunney, T.D.; Behjati, S.; Katan, M. Dysfunction of phospholipase Cγ in immune disorders and cancer. *Trends Biochem. Sci.* **2014**, *39*, 603–611. [CrossRef] [PubMed]

22. Tang, W.; Zhou, Y.; Sun, D.; Dong, L.; Xia, J.; Yang, B. Oncogenic role of PLCG1 in progression of hepatocellular carcinoma. *Hepatol. Res.* **2019**, *49*, 559–569. [CrossRef] [PubMed]

23. Wu, C.; Ma, M.H.; Brown, K.R.; Geisler, M.; Li, L.; Tzeng, E.; Jia, C.Y.H.; Jurisica, I.; Li, S.S.C. Systematic identification of SH3 domain-mediated human protein-protein interactions by peptide array target screening. *Proteomics* **2007**, *7*, 1775–1785. [CrossRef] [PubMed]

24. Huang, P.S.; Davis, L.; Huber, H.; Goodhart, P.J.; Wegrzyn, R.E.; Oliff, A.; Heimbrook, D.C. An SH3 domain is required for the mitogenic activity of microinjected phospholipase C-γ1. *FEBS Lett.* **1995**, *358*, 287–292. [CrossRef]

25. Smith, M.R.; Liu, Y.L.; Kim, S.R.; Bae, Y.S.; Kim, C.G.; Kwon, K.S.; Rhee, S.G.; Kung, H.F. PLCγ1 Src homology domain induces mitogenesis in quiescent NIH 3T3 fibroblasts. *Biochem. Biophys. Res. Commun.* **1996**, *222*, 186–193. [CrossRef]

26. Seedorf, K.; Kostka, G.; Lammers, R.; Bashkin, P.; Daly, R.; Burgess, W.H.; Van der Bliek, A.M.; Schlessinger, J.; Ullrich, A. Dynamin binds to SH3 domains of phospholipase Cγ and GRB-2. *J. Biol. Chem.* **1994**, *269*, 16009–16014. [PubMed]

27. Kitchen, D.B.; Decornez, H.; Furr, J.R.; Bajorath, J. Docking and scoring in virtual screening for drug discovery: Methods and applications. *Nat. Rev. Drug Discov.* **2004**, *3*, 935–949. [CrossRef]

28. Kwofie, S.K.; Dankwa, B.; Enninful, K.S.; Adobor, C.; Broni, E.; Ntiamoah, A.; Wilson, M.D. Molecular docking and dynamics simulation studies predict munc18b as a target of mycolactone: A plausible mechanism for granule exocytosis impairment in Buruli Ulcer Pathogenesis. *Toxins (Basel)* **2019**, *11*, 181. [CrossRef]

29. Wade, R.C.; Salo-Ahen, O.M.H. Molecular Modeling in Drug Design. *Molecules* **2019**, *24*, 321. [CrossRef]

30. Wang, S.Q.; Du, Q.S.; Huang, R.B.; Zhang, D.W.; Chou, K.C. Insights from investigating the interaction of oseltamivir (Tamiflu) with neuraminidase of the 2009 H1N1 swine flu virus. *Biochem. Biophys. Res. Commun.* **2009**, *386*, 432–436. [CrossRef]

31. Li, X.B.; Wang, S.Q.; Xu, W.R.; Wang, R.L.; Chou, K.C. Novel inhibitor design for hemagglutinin against H1N1 influenza virus by core hopping method. *PLoS ONE* **2011**, *6*, e28111. [CrossRef] [PubMed]

32. Ma, Y.; Wang, S.Q.; Xu, W.R.; Wang, R.L.; Chou, K.C. Design novel dual agonists for treating type-2 diabetes by targeting peroxisome proliferator-activated receptors with core hopping approach. *PLoS ONE* **2012**, *7*, e38546. [CrossRef] [PubMed]

33. Franchini, S.; Battisti, U.M.; Prandi, A.; Tait, A.; Borsari, C.; Cichero, E.; Fossa, P.; Cilia, A.; Prezzavento, O.; Ronsisvalle, S.; et al. Scouting new sigma receptor ligands: Synthesis, pharmacological evaluation and molecular modeling of 1,3-dioxolane-based structures and derivatives. *Eur. J. Med. Chem.* **2016**, *112*, 1–19. [CrossRef] [PubMed]

34. Franchini, S.; Manasieva, L.; Sorbi, C.; Battisti, U.; Fossa, P.; Cichero, E.; Denora, N.; Iacobazzi, R.; Cilia, A.; Pirona, L.; et al. Synthesis, biological evaluation and molecular modeling of 1-oxa-4-thiaspiro- and 1,4-dithiaspiro[4.5]decane derivatives as potent and selective 5-HT1A receptor agonists. *Eur. J. Med. Chem.* **2017**, *125*, 435–452. [CrossRef] [PubMed]

35. Ghamari, N.; Zarei, O.; Reiner, D.; Dastmalchi, S.; Stark, H.; Hamzeh-Mivehroud, M. Histamine H3 receptor ligands by hybrid virtual screening, docking, molecular dynamics simulations, and investigation of their biological effects. *Chem. Biol. Drug Des.* **2019**, *93*, 832–843. [CrossRef] [PubMed]

36. Selvakumar, J.N.; Chandrasekaran, S.D.; Doss, G.P.C.; Kumar, T.D. Inhibition of the ATPase Domain of Human Topoisomerase IIa on HepG2 Cells by 1, 2-benzenedicarboxylic Acid, Mono (2-ethylhexyl) Ester: Molecular Docking and Dynamics Simulations. *Curr. Cancer Drug Targets* **2019**, *19*, 495–503. [CrossRef] [PubMed]

37. Gaulton, A.; Bellis, L.J.; Bento, A.P.; Chambers, J.; Davies, M.; Hersey, A.; Light, Y.; McGlinchey, S.; Michalovich, D.; Al-Lazikani, B.; et al. ChEMBL: A large-scale bioactivity database for drug discovery. *Nucleic Acids Res.* **2012**, *40*, D1100–D1107. [CrossRef] [PubMed]

38. Dixon, S.L.; Smondyrev, A.M.; Knoll, E.H.; Rao, S.N.; Shaw, D.E.; Friesner, R.A. PHASE: A new engine for pharmacophore perception, 3D QSAR model development, and 3D database screening: 1. Methodology and preliminary results. *J. Comput. Aided. Mol. Des.* **2006**, *20*, 647–671. [CrossRef] [PubMed]

39. Dixon, S.L.; Smondyrev, A.M.; Rao, S.N. PHASE: A novel approach to pharmacophore modeling and 3D database searching. *Chem. Biol. Drug Des.* **2006**, *67*, 370–372. [CrossRef]

40. *Schrödinger Release 2018-2: Phase*; Schrödinger, LLC: New York, NY, USA, 2018.

41. Madhavi Sastry, G.; Adzhigirey, M.; Day, T.; Annabhimoju, R.; Sherman, W. Protein and ligand preparation: Parameters, protocols, and influence on virtual screening enrichments. *J. Comput. Aided. Mol. Des.* **2013**, *27*, 221–234. [CrossRef]

42. *Schrödinger Release 2018-2: Protein Preparation Wizard*; Schrödinger, LLC: New York, NY, USA, 2016.

43. *Schrödinger Release 2018-2: LigPrep*; Schrödinger, LLC: New York, NY, USA, 2018.

44. Shelley, J.C.; Cholleti, A.; Frye, L.L.; Greenwood, J.R.; Timlin, M.R.; Uchimaya, M. Epik: A software program for pKa prediction and protonation state generation for drug-like molecules. *J. Comput. Aided. Mol. Des.* **2007**, *21*, 681–691. [CrossRef] [PubMed]

45. Greenwood, J.R.; Calkins, D.; Sullivan, A.P.; Shelley, J.C. Towards the comprehensive, rapid, and accurate prediction of the favorable tautomeric states of drug-like molecules in aqueous solution. *J. Comput. Aided. Mol. Des.* **2010**, *24*, 591–604. [CrossRef] [PubMed]

46. *Schrödinger Release 2018-2: Epik*; Schrödinger, LLC: New York, NY, USA, 2018.

47. Friesner, R.A.; Banks, J.L.; Murphy, R.B.; Halgren, T.A.; Klicic, J.J.; Mainz, D.T.; Repasky, M.P.; Knoll, E.H.; Shelley, M.; Perry, J.K.; et al. Glide: A new approach for rapid, accurate docking and scoring. 1. Method and assessment of docking accuracy. *J. Med. Chem.* **2004**, *47*, 1739–1749. [CrossRef] [PubMed]

48. Halgren, T.A.; Murphy, R.B.; Friesner, R.A.; Beard, H.S.; Frye, L.L.; Pollard, W.T.; Banks, J.L. Glide: A new approach for rapid, accurate docking and scoring. 2. Enrichment factors in database screening. *J. Med. Chem.* **2004**, *47*, 1750–1759. [CrossRef] [PubMed]

49. Friesner, R.A.; Murphy, R.B.; Repasky, M.P.; Frye, L.L.; Greenwood, J.R.; Halgren, T.A.; Sanschagrin, P.C.; Mainz, D.T. Extra precision glide: Docking and scoring incorporating a model of hydrophobic enclosure for protein-ligand complexes. *J. Med. Chem.* **2006**, *49*, 6177–6196. [CrossRef] [PubMed]

50. *Schrödinger Release 2018-2: Glide*; Schrödinger, LLC: New York, NY, USA, 2018.

51. Sherman, W.; Day, T.; Jacobson, M.P.; Friesner, R.A.; Farid, R. Novel procedure for modeling ligand/receptor induced fit effects. *J. Med. Chem.* **2006**, *49*, 534–553. [CrossRef]

52. Sherman, W.; Beard, H.S.; Farid, R. Use of an induced fit receptor structure in virtual screening. *Chem. Biol. Drug Des.* **2006**, *67*, 83–84. [CrossRef]

53. *Schrödinger Release 2018-2: Induced Fit Docking protocol; Glide*; Schrödinger, LLC: New York, NY, USA, 2016.

54. Jorgensen, W.L.; Maxwell, D.S.; Tirado-Rives, J. Development and testing of the OPLS all-atom force field on conformational energetics and properties of organic liquids. *J. Am. Chem. Soc.* **1996**, *118*, 11225–11236. [CrossRef]

55. *Schrödinger Release 2018-2: Prime*; Schrödinger, LLC: New York, NY, USA, 2018.

56. Case, D.A.; Ben-Shalom, I.Y.; Brozell, S.R.; Cerutti, D.S.; Cheatham, T.E.I.; Cruzeiro, V.W.D.; Darden, T.A.; Duke, R.E.; Ghoreishi, D.; Gilson, M.K.; et al. *Amber 2018*; University of California: San Francisco, CA, USA, 2018.

57. Wang, J.; Wolf, R.M.; Caldwell, J.W.; Kollman, P.A.; Case, D.A. Development and testing of a general Amber force field. *J. Comput. Chem.* **2004**, *25*, 1157–1174. [CrossRef]

58. Hornak, V.; Abel, R.; Okur, A.; Strockbine, B.; Roitberg, A.; Simmerling, C. Comparison of multiple amber force fields and development of improved protein backbone parameters. *Proteins Struct. Funct. Genet.* **2006**, *65*, 712–725. [CrossRef]

59. Jakalian, A.; Jack, D.B.; Bayly, C.I. Fast, efficient generation of high-quality atomic charges. AM1-BCC model: II. Parameterization and validation. *J. Comput. Chem.* **2002**, *21*, 132–146. [CrossRef]

60. Mark, P.; Nilsson, L. Structure and dynamics of the TIP3P, SPC, and SPC/E water models at 298 K. *J. Phys. Chem. A* **2001**, *105*, 9954–9960. [CrossRef]

61. Darden, T.; York, D.; Pedersen, L. Particle mesh Ewald: An N·log(N) method for Ewald sums in large systems. *J. Chem. Phys.* **1993**, *98*, 10089–10092. [CrossRef]

62. Forester, T.R.; Smith, W. SHAKE, rattle, and roll: Efficient constraint algorithms for linked rigid bodies. *J. Comput. Chem.* **1998**, *19*, 102–111. [CrossRef]

63. Roe, D.R.; Cheatham, T.E. PTRAJ and CPPTRAJ: Software for processing and analysis of molecular dynamics trajectory data. *J. Chem. Theory Comput.* **2013**, *9*, 3084–3095. [CrossRef] [PubMed]

64. Humphrey, W.; Dalke, A.; Schulten, K. VMD: Visual molecular dynamics. *J. Mol. Graph.* **1996**, *14*, 33–38. [CrossRef]

65. Genheden, S.; Ryde, U. The MM/PBSA and MM/GBSA methods to estimate ligand-binding affinities. *Expert Opin. Drug Discov.* **2015**, *10*, 449–461. [CrossRef]

 International Journal of
Molecular Sciences

Article

Application of the Movable Type Free Energy Method to the Caspase-Inhibitor Binding Affinity Study

Song Xue [1], Hao Liu [2] and Zheng Zheng [3],* 

[1] School of Statistics and Mathematics, Zhongnan University of Law and Economics, Wuhan 430073, China; songxue@outlook.com
[2] School of Mechanical and Electronic Engineering, Wuhan University of Technology, Wuhan 430070, China; haoliulh@gmail.com
[3] School of Chemistry, Chemical Engineering and Life Science, Wuhan University of Technology, Wuhan 430070, China
* Correspondence: laozhengzz@gmail.com; Tel.: +86-135-5285-6384

Received: 30 August 2019; Accepted: 25 September 2019; Published: 29 September 2019

Abstract: Many studies have provided evidence suggesting that caspases not only contribute to the neurodegeneration associated with Alzheimer's disease (AD) but also play essential roles in promoting the underlying pathology of this disease. Studies regarding the caspase inhibition draw researchers' attention through time due to its therapeutic value in the treatment of AD. In this work, we apply the "Movable Type" (MT) free energy method, a Monte Carlo sampling method extrapolating the binding free energy by simulating the partition functions for both free-state and bound-state protein and ligand configurations, to the caspase-inhibitor binding affinity study. Two test benchmarks are introduced to examine the robustness and sensitivity of the MT method concerning the caspase inhibition complexing. The first benchmark employs a large-scale test set including more than a hundred active inhibitors binding to caspase-3. The second benchmark includes several smaller test sets studying the relative binding free energy differences for minor structural changes at the caspase-inhibitor interaction interfaces. Calculation results show that the RMS errors for all test sets are below 1.5 kcal/mol compared to the experimental binding affinity values, demonstrating good performance in simulating the caspase-inhibitor complexing. For better understanding the protein-ligand interaction mechanism, we then take a closer look at the global minimum binding modes and free-state ligand conformations to study two pairs of caspase-inhibitor complexes with (1) different caspase targets binding to the same inhibitor, and (2) different polypeptide inhibitors targeting the same caspase target. By comparing the contact maps at the binding site of different complexes, we revealed how small structural changes affect the caspase-inhibitor interaction energies. Overall, this work provides a new free energy approach for studying the caspase inhibition, with structural insight revealed for both free-state and bound-state molecular configurations.

Keywords: caspase inhibition; protein-ligand binding free energy; Monte Carlo sampling; docking and scoring; molecular conformational sampling

---

1. Introduction

Alzheimer disease (AD) is a neurodegenerative disorder characterized by the neuronal and synaptic loss as well as the accumulation of β-amyloid plaques and neurofibrillary tangles (NFTs) within selective brain regions. Yet its cause, time course or mechanisms are still not well understood [1–4]. Scientists have proven that the programmed cell death pathway, also known as apoptosis, plays a significant role in the pathogenesis of age-related neurodegenerative diseases, particularly in AD [2,5–7]. Caspases, a family of serine-aspartyl proteases, are involved in the initiation and execution of apoptosis.

They are known to exist in our cells as inactive precursors which kill the cell once activated and lead to the proteolytic cleavage of several neuronal proteins including tau, APP, presenilin (PS1, PS2), actin, fodrin, etc. Therefore, caspases are believed to be critically related to the pathogenesis of AD [4,8–11]. Many research results have been published to elucidate the correlation between AD pathogenesis and caspases family members, mostly caspases-2, 3, 6, 7, 8 and 9 [2,12–19]. These studies suggest that preventing caspase activation may be a promising therapeutic for the treatment of AD. The activation or the activity of the caspases can be regulated in two ways: (1) specific molecules such as Bcl-2, FLIP or IAPs can be used to control the processing and activation of a caspase; (2) a number of molecules that directly interact with a caspase can be used to inhibit the proteases that have already been activated. These molecules are called caspase inhibitors [20–24]. Various caspase inhibitors, including small molecules, peptidomimetic and peptide compounds, have been designed to study the relationship between caspases and other factors involved in apoptosis.

Structure-based drug design using high-performance computers have long played important roles in the de novo drug/biomolecule discovery studies. The long-pursued essential of structure-based drug design is the estimation of the free energy change associated with the binding process of a ligand to a biochemical system, for which the calculation speed and accuracy are both crucial [25–27]. A number of free energy estimation methods have been developed, including end-point methods, pathway-based free energy calculations, and pathway-independent free energy methods. The end-point methods for free energy estimation (e.g., docking, molecular mechanics combined with the Poisson–Boltzmann or generalized Born and surface area continuum solvation (MMPBSA or MMGBSA)) are relatively fast, but the single static structure which they usually rely on often leads to the neglect of the receptor flexibility and thus compromise the calculation accuracy [28–31]. The pathway-based free energy methods, which can be broadly categorized into alchemical and potential of mean force approaches, are usually computationally expensive due to the extensive sampling required to estimate the binding free energies [32–34]. The alchemical approaches use the thermodynamic cycle built with nonphysical intermediate states to compute the free energy differences between the end states. The two most commonly used alchemical free energy methods are Free energy perturbation (FEP) [35] and thermodynamic integration (TI) [36]. The potential of mean force (PMF) approach [37–39], with umbrella sampling coupled with the WHAM (weighted histogram analysis method) analysis, is one of the most widely adopted PMF approaches [40]. Other than the high computational cost caused by the intensive sampling, the pathway free energy methods are also limited by simulation time scales. Constitutionally, the underlying force field has a powerful hold on the accuracy of all free energy estimation methods, leaving improvements in all these methods an active area of research [41–43]. On the other hand, the pathway-independent free energy methods, e.g., Monte Carlo free energy sampling methods, use Markov model for the molecular configurational-state sampling. Such methods could potentially gain significant speed benefits from parallel computing according to their stochastic sampling protocols, which also avoids the difficulty of crossing the energy barriers during simulations in the pathway-based methods. However, to generate a converged energy ensemble takes no less computational effort compared to the pathway-based methods. Capturing the significant configurational states are crucial for the pathway-independent free energy methods, which require thorough and careful sampling against the energy landscape.

In this research, we used the Movable Type (MT) free energy method, a novel Monte Carlo free energy algorithm developed by our group to evaluate the binding affinity between a variety of caspase inhibitors and their caspase targets [44]. By comparing the binding free energies and the predicted significant binding modes calculated by our simulation model to those obtained from experiments, we could validate the accuracy of our model against this particular protein target family, and provide potential theoretical support for the future development of the therapeutic intervention for AD.

2. Results and Discussion

The goals of this research are (1) to examine the accuracy of Movable Type free energy method in calculating the binding free energy between different caspase targets and various inhibitors, and (2) to apply the MT method to the structural analysis of the caspase-inhibitor binding mechanism. The results could provide theoretical support to proceed further study the feasibility of applying the Movable Type Free Energy Method to design caspase ligand inhibitors, which are closely related to Alzheimer's disease.

Two different test benchmarks were introduced in this work. First, a relatively large test set was studied to obtain a general picture of the MT method's performance to differentiate the binding affinities of a large variety of ligand structures binding to the caspase-3 protein target (the caspase target having the most significant number of ligands with known binding affinities). Then we performed a series of relative binding free energy reproduction studies to carefully examine the MT binding affinity prediction regarding (1) ligands of different structural categories bound to specific targets, and (2) ligands from the same structural category bound to different caspase targets, for more detailed computational study of the caspase-ligand bindings.

2.1. Large-Scale Validation Benchmark

The first benchmark includes structures and IC50 data (which can be converted to binding free energy via approximation) of 113 small molecular ligands bound to the caspase-3 target that are proven to have binding affinities, published on DUD-E data website (http://dude.docking.org/). The DUD-E website provides several hundred structures of small molecules that actively bind to caspase-3. After screening, redundancy structures, as well as structures with high molecular weights (MW) (>1000 Da) or high degrees of freedom (>1000 rotatable bonds) were abandoned, with the rest 113 active ligands forming the test set used in our validation. Ligand structures were prepared by adding the missing hydrogen atoms, Missing residues at the caspase-3 target protein were added and locally optimized before the calculation. The active compounds' IC50 data collected from the DUD-E website were transferred to pIC50 values and further approximated to the binding free energies by assigning the unit of energy:

$$\Delta G_{binding} = RT \ln K_d \approx RT \ln IC50 = -RT \times \frac{e}{10} \times pIC50 \tag{1}$$

where R is the gas constant and T is temperature in Kelvin, which is set to 298.15 K in this work; e is the base of the natural logarithm.

IC50 is strongly related to the inhibitor's binding affinity, and also affected by other factors as the substrate's and receptor's concentrations. The inhibitor's binding affinities can be approximated as pIC50 values when the substrate's concentration is very small. On one hand, IC50 data are more easily accessible compared to K_i or K_d data from the public databases [45], being popular for the large-scaling binding affinity prediction evaluations provided by many widely used databases like BindingDB [45,46], DUD-E [47] and ChEMBL [48], etc. On the other hand, not all experimental IC50 values are comparable to the binding affinity data if without small enough substrate concentrations [49], plus that different experimental IC50 values have been found regarding the same protein-ligand complex system [50], indicating reliability issues for using the public databases in the calculation evaluations. Despite the aforementioned issues, IC50 data are still broadly used in the virtual screening and binding affinity simulation studies [51–53], partly because of the limited accessible K_i or K_d data, and also because the substrate's or receptor's concentration-related terms can be cancelled out (Equation (2)) when comparing the relative binding affinities of those protein-ligand complex systems with the same mechanism of inhibition e.g., virtual screening study targeting the same receptor's binding site (Equation (3)).

$$\frac{K_{i,1}}{K_{i,2}} = \frac{IC_{50,1}}{IC_{50,2}} \tag{2}$$

$$\Delta\Delta G_{1,2} = \Delta G_1 - \Delta G_2 = (-RT \ln K_{i,1}) - (-RT \ln K_{i,2}) = (-RT \ln IC_{50,1}) - (-RT \ln IC_{50,2}) \qquad (3)$$

The MT protocol was utilized to perform the virtual screening. The calculation results were shown in Figure 1 together with the experimental $RT\ln IC50$ data generated using Equation (1). As active compounds, all the ligands in this test set are relatively tight binders, with the binding affinity distributed between −8 to −14 kcal/mol and mostly ranged between −8 to −12 kcal/mol. Statistics of this calculation approach showed an RMSE as 0.746 kcal/mol, the r^2 coefficient as 0.552 and Kendall's tau correlation as 0.506, revealing a good prediction accuracy and ranking capability of the MT method against the large-scaling caspase-3 target-ligand virtual screening test set (Figure 1). Introducing the first test benchmark revealed a general picture of the binding affinity prediction using the MT method against a large number of active small molecules, with diverse structural features, bound to the caspase-3 target. Further explorations including relative binding affinity difference study referencing minor structural changes and structural based protein-ligand interaction interface analysis were also carried out to examine the reliability of the MT protocol against the caspase-ligand binding prediction.

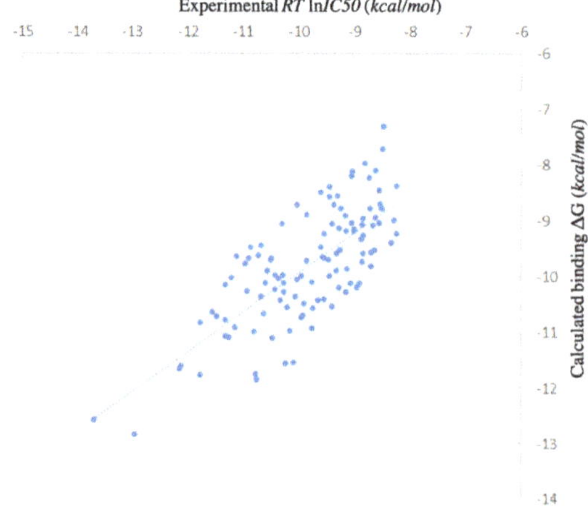

Figure 1. Scattered plot comparing binding free energy calculated by Movable Type Method to experimental data for the DUD-E CASP3 test set (Table S1).

2.2. Test Benchmark Studying the Binding ∆∆G Regarding the Structural Changes at the Binding Interface

In the second test benchmark, we employed a series of smaller test sets with high quality protein-ligand crystal structures, and carefully categorized ligands according to their structural similarities, so that we can further explore the binding affinity prediction accuracy by using the MT method, its sensitivity against local structural changes at the protein-ligand interaction interfaces, and even more, the potency of applying this method to the inverse docking study related to the caspase inhibitors.

In this test benchmark, protein-ligand complex crystal structures were selected from the Protein Data Bank (PDB) and categorized into three test sets according to the ligand structural features. The first test set aimed to study the relative binding free energy changes of different ligands bound to the same protein target. Given the same target and same binding site residue environment, it was important to explore the capability of the MT method to differentiate the binding affinity against minor to major changes concerning the ligand structures. Caspase-3, as one of the most important AD related target, was selected as the protein target in this test benchmark as well for the relative binding affinity study.

16 caspase-3 inhibitors were selected from the Protein Data Bank and categorized into two sub-groups based on their structural characteristics: inhibitors with no amino acid structures while having MWs less than 500 Da were selected to the small molecule inhibitors sub-group; inhibitors containing polypeptide backbones with natural or unnatural amino acids were classified to the

peptidomimetic inhibitors sub-group. The results of applying the MT method to calculate the binding free energy were listed in Tables 1 and 2 below:

Table 1. Comparison of the binding free energy calculated by Moveable Type to that obtained from experiment for the caspase-3—small molecule inhibitor test set.

PDB ID	Ligand	Ligand Mass (Da)	Experimental ΔG (kcal/mol)	Calculated ΔG (kcal/mol)
3h0e	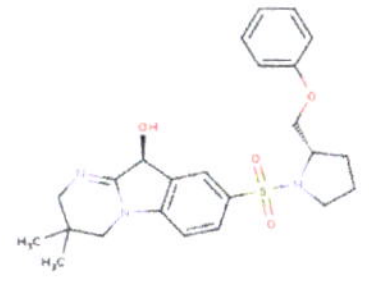	455.57	−11.11	−9.15
1gfw	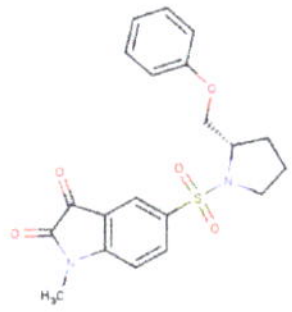	400.45	−10.66	−10.34
3dei	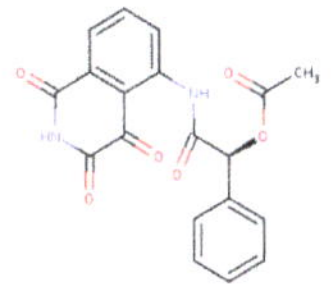	366.32	−10.08	−8.56
3dej	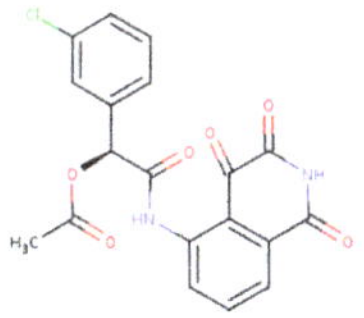	400.77	−10.90	−8.46
3dek	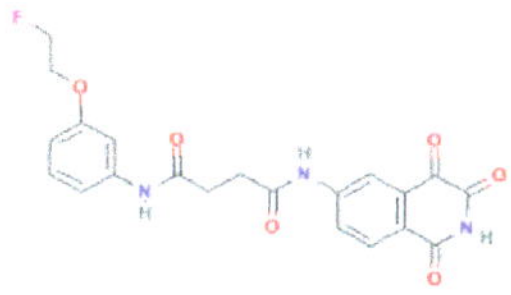	427.38	−10.32	−10.37
1nms	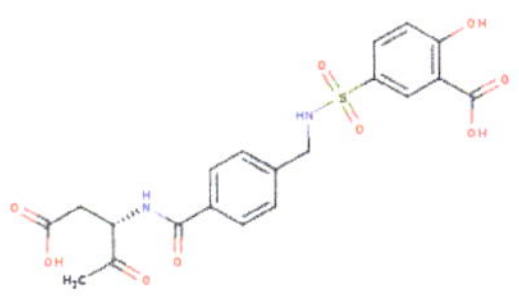	464.45	−9.13	−9.46
1re1	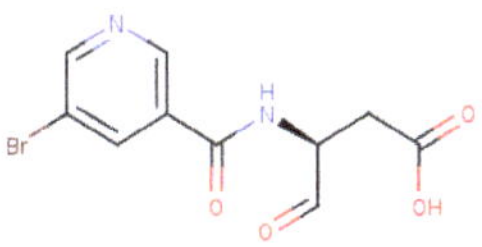	301.09	−7.099	−7.133
1rhm	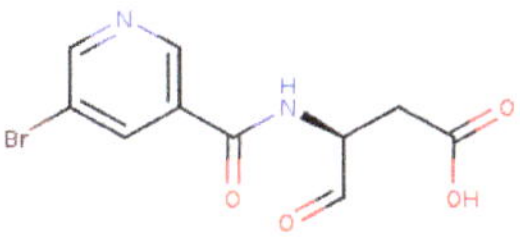	301.09	−7.849	−7.728

Table 2. Comparison of the binding free energy calculated by Moveable Type to that obtained from experiment for the caspase-3—peptidomimetic molecule inhibitor test set.

PDB ID	Ligand	Ligand Mass (Da)	Experimental ΔG (kcal/mol)	Calculated ΔG (kcal/mol)
1rhu	5,6,7 tricyclic peptidomimetic	638.69	−11.61	−10.88
1rhr	Cinnamic acid methyl ester	651.14	−11.04	−10.79
1rhj	Pryazinone	574.69	−10.96	−11.02
4jje	ACE-1MH-ASP-B3L-HLX-1U8 (Unnatural amino acid peptides)	838.94	−10.41	−10.90
2h5i	Ac-DEVD-Cho	504.49	−12.11	−11.18
2h5j	Ac-DMQD-Cho	535.57	−10.779	−11.06
4jr0	Ac-DEVD-CMK	552.96	−11.17	−11.26
3gjt	Ac-IEPD (Diverse P4 Residues in Peptides)	498.13	−9.23	−9.44

The small molecule subgroup contains ligands with more spread-out binding affinities while inhibitors in the peptidomimetic inhibitor subgroups are all tight binders. Binding affinity predictions using the MT method were illustrated in Figure 2 to compare with the experimental data. Against the small molecule subgroup, the MT method reproduced an RMSE as 1.242 kcal/mol, r^2 correlation coefficient as 0.501, and Kendall's tau as 0.357. Regarding the peptidomimetic inhibitor subgroup, the MT calculation results had an RMSE as 0.479 kcal/mol, r^2 coefficient as 0.655, and Kendall's tau as 0.444 compared to the experimental data. Calculation against the peptidomimetic inhibitor subgroup were generally better than the small molecule subgroup. By merging the two subgroups, we also looked at the MT calculation performance against the total caspase3-ligands test set. For all the 16 different ligands bound to the caspase3 target, we generated an RMSE as 0.920 kcal/mol, r^2 coefficient as 0.647, and Kendall's tau as 0.559.

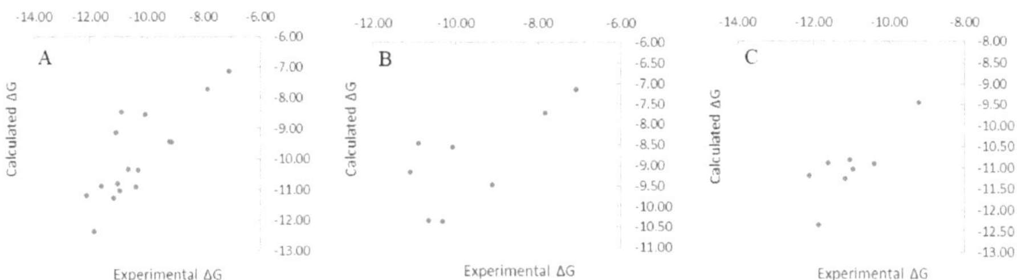

Figure 2. Scattered plots comparing the binding free energy calculated by Moveable Type to experimental data for the caspase-3-Inhibitor test set. (**A**) All the test cases. (**B**) small molecule inhibitors sub-group. (**C**) peptidomimetic inhibitor sub-group.

In the caspase-3-Inhibitor test set, the ligands' MWs varied from 301.09 to 838.94 Da, with an r^2 correlation as 0.314 with the binding affinity distribution, compared to the MT calculation results whose r^2 coefficient as 0.647 regarding the experimental data. The MT method is not ligands' MW dependent, according to this validation. Regarding this test set, the absolute errors of all the MT calculation results were lower than 2.5 kcal/mol for all the 16 complexes, 15 predictions had the absolute errors lower than 2 kcal/mol; 13 predictions had the absolute errors lower than 1 kcal/mol. A generally good binding affinity prediction against the caspase-3-Inhibitor test set were revealed by using the MT free energy protocol.

Hereby we used one example, namely 1gfw, to illustrate the sampled significant ligand's conformations in the free state and the docked poses in bound state, and the calculated ensemble energies in both free and bound state, to further demonstrate how the MT computational protocol worked.

1gfw contains a relatively small ligand with 5 heavy-heavy atom rotatable bonds. The MT-CS conformational search program generated 134 distinguished conformers and calculated their conformational energies in the solution phase by employing the KMTSIM solvation model. The top 9 ligand conformers according to their energy ranking were shown in Figure 3, with their energy distribution shown in Figure 4. The free-state ligand's partition function, Z_L was in-turn calculated using Equation (6). Z_L was a very big number as the sum of all the ligand's conformational local partition functions, which was shown as $-RT \log (Z_L)$ in this work for better revealing its physical meaning. The MT-CS calculation had $-RT \log (Z_L) = -3.99$ kcal/mol, representing the ensemble energy of the free-state ligand's conformations, an energy barrier that the binding process had to overcome.

The heatmap docking method generated 115 unique docked poses for this protein-ligand complex. The best docked ligand pose had a structural RMSD as 2.08 Å compared to the ligand's crystal structure. We showed the top 9 docked complex poses in Figure 5, and the protein-ligand binding interaction energies in Figure 6. Z_{PL} was calculated using Equation (7) summing all the complexes' configurational local partition functions. $-RT \log (Z_{PL}) = -14.33$ kcal/mol was generated as the ensemble energy of the complex considering all the 115 binding conformations in the solution phase. So that we derived the final binding free energy using Equation (8). The $\Delta G_{binding}$ was then calculated as -14.33 kcal/mol $- (-3.99$ kcal/mol$) = -10.34$ kcal/mol.

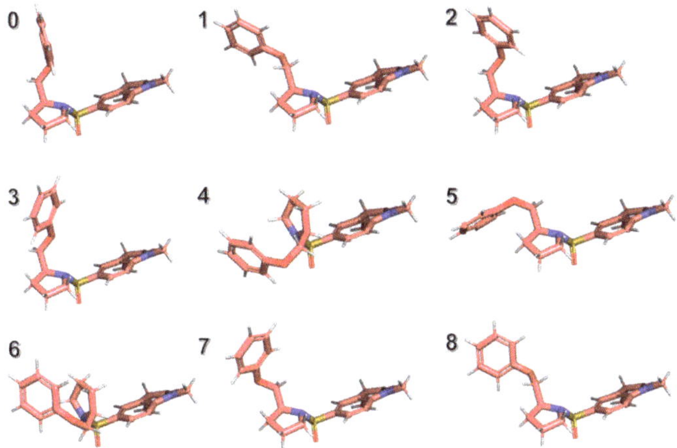

Figure 3. The top 9 significant ligand's conformations for the ligand in its free solution phase with respect to the 1gfw caspase-3-Inhibitor complex (from conformer 0 to conformer 9), generated by using the MT protocol.

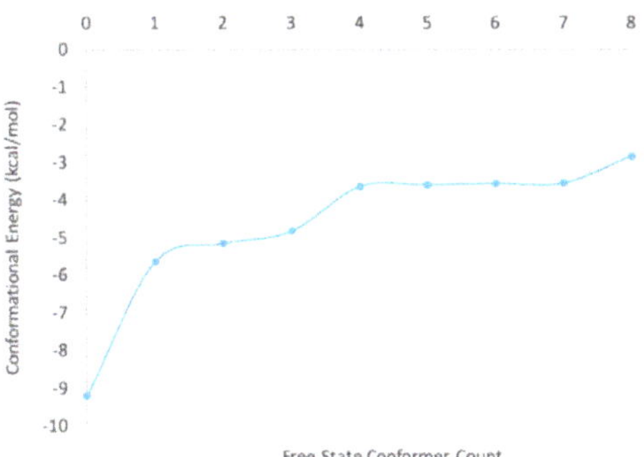

Figure 4. Energy distribution of the top 9 sampled ligand conformations ranked based on the conformational energy in the solution phase.

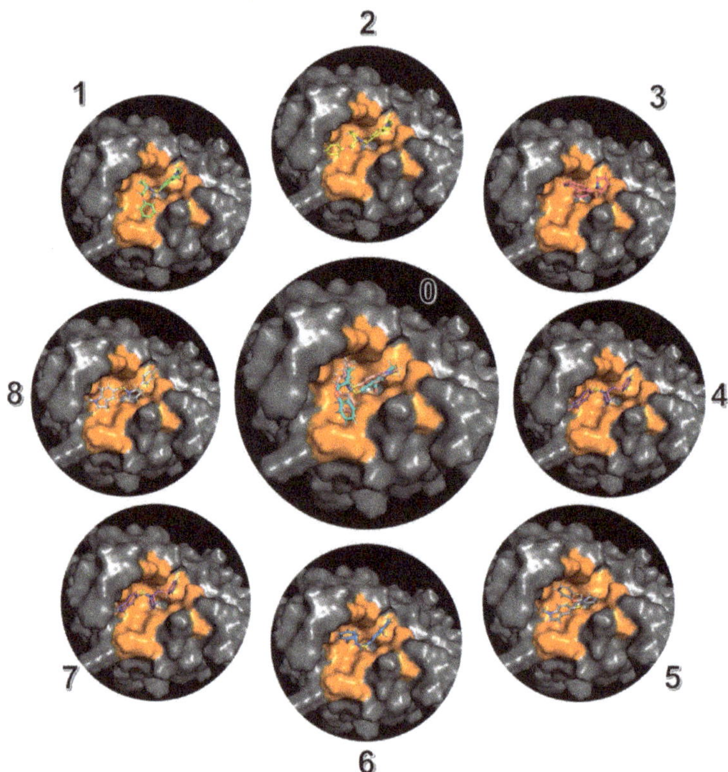

Figure 5. The top 9 significant binding modes for the 1gfw caspase-3-inhibitor complex, indexed from 0 to 9. All the binding modes are generated by using the heatmap docking method. The best docked ligand pose (in cyan) is shown together with the crystal ligand (in pink) in Figure 5-0 in the middle, with a structural RMSD as 2.08 Å.

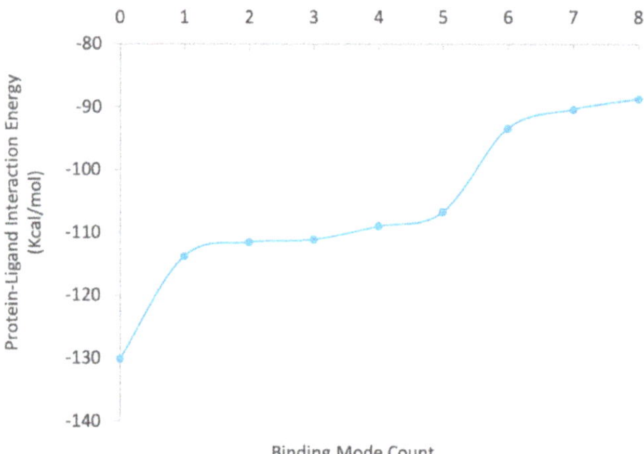

Figure 6. Protein-ligand nonbonding interaction energy distribution of the top 9 sampled binding modes ranked based on the protein-ligand nonbonding interaction energy.

Given the success of the first test set, we were encouraged to expand our study on other caspase targets. Polypeptide inhibitors were found with better selectivity and more effective compared to the small molecular inhibitors against the caspase targets, which gradually drew researchers' attention through time. In this work, we studied the polypeptide inhibitors with similar structures binding to different caspase targets, to explore the performance of the MT method reproducing the small relative binding affinity differences among the test cases.

We collected the crystal structures and binding affinity data of 15 different caspase-polypeptide inhibitor complexes from the Protein Data Bank. The MT protocol was applied to reproduce the binding affinities and significant binding modes reproductions. The calculation results agreed quite well with the experimental data and generated a RMSE as 0.733 kcal/mol, an r^2 coefficient as 0.752, and a Kendall's tau as 0.651 (Table 3). In the first and second test benchmarks, we focused on different ligands binding to the same caspase target. Within this test set, we particularly examined the cases with the same inhibitor binding to different caspase targets.

Table 3. Binding free energy calculation results by using the MT protocol against the caspase-polypeptide complexing test set.

PDB ID	Caspase Target	Peptide Ligand	Experimental ΔG (kcal/mol)	Calculated ΔG (kcal/mol)
2h5j	caspase-3	Ac-DMQD-Cho	−10.78	−11.06
2ql5	caspase-7	Ac-DMQD-Cho	−11.04	−12.72
2ql9	caspase-7	Ac-DQMD-Cho	−12.30	−12.51
2qlf	caspase-7	Ac-DNLD-Cho	−12.06	−12.26
2qlb	caspase-7	Ac-EMSD-Cho	−8.03	−8.46
2ql7	caspase-7	Ac-IEPD-Cho	−8.53	−8.16
1f1j	caspase-7	Ac-DEVD-Cho	−11.99	−10.74
2h5i	caspase-3	Ac-DEVD-Cho	−12.11	−11.18
4jr0	caspase-3	Ac-DEVD-CMK	−11.17	−11.26
3r7b	caspase-2	Ac-DVAD-Cho	−8.38	−8.42
3r5j	caspase-2	Ac-ADVAD-Cho	−9.48	−9.72
3r6g	caspase-2	Ac-VDVAD-Cho	−10.36	−10.68
3gjt	caspase-3	Ac-IEPD	−9.23	−9.44
1f9e	caspase-8	Phq-DEVD	−11.86	−10.49
4jje	caspase-3	Ac-1MH-ASP-B3L-HLX-1U8	−10.41	−10.90

Hereby we looked at two pairs of complex structures as representative examples, to examine how the small structural differences at the binding interfaces affecting the binding affinities between the caspase targets and polypeptide inhibitors.

First, we compared the calculation results between 2h5i and 1f1j, two complexes with the same peptide ligand, Ac-DEVD-Cho, targeting different caspase receptors, caspase-3, and caspase-7. The global minimum binding modes for both of the complexes provided us a clear view of their protein-ligand interaction maps. By using the MT protocol, the global minimum binding mode for the caspase-3-Ac-DEVD-Cho complex had a structural RMSD as 1.17 Å, and the global minimum binding mode for the caspase-7-Ac-DEVD-Cho complex had a structural RMSD as 1.44 Å, both compared to their corresponding crystal structures.

Both caspase-3 and caspase-7 targets had clip-shaped binding sites with similar volumes occupied by the polypeptide inhibitor, Ac-DEVD-Cho, according to the highlighted area in Figures 7 and 8. Both binding sites used short amino acid chains to form a series of backbone-backbone hydrogen bonds stabilizing the polypeptide inhibitor, i.e., S205, R207 and S209 at the caspase-3 binding site formed 4 hydrogen bonds with the Asp, Val, Glu backbone residues and the acetyl capping group of the polypeptide inhibitor respectively; S231, R233, and Q276 forms 4 hydrogen bonds with the Asp, Val, Glu, and Asp backbone residues as well. W206 and Y204 from caspase-3 applied bulky aromatic side-chain structures to limit the flexibility of the polypeptide inhibitor by holding its Valine side chain in between. Similarly, caspase-7 used the indole side chain of W232 and the phenol side chain of Y230 to drag the ligand's valine side chain by forming a C-H/π interaction. Several other residues at the protein's clip-shaped binding site also stabilize the target-inhibitor complex by forming hydrogen bonds with the side chain and capping groups of Ac-DEVD-Cho. At the caspase-3 binding site, W214, S249, and N208 formed hydrogen bonds with the carboxyl side chain of the acetyl capped aspartic acid residue; R207 formed a hydrogen bond with the carboxyl group from the glutamic acid side chain; R64, Q161 and R207 formed hydrogen bonds with the aldehyde capped aspartic acid side chain; and G122 formed a hydrogen bond with the aldehyde capping group on the ligand. On the other hand, at the caspase-7 binding site, S234, W240 and Q276 formed hydrogen bonds with the carboxyl side chain of the inhibitor's acetyl capped aspartic acid residue; N88 formed a hydrogen bond with the carboxyl group from the glutamic acid side chain; R87, Q184 and R233 formed hydrogen bonds with the aldehyde capped aspartic acid side chain; and R87 also formed a hydrogen bond with the aldehyde capping group on the ligand.

With quite similar interaction maps, the MT protocol generated very close protein-ligand interaction energies of these two global-minimum binding modes. The caspase-3-Ac-DEVD-Cho binding mode had −163.92 kcal/mol for the protein-ligand interface contact energy and the caspase-7-Ac-DEVD-Cho binding mode had −159.73 kcal/mol as its own. It also led to quite similar binding affinity predictions, with −11.18 kcal/mol for 2h5i and −10.74 kcal/mol for 1f1j.

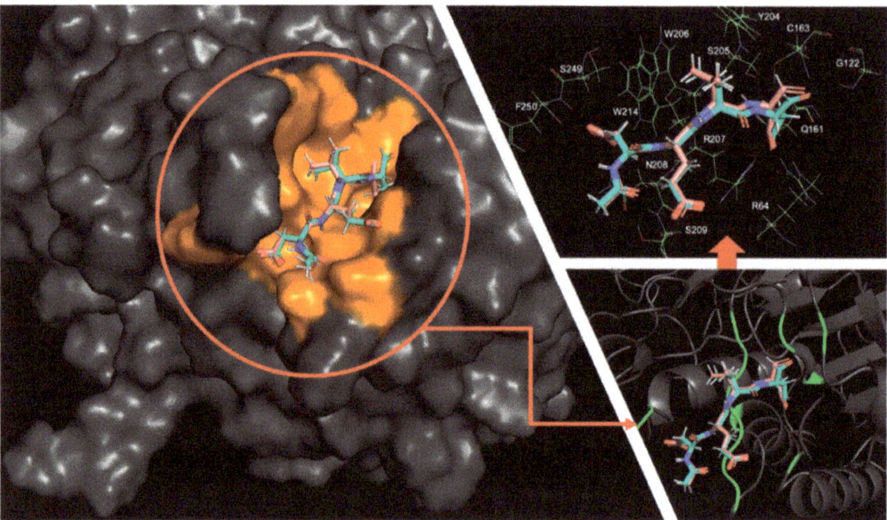

Figure 7. The global minimum docked pose (cyan) together with the crystal ligand conformation (pink) for Ac-DEVD-Cho bound to caspase-3. The orange surface on the left shows the area of interaction interface at the caspase-3 binding site. The green ribbon on the bottom right shows the locations of the residues having close contact (within 3 Å) with the global minimum docked pose. All residues having close contact with the global minimum Ac-DEVD-Cho pose are shown in the picture on the top right.

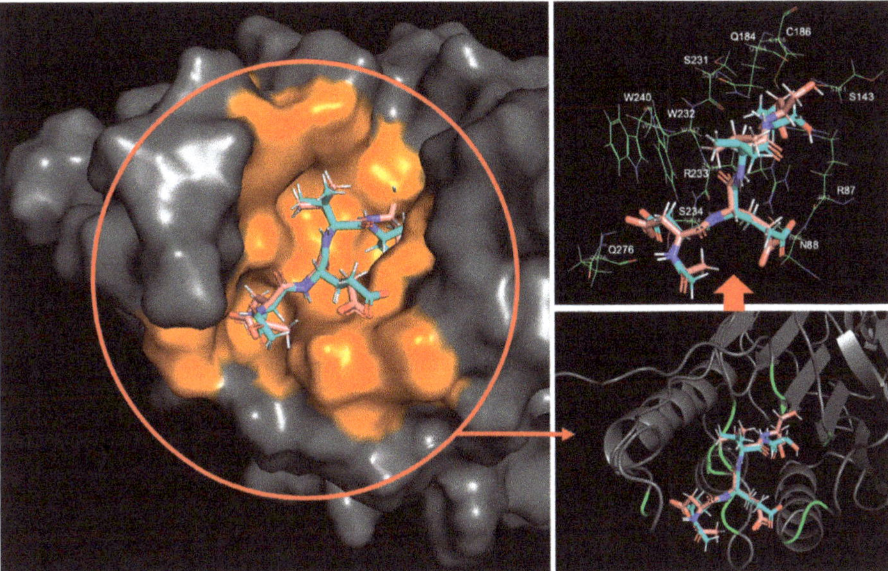

Figure 8. The global minimum docked pose (cyan) together with the crystal ligand conformation (pink) for Ac-DEVD-Cho bound to caspase-7. The orange surface on the left shows the area of interaction interface at the caspase-7 binding site. The green ribbon on the bottom right shows the locations of the residues having close contact (within 3 Å) with the global minimum docked pose. All residues having close contact with the global minimum Ac-DEVD-Cho pose are shown in the picture on the top right.

Another comparison study focused on the two complexes with the PDBID 2ql9 and 2qlb, using the same target protein: caspase-7, binding to two different polypeptide inhibitors: Ac-DQMD-Cho and Ac-ESMD-Cho. Similarly, in both cases, the caspase target provided a short amino acid chain to seize the peptide inhibitor by a series of hydrogen bonds. S231, W232, R233 and S234 formed four hydrogen bonds with both of the peptide inhibitors' backbone structures respectively. Also, the caspase-7 receptor prepared Y230, W232 and F282 with their aromatic side chains to stabilize the two inhibitors with the C-H/π interactions. Meanwhile, by introducing the R87, Q184 residues to form hydrogen bonds with the carboxyl groups from the glutamic acid, and the aldehyde capping groups respectively, and by using the Q276 residue to form a hydrogen bond with the acetyl capping groups, the caspase-7 receptor further locked both of the peptide inhibitors at the binding site (Figure 9).

The main reason causing the interaction energy difference for the two inhibitors lay in that the glutamine residue from Ac-DQMD-Cho formed two more hydrogen bonds with the amide group on the R233 residue from the caspase-7 binding site. On the other hand, the side chain of the serine from Ac-ESMD-Cho was too short to stretch out to form such hydrogen bonds. It resulted in the ~10 kcal/mol interaction energy difference between these two global minimum binding modes, with the protein-ligand contact energy as −201.01 kcal/mol for 2ql9 and that as −190.78 kcal/mol for 2qlb. On the other hand, the free-state ligand's ensemble energy for Ac-DQMD-Cho was −12.217 kcal/mol and that for Ac-ESMD-Cho was −9.18 kcal/mol. It showed that Ac-DQMD-Cho was slightly more favored in the water-solvated free state than Ac-ESMD-Cho, also indicating that the more flexible structure of Ac-DQMD-Cho restored larger configurational entropy compared to Ac-ESMD-Cho. However, the slightly increased protein-ligand complexing barrier for Ac-DQMD-Cho did not stop it from earning ~4 kcal/mol more preferred binding free energy compared to Ac-ESMD-Cho.

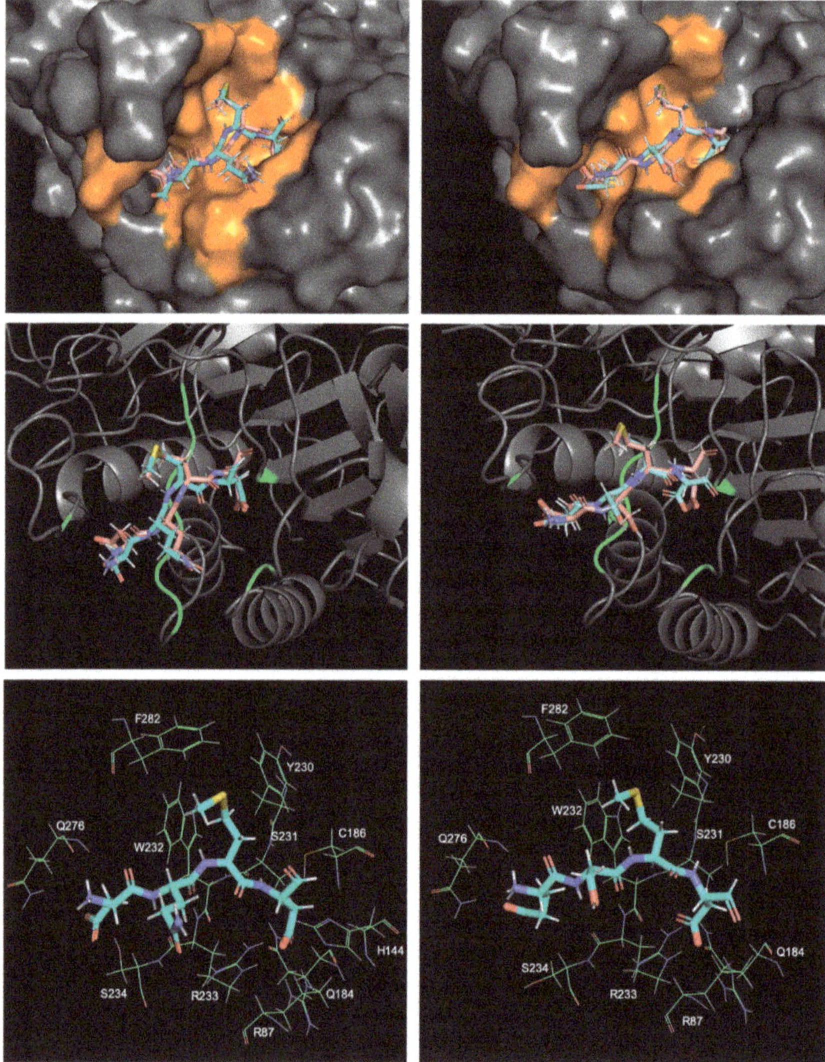

Figure 9. Illustrations of the caspase-7 binding site with Ac-DQMD-Cho (pictures on the left) and Ac-ESMD-Cho (pictures on the right). The global minimum docked pose (cyan) are shown together with the crystal ligand conformation (pink). Orange regions on both binding sites shows the difference of the contacts areas. The green ribbons also indicate more residues from the caspase-7 binding site having significant contact with Ac-DQMD-Cho compared to Ac-ESMD-Cho. Pictures on the bottom show that the glutamine residue from Ac-DQMD-Cho forms extra hydrogen bonds to the R233 residue at the caspase-7 binding site, while no hydrogen bond can be found at the same location for the caspase-7-Ac-ESMD-Cho complex.

3. Materials and Methods

The MT method was first developed in our lab in 2013 [44]. Further refinement was later on published in 2018 [54]. Since the detailed illustrations, thorough validations and calculation comparisons with other top-notch methods can be found in our previous publications, and our focus in this work is the MT method validation and application regarding the caspase inhibition instead of a methodology demonstration, only a brief introduction of this method was included in this paper.

The MT method simplifies the molecular energy state simulation and reduces the computational complexity by separating the sampling of the molecular states into samplings of independent atom pairwise contacts during molecular movements. In a molecular system, each atom possesses independent degree of freedom for its movement, hence the free energy change of a molecule can be simulated using the free energy changes of all the atoms in this molecular system. Given that all atoms are allowed a small movement range, the MT method assumes that every pairwise work on atom A from another atom i is independent from each other. Since every atom, including atom A and every atom i, possesses its own moving degrees of freedom, all the atom A-i pairwise energy states can be extrapolated using the E_{Ai} vector, where τ^0_{Ai} represents the atom A-i relative coordinate from the input structure, and $\Delta\tau$ is their geometric deviation step unit with a sampling range ($\pm n\Delta\tau$).

$$
E_{Ai} = \begin{bmatrix}
E_{Ai}\left(\tau^0_{Ai} - n\Delta\tau\right) \\
\vdots \\
E_{Ai}\left(\tau^0_{Ai} + \Delta\tau\right) \\
E_{Ai}\left(\tau^0_{Ai}\right) \\
E_{Ai}\left(\tau^0_{Ai} - \Delta\tau\right) \\
\vdots \\
E_{Ai}\left(\tau^0_{Ai} - n\Delta\tau\right)
\end{bmatrix}
\tag{4}
$$

All energy states for atom A from the atom A-i relative coordinate change can be generated using the reversible work on atom A from atom i during their movement. The atom pairwise reversible work is calculated as the sum of the work on three orthogonal directions (x, y and z directions) with respect to all the atom A-i pairwise energy changes. Equation (5) illustrated the calculation of the pairwise reversible work regarding atom A along the x axis.

$$
E^x_A = \sum_i^{N-1} F_{Ai} \times \cos(\theta_{Ai})\Delta r
$$

$$
= \left(\begin{bmatrix} F_{A\alpha}\left(r^1_{A\alpha}\right) \\ F_{A\alpha}\left(r^2_{A\alpha}\right) \\ \vdots \\ F_{A\alpha}\left(r^n_{A\alpha}\right) \end{bmatrix} \times \begin{bmatrix} \cos\left(\theta^1_{A\alpha}\right) \\ \cos\left(\theta^2_{A\alpha}\right) \\ \vdots \\ \cos\left(\theta^n_{A\alpha}\right) \end{bmatrix} + \begin{bmatrix} F_{A\beta}\left(r^1_{A\beta}\right) \\ F_{A\beta}\left(r^2_{A\beta}\right) \\ \vdots \\ F_{A\beta}\left(r^n_{A\beta}\right) \end{bmatrix} \times \begin{bmatrix} \cos\left(\theta^1_{A\beta}\right) \\ \cos\left(\theta^2_{A\beta}\right) \\ \vdots \\ \cos\left(\theta^n_{A\beta}\right) \end{bmatrix} + \begin{bmatrix} F_{A\gamma}\left(r^1_{A\gamma}\right) \\ F_{A\gamma}\left(r^2_{A\gamma}\right) \\ \vdots \\ F_{A\gamma}\left(r^n_{A\gamma}\right) \end{bmatrix} \times \begin{bmatrix} \cos\left(\theta^1_{A\gamma}\right) \\ \cos\left(\theta^2_{A\gamma}\right) \\ \vdots \\ \cos\left(\theta^n_{A\gamma}\right) \end{bmatrix} \ldots \right)\Delta r
\tag{5}
$$

where F_{Ai} represents the vector of forces regarding atom A-i with their pairwise distances ranged from r^1_{Ai} to r^n_{Ai}; θ_{Ai} is the collection of angles of inclination of all the i-A vector (Figure 10) regarding the x axis. Δr is the sampling step unit. $F_{Ai} \times \cos(\theta_{Ai})\Delta r$ generates all the atom A-i pairwise energy states. E^x_A is the ensemble of work on atom A from all the surrounding atoms along the x axis.

In this work, we set the distance sampling range (r^1_{Ai} to r^n_{Ai}) as 1 Å, angle sampling range (θ^1_{Ai} to θ^n_{Ai}) as 30 degrees, and Δr as 0.005 Å. The reversible work on atom A is calculated in all the three orthogonal directions and summed as in Equation (6).

$$
E_A = E^x_A + E^y_A + E^z_A
\tag{6}
$$

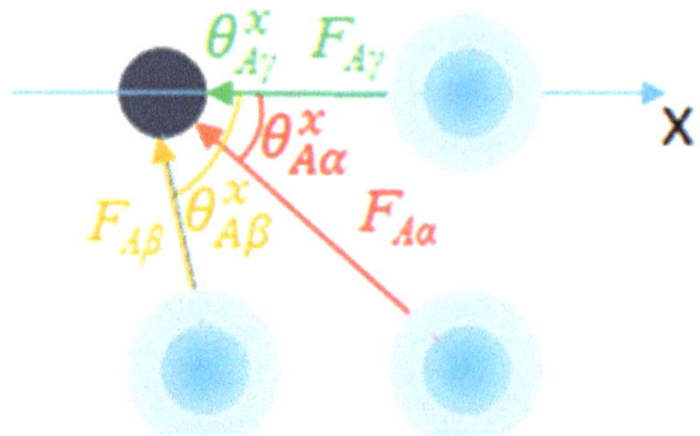

Figure 10. With respect to atom A (dark blue dot on the upper left corner), all atom pairwise contacts on this atom are independent from each other. With atom A moving along the x axis, the ensemble energy.

The partition function (Z) for atom A can be generated as:

$$Z_A = \exp\left(-\frac{1}{RT}E_A\right) \tag{7}$$

For all atoms in the selected molecular system, the atomic ensemble energies are calculated separately to ensure that the molecular local partition function can be numerically calculated for each atomic movement in its given sampling range.

$$Z_M = Z_A \times Z_B \times Z_C \times \cdots \tag{8}$$

By feeding the MT protocol with multiple molecular configurations, local molecular partition functions Z_M can be calculated using Equation (8) for estimation of the free energy. Regarding the protein-ligand binding affinity study, conformations for both free and bound states are generated using the Monte Carlo sampling protocols followed by local minimizations. The free state molecular system includes unbound ligand and protein in the solution phase. Z_L and Z_P are their corresponding partition functions which are necessary for the binding free energy calculation. On the other hand, the bound state molecular system includes the protein ligand molecules in the complex form in the solution. Z_{PL} is the bound state partition function containing all the protein-ligand binding mode energy states. In the present study we only performed the ligand conformational sampling and the protein-ligand binding mode sampling by considering the flexibility of the ligand structures and the protein binding site residues while keeping the rest of the protein geometry fixed. The protein conformational sampling is skipped because (1) the massive degrees of freedom associated with inclusion of protein flexibility will significantly increase the computational burden, while (2) having limited contributions to the computational accuracies regarding relative binding affinities studies using identical or similar protein target, due to that the Z_{PL} values are very similar among all the test cases.

In-house programs developed in our group are introduced to perform such tasks. For the free-state calculation, the MT-CS conformational search program [55] was introduced to generate significant free-state molecular conformations with reference to the molecular flexibility. The MT-CS conformational search program generated ligand conformers using a torsion library with pre-calculated torsion energies using the GARF energy model [56], the solvation free energy was calculated using the KMTISM model [57]. The MT protocol was then applied to each ligand conformer to estimate the local partition function Z_L. The ligand's total partition function was then generated using all the MT-CS sampled configurational ensemble energies in Equation (9).

$$Z_L = \sum_{\alpha}^{N_{L\ conformation}} Z_L^\alpha = Z_L^1 + Z_L^2 + \cdots + Z_L^N \tag{9}$$

The Heatmap docking program [44] was employed for the bound state configuration sampling in this work. The bound-state protein-ligand complex ensemble energy is calculated using the same protocol by summing all the local partition functions.

$$Z_{PL} = \sum_{\alpha}^{N_{PL\ conformation}} Z_{PL}^{\alpha} = Z_{PL}^1 + Z_{PL}^2 + \cdots + Z_{PL}^N \tag{10}$$

The binding free energy change was then estimated by using the ratio of partition functions in bound and free states. The whole calculation protocol is also illustrated in Figure 11.

$$\Delta G_{binding} \approx -RT \log\left(\frac{Z_{PL}}{Z_L}\right) \tag{11}$$

In this work, we utilized the MT free energy protocol briefed above for the caspase-inhibitor binding affinity study. All related codes and data can be obtained by contacting the authors for validation and review purpose only.

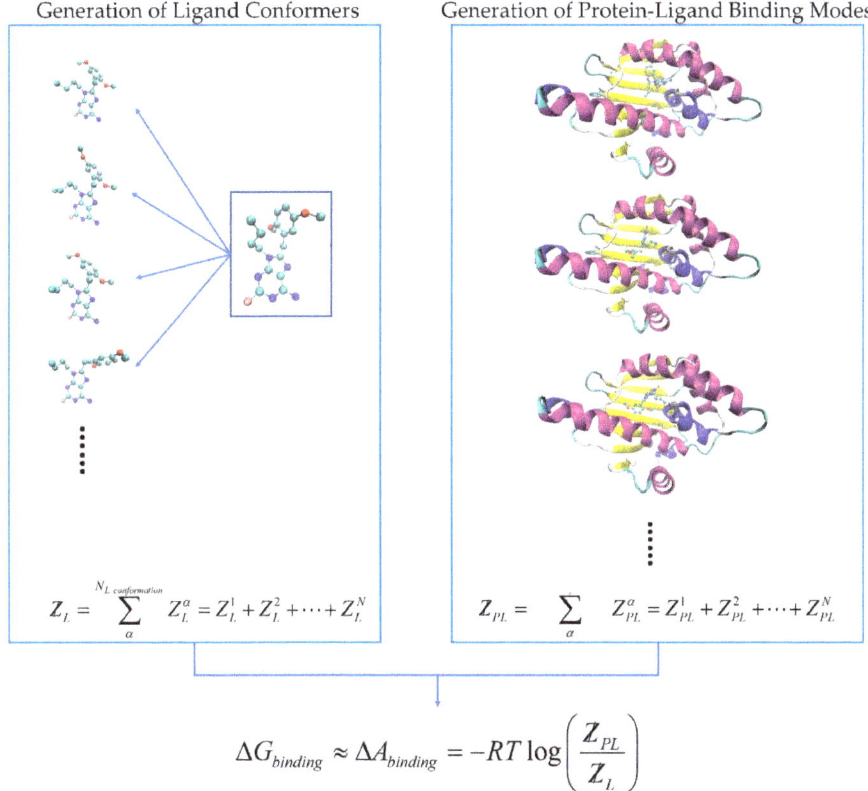

$$Z_L = \sum_{\alpha}^{N_{L\ conformation}} Z_L^{\alpha} = Z_L^1 + Z_L^2 + \cdots + Z_L^N$$

$$Z_{PL} = \sum_{\alpha} Z_{PL}^{\alpha} = Z_{PL}^1 + Z_{PL}^2 + \cdots + Z_{PL}^N$$

$$\Delta G_{binding} \approx \Delta A_{binding} = -RT \log\left(\frac{Z_{PL}}{Z_L}\right)$$

Figure 11. A flow chart for the ensemble energy calculation protocol employed in this work. Both bound-state and free-state ensembles were generated using the programs indicated (Heatmap docking and MT-CS). The final free energy change was then calculated using the ratio of partition functions Z_{PL}/Z_L.

4. Conclusions

We applied our newly developed Movable Type free energy protocol to the caspase-inhibitor complexing study. Using a Monte Carlo sampling approach, the MT method generated the significant binding modes and calculated the binding free energies using the ratio of the partition functions referencing the bound state and free state protein-ligand systems. Both the large-scaling and carefully set-up small test sets were introduced to provide a comprehensive study regarding the robustness and sensitivity of the MT protocol against such complexing systems. Results revealed good agreements of the calculation predictions with the experimental binding affinities and the global minimum binding modes. Through detailed case studies, we further illustrated the MT protocol mechanism for the free energy extrapolation using a Monte Carlo based sampling method. Moreover, we also took a close look at the global minimum binding mode structures to study how minor changes in the interaction interfaces affecting the binding affinities and how with different interaction interfaces achieved similar binding affinities. Generally, this work provided us useful computational information for the binding affinity prediction using the MT protocol. Future studies including computation-experiment combinatorial research can be expected for the structural based caspase inhibitor design. We also plan to apply the MT protocol to the caspase inhibitor-related inverse docking study.

Supplementary Materials: Supplementary materials can be found at http://www.mdpi.com/1422-0067/20/19/4850/s1.

Author Contributions: Conceptualization, Z.Z.; methodology, Z.Z.; software, S.X. and Z.Z.; validation, S.X. and H.L.; formal analysis, S.X.; investigation, S.X.; resources, Z.Z.; data curation, S.X. and H.L.; writing—original draft preparation, S.X.; writing—review and editing, S.X. H.L. and Z.Z.; visualization, S.X.; supervision, Z.Z.; project administration, Z.Z.

Funding: This research received no external funding.

Acknowledgments: We thank Taolei Sun from Wuhan University of Technology for useful discussions of the caspase inhibition mechanism and the computational methodology application.

Conflicts of Interest: The authors declare no conflict of interest.

Abbreviations

AD	Alzheimer's Disease
MT	Movable Type
NFTs	Neurofibrillary tangles

References

1. Rohn, T.T.; Head, E.; Su, J.H.; Anderson, A.J.; Bahr, B.A.; Cotman, C.W.; Cribbs, D.H. Correlation between caspase activation and neurofibrillary tangle formation in Alzheimer's disease. *Am. J. Pathol.* **2001**, *158*, 189–198. [CrossRef]
2. Stadelmann, C.; Deckwerth, T.L.; Srinivasan, A.; Bancher, C.; Brück, W.; Jellinger, K.; Lassmann, H. Activation of caspase-3 in single neurons and autophagic granules of granulovacuolar degeneration in Alzheimer's disease. Evidence for apoptotic cell death. *Am. J. Pathol.* **1999**, *155*, 1459–1466. [CrossRef]
3. Gervais, F.G.; Xu, D.; Robertson, G.S.; Vaillancourt, J.P.; Zhu, Y.; Huang, J.; LeBlanc, A.; Smith, D.; Rigby, M.; Shearman, M.S.; et al. Involvement of caspases in proteolytic cleavage of Alzheimer's amyloid-beta precursor protein and amyloidogenic A beta peptide formation. *Cell* **1999**, *97*, 395–406. [CrossRef]
4. Cotman, C.W.; Poon, W.W.; Rissman, R.A.; Blurton-Jones, M. The role of caspase cleavage of tau in Alzheimer disease neuropathology. *J. Neuropathol. Exp. Neurol.* **2005**, *64*, 104–112. [CrossRef] [PubMed]
5. Roth, K.A. Caspases, apoptosis, and Alzheimer disease: Causation, correlation, and confusion. *J. Neuropathol. Exp. Neurol.* **2001**, *60*, 829–838. [CrossRef] [PubMed]
6. Obulesu, M.; Lakshmi, M.J. Apoptosis in Alzheimer's disease: An understanding of the physiology, pathology and therapeutic avenues. *Neurochem. Res.* **2014**, *39*, 2301–2312. [CrossRef] [PubMed]
7. Cribbs, D.H.; Poon, W.W.; Rissman, R.A.; Blurton-Jones, M. Caspase-Mediated Degeneration in Alzheimer's Disease. *Am. J. Pathol.* **2004**, *165*, 353–355. [CrossRef]

8. Kim, T.W.; Pettingell, W.H.; Jung, Y.K.; Kovacs, D.M.; Tanzi, R.E. Alternative cleavage of Alzheimer-associated presenilins during apoptosis by a caspase-3 family protease. *Science* **1997**, *277*, 373–376. [CrossRef] [PubMed]

9. Gamblin, T.C.; Chen, F.; Zambrano, A.; Abraha, A.; Lagalwar, S.; Guillozet, A.L.; Lu, M.; Fu, Y.; Garcia-Sierra, F.; LaPointe, N.; et al. Caspase cleavage of tau: Linking amyloid and neurofibrillary tangles in Alzheimer's disease. *Proc. Natl. Acad. Sci. USA* **2003**, *100*, 10032–10037. [CrossRef] [PubMed]

10. Rissman, R.A.; Poon, W.W.; Blurton-Jones, M.; Oddo, S.; Torp, R.; Vitek, M.P.; LaFerla, F.M.; Rohn, T.T.; Cotman, C.W. Caspase-cleavage of tau is an early event in Alzheimer disease tangle pathology. *J. Clin. Investig.* **2004**, *114*, 121–130. [CrossRef] [PubMed]

11. Binder, L.I.; Guillozet-Bongaarts, A.L.; Garcia-Sierra, F.; Berry, R.W. Tau, tangles, and Alzheimer's disease. *Biochim. Biophys. Acta Mol. Basis Dis.* **2005**, *1739*, 216–223. [CrossRef] [PubMed]

12. Leroy, K.; Ando, K.; Laporte, V.; Dedecker, R.; Suain, V.; Authelet, M.; Héraud, C.; Pierrot, N.; Yilmaz, Z.; Octave, J.N.; et al. Lack of tau proteins rescues neuronal cell death and decreases amyloidogenic processing of APP in APP/PS1 mice. *Am. J. Pathol.* **2012**, *181*, 1928–1940. [CrossRef] [PubMed]

13. Guo, H.; Albrecht, S.; Bourdeau, M.; Petzke, T.; Bergeron, C.; LeBlanc, A.C. Active caspase-6 and caspase-6-cleaved tau in neuropil threads, neuritic plaques, and neurofibrillary tangles of Alzheimer's disease. *Am. J. Pathol.* **2004**, *165*, 523–531. [CrossRef]

14. Rohn, T.T.; Rissman, R.A.; Davis, M.C.; Kim, Y.E.; Cotman, C.W.; Head, E. Caspase-9 activation and caspase cleavage of tau in the Alzheimer's disease brain. *Neurobiol. Dis.* **2002**, *11*, 341–354. [CrossRef] [PubMed]

15. LeBlanc, A.; Liu, H.; Goodyer, C.; Bergeron, C.; Hammond, J. Caspase-6 role in apoptosis of human neurons, amyloidogenesis, and Alzheimer's disease. *J. Biol. Chem.* **1999**, *274*, 23426–23436. [CrossRef] [PubMed]

16. Klaiman, G.; Petzke, T.L.; Hammond, J.; Leblanc, A.C. Targets of caspase-6 activity in human neurons and Alzheimer disease. *Mol. Cell Proteom.* **2008**, *7*, 1541–1555. [CrossRef] [PubMed]

17. Horowitz, P.M.; Patterson, K.R.; Guillozet-Bongaarts, A.L.; Reynolds, M.R.; Carroll, C.A.; Weintraub, S.T.; Bennett, D.A.; Cryns, V.L.; Berry, R.W.; Binder, L.I. Early N-terminal changes and caspase-6 cleavage of tau in Alzheimer's disease. *J. Neurosci.* **2004**, *24*, 7895–7902. [CrossRef]

18. Su, J.H.; Zhao, M.; Anderson, A.J.; Srinivasan, A.; Cotman, C.W. Activated caspase-3 expression in Alzheimer's and aged control brain: Correlation with Alzheimer pathology. *Brain Res.* **2001**, *898*, 350–357. [CrossRef]

19. Rohn, T.T.; Head, E.; Nesse, W.H.; Cotman, C.W.; Cribbs, D.H. Activation of caspase-8 in the Alzheimer's disease brain. *Neurobiol. Dis.* **2001**, *8*, 1006–1016. [CrossRef]

20. Pate, K.M.; Rogers, M.; Reed, J.W.; van der Munnik, N.; Vance, S.Z.; Moss, M.A. Anthoxanthin Polyphenols Attenuate Aβ Oligomer-induced Neuronal Responses Associated with Alzheimer's Disease. *CNS Neurosci. Ther.* **2017**, *23*, 135–144. [CrossRef]

21. Pérez, M.J.; Vergara-Pulgar, K.; Jara, C.; Cabezas-Opazo, F.; Quintanilla, R.A. Caspase-Cleaved Tau Impairs Mitochondrial Dynamics in Alzheimer's Disease. *Mol. Neurobiol.* **2018**, *55*, 1004–1018. [CrossRef] [PubMed]

22. Forner, S.; Baglietto-Vargas, D.; Martini, A.C.; Trujillo-Estrada, L.; LaFerla, F.M. Synaptic Impairment in Alzheimer's Disease: A Dysregulated Symphony. *Trends Neurosci.* **2017**, *40*, 347–357. [CrossRef] [PubMed]

23. Chu, J.; Lauretti, E.; Praticò, D. Caspase-3-dependent cleavage of Akt modulates tau phosphorylation via GSK3β kinase: Implications for Alzheimer's disease. *Mol. Psychiatry.* **2017**, *22*, 1002–1008. [CrossRef] [PubMed]

24. Rohn, T.T. The role of caspases in Alzheimer's disease; potential novel therapeutic opportunities. *Apoptosis* **2010**, *15*, 1403–1409. [CrossRef] [PubMed]

25. Gilson, M.K.; Zhou, H.X. Calculation of protein-ligand binding affinities. *Annu. Rev. Bioph. Biom.* **2007**, *36*, 21–42. [CrossRef] [PubMed]

26. Michel, J.; Essex, J.W. Prediction of protein-ligand binding affinity by free energy simulations: Assumptions, pitfalls and expectations. *J. Comput. Aid. Mol Des.* **2010**, *24*, 639–658. [CrossRef]

27. Shirts, M.R.; Mobley, D.L.; Chodera, J.D. Alchemical Free Energy Calculations: Ready for Prime Time? *Ann. Rep. Comp. Chem.* **2007**, *3*, 41–59.

28. Halperin, I.; Ma, B.Y.; Wolfson, H.; Nussinov, R. Principles of docking: An overview of search algorithms and a guide to scoring functions. *Proteins* **2002**, *47*, 409–443. [CrossRef]

29. Yuriev, E.; Ramsland, P.A. Latest developments in molecular docking: 2010-2011 in review. *J. Mol. Recognit.* **2013**, *26*, 215–239. [CrossRef]

30. Kollman, P.A.; Massova, I.; Reyes, C.; Kuhn, B.; Huo, S.; Chong, L.; Lee, M.; Lee, T.; Duan, Y.; Wang, W.; et al. Calculating structures and free energies of complex molecules: Combining molecular mechanics and continuum models. *Acc. Chem. Res.* **2000**, *33*, 889–897. [CrossRef]

31. Genheden, S.; Ryde, U. The MM/PBSA and MM/GBSA methods to estimate ligand-binding affinities. *Expert Opin. Drug Discov.* **2015**, *10*, 449–461. [CrossRef] [PubMed]

32. Chodera, J.D.; Mobley, D.L.; Shirts, M.R.; Dixon, R.W.; Branson, K.; Pande, V.S. Alchemical free energy methods for drug discovery: Progress and challenges. *Curr. Opin. Struc. Biol.* **2011**, *21*, 150–160. [CrossRef] [PubMed]

33. Christ, C.D.; Mark, A.E.; van Gunsteren, W.F. Feature Article Basic Ingredients of Free Energy Calculations: A Review. *J. Comput. Chem.* **2010**, *31*, 1569–1582. [PubMed]

34. Woo, H.J.; Roux, B. Calculation of absolute protein-ligand binding free energy from computer simulations. *Proc. Natl. Acad. Sci. USA* **2005**, *102*, 6825–6830. [CrossRef] [PubMed]

35. Zwanzig, R.W. High-Temperature Equation of State by a Perturbation Method.1. Nonpolar Gases. *J. Chem. Phys.* **1954**, *22*, 1420–1426. [CrossRef]

36. Kirkwood, J.G. Statistical mechanics of fluid mixtures. *J. Chem. Phys.* **1935**, *3*, 300–313. [CrossRef]

37. Buch, I.; Sadiq, S.K.; De Fabritiis, G. Optimized Potential of Mean Force Calculations for Standard Binding Free Energies. *J. Chem. Theory Comput.* **2011**, *7*, 1765–1772. [CrossRef] [PubMed]

38. Essex, J.W.; Severance, D.L.; TiradoRives, J.; Jorgensen, W.L. Monte Carlo simulations for proteins: Binding affinities for trypsin-benzamidine complexes via free-energy perturbations. *J. Phys. Chem. B* **1997**, *101*, 9663–9669. [CrossRef]

39. Gumbart, J.C.; Roux, B.; Chipot, C. Standard Binding Free Energies from Computer Simulations: What Is the Best Strategy? *J. Chem. Theory Comput.* **2013**, *9*, 794–802. [CrossRef]

40. Roux, B. The Calculation of the Potential of Mean Force Using Computer-Simulations. *Comput. Phys. Commun.* **1995**, *91*, 275–282. [CrossRef]

41. Jorgensen, W.L. Free-Energy Calculations—a Breakthrough for Modeling Organic-Chemistry in Solution. *Accounts Chem. Res.* **1989**, *22*, 184–189. [CrossRef]

42. Klimovich, P.V.; Shirts, M.R.; Mobley, D.L. Guidelines for the analysis of free energy calculations. *J. Comput. Aid. Mol Des.* **2015**, *29*, 397–411. [CrossRef] [PubMed]

43. Shirts, M.R.; Mobley, D.L. An introduction to best practices in free energy calculations. *Methods Mol. Biol.* **2013**, *924*, 271–311. [PubMed]

44. Zheng, Z.; Ucisik, M.N.; Merz, K.M., Jr. The Movable Type Method Applied to Protein–Ligand Binding. *J. Chem. Theory Comput.* **2013**, *9*, 5526–5538. [CrossRef]

45. Liu, T.; Lin, Y.; Wen, X.; Jorissen, R.N.; Gilson, M.K. BindingDB: A web-accessible database of experimentally determined protein-ligand binding affinities. *Nucleic Acids Res.* **2007**, *35*, D198–D201. [CrossRef] [PubMed]

46. Gilson, M.K.; Liu, T.; Baitaluk, M.; Nicola, G.; Hwang, L.; Chong, J. BindingDB in 2015: A public database for medicinal chemistry, computational chemistry and systems pharmacology. *Nucleic Acids Res.* **2016**, *44*, D1045–D1053. [CrossRef] [PubMed]

47. Mysinger, M.M.; Carchia, M.; Irwin, J.J.; Shoichet, B.K. Directory of Useful Decoys, Enhanced (DUD-E): Better Ligands and Decoys for Better Benchmarking. *J. Med. Chem.* **2012**, *55*, 6582–6594. [CrossRef]

48. Gaulton, A.; Bellis, L.J.; Bento, A.P.; Chambers, J.; Davies, M.; Hersey, A.; Light, Y.; McGlinchey, S.; Michalovich, D.; Al-Lazikani, B.; et al. ChEMBL: A large-scale bioactivity database for drug discovery. *Nucleic Acids Res.* **2012**, *40*, D1100–D1107. [CrossRef]

49. Cheng, Y.; Prusoff, W.H. Relationship between the inhibition constant (K_I) and the concentration of inhibitor which causes 50 per cent inhibition (I_{50}) of an enzymatic reaction. *Biochem. Pharm.* **1973**, *22*, 3099–3108.

50. Kalliokoski, T.; Kramer, C.; Vulpetti, A.; Gedeck, P. Comparability of mixed IC_{50} data—a statistical analysis. *PLoS ONE* **2013**, *8*, e61007. [CrossRef]

51. Kastritis, P.L.; Rodrigues, J.P.G.L.M.; Bonvin, A.M.J.J. HADDOCK$_{2P2I}$: A Biophysical Model for Predicting the Binding Affinity of Protein–Protein Interaction Inhibitors. *J. Chem. Inf. Model.* **2014**, *54*, 826–836. [CrossRef] [PubMed]

52. Sakkiah, S.; Thangapandian, S.; John, S.; Kwon, Y.J.; Lee, K.W. 3D QSAR pharmacophore based virtual screening and molecular docking for identification of potential HSP90 inhibitors. *Eur. J. Med. Chem.* **2010**, *45*, 2132–2140. [CrossRef] [PubMed]

53. Schneider, G.; Odile, C.; Hilfiger, L.; Schneider, P.; Kirsch, S.; Böhm, H.; Neidhart, W. Virtual Screening for Bioactive Molecules by Evolutionary De Novo Design. *Angew. Chem. Int. Ed.* **2000**, *39*, 4130–4133. [CrossRef]
54. Bansal, N.; Zheng, Z.; Song, L.F.; Pei, J.; Merz, K.M., Jr. The Role of the Active Site Flap in Streptavidin/Biotin Complex Formation. *J. Am. Chem. Soc.* **2018**, *140*, 5434–5446. [CrossRef] [PubMed]
55. Pan, L.; Zheng, Z.; Wang, T.; Merz, K.M., Jr. A Free Energy Based Conformational Search Algorithm Using the "Movable Type" Sampling Method. *J. Chem. Theory Comput.* **2015**, *11*, 5853–5864. [CrossRef] [PubMed]
56. Zheng, Z.; Pei, J.; Bansal, N.; Liu, H.; Song, L.F.; Merz, K.M., Jr. Generation of Pairwise Potentials Using Multi-Dimensional Data Mining. *J. Chem. Theory Comput.* **2018**, *14*, 5045–5067. [CrossRef] [PubMed]
57. Zheng, Z.; Wang, T.; Li, P.; Merz, K.M., Jr. KECSA-Movable Type Implicit Solvation Model (KMTISM). *J. Chem. Theory Comput.* **2014**, *11*, 667–682. [CrossRef]

International Journal of
Molecular Sciences

Article

DockNmine, a Web Portal to Assemble and Analyse Virtual and Experimental Interaction Data

Ennys Gheyouche , Romain Launay, Jean Lethiec, Antoine Labeeuw, Caroline Roze,
Alan Amossé and Stéphane Téletchéa *

UFIP, Université de Nantes, UMR CNRS 6286, 2 rue de la Houssinière, 44322 Nantes, France;
ennys.gheyouche@univ-nantes.fr (E.G.); romain.launay1@etu.univ-nantes.fr (R.L.); jean.lethiec@gmail.com (J.L.);
antoine.labeeuw@etu.univ-nantes.fr (A.L.); caroline@affilogic.com (C.R.); alan.amosse@gmail.com (A.A.)
* Correspondence: stephane.teletchea@univ-nantes.fr; Tel.: +33-251-125-636

Received: 9 September 2019; Accepted: 7 October 2019 ; Published: 12 October 2019

Abstract: Scientists have to perform multiple experiments producing qualitative and quantitative data to determine if a compound is able to bind to a given target. Due to the large diversity of the potential ligand chemical space, the possibility of experimentally exploring a lot of compounds on a target rapidly becomes out of reach. Scientists therefore need to use virtual screening methods to determine the putative binding mode of ligands on a protein and then post-process the raw docking experiments with a dedicated scoring function in relation with experimental data. Two of the major difficulties for comparing docking predictions with experiments mostly come from the lack of transferability of experimental data and the lack of standardisation in molecule names. Although large portals like PubChem or ChEMBL are available for general purpose, there is no service allowing a formal expert annotation of both experimental data and docking studies. To address these issues, researchers build their own collection of data in flat files, often in spreadsheets, with limited possibilities of extensive annotations or standardisation of ligand descriptions allowing cross-database retrieval. We have conceived the dockNmine platform to provide a service allowing an expert and authenticated annotation of ligands and targets. First, this portal allows a scientist to incorporate controlled information in the database using reference identifiers for the protein (Uniprot ID) and the ligand (SMILES description), the data and the publication associated to it. Second, it allows the incorporation of docking experiments using forms that automatically parse useful parameters and results. Last, the web interface provides a lot of pre-computed outputs to assess the degree of correlations between docking experiments and experimental data.

Keywords: protein–ligand analysis; drug discovery and design; structure–activity relationships

---

1. Introduction

There is a booming demand to develop precision medicine products, that is, to design new drugs targeting regular or pathological protein variants [1–3]. These targeted strategies are very promising for the treatment of cancers and other diseases but are very challenging to set up [4]. First, it is necessary to assemble the knowledge on biological processes of interest, in order to identify which protein should be specifically targeted by new drugs [5]. Second, a review of known ligands, be them agonists or antagonists, is essential to identify key binding motifs [6]. Last, when possible, one needs to gain as much as possible insight into the protein three-dimensional structure obtained by crystallography or NMR and the allostery associated with the protein [7].

It takes time and expertise to get a broad overview of the protein to target and of its specific modifications related to a disease. The limitations in this process comes from the immense gap of knowledge that individuals in one laboratory can apprehend, in comparison with the ocean of data available in scientific literature. Fortunately, in order to link the experimental activities of various small chemical entities with their (protein) targets into databases, there is a strong ongoing effort to organise these data logically by human experts, a process called curation. Once set up, these databases can be queried via their web interface but also queried using dedicated programmatic access for batch data retrieval [8,9].

An important limitation for gathering experimental results for a target comes from the standardisation of experimental data, of target names and of small chemical entities. For example, it is common practice to reference a molecule by a common name in a given laboratory, to use a chemical name or to name it based on the biological process interrupted by the drug. These difference in protein nomenclature is visible for the Tartget Of Rapamycin (mTOR) where the protein itself is referenced as mTOR [10] but the protein name in Uniprot is Serine/threonine-protein kinase mTOR. This variety of definitions for small chemical entities and protein targets is not important when people are working closely together but this renders the comparison of data very difficult between laboratories.

Before being publicly available, either published or patented, compounds synthesized or in tests have to stay private. In the meantime, the teams of biologists, chemists and chemoinformaticians/structural biochemists need to collaborate to bring together their results in a comprehensive and efficient way. This requirement of privacy and collaborative methodology starts to be critical when the collaborators are split in different and sometimes geographically dispersed teams.

We have set up the dockNmine portal (http://www.ufip.univ-nantes.fr/tools/docknmine/) to ease the data management, exchange and analysis of project-based medicinal chemistry studies. The portal allows to manage private experimental data and private docking studies but also makes use of public data when possible for homogenising proteins description and small molecules activities. We now describe dockNmine organisation and implementation.

2. Results

The dockNmine home page is divided into six independent services to ease a logical workflow for processing docking and experimental data (Figure 1A). After a broad overview of dockNmine organisation, a detailed explanation of its services is provided in dedicated sections.

2.1. DockNmine Overview

The portal philosophy is directly inspired from funded-based projects, therefore it is designed to isolate independent and confidential data from different users. This management by project allows to assemble ligands, protein(s), docking and experimental data into a coherent ensemble, via dedicated feature-control checks. Once registered and connected, the user can start a new project or join an existing one. In both cases, either a private project or a shared project within a small group, the connected used can start to organise his computational data using dockNmine services. We have set up as an example a new project called "Target" created by the user "demo" by following the workflow presented in Figure 1B.

The user can now use the Target tab (Figure 1A3) to add the description of its target from a unique identifier, for example the Solute carrier family 2, facilitated glucose transporter member 1, whose uniprot ID is P11166. A request is triggered on uniprot to retrieve its protein name, its description and other parameters (Figure 2B). This protein is automatically added to the main project as its first target. This transporter allows the exchange of glucose and is important for glucose supply in brain and other organs [11–13]. The beta-D-glycopyranose is referenced under the ID 64689 in PubChem, this ligand is added via the form provided under the Ligand tab (Figure 1A4) using the "Add a ligand" link. The dockNmine request triggers a query on PubChem to download reference ligand information and the file containing its three-dimensional structure in sdf format. This file is processed using rdkit to generate additional descriptors. The summary page for known ligands shows the result of this process (Figure 1D). Since the crystal structure of GLUT1 contains another ligand (B-nonylglucoside), this ligand was also incorporated into dockNmine. These ligands were screened on the GLUT1 structure (PDB ID: 4PYP) using vina (see Supplementary Materials) and the output of vina is incorporated using the Docking service. To upload a complete docking result, its structure file has to be previously uploaded to the target but this step is more extensively described below (Figure 2D). Once all mandatory parameters are present, the docking information is added to dockNmine (Figure 3).

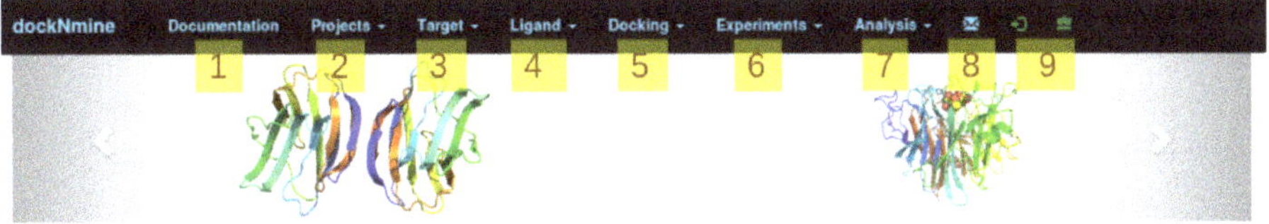

The dockNmine portal aims at gathering public and private data into a unique service. Automated queries on protein targets and ligand definitions are performed to Uniprot, PubChem and ChemBL to enhance the results of pre-computed docking experiments. When available, public data are automatically added to the docking results to produce state-of-the-art protein-ligand binding analysis such as ROC curves or enrichment analysis. Users can also upload their own private data to analyze them automatically. Data access is controlled by project membership, user role and object-based restrictions.

A

B

Figure 1. Overview of dockNmine. (**A**) (1) Link to the documentation of each service; (2–7) access to each service independently; (8) contact link; (9) login and registration links. A simple demonstration of the functionalities is accessible upon connection using the demo account using the log-in glyphicon (login: *demo*, password: *demo*) or by registering upon clicking on the briefcase; (**B**) Once connected, the user can create or join a project where all his data will be assembled (rounded-corner rectangle), he shall then add required protein and ligands (parallelogram) and link them to experimental data and docking results (diamond).

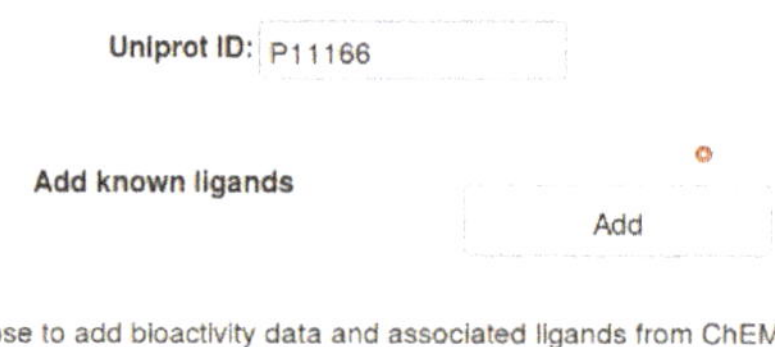

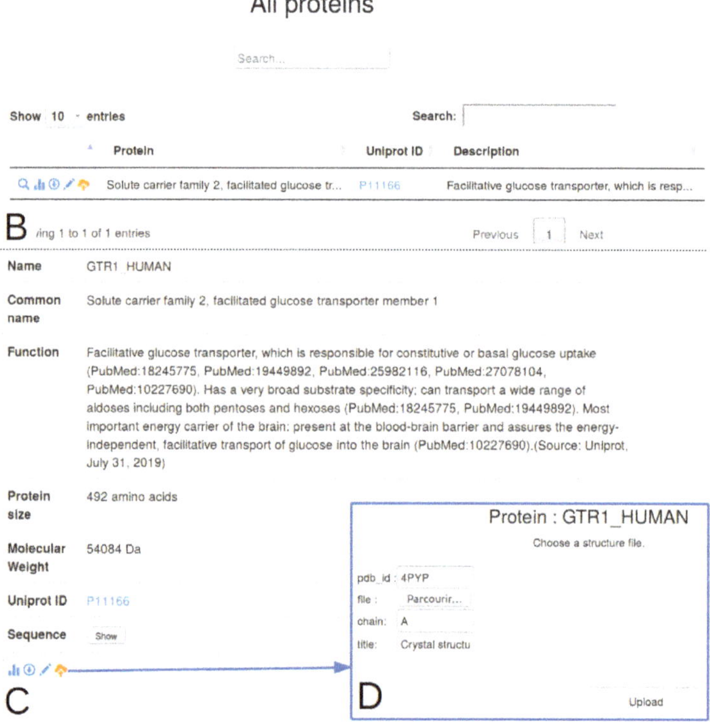

Figure 2. Automated target management in dockNmine for Solute carrier family 2, facilitated glucose transporter member 1. (**A**) The screenshot presents a request for the retrieval of data for the Uniprot ID P11166 and of known ligands from ChEMBL; (**B**) A condensed view of the targets for the project is provided. Some glyphicons are provided to see the details of the entry (magnifier icon), to get detailed statistics (histogram), to download existing data in csv format (circled arrow), to add a comment (pen) or to add a three-dimensional structure for the target (orange arrow towards a cloud); (**C**) Detail of a given entry; (**D**) If required, the user can upload one or many structures for the target. As structure files can be processed in virtual screening experiments, only the structure file is mandatory, all other fields being optional.

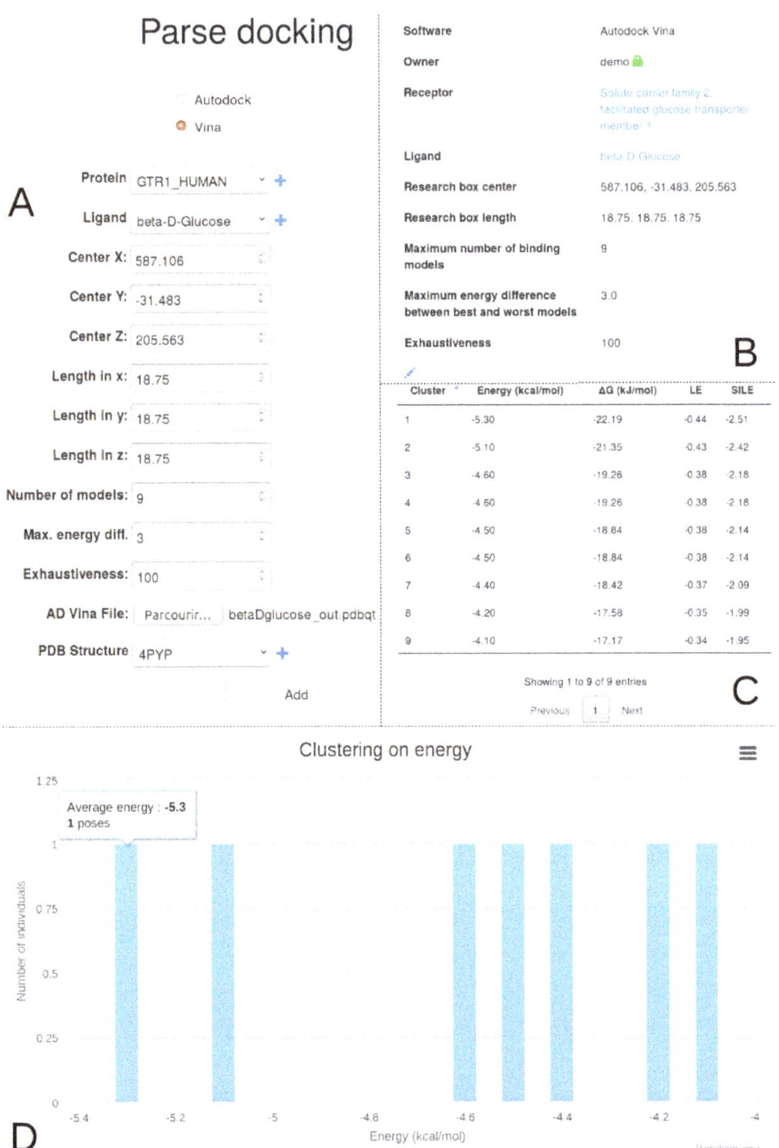

Cluster	Energy (kcal/mol)	ΔG (kJ/mol)	LE	SILE
1	-5.30	-22.19	-0.44	-2.51
2	-5.10	-21.35	-0.43	-2.42
3	-4.60	-19.26	-0.38	-2.18
4	-4.60	-19.26	-0.38	-2.18
5	-4.50	-18.84	-0.38	-2.14
6	-4.50	-18.84	-0.38	-2.14
7	-4.40	-18.42	-0.37	-2.09
8	-4.20	-17.58	-0.35	-1.99
9	-4.10	-17.17	-0.34	-1.95

Figure 3. Vina docking results import for the beta-D-glucose docked in the glucose transporter. (**A**) Upon selection of the docking method, a dedicated form allows to link protein, ligand and docking results. Detailed docking parameters must be provided to allow a further comparison of docking profiles between experiments. If required, the plus glyphicon allows to add a target, a ligand or a target structure prior to entering the docking results; (**B**) Detail of the docking analysis. This view indicates the principal features of the docking method; (**C**) After docking processing, the cluster energy of vina is transformed into kJ/mol, LE and SILE automatically, to ease comparison against other experiments or other ligands; (**D**) Interactive graph depicting the discrete cluster and associated energy. This graph, which can be easily downloaded as an image, allows a rapid overview of the docking energy dispersion for the ligand.

The validity of a virtual screening must be assessed by comparing its results with experimental data. Some experimental data are already openly aggregated by entities like BindingBD or ChEMBL but this information can be retrieved automatically during the target addition to dockNmine. Though, there is a important amount of data impossible to share immediately, mostly coming from internal experiments in the laboratory. The Experiment service allows to include data from the laboratory and its partners without the need to disclose it too early. We have included into the demo account a piperazin derivative where a cellular IC50 was determined of 50 nM is referenced in Binding DB (ID 50155745) from the work of Siebeneicher and co-workers [14]. The structure file for this compound was incorporated from PubChem using the identifier 1387075.

2.2. Target Management

A more extensive explanation of the incorporation process and data management is provided in this section for the addition of a given target. As indicated above, we study the Solute carrier family 2, facilitated glucose transporter member 1 (Uniprot ID P11166). Within a given project, due to the variety of disciplines involved, scientists may refer to a target using an acronym, its gene name or a common name. These names are prone to change so we have provided a simplified way to incorporate a given target from its reference uniprot identifier, as shown in Figure 2. By clicking on Target-> Add a protein, dockNmine retrieves automatically from Uniprot the target name, common name, function, size, molecular weigth and sequence. A checkbox is also provided to grab bioactivities of ChEMBL compounds for the target using ChEMBL web services [15,16] but the addition of ligands is processed asynchronously to allow a better experience in the interface. The process can be sufficient if only experimental data are to be added for a target but if docking results need to be incorporated, then a reference structure file is required. This addition is accessible by clicking on the orange glyphicon (Figure 2B,C). Since the structure file often needs to be extensively processed for docking [6], only the resulting structure file in pdb or cif format is required. We have uploaded for the purpose of demonstration the unprocessed crystal structure of the glucose transporter resolved by Deng and colleagues [11]. Using the target management service, the user is able to incorporate most important data in few minutes.

2.3. Ligand Import And Management

There are two ways to manage ligand incorporation into the database—(i) by adding one ligand at a time via the dedicated form, (ii) by uploading multiple ligands from a file, which will be regrouped into a library.

The single ligand form (Figure 4A) allows to incorporate ligand information from the Protein Data Bank (PDB) [17], PubChem [8] or ChEMBL [9]. The crystal structure of the glucose transporter contains a glycosyl analogue, b-nonyglucoside, used to trap the protein in an inward-open conformation [11]. This ligand has the identifier BNG in the PDB. When selecting the PDB input format and searching for BNG, a query on the PDB is performed to retrieve the ligand name, chemical formula, SMILES notation, molecular weight and InChiKey. The ideal three-dimensional coordinates of the molecule is downloaded in sdf format and added as the reference conformation for the ligand. By default the visibility of any ligand entered using the single ligand form is restricted to the project members only.

If multiple ligands are to be added rapidly, that is, from a commercial supplier, another possibility is to create a dedicated library from a multiple-compound sdf file (Figure 4C). For example, Siebeneicher and colleagues [14] have determined a series of GLUT inhibitors involving piperazine derivatives. These results are available in ChEMBL in the document report card CHEMBL3779893. Five compounds were downloaded in sdf format from the ChEMBL report (Table 1) and assembled into a single sdf file. This library was incorporated with a free text name using a file upload form (Figure 4C), resulting into the addition of the five new ligands into the database, joined into the "GLUT inhibitors 2016" library. To stress the system, the incorporation of a larger library of >11,000 compounds was assessed (data not shown). In this situation, the rate for ligand processing was about 100 ligands/s.

The library facility can also be used to regroup existing individual ligands into a coherent ensemble (Figure 4E,F). This approach allows to delineate sub ensemble of ligands for easier data extraction and analysis but has no strong dependence on the classified ligands. It is therefore easy to remove a library, this will not cascade to the removal of ligands or of ligands data.

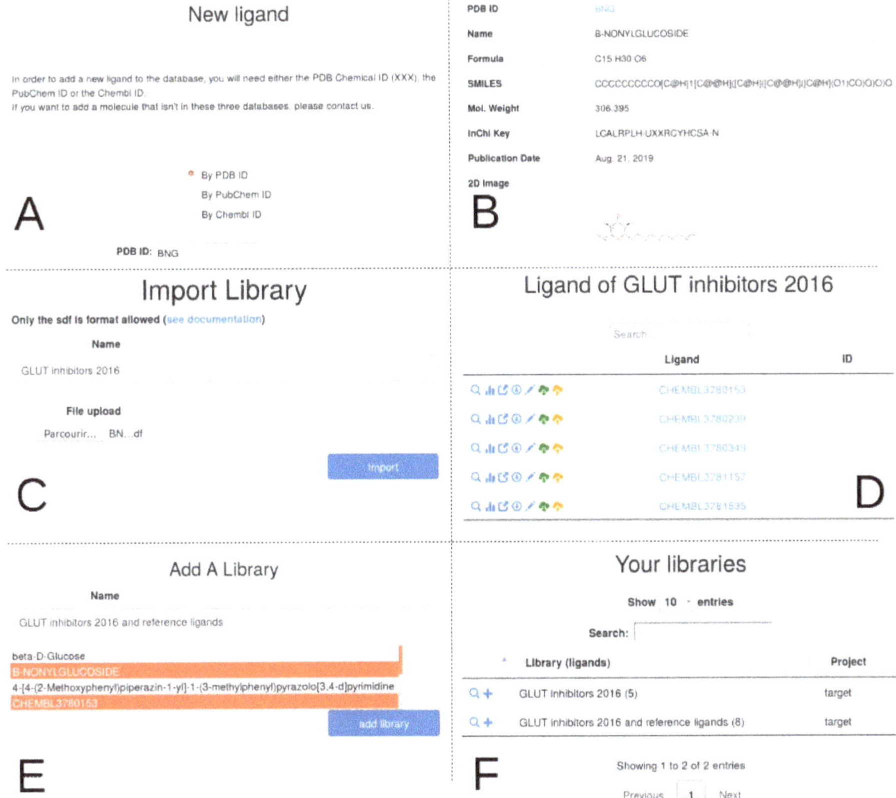

Figure 4. Additions of a single ligand or of multiple ligands into a library. (**A**) The form allows to add a single ligand from either the Protein Data Bank (PDB), PubChem or ChEBML. The query using the PDB request is shown here; (**B**) After a short period, the details of the added ligand can be accessed, if available, a 2D depiction of the molecule is displayed; (**C**) For more extensive data incorporation, the simplest way is to add a library and alongside a valid sdf file; (**D**) In this case all ligands available in the sdf file are processed, de-duplicated and added to the library; (**E**) If necessary, a small subset of ligands can be arranged in another library; (**F**) This new library will be referenced.

Table 1. Selected piperazine derivatives from the work of Siebeneicher and colleagues [14] and their reference in ChEMBL. The vina free energy value is indicated for the best cluster, see Supplementary Materials for calculation details.

Compound ID	IC50 (nM)	ChEMBL ID	PubChem ID	Vina Energy (kcal/mol)
13	1	3780239	72547759	−6.2
3	25	3781157	1977736	−7.4
66	80	3781535	127030174	−5.3
63	510	3780349	52149799	−6.3
41	44,000	3780153	127030188	−5.3

In both ligand import processes, it is important to not duplicate ligands even if different names are found, coming from the diversity of upstream sources. Instead of relying on ligand names, we have chosen a more robust approach by comparing the InChiKey [18] of new ligands to existing ones. If the InChiKey is not available, a Morgan fingerprint [19] is computed using rkdit [20] and used for comparison.

2.4. Docking Import And Management

The docking process needs a large amount of computational power to screen libraries of ligands against a given target. We envision to provide the possibility of performing some dockings through the interface of dockNmine in future revisions for a limited set of molecules but the potential ressources required are not yet available. Up to now, it is however possible to record already existing vina or autodock [21–23] results. This process already allows to standardize ligand import and management and to perform basic analysis. The course of action for pre-computed docking data is detailed hereafter.

The user needs to define the target, the ligand and the structure file used for docking. Depending on the docking software, autodock or vina, different information needs to be provided. The completed input form is presented in the Figure 3A. It is possible to not enter all docking parameters but it is recommended to add them all in order to be able to compare different docking experiments for the same partners (target+structure and ligand). Upon form submission, the pdbqt vina output file is parsed to extract the cluster number and its associated energy for each pose. Once processed, the user is redirected to the docking list page where he can inspect the incorporated docking by clicking on the magnifier icon.

This magnifier icon redirects to the details of the processed data (Figure 3B–D) where not only the docking parameters are listed but also the extracted energy by cluster (and/or pose if autodock vina is used). The cluster energy is transformed into kJ to allow a rapid comparison with other experiments and two indicators are also computed—Ligand Efficiency (LE) and Size-Independent Ligand Efficiency (SILE) [24]. These indicators were developed to compensate the tendency of large ligands to obtain better docking scores based on the ligand size rather than being effectively more active experimentally. These two measures are nowadays discussed or further explored [25] but we have chosen to not provide additional indicators in the table since more advanced features are available under the analysis tab.

This single-step ligand incorporation mechanism is perfectly fit to compare ligands or docking parameters for a small amount of compounds. Within minutes, it is possible to arrange properly and formally the docking data without expert needs. After the output ligand file is uploaded in pdbqt format, all other steps are automated to ease the user experience.

2.5. Experimental Data

Although virtual screening may be useful for finding the needle in a haystack, it is important to rely on experimental validation to assess the predictive power of calculations. The experimental tab allows to add to individual ligand experimental results for six different experiments—(i) IC50; (ii) hemaglutination; (iii) isothermal titration calorimetry (ITC); (iv) surface plasmon resonance (SPR); (v) fluorescence anisotropy; (vi) affinity chromatography (Figure 5). This initial list can be easily extended to adjust to a specific method but should already be generic enough to register most of experimental data. A large comment box allows to indicate the method in detail and/or the reference study. This free-text addition is important to keep track of a given laboratory result prior to publication, altogether with better procedures for chemical names and entities across partners of the project.

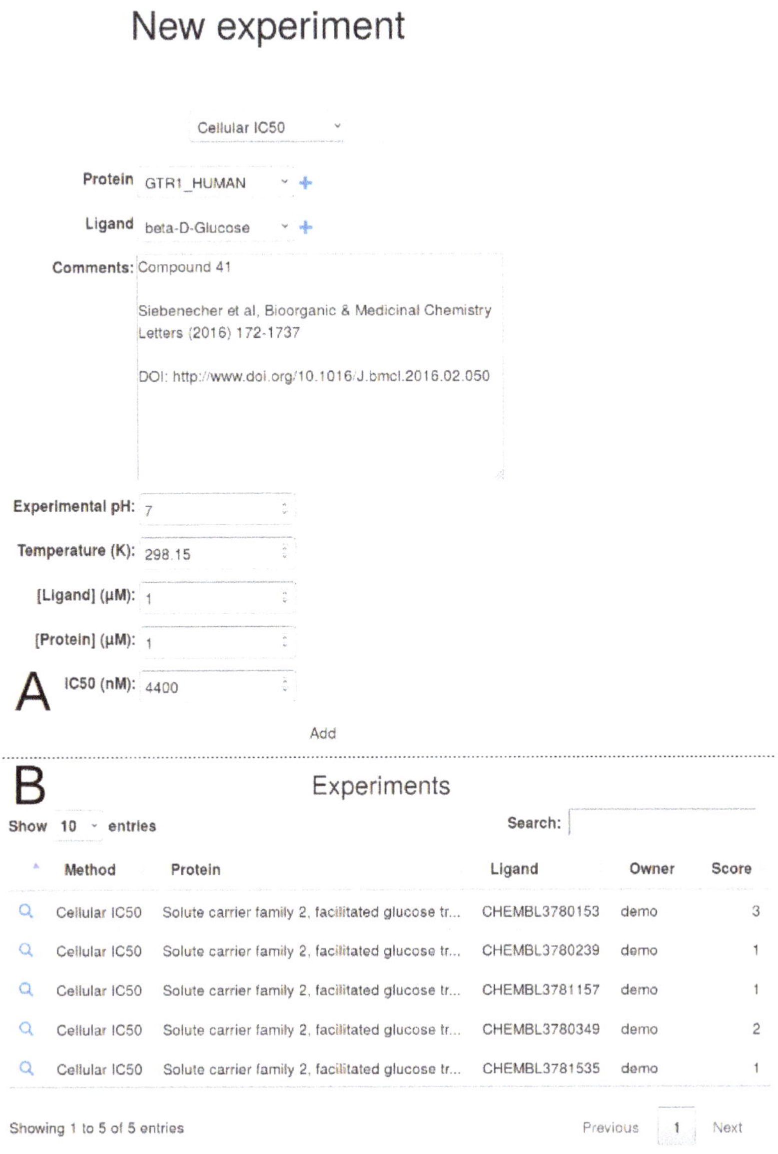

Figure 5. Experimental data addition for a selection of ligands from the study of Seibeneicher and co-workers [14]. (**A**) After IC50 selection from the drop-down menu, a method specific form is shown to the user. Pre-defined valued are provided for pH, temperature, target and ligand concentrations since they are seldomly used. The user can complete the free text box to indicate the data origin, either being from literature of from private laboratory experiments; (**B**) Upon form validation, all experimental data are listed, with an auto-computed normalised score important for comparison with virtual predictions.

2.6. Library Analysis

Virtual screening studies theoretically provide ligand-binding predictions in close agreement with known experimental data. It is however difficult to compare virtual results directly with experimental values. First, the free energy of binding is often tuned and estimated from a limited set of ligands (autodock, vina), which may be largely unrelated to the study of interest. Second, not a lot of study provide direct measure of ligand-protein interaction with a defined kD or ki. Third, even if the binding of a chemical entity is a direct measure, for instance using surface plasmon resonance (SPR) or thermocalorimetry (ITC) methods, the binding consequence can lead to different definitions, the ligand being classified at least as an agonist or antagonist, not counting the partial or reverse definitions and receptor allostery [26]. We have incorporated interactive indicators to provide a better insight into the predictive power of docking experiments (Figure 6). The Receiver Operating Characteristic analysis (ROC) [27] is useful in rapid evaluation of the docking performance by comparing True and False prediction rates. ROC curves are provided per ligand (Figure 6A) or synthetically for a given target in the detailed target page. For a more expert analysis of docking performance against experimental results, advanced features are displayed under the analysis menu. These analysis based on clinical epidemiology were defined by Empereur-Mot and his colleagues [28] and a demo is provided online (http://stats.drugdesign.fr/). For both analysis, it is important to rank the ligand according to the virtual and experimental results. We have defined three classes for ranking ligands—(i) good; (ii) intermediate; and (iii) bad. For docking results, the good category is reported if the autodock or vina free energy (kcal/mol) of the best cluster is ≤ -10, an intermediate ligand lies in the interval $-10 >$ energy ≤ -6.5 and a bad ligand has a lower free energy of binding. The corresponding thresholds for a IC50 (nM) result are ≤ 100 for a good ligand, >100 and ≤ 1000 for an intermediate ligand, and above 1000 for a bad ligand. Pre-defined thresholds are also provided for each experimental result which can be incorporated in dockNmine. The experimental and virtual categories are then compared to indicate if there is an agreement between predictions and results. This information is then transformed automatically for being displayed in the dedicated graphs.

2.7. Access Controls

Any user can freely discover dockNmine without authentication but a demonstration account is provided to evaluate all of its functionalities. Once connected or registered using the log-in or the briefcase glyphicons (Figure 1A), the user is attributed a **Project Manager** role (Table 2) and can therefore create a new project, mandatory to start adding data in dockNmine. Once set up, the Project Manager can share the project credentials with collaborators to allow them to join the project. The access controls systems is a combination of Guardian rules and of permissions offered by Django's internal group management system. A lot of predefined permissions can thus be finely tuned to Create, Read, Update and Delete data (CRUD) on any object or data in the database. Since this granularity may be hard to apprehend, we have set up pre-defined roles (Table 2) where these permissions are clearly split into viewers (Project Member) and editors roles (Project Manager) with a clear separation of privileges. To ease dockNmine usage, by default all new accounts have a Manager role but it is recommended to restrict this default role to **Member** for some collaborators in order to avoid errors. Only the principal dockNmine administrator has full control over the service and can adjust permissions for users, objects and data.

Before adding any information, connected users have to select an existing project. This mechanism allows to automatically define data visibility and origin, as all members of the project can see targets, ligands, dockings, experiments and their automated analysis. This per-project access allows to restrict data visibility to project members and accross projects. Pre-defined Manager and Member roles are indicated in the login page to demonstrate the difference between them. With a Member role, the user can add a ligand to the library but cannot upload a specific structure file. With a Manager role, the user can upload its own sdf file, carefully prepared using an external software or web service.

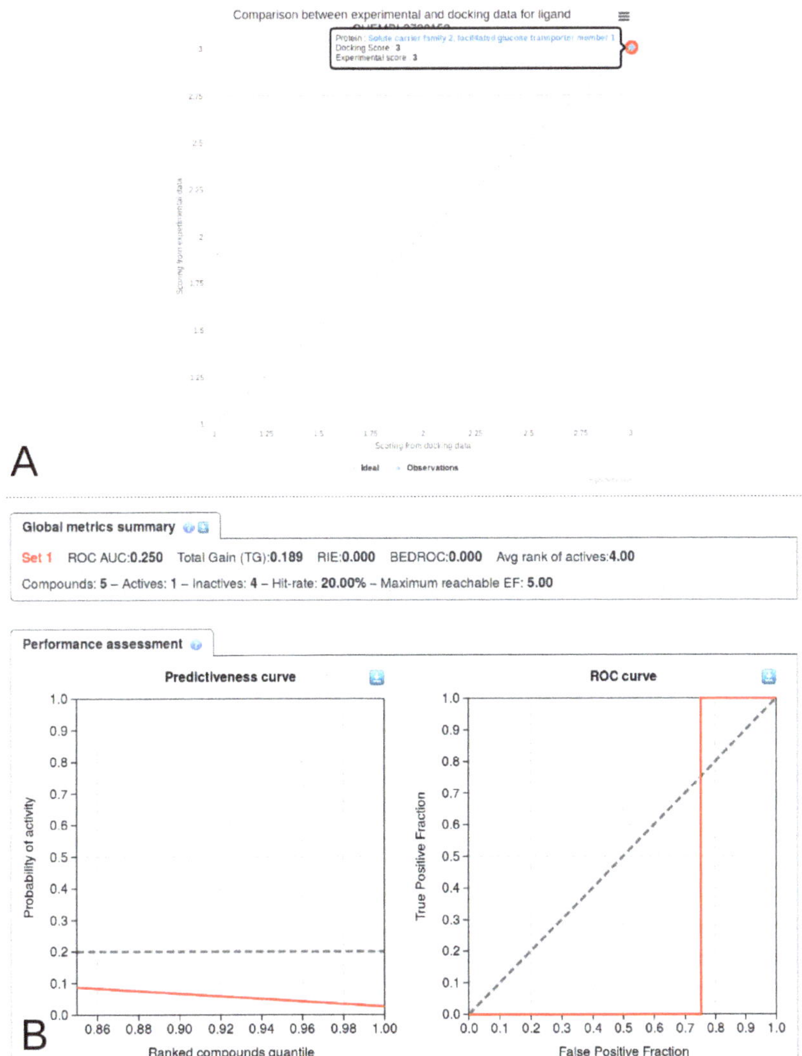

Figure 6. Analysis of ligand classification using reference methods. Experimental data were taken from the work of Seibeneicher and co-workers [14], the docking results were computed for this study. (**A**) Single ligand analysis for CHEMBL3780153. Both the experimental and docking values allow to classify it as a good ligand; (**B**) A more complete analysis of the overall virtual screening allows to evaluate the ongoing project evolution.

Table 2. CRUD permissions management in dockNmine.

	Project	Target	Ligand	Experimental Method	Docking	Library
SuperUser	CRUD	CRUD	CRUD	CRUD	CRUD	CRUD
Manager	CRU	CRU	CRU	CRU	CRU	CRU
Member	R	R	CR	CR	CR	CR
Anonymous	R	R	R	R	R	R

2.8. Extending Docknmine

This web server is meant to assemble various docking experiments and to compare it with existing or *in-house* experimental data. Since there is a large variety of virtual and experimental methods, it is possible to update dockNmine to take them into account. Some expertise is required in Django development so users are advised to first contact the corresponding author.

3. Discussion

One of the biggest challenges when trying to reach the precision medicine objectives, the goal to provide a per-individual efficient treatment, is the need to take into account little variation in protein sequences in order to predict if these mutations will lead to dramatic changes in the protein structure [29]. This research field, also known as structural genomics [30,31], is developing rapidly though many technological obstacles still need to be leveraged to be applied blindly [32]. In order to illustrate how dockNmine could properly be used to integrate results from these approaches, we have detailed its independent services allowing to assemble existing public and private knowledge into logically organised comprehensive data sets.

3.1. Single Protein Analysis

We chose the GLUT1 receptor (Solute carrier family 2, facilitated glucose transporter member 1) as an example. We indicated how to add this protein entry into the portal and how to add virtual and publicly available experimental data from databases and scientific literature [9,14]. We provided original docking data (see Supplementary Materials for details) to exemplify how one should make use of the portal. If one was willing to reproduce these steps, much more attention would have to be paid to the virtual screening experiments for getting relevant results [6]. Since these steps require manual expertise and computational time, we have not allowed docking computation to be performed directly within the portal. This situation may change but there are already efficient and popular solutions available for users interested in performing large virtual screening studies [33–36]. One of the critical steps for setting up a virtual screening approach is related to ligand preparation, classification and ranking. There are many challenges for each of these steps but again a lot of reliable solutions exist [37,38]. In the end, users can also make use of commercial software which provide a lot of facilities for setting up virtual screening studies, by pipe-lining all these steps silently.

3.2. Multiple Proteins Analysis

In order to get a broader overview of the GLUT family response to different molecules, we have processed two other glucose transporters, GLUT2 and GLUT3, for which experimental binding data are also provided in the reference article [14]. The crystallographic structure of GLUT3 was determined by Deng and colleagues (PDB ID: 4ZW9) [39], the model of GLUT2 was downloaded from Swiss-Model (automatically computed from the structure 4ZXC) [40]. We selected a crystallographic structure and a publicly available model to compute docking energies with vina for illustrating how one could compare his experimental data with predictions without advanced expertise on computational protein modelling. The incorporation of these proteins, of virtual screening data and of experimental information were processed as previously described. These results are available by ligand (Figure 7A) or by target (Figure 7B) from the drop-down options of the Analysis tab. These views present individual graph and table for dockings and experiments. For the ligand CHEMBL3780153, the experimental IC50 binding values are 4400 nM for GLUT1 and 2200 nM for GLUT3, no data being available for GLUT2. By selecting the GLUT2 receptor from the menu, the predicted binding value of the best pose is -8.3 kcal/mol, ranking it as a medium binder on the target according to our ranking procedure. Since the docking value allows to rank this ligand in the lowest binding category (rank 3) for both GLUT1 and GLUT3, in agreement with experimental data, this rapid comparison suggests a better recognition of CHEMBL3780153 by GLUT2. This simple comparison of docking and experimental results could be exact but the user is warned about the limited predictive power of the included dataset, with a negative predictiveness and an Area Under the Curve (AUC) under 0.5 for GLUT1 (Figure 6B).

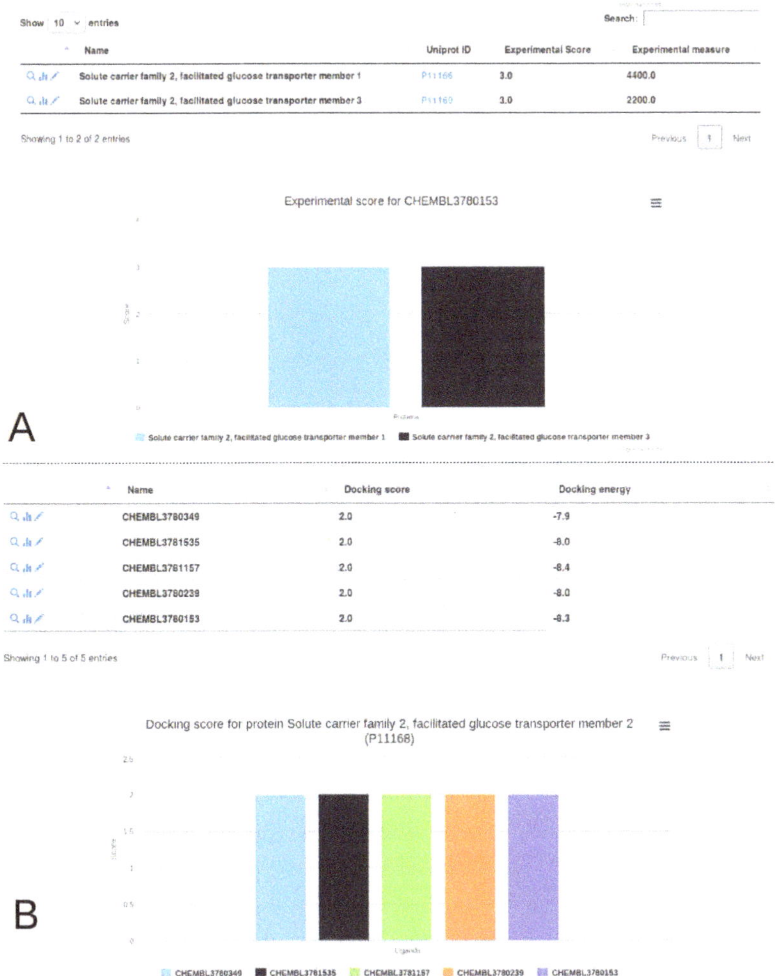

Figure 7. Comparison of experimental and virtual data for GLUT proteins and their ligands. Experimental data from Seibeneicher and co-workers [14], the docking results were computed for this study. (**A**) Comparison of ligand results for CHEMBL3780153. The docking was performed on all proteins but experimental values are only available for GLUT1 and GLUT3; (**B**) Tabular results and graphical representation of docking results for GLUT2.

To move towards precision medicine, dockNmine can act as a central gathering portal to add much more experimental and docking information. In the case of the GLUT members, this would required adding docking and experimental data for the 14 members of the SLC2 family, from the recently published study in Reference [41,42].

3.3. Advanced Analysis

The provided analysis pages are simple and standard methods for comparing docking and experimental data. Since these views may not be sufficient, we offer to download the complete data set in the detailed target and ligand pages, by clicking on the download glyphicon. In this case, docking and experimental data are arranged into a convenient csv file for further processing in any spreadsheet.

The existing portal already allows to assemble a lot of knowledge seamlessly. As more information may be required for further deciphering protein-ligand interaction, for instance for advanced machine learning processing [43], new features shall be incorporated [44]. Some of future improvements may comprise the addition of direct docking computation from the interface, advanced protein-ligand analysis [45] and visualisation [46] and other functionalities demanded by users.

4. Materials and Methods

4.1. Server Design, Implementation And Security

The django framework (https://www.djangoproject.com/) was used to arrange data into dedicated classes. Access controls are ensured by django's built-in system supplemented by the guardian module to provide per-object control. Each page in the interface is submitted to permission validation ensured by a dedicated decorator developed specifically for this purpose. User interaction and interactive displays are provided using Bootstrap (http://getbootstrap.com/) and jQuery (http://jquery.com/). The specific protein-ligand global statistics analysis is derived from the work of Empereur-Mot and collaborators [28].

4.2. Data Retrieval And Processing

Queries on external databases are executed using the python3 requests module. When available, public API are used like for ChEMBL for instance [9], otherwise simple HTTP request are performed. Queries are then processed using biopython [47] for proteins and rdkit [20] for small chemical entities.

5. Conclusions

With the need to address large-scale, diverse and targeted protein-ligand interaction predictions, it is essential to be able to quickly assemble public and private experimental and virtual data. The dockNmine portal aims at providing the first component for this ambitious goal; it is freely accessible at the http://www.ufip.univ-nantes.fr/tools/docknmine/.

Supplementary Materials: The following are available online at http://www.mdpi.com/1422-0067/20/20/5062/s1, Supplemental Methods S1: Receptor and ligand setup.

Author Contributions: Conceptualization, S.T.; methodology, E.G., A.L., R.L. and S.T.; software, J.L., C.R., A.A., A.L.; validation, R.L., E.G. and S.T.; formal analysis, C.R. and A.A.; investigation, S.T.; resources, S.T.; data curation, C.R. and A.A.; writing–original draft preparation, R.L. and S.T.; writing–review and editing, E.G. and S.T.; visualization, R.L. and S.T.; supervision, S.T.; project administration, S.T.; funding acquisition, S.T.

Funding: This research is part of the PIRAMID and TROPIC projects which received financial support from the ANR grant "Programme d'investissements d'Avenir" (ANR-16-IDEX-0007), from Région Pays de La Loire and from Nantes Métropole.

Acknowledgments: The authors thank 'Centre de Calcul Intensif des Pays de la Loire (CCIPL)" for computational ressources.

Conflicts of Interest: The authors declare no conflict of interest.

Abbreviations

The following abbreviations are used in this manuscript:

AUC	Area Under the Curve
CADD	Computer-Aided Drug Design
CRUD	Create, Read, Update, Delete

ITC	isothermal titration calorimetry
LE	Ligand Efficiency
NMR	Nuclear Magnetic Resonance
PDB	Protein Data Bank
ROC	Receiver Operating Characteristics
SILE	Size-Independent Ligand Efficiency
SPR	Surface Plasmon Resonance

References

1. Pacanowski, M.; Huang, S.M. Precision Medicine. *Clin. Pharmacol. Ther.* **2016**, *99*, 124–129. [CrossRef]
2. Schütte, M.; Ogilvie, L.A.; Rieke, D.T.; Lange, B.M.H.; Yaspo, M.L.; Lehrach, H. Cancer Precision Medicine: Why More Is More and DNA Is Not Enough. *Public Health Genom.* **2017**, *20*, 70–80. [CrossRef] [PubMed]
3. Marzagalli, M.; Raimondi, M.; Fontana, F.; Montagnani Marelli, M.; Moretti, R.M.; Limonta, P. Cellular and molecular biology of cancer stem cells in melanoma: Possible therapeutic implications. *Semin. Cancer Biol.* **2019**. [CrossRef]
4. Zaman, A.; Wu, W.; Bivona, T.G. Targeting Oncogenic BRAF: Past, Present, and Future. *Cancers* **2019**, *11*, 1197. [CrossRef] [PubMed]
5. Villoutreix, B.O.; Eudes, R.; Miteva, M.A. Structure-Based Virtual Ligand Screening: Recent Success Stories. *Comb. Chem. High Throughput Screen.* **2009**, *12*, 1000–10016. [CrossRef] [PubMed]
6. Forli, S. Charting a Path to Success in Virtual Screening. *Molecules* **2015**, *20*, 18732–18758. [CrossRef]
7. Rognan, D. Proteome-scale docking: Myth and reality. *Drug Discov. Today Technol.* **2013**, *10*, e403–e409. [CrossRef]
8. Kim, S.; Chen, J.; Cheng, T.; Gindulyte, A.; He, J.; He, S.; Li, Q.; Shoemaker, B.A.; Thiessen, P.A.; Yu, B.; et al. PubChem 2019 update: Improved access to chemical data. *Nucleic Acids Res.* **2019**, *47*, D1102–D1109. [CrossRef]
9. Gaulton, A.; Hersey, A.; Nowotka, M.; Bento, A.P.; Chambers, J.; Mendez, D.; Mutowo, P.; Atkinson, F.; Bellis, L.J.; Cibrián-Uhalte, E.; et al. The ChEMBL database in 2017. *Nucleic Acids Res.* **2017**, *45*, D945–D954. [CrossRef]
10. Guertin, D.A.; Sabatini, D.M. Defining the Role of mTOR in Cancer. *Cancer Cell* **2007**, *12*, 9–22. [CrossRef]
11. Deng, D.; Xu, C.; Sun, P.; Wu, J.; Yan, C.; Hu, M.; Yan, N. Crystal structure of the human glucose transporter GLUT1. *Nature* **2014**, *510*, 121–125. [CrossRef] [PubMed]
12. Galochkina, T.; Chong, M.N.F.; Challali, L.; Abbar, S.; Etchebest, C. New insights into GluT1 mechanics during glucose transfer. *Sci. Rep.* **2019**, *9*, 1–14. [CrossRef] [PubMed]
13. Téletchéa, S.; Santuz, H.; Léonard, S.; Etchebest, C. Repository of Enriched Structures of Proteins Involved in the Red Blood Cell Environment (RESPIRE). *PLoS ONE* **2019**, *14*, e0211043. [CrossRef] [PubMed]
14. Siebeneicher, H.; Bauser, M.; Buchmann, B.; Heisler, I.; Müller, T.; Neuhaus, R.; Rehwinkel, H.; Telser, J.; Zorn, L. Identification of novel GLUT inhibitors. *Bioorganic Med. Chem. Lett.* **2016**, *26*, 1732–1737. [CrossRef] [PubMed]
15. Davies, M.; Nowotka, M.; Papadatos, G.; Dedman, N.; Gaulton, A.; Atkinson, F.; Bellis, L.; Overington, J.P. ChEMBL web services: Streamlining access to drug discovery data and utilities. *Nucleic Acids Res.* **2015**, *43*, W612–W620. [CrossRef]
16. Nowotka, M.M.; Gaulton, A.; Mendez, D.; Bento, A.P.; Hersey, A.; Leach, A. Using ChEMBL web services for building applications and data processing workflows relevant to drug discovery. *Expert Opin. Drug Discov.* **2017**, *12*, 757–767. [CrossRef]
17. Burley, S.K.; Berman, H.M.; Bhikadiya, C.; Bi, C.; Chen, L.; Costanzo, L.D.; Christie, C.; Duarte, J.M.; Dutta, S.; Feng, Z.; et al. Protein Data Bank: The single global archive for 3D macromolecular structure data. *Nucleic Acids Res.* **2019**, *47*, D520–D528. [CrossRef]
18. Heller, S.R.; McNaught, A.; Pletnev, I.; Stein, S.; Tchekhovskoi, D. InChI, the IUPAC International Chemical Identifier. *J. Cheminform.* **2015**, *7*, 23. [CrossRef]
19. Rogers, D.; Hahn, M. Extended-Connectivity Fingerprints. *J. Chem. Inf. Model.* **2010**, *50*, 742–754. [CrossRef]
20. RDKit: Open-Source Cheminformatics. Available online: http://www.rdkit.org (accessed on 28 August 2019).

21. Morris, G.M.; Huey, R.; Lindstrom, W.; Sanner, M.F.; Belew, R.K.; Goodsell, D.S.; Olson, A.J. AutoDock4 and AutoDockTools4: Automated docking with selective receptor flexibility. *J. Comput. Chem.* **2009**, *30*, 2785–2791. [CrossRef]

22. Trott, O.; Olson, A.J. AutoDock Vina: Improving the speed and accuracy of docking with a new scoring function, efficient optimization, and multithreading. *J. Comput. Chem.* **2009**, *31*, 455–461. [CrossRef] [PubMed]

23. Cosconati, S.; Forli, S.; Perryman, A.L.; Harris, R.; Goodsell, D.S.; Olson, A.J. Virtual screening with AutoDock: Theory and practice. *Expert Opin. Drug Discov.* **2010**, *5*, 597–607. [CrossRef] [PubMed]

24. Nissink, J.W.M. Simple Size-Independent Measure of Ligand Efficiency. *J. Chem. Inf. Model.* **2009**, *49*, 1617–1622. [CrossRef] [PubMed]

25. Kenny, P.W. The nature of ligand efficiency. *J. Cheminform.* **2019**, *11*, 8. [CrossRef] [PubMed]

26. Vecchio, E.A.; Baltos, J.A.; Nguyen, A.T.N.; Christopoulos, A.; White, P.J.; May, L.T. New paradigms in adenosine receptor pharmacology: Allostery, oligomerization and biased agonism. *Br. J. Pharmacol.* **2018**, *175*, 4036–4046. [CrossRef] [PubMed]

27. Triballeau, N.; Acher, F.; Brabet, I.; Pin, J.P.; Bertrand, H.O. Virtual Screening Workflow Development Guided by the "Receiver Operating Characteristic" Curve Approach. Application to High-Throughput Docking on Metabotropic Glutamate Receptor Subtype 4. *J. Med. Chem.* **2005**, *48*, 2534–2547. [CrossRef]

28. Empereur-mot, C.; Guillemain, H.; Latouche, A.; Zagury, J.F.; Viallon, V.; Montes, M. Predictiveness curves in virtual screening. *J. Cheminform.* **2015**, *7*, 52. [CrossRef]

29. Lam, S.D.; Das, S.; Sillitoe, I.; Orengo, C. An overview of comparative modelling and resources dedicated to large-scale modelling of genome sequences. *Acta Crystallogr. Sect. D* **2017**, *73*, 628–640. [CrossRef]

30. Goodwin, S.; McPherson, J.D.; McCombie, W.R. Coming of age: Ten years of next-generation sequencing technologies. *Nat. Rev. Genet.* **2016**, *17*, 333–351. [CrossRef]

31. Marschall, T.; Marz, M.; Abeel, T.; Dijkstra, L.; Dutilh, B.E.; Ghaffaari, A.; Kersey, P.; Kloosterman, W.P.; Mäkinen, V.; Novak, A.M.; et al. Computational pan-genomics: Status, promises and challenges. *Brief. Bioinform.* **2018**, *19*, 118–135. [CrossRef]

32. Kuhlman, B.; Bradley, P. Advances in protein structure prediction and design. *Nat. Rev. Mol. Cell Biol.* **2019**, 1–17. [CrossRef] [PubMed]

33. Irwin, J.J.; Shoichet, B.K.; Mysinger, M.M.; Huang, N.; Colizzi, F.; Wassam, P.; Cao, Y. Automated Docking Screens: A Feasibility Study. *J. Med. Chem.* **2009**, *52*, 5712–5720. [CrossRef] [PubMed]

34. Coleman, R.G.; Sharp, K.A. Protein Pockets: Inventory, Shape, and Comparison. *J. Chem. Inf. Model.* **2010**, *50*, 589–603. [CrossRef] [PubMed]

35. Bullock, C.; Cornia, N.; Jacob, R.; Remm, A.; Peavey, T.; Weekes, K.; Mallory, C.; Oxford, J.T.; McDougal, O.M.; Andersen, T.L. DockoMatic 2.0: High Throughput Inverse Virtual Screening and Homology Modeling. *J. Chem. Inf. Model.* **2013**, *53*, 2161–2170. [CrossRef] [PubMed]

36. Dallakyan, S.; Olson, A.J. Small-Molecule Library Screening by Docking with PyRx. In *Chemical Biology: Methods and Protocols*; Hempel, J.E., Williams, C.H., Hong, C.C., Eds.; Methods in Molecular Biology; Springer: New York, NY, USA, 2015; pp. 243–250. [CrossRef]

37. Backman, T.W.H.; Cao, Y.; Girke, T. ChemMine tools: An online service for analyzing and clustering small molecules. *Nucleic Acids Res.* **2011**, *39*, W486–W491. [CrossRef] [PubMed]

38. Capuzzi, S.J.; Kim, I.S.J.; Lam, W.I.; Thornton, T.E.; Muratov, E.N.; Pozefsky, D.; Tropsha, A. Chembench: A Publicly Accessible, Integrated Cheminformatics Portal. *J. Chem. Inf. Model.* **2017**, *57*, 105–108. [CrossRef]

39. Deng, D.; Sun, P.; Yan, C.; Ke, M.; Jiang, X.; Xiong, L.; Ren, W.; Hirata, K.; Yamamoto, M.; Fan, S.; Yan, N. Molecular basis of ligand recognition and transport by glucose transporters. *Nature* **2015**, *526*, 391–396. [CrossRef]

40. Waterhouse, A.; Bertoni, M.; Bienert, S.; Studer, G.; Tauriello, G.; Gumienny, R.; Heer, F.T.; de Beer, T.A.P.; Rempfer, C.; Bordoli, L.; et al. SWISS-MODEL: Homology modelling of protein structures and complexes. *Nucleic Acids Res.* **2018**, *46*, W296–W303. [CrossRef]

41. Schmidl, S.; Iancu, C.V.; Choe, J.Y.; Oreb, M. Ligand Screening Systems for Human Glucose Transporters as Tools in Drug Discovery. *Front. Chem.* **2018**, *6*, 183. [CrossRef]

42. Holman, G.D. Chemical biology probes of mammalian GLUT structure and function. *Biochem. J.* **2018**, *475*, 3511–3534. [CrossRef]

43. Pedregosa, F.; Varoquaux, G.; Gramfort, A.; Michel, V.; Thirion, B.; Grisel, O.; Blondel, M.; Prettenhofer, P.; Weiss, R.; Dubourg, V.; et al. Scikit-learn: Machine Learning in Python. *J. Mach. Learn. Res.* **2011**, *12*, 2825–2830.

44. Wójcikowski, M.; Zielenkiewicz, P.; Siedlecki, P. Open Drug Discovery Toolkit (ODDT): A new open-source player in the drug discovery field. *J. Cheminform.* **2015**, *7*, 26. [CrossRef] [PubMed]

45. Salentin, S.; Schreiber, S.; Haupt, V.J.; Adasme, M.F.; Schroeder, M. PLIP: Fully automated protein–ligand interaction profiler. *Nucleic Acids Res.* **2015**, *43*, W443–W447. [CrossRef] [PubMed]

46. Li, H.; Leung, K.S.; Nakane, T.; Wong, M.H. iview: An interactive WebGL visualizer for protein-ligand complex. *BMC Bioinform.* **2014**, *15*, 56. [CrossRef] [PubMed]

47. Cock, P.J.A.; Antao, T.; Chang, J.T.; Chapman, B.A.; Cox, C.J.; Dalke, A.; Friedberg, I.; Hamelryck, T.; Kauff, F.; Wilczynski, B.; et al. Biopython: Freely available Python tools for computational molecular biology and bioinformatics. *Bioinformatics* **2009**, *25*, 1422–1423. [CrossRef]

International Journal of
Molecular Sciences

Article

Virtual Screening Using Pharmacophore Models Retrieved from Molecular Dynamic Simulations

Pavel Polishchuk [1],* , Alina Kutlushina [1], Dayana Bashirova [2], Olena Mokshyna [1] and Timur Madzhidov [2]

[1]   Institute of Molecular and Translational Medicine, Faculty of Medicine and Dentistry, Palacky University and University Hospital in Olomouc, Hnevotinska 5, 77900 Olomouc, Czech Republic; alina.kutlushina@upol.cz (A.K.); olena.mokshyna@upol.cz (O.M.)

[2]   A.M. Butlerov Institute of Chemistry, Kazan Federal University, Kremlyovskaya Str. 18, 420008 Kazan, Russia; dayana.bashirova@yandex.ru (D.B.); timur.madzhidov@kpfu.ru (T.M.)

*   Correspondence: pavlo.polishchuk@upol.cz; Tel.: +420-585632298

Received: 2 November 2019; Accepted: 18 November 2019; Published: 20 November 2019

Abstract: Pharmacophore models are widely used for the identification of promising primary hits in compound large libraries. Recent studies have demonstrated that pharmacophores retrieved from protein-ligand molecular dynamic trajectories outperform pharmacophores retrieved from a single crystal complex structure. However, the number of retrieved pharmacophores can be enormous, thus, making it computationally inefficient to use all of them for virtual screening. In this study, we proposed selection of distinct representative pharmacophores by the removal of pharmacophores with identical three-dimensional (3D) pharmacophore hashes. We also proposed a new conformer coverage approach in order to rank compounds using all representative pharmacophores. Our results for four cyclin-dependent kinase 2 (CDK2) complexes with different ligands demonstrated that the proposed selection and ranking approaches outperformed the previously described common hits approach. We also demonstrated that ranking, based on averaged predicted scores obtained from different complexes, can outperform ranking based on scores from an individual complex. All developments were implemented in open-source software pharmd.

Keywords: pharmacophore; molecular dynamics; virtual screening

1. Introduction

Pharmacophore models are widely used in the early stages of drug development to identify potential hits in large datasets. These models encode spatial arrangements of features which are important for protein–ligand interactions and can be derived from available three-dimensional (3D) structures of protein-ligand complexes. The X-ray structures of complexes from the Protein Data Bank [1] are usually used for structure-based pharmacophore modeling. However, X-ray structures represent only a static view and can fail to describe the complexity of ligand–protein interactions. Protein-ligand complexes are inherently flexible species and their dynamic behavior greatly determines protein-ligand recognition. Molecular dynamics (MD) is a well-established approach for the simulation of the flexibility of large molecular systems and is widely used for the investigation of protein-ligand complexes' dynamic behavior. MD simulations act as a rich source of information about studied systems, and thus can be used for drug design purposes. In particular, ensemble docking [2,3] employs individual snapshots of MD trajectory.

In several recent studies, researchers applied pharmacophore modeling for MD trajectory analysis. Choudhury et al. derived models from snapshots of a 40 ns trajectory and validated them on the external set of known active and inactive compounds to select the most reasonable pharmacophores [4]. They obtained only eight pharmacophores by selecting snapshots every 5 ns of the trajectory. Such amounts of pharmacophores are not only unrepresentative, but also this approach is applicable only if there are enough data on known active and inactive compounds for model validation and selection (because a priori is impossible to estimate the usefulness of models for virtual screening). Other researchers have clustered MD trajectories to select representative pharmacophore models [5,6] which reduced computational complexity due to fewer models. However, such approaches depend on a chosen clustering algorithm and its tuning parameters and can overlook some less populated states, which might be important for ligand-receptor recognition. Each of these approaches also requires datasets of known compounds to validate and select the most appropriate and accurate models.

Recently, Wieder et al. proposed the "common hits approach" (CHA) which requires no information about known ligands to validate and select predictive pharmacophore models [7]. They proposed the use of all representative pharmacophore models retrieved from a single MD trajectory of a protein-ligand complex to rank compounds according to the number of matched models. They demonstrated high performance of the CHA on a number of protein-ligand complexes. Nevertheless, the proposed selection procedure of representative pharmacophore models in that study has some weaknesses. The authors retrieved 20,000 MD trajectory snapshots and the corresponding number of pharmacophore models. To select representative pharmacophore models they grouped all models according to the number and types of pharmacophore features. The energy of ligand conformations corresponding to each pharmacophore model was calculated with the Merck Molecular Force Field (MMFF). A conformer with median energy was identified within each group and the corresponding pharmacophore model was selected as representative. The spatial arrangement of features was ignored because pharmacophore models were grouped only by the type and the number of pharmacophore features. Therefore, dissimilar pharmacophores with substantially different geometry but the same set of features can get to the same group, which will not correspond to a single representative model.

Nevertheless, there is a need to develop a stable approach capable of selecting representative pharmacophore models with a minimal number of tuning parameters. In this study, we used previously developed 3D pharmacophore hashes [8] which were able to identify identical pharmacophore models within a given binning step. A 3D pharmacophore hash is a unique identifier of a pharmacophore that takes into account distances between features and their spatial arrangement, including stereoconfiguration. A binning step, the only sensitive tuning parameter, was used for discretization of interfeature distances to enable fuzzy matching of pharmacophores by calculated hashes. The removal of pharmacophores with duplicated hashes reduced the whole set of pharmacophore models retrieved from the MD trajectory to a subset of representative ones, further used for virtual screening.

We also proposed a new approach of compound ranking, called the "conformers coverage approach" (CCA). Similar to the common hits approach, it uses all representative pharmacophore models, and therefore does not require validation and selection of individual models based on sets of known active and inactive compounds. Supposedly, if a greater number of existing compound conformers can fit protein conformational states, the more favorable binding would be observed as flexibility of a compound that better corresponds to flexibility of a protein (thus the ligand would lose fewer degrees of freedom upon a binding event and less binding entropy decrease may be observed). In the case when multiple complexes of a protein with different ligands are available, a consensus ranking can be performed by averaging CCA scores across different complexes. We also demonstrated the influence of pharmacophore model complexity represented by the number of features on virtual screening performance. The proposed approach to retrieve pharmacophore models from MD trajectory and virtual screening was implemented in open-source software available on GitHub (https://github.com/ci-lab-cz/pharmd).

2. Materials and Methods

2.1. Protein Target and Compound Dataset

We chose cyclin-dependent kinase 2 (CDK2) as a protein target due to the abundance of X-ray structures of this protein in complexes with small molecules. Among the available PDB X-ray complexes, we selected four with high affinity inhibitors (2C6O, 2FVD, 2XMY, and 5D1J) (Figure 1). The corresponding dataset of known inhibitors and decoys from the DUD-E dataset [9] was used for the validation of developed pharmacophore models. After a thorough check, all duplicates were removed from the validation set. In addition, the ligands presented in the selected four complexes were removed from the DUD-E dataset to avoid overestimation of model performance. The final dataset contained 473 active compounds and 27,853 decoys. For compounds with an undefined configuration of stereocenters or double bonds, all possible stereoisomers were enumerated. For each stereoisomer of a compound, up to 100 conformers were generated, within the energy gap 100 kcal/mol from the lowest energy conformer, using the MMFF force field implemented in RDKit [10]. Such a large energy gap was deliberately chosen to cover larger conformational space because large and flexible compounds with polar or charged groups can form bent conformations, where oppositely charged groups are close together. In addition, we considered conformers with a root-mean-square distance less than 0.5 Å as duplicates and removed them.

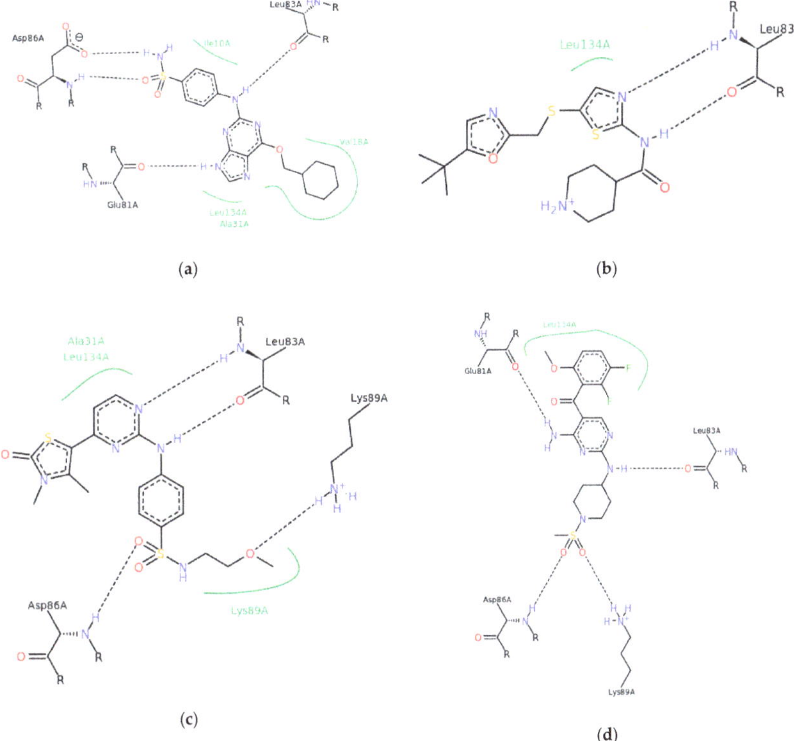

Figure 1. Protein–ligand interaction charts of four selected cyclin-dependent kinase 2 (CDK2) complexes. (a) IC_{50} = 5–8.1 nM [11–13] 2C6O; (b) IC_{50} = 38–46 nM [14,15] 5D1J; (c) K_i = 0.11 nM [16] 2XMY; (d) K_i = 3 nM [17] 2FVD.

2.2. Molecular Dynamic Simulations

All molecular dynamics simulations were done using GROMACS 2016 with GPU support [18,19]. First, protein and ligand topologies were prepared. In the original 2XMY structure from PDB, two possible ligand structures are overlapped. They have extremely low RMSD, therefore, we arbitrarily chose structure A. For protein topology generation, we used the Amber99SB-ILDN force field [20]. The ligand topologies were prepared with an Antechamber 17.3 [21,22] using GAFF2 force field parameters and checked using parmchk utility and manually.

Each protein-ligand system was placed in a dodecahedron water cell with a minimal distance to the cell wall of 1 Å. The TIP3P [23] model was used for water description. The maximum number of steps for energy minimization was 50,000, but for all four complexes, the steepest descent converged at approximately 1000 steps. After energy minimization, each system was NVT and NPT equilibrated (100 ps per each equilibration) following guidelines published by Justin Lemkul [24]. Then, equilibrated protein-ligand complexes were simulated under NPT ensemble with a V-rescale thermostat and a Parrinello-Rahman barostat at 310 K for 50 ns with 2 fs time step. The temperature (during NVT equilibration) and density (during NPT equilibration) were carefully monitored and found acceptably stable. The simulations' convergence was analyzed using RMSD and gyration radius plots, as well as by temperature and density as additional parameters (see Supplementary Materials).

2.3. Pharmacophore Model Retrieval

Individual snapshots of each 20 ps of MD trajectory of a protein-ligand complex were extracted to PDB files using MDTraj library [25]. A total of 2500 snapshots were retrieved from the MD trajectory of each complex. Water molecules were removed because we were interested in the identification of only direct ligand–protein interactions, and also this significantly sped up pharmacophore recognition. From each snapshot, we retrieved a pharmacophore model by the identification of hydrogen bonds, hydrophobic and aromatic interaction centers between protein and ligand, using PLIP library [26]; electrostatic interactions were identified as short contacts (less than 3.8 Å) between the side chains of charged amino acids (Glu, Asp, Lys, Arg, and His) and oppositely charged ligands. The assignment of pharmacophore features was refined to satisfy pharmacophore feature patterns implemented in pmapper [8] which was further used to derive 3D pharmacophore hashes from individual pharmacophore models. The binning step and tolerance for the calculation of 3D pharmacophore hashes were set to default 1 Å and 0, respectively. The former represents how models tolerate deviation of distances between features, and the later represents the tolerance to deviation of pharmacophore quadruplets from planarity to calculate stereoconfiguration of a pharmacophore [8]. In our previous study, we observed an extremely weak effect from changing the tolerance parameter, and therefore it was set to 0. The pharmacophore features (such as H-bond donors and acceptors) were undirected due to shortcomings of the current pmapper implementation. Because models with undirected features are less specific, this could lead to somewhat lower hit rates and enrichment. More details about the computing of 3D pharmacophore hashes can be obtained by referring to our previous publication [8].

2.4. Virtual Screening With Ensembles of MD-Based Pharmacophore Models

We reduced the number of considered pharmacophore models in each ensemble to representative ones by removing duplicates, i.e., pharmacophore models with identical hashes. Individual representative pharmacophore models were screened on the DUD-E dataset. Finally, compounds were ranked according to two strategies, common hits approach (CHA) and conformers coverage approach (CCA) (Figure 2). Within the CHA strategy proposed by Wieder at el., compounds were ranked according to the percentage of representative pharmacophore models matching at least one compound conformer, which was equivalent to the number of models matching the given compound from the original study [7]. It was suggested that active compounds should have a greater number of matched models. Within the proposed CCA strategy, compounds were ranked according to the percentage of conformers matching at least one representative pharmacophore model. We suggested that the compounds in which conformers fit more frequently to the pharmacophores observed within the MD simulations of a protein-ligand complex could have a more favorable binding due to less decrease in binding entropy. The more ligand conformers could fit the observed conformational states of a protein, the fewer degrees of freedom of a ligand would be lost upon a binding event.

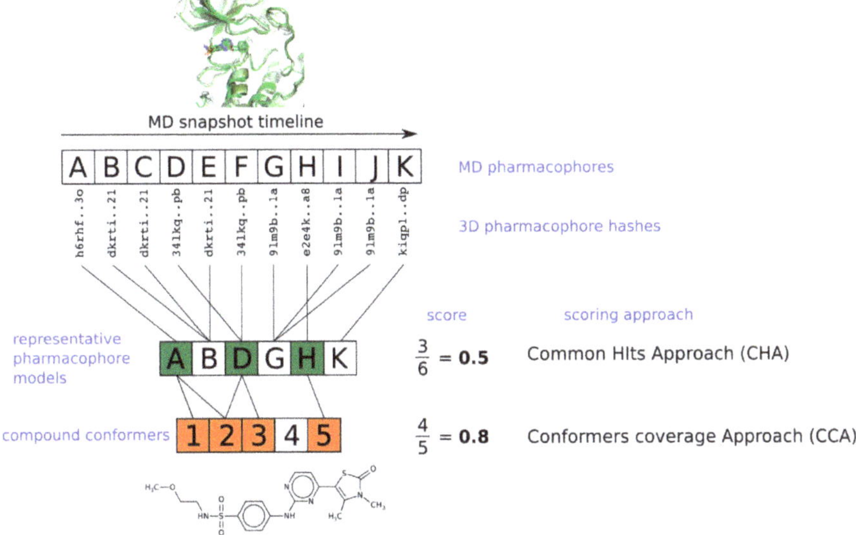

Figure 2. Compound scoring schemes based on the proposed conformers coverage approach and the previously developed common hits approach. Distinct representative pharmacophore models were selected among all molecular dynamics (MD) pharmacophores based on their three-dimensional (3D) pharmacophore hashes.

To estimate screening performance, we calculated the precision, Equation (1), and the enrichment factor, Equation (2) which are the most important screening parameters, as the models should result in the lowest possible number of false positives and demonstrate enrichment over random selection. Enrichment was calculated at different percentages of selected compounds as follows: 0.25%, 0.5%, 1%, 2%, 5%, 10%, and 100%. Basically, we selected the specified percentage of compounds and all compounds having a score identical to the last compound in the list, and therefore the actual number of compounds could be greater than the given percentage. In addition, if the compounds retrieved by pharmacophore models was fewer than the given percentage we used only the retrieved compounds to calculate the statistics because the remaining compounds could not be ranked reasonably. All enumerated stereoisomers of a compound were treated as a single compound during virtual screening.

$$\text{precision} = TP/(TP + FP) \tag{1}$$

$$\text{enrichment factor} = \text{precision}/\text{baseline precision} \tag{2}$$

where TP is a number of true hits retrieved by a model, FP is a number of decoys retrieved by a model, baseline precision is calculated according to Equation (1) where all hits were considered as true positives and all decoys as false positives. The baseline precision for the DUD-E dataset was 0.0167.

3. Results and Discussion

A total of 2500 frames were extracted from each MD trajectory of four complexes and the corresponding number of structure-based pharmacophore models was derived. Three-dimensional pharmacophore hashes were calculated for each pharmacophore to identify highly similar ones. By design, the pharmacophores with identical hashes should have a root-mean-square distance (RMSD) within the chosen binning step. In order to verify this, we aligned pairs of pharmacophore models with identical sets of features and calculated best-fit RMSD values. As expected, pharmacophores having identical hashes have a distribution of RMSD values from 0 to 0.93 Å across all four protein targets, whereas RMSD values for pairs of pharmacophores having different hashes were distributed in a wider range, from 0.01 to 4.96 Å (Figure 3). This indicates an important feature, i.e., identical 3D pharmacophore hashes always correspond to similar pharmacophores, however, similar pharmacophores do not always have identical hashes. This means that by removing pharmacophores with identical hashes we achieved the main purpose of reducing the number of pharmacophores, although keeping some redundancy among remaining representative pharmacophores.

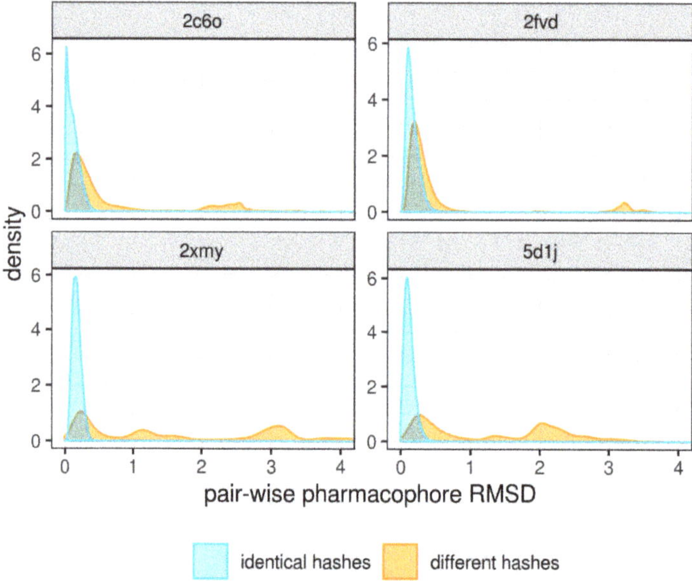

Figure 3. Gaussian kernel density of distribution of root-mean-square deviation values for the best fit between pairs of pharmacophores with identical and different hashes.

Elimination of pharmacophores with duplicated hashes substantially reduced the number of pharmacophores for 2C6O, 2FVD, and 5D1J targets to 13.5%, 17.6%, and 27.3%, respectively. The pharmacophores retrieved for 2XMY target were the most diverse and the number of distinct pharmacophore hashes was high, 80.3% (Figure 4). This can be explained by higher flexibility of a 2XMY ligand and more complex pharmacophore models for 2XMY with a greater number of features than pharmacophores for other complexes.

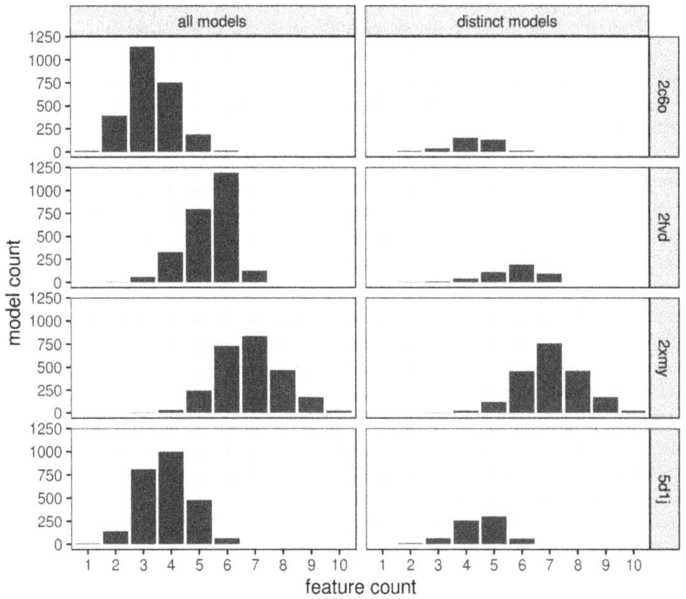

Figure 4. Distribution of all pharmacophores and pharmacophores having distinct 3D pharmacophore hashes according to their feature count.

We investigated the issue of how well the generated conformers of cocrystallized ligands reproduce binding modes represented by corresponding MD pharmacophores. Conformers of four cocrystallized ligands were generated using the same setup as for the DUD-E dataset. The most complex pharmacophores matched by 2C6O ligand were 16 four-feature pharmacophores. However, in general the complexity of corresponding MD pharmacophores was relatively low with only a few six-feature pharmacophores available (Figure 4). The 2FVD ligand matched five seven-feature pharmacophores which were the most complex among the available MD pharmacophores. It also matched 11 six-feature and 40 five-feature pharmacophores. The 2XMY ligand could match three six-feature and four five-feature pharmacophores, whereas the most complex MD pharmacophores contained 10 features. This can be explained by the high flexibility of a ligand that makes it difficult to find a conformer that perfectly matches such complex pharmacophores. The 5D1J ligand matched one five-feature and 13 four-feature pharmacophores, whereas the most complex MD pharmacophores had six features. These results indicate that the chosen setup of conformer generation can reproduce binding modes of cocrystallized ligands quite well.

To create baseline models, we retrieved pharmacophore models from initial PDB complexes using the same procedure. Only a four-feature PDB pharmacophore identified for 5D1J target had a hash identical to one of those retrieved from the MD simulations. The pharmacophore models for the remaining three complexes had a higher number of features relative to pharmacophores observed in the MD simulations (Figure 4). For the 2FVD complex, the PDB pharmacophore contained seven features, the number of representative MD pharmacophores of the same complexity was only 20%, and more complex pharmacophores were not observed. The 2XMY PDB pharmacophore consisted of nine features whereas only 9.7% of representative MD pharmacophores had the same or higher complexity. The PDB pharmacophore from the 2C6O complex had six features and there were just 15 representative MD pharmacophores of the same complexity (4.4%). On the one hand, due to a large number of features, the PDB pharmacophores extracted from the 2C6O, 2FVD, and 2XMY complexes were too specific and failed to retrieve any hits from the DUD-E dataset. On the other hand, the four-feature 5D1J PDB pharmacophore model was too loose and retrieved 10 hits and 818 non-hits that resulted in a poor enrichment equal to 0.72. The observation that PDB pharmacophore models can result in poor performance agrees with previous studies of other authors [4,7].

One expects that more complex pharmacophore models are more specific, which results in fewer retrieved hits and improves the chances of finding true hits. To estimate how the complexity of selected models affects virtual screening performance, we calculated enrichment based on all hits retrieved by any representative MD pharmacophores having at least a specified number of features (corresponds to $EF_{100\%}$). As expected, virtual screening based on all models with very simple pharmacophores having one, two or three features resulted in the lowest performance in all four cases. As the minimal complexity increased from a four-feature to a seven-feature model, enrichment improved and the number of retrieved hits significantly decreased (Table 1).

Table 1. The overall number of compounds retrieved from the DUD-E dataset by representative MD pharmacophore models of different minimum complexity.

PDB	Minimum Number of Pharmacophore Features in Models	Number of Representative Models	Number of Retrieved Compounds	TP/FP	$EF_{100\%}$ [1]
2C6O	1	338	27,884 (98.6%)	471/27,413	1.01
	4	295	8109 (28.7%)	178/7931	1.31
	5	143	291 (1.03%)	32/259	6.58
2FVD	1	440	25262 (89.3%)	430/24,832	1.02
	4	431	7745 (27.4%)	180/7565	1.39
	5	390	205 (0.73%)	22/183	6.42
	6	282	2 (0.007%)	2/0	59.79
2XMY	1	2009	14,877 (52.6%)	337/14,540	1.35
	4	2008	10,470 (37.0%)	300/10,170	1.71
	5	1988	707 (2.5%)	88/619	7.44
	6	1868	33 (0.117%)	24/9	43.48
	7	1411	1 (0.004%)	1/0	59.79
5D1J	1	683	27,884 (98.6%)	471/27,413	1.01
	4	609	15,312 (54.1%)	270/15,042	1.05
	5	356	116 (0.41%)	9/107	4.64

[1] Enrichment factor calculated for all retrieved hits.

Due to the poor performance of the simple models, we used ensembles of models having at least four pharmacophore features while comparing two ranking strategies, CHA and CCA. In almost all cases, CCA demonstrated higher early enrichment factors than CHA (Figure 5). For example, for ensembles consisting of at least five-feature models, enrichment at 0.25% was 6.27 and 10.25 (2C6O), 4.98 and 10.5 (2FVD), 22.7 and 35.0 (2XMY), and 4.64 and 4.23 (5D1J) for CHA and CCA, respectively. A similar trend is observed for other percentages of selected compounds and model ensembles (Figure 5).

The ensembles of pharmacophore models consisting of a greater number of distinct complex models, such as in the case of 2XMY complex, result in better virtual screening performance. Early enrichment calculated within CCA at 0.25% and 0.5% was 43.5 and 35.0 for ensembles consisting of at least five- and six-feature models, respectively. This may suggest that complexes of ligands with a greater number of interactions are more preferable for virtual screening, if available.

We expected that using undirected features for H-bond donors and acceptors, as well as for aromatic features, would reduce virtual screening performance. Models with undirected features were less specific, and thus could retrieve more false positives. We did not investigate this issue explicitly because directed features were not implemented in the pmapper software. However, rather high early enrichments were achieved for many complexes, with up to 43.5 $EF_{0.25\%}$ for 2XMY. This indicates that directed features may sometimes be inessential. The only exception was the 5FVD complex in which we achieved only moderate enrichment for the ensemble consisting of at least five-feature models, $EF_{0.25\%}$ was 4.64. But this also may result from a specific binding mode of a ligand in the 5D1J complex not matched by other active compounds from the DUD-E dataset.

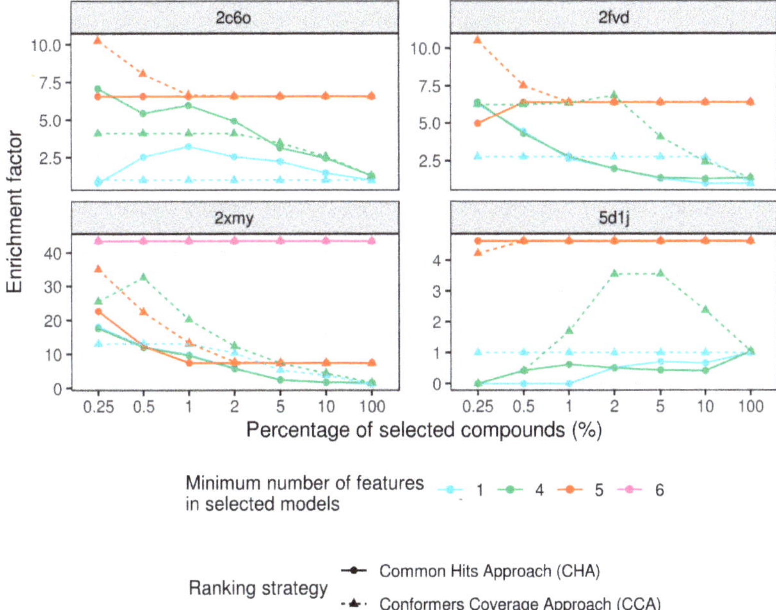

Figure 5. Enrichment factor for two ranking strategies at different complexity of selected models.

In the case of several available X-ray structures, it is possible to combine predictions to improve screening accuracy. Compounds were ranked in descending order according to their average CCA scores calculated for different protein targets. We used only pharmacophore ensembles including models with at least four and five features because simpler models resulted in poor performance. More complex models were unavailable for all studied complexes. The consensus of four complexes demonstrated good performance in both cases with $EF_{0.25\%}$ being 24.8 and 22.1 (Figure 6). However, such high performance was mainly determined by high performance of the ensemble of pharmacophore models extracted from the MD trajectory of the 2XMY complex and the consensus ranking based on four complexes did not outperform the one for the 2XMY complex. Therefore, we evaluated consensus performance based on the average CCA scores among only three model ensembles with poorer performance (2C6O, 2FVD, and 5D1J). A substantial improvement was observed for the models having at least four features. The enrichment factor at 0.25% reached 17.9 for a consensus ranking, whereas it was 4.09, 6.26, and 0 for individual model ensembles of 2C6O, 2FVD, and 5D1J complexes, respectively. The improvement of consensus ranking, based on the output of ensembles comprising at least five-feature models, was less apparent as compared with individual model ensembles. The $EF_{0.25\%}$ was 11.1 for consensus ranking and 10.2 for 2C6O, 10.5 for 2FVD, and 4.2 for 5D1J. These results encourage the application of consensus ranking whenever possible because it decreases bias introduced by individual model ensembles and gives more robust output.

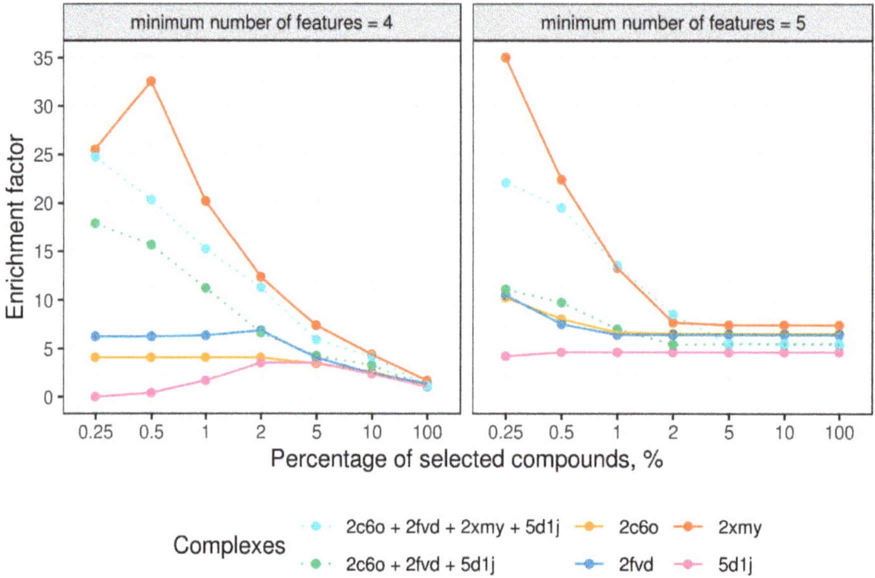

Figure 6. Enrichment factors for single pharmacophore ensembles and for consensus predictions made by averaging the scores of single compounds calculated for individual model ensembles within the conformers coverage approach.

4. Conclusions

In this study, we demonstrated the advantages of "dynamic pharmacophores", i.e., sets of pharmacophores extracted from snapshots of molecular dynamics trajectories, for virtual screening of biologically active compounds. The previously developed 3D pharmacophore hashes were successfully applied to identify identical pharmacophores and reduce the number of retrieved MD pharmacophores to representative ones. This approach omits complex calculations such as pharmacophore models clustering and selection of representative models. Since it linearly scales with a number of pharmacophore models, it can be freely applied to any number of snapshots. The 3D pharmacophore hash generation requires only one tuning parameter, i.e., binning step, which determines the fuzziness of hashes obtained. As we demonstrated, all pharmacophores with identical hashes had a pairwise root-mean-square distance less than the chosen binning step 1 Å.

We also proposed a new ranking approach, conformers coverage approach, based on the percentage of compound conformers matching representative pharmacophores from an ensemble of MD pharmacophores of an individual protein-ligand complex. Apparently, compounds with a high percentage of fitted conformers could lose fewer degrees of freedom upon binding and, as a consequence, binding entropy could be more favorable. We demonstrated that conformers coverage approach outperforms the previously proposed common hits approach on four selected protein-ligand complexes that supported the validity of the proposed approach. More rigorous validation on a larger number of complexes would be desirable.

As we observed, the usage of more complex pharmacophores with more features results in a higher performance of virtual screening. Models with three or fewer features are not recommended for virtual screening due to poor performance. As expected, there is a trade-off between the accuracy of predictions and the number of retrieved hits. Models of higher complexity result in higher enrichment values and less retrieved hits. Therefore, one should choose a model complexity depending on the particular goals of a study, but models should have at least four features and more.

Developed tools for the extraction of snapshots from a MD trajectory, assignment of pharmacophore features based on protein-ligand complex geometry, calculation of pharmacophore hashes, and virtual screening of compounds are freely available on GitHub (https://github.com/ci-lab-cz/pharmd).

Supplementary Materials: Supplementary materials can be found at http://www.mdpi.com/1422-0067/20/23/5834/s1.

Author Contributions: Conceptualization, P.P.; data curation, A.K.; investigation, P.P., A.K., and O.M.; methodology, P.P. and T.M.; project administration, P.P.; software, P.P. and D.B.; supervision, P.P.; validation, A.K.; visualization, P.P. and O.M.; writing—original draft, P.P.; writing—review and editing, P.P., O.M., and T.M.

Funding: This research was funded by the Ministry of Education, Youth and Sports of the Czech Republic, agreement number MSMT-5727/2018-2 and by the Ministry of Science and Higher Education of the Russian Federation, agreement No. 14.587.21.0049 (unique identifier RFMEFI58718X0049).

Conflicts of Interest: The authors declare no conflict of interest.

References

1. Berman, H.M.; Westbrook, J.; Feng, Z.; Gilliland, G.; Bhat, T.N.; Weissig, H.; Shindyalov, I.N.; Bourne, P.E. The Protein Data Bank. *Nucleic Acids Res.* **2000**, *28*, 235–242. [CrossRef]

2. Hritz, J.; de Ruiter, A.; Oostenbrink, C. Impact of Plasticity and Flexibility on Docking Results for Cytochrome P450 2D6: A Combined Approach of Molecular Dynamics and Ligand Docking. *J. Med. Chem.* **2008**, *51*, 7469–7477. [CrossRef]

3. Campbell, A.J.; Lamb, M.L.; Joseph-McCarthy, D. Ensemble-Based Docking Using Biased Molecular Dynamics. *J. Chem. Inf. Modeling* **2014**, *54*, 2127–2138. [CrossRef]

4. Choudhury, C.; Priyakumar, U.D.; Sastry, G.N. Dynamics Based Pharmacophore Models for Screening Potential Inhibitors of Mycobacterial Cyclopropane Synthase. *J. Chem. Inf. Modeling* **2015**, *55*, 848–860. [CrossRef]

5. Sohn, Y.-s.; Park, C.; Lee, Y.; Kim, S.; Thangapandian, S.; Kim, Y.; Kim, H.-H.; Suh, J.-K.; Lee, K.W. Multi-conformation dynamic pharmacophore modeling of the peroxisome proliferator-activated receptor γ for the discovery of novel agonists. *J. Mol. Graph. Model.* **2013**, *46*, 1–9. [CrossRef]

6. Spyrakis, F.; Benedetti, P.; Decherchi, S.; Rocchia, W.; Cavalli, A.; Alcaro, S.; Ortuso, F.; Baroni, M.; Cruciani, G. A Pipeline To Enhance Ligand Virtual Screening: Integrating Molecular Dynamics and Fingerprints for Ligand and Proteins. *J. Chem. Inf. Modeling* **2015**, *55*, 2256–2274. [CrossRef] [PubMed]

7. Wieder, M.; Garon, A.; Perricone, U.; Boresch, S.; Seidel, T.; Almerico, A.M.; Langer, T. Common Hits Approach: Combining Pharmacophore Modeling and Molecular Dynamics Simulations. *J. Chem. Inf. Modeling* **2017**, *57*, 365–385. [CrossRef] [PubMed]

8. Kutlushina, A.; Khakimova, A.; Madzhidov, T.; Polishchuk, P. Ligand-Based Pharmacophore Modeling Using Novel 3D Pharmacophore Signatures. *Molecules* **2018**, *23*, 3094. [CrossRef] [PubMed]

9. Mysinger, M.M.; Carchia, M.; Irwin, J.J.; Shoichet, B.K. Directory of Useful Decoys, Enhanced (DUD-E): Better Ligands and Decoys for Better Benchmarking. *J. Med. Chem.* **2012**, *55*, 6582–6594. [CrossRef] [PubMed]

10. Landrum, G. RDKit: Open-Source Cheminformatics Software. Available online: https://www.rdkit.org (accessed on 2 November 2019).

11. Sayle, K.L.; Bentley, J.; Boyle, F.T.; Calvert, A.H.; Cheng, Y.; Curtin, N.J.; Endicott, J.A.; Golding, B.T.; Hardcastle, I.R.; Jewsbury, P.; et al. Structure-Based design of 2-Arylamino-4-cyclohexylmethyl-5-nitroso-6-aminopyrimidine inhibitors of cyclin-Dependent kinases 1 and 2. *Bioorganic Med. Chem. Lett.* **2003**, *13*, 3079–3082. [CrossRef]

12. Pratt, D.J.; Bentley, J.; Jewsbury, P.; Boyle, F.T.; Endicott, J.A.; Noble, M.E.M. Dissecting the Determinants of Cyclin-Dependent Kinase 2 and Cyclin-Dependent Kinase 4 Inhibitor Selectivity. *J. Med. Chem.* **2006**, *49*, 5470–5477. [CrossRef] [PubMed]

13. Coxon, C.R.; Anscombe, E.; Harnor, S.J.; Martin, M.P.; Carbain, B.; Golding, B.T.; Hardcastle, I.R.; Harlow, L.K.; Korolchuk, S.; Matheson, C.J.; et al. Cyclin-Dependent Kinase (CDK) Inhibitors: Structure–Activity Relationships and Insights into the CDK-2 Selectivity of 6-Substituted 2-Arylaminopurines. *J. Med. Chem.* **2017**, *60*, 1746–1767. [CrossRef] [PubMed]

14. Choong, I.C.; Serafimova, I.; Fan, J.; Stockett, D.; Chan, E.; Cheeti, S.; Lu, Y.; Fahr, B.; Pham, P.; Arkin, M.R.; et al. A diaminocyclohexyl analog of SNS-032 with improved permeability and bioavailability properties. *Bioorganic Med. Chem. Lett.* **2008**, *18*, 5763–5765. [CrossRef]

15. Fan, J.; Fahr, B.; Stockett, D.; Chan, E.; Cheeti, S.; Serafimova, I.; Lu, Y.; Pham, P.; Walker, D.H.; Hoch, U.; et al. Modifications of the isonipecotic acid fragment of SNS-032: Analogs with improved permeability and lower efflux ratio. *Bioorganic Med. Chem. Lett.* **2008**, *18*, 6236–6239. [CrossRef] [PubMed]

16. Wang, S.; Griffiths, G.; Midgley, C.A.; Barnett, A.L.; Cooper, M.; Grabarek, J.; Ingram, L.; Jackson, W.; Kontopidis, G.; McClue, S.J.; et al. Discovery and Characterization of 2-Anilino-4- (Thiazol-5-yl)Pyrimidine Transcriptional CDK Inhibitors as Anticancer Agents. *Chem. Biol.* **2010**, *17*, 1111–1121. [CrossRef] [PubMed]

17. Chu, X.-J.; DePinto, W.; Bartkovitz, D.; So, S.-S.; Vu, B.T.; Packman, K.; Lukacs, C.; Ding, Q.; Jiang, N.; Wang, K.; et al. Discovery of [4-Amino-2-(1-methanesulfonylpiperidin-4-ylamino)pyrimidin-5-yl](2,3-difluoro-6-methoxyphenyl)methanone (R547), A Potent and Selective Cyclin-Dependent Kinase Inhibitor with Significant in Vivo Antitumor Activity. *J. Med. Chem.* **2006**, *49*, 6549–6560. [CrossRef]

18. Abraham, M.J.; Murtola, T.; Schulz, R.; Páll, S.; Smith, J.C.; Hess, B.; Lindahl, E. GROMACS: High performance molecular simulations through multi-level parallelism from laptops to supercomputers. *SoftwareX* **2015**, *1–2*, 19–25. [CrossRef]

19. Abraham, M.J.; van der Spoel, D.; Lindahl, E.; Hess, B.; The GROMACS Development Team. GROMACS User Manual Version 2016. Available online: www.gromacs.org (accessed on 18 November 2019).

20. Ponder, J.W.; Case, D.A. Force Fields for Protein Simulations. In *Advances in Protein Chemistry*; Academic Press: Cambridge, MA, USA, 2003; Volume 66, pp. 27–85.

21. Wang, J.; Wang, W.; Kollman, P.A.; Case, D.A. Automatic atom type and bond type perception in molecular mechanical calculations. *J. Mol. Graph. Model.* **2006**, *25*, 247–260. [CrossRef]

22. Wang, J.; Wolf, R.M.; Caldwell, J.W.; Kollman, P.A.; Case, D.A. Development and testing of a general amber force field. *J. Comput. Chem.* **2004**, *25*, 1157–1174. [CrossRef]

23. Jorgensen, W.L.; Chandrasekhar, J.; Madura, J.D.; Impey, R.W.; Klein, M.L. Comparison of simple potential functions for simulating liquid water. *J. Chem. Phys.* **1983**, *79*, 926–935. [CrossRef]

24. Lemkul, J. From Proteins to Perturbed Hamiltonians: A Suite of Tutorials for the GROMACS-2018 Molecular Simulation Package [Article v1. 0]. *Living J. Comput. Mol. Sci.* **2018**, *1*, 5068. [CrossRef]

25. McGibbon, R.T.; Beauchamp, K.A.; Harrigan, M.P.; Klein, C.; Swails, J.M.; Hernández, C.X.; Schwantes, C.R.; Wang, L.P.; Lane, T.J.; Pande, V.S. MDTraj: A Modern Open Library for the Analysis of Molecular Dynamics Trajectories. *Biophys. J.* **2015**, *109*, 1528–1532. [CrossRef] [PubMed]

26. Salentin, S.; Schreiber, S.; Haupt, V.J.; Adasme, M.F.; Schroeder, M. PLIP: Fully automated protein–ligand interaction profiler. *Nucleic Acids Res.* **2015**, *43*, W443–W447. [CrossRef] [PubMed]

International Journal of
Molecular Sciences

Article

3D-PP: A Tool for Discovering Conserved Three-Dimensional Protein Patterns

Alejandro Valdés-Jiménez [1,2], Josep-L. Larriba-Pey [3], Gabriel Núñez-Vivanco [1,*] and Miguel Reyes-Parada [4,5,*]

1   Center for Bioinformatics, Simulations and Modelling, Universidad de Talca, 3460000 Talca, Chile
2   PhD Program on Computer Architecture, Universitat Politécnica de Catalunya, 08034 Barcelona, Spain
3   DAMA-UPC, Universitat Politécnica de Catalunya BarcelonaTech, 08034 Barcelona, Spain
4   Facultad de Ciencias de la Salud, Universidad Autonóma de Chile, 3467987 Talca, Chile
5   School of Medicine, Faculty of Medical Sciences, Universidad de Santiago de Chile, 9170022 Santiago, Chile
*   Correspondence: ganunez@utalca.cl (G.N.-V.); miguel.reyes@usach.cl (M.R.-P.)

Received: 11 May2019; Accepted: 20 June 2019; Published: 28 June 2019

Abstract: Discovering conserved three-dimensional (3D) patterns among protein structures may provide valuable insights into protein classification, functional annotations or the rational design of multi-target drugs. Thus, several computational tools have been developed to discover and compare protein 3D-patterns. However, most of them only consider previously known 3D-patterns such as orthosteric binding sites or structural motifs. This fact makes necessary the development of new methods for the identification of all possible 3D-patterns that exist in protein structures (allosteric sites, enzyme-cofactor interaction motifs, among others). In this work, we present 3D-PP, a new free access web server for the discovery and recognition all similar 3D amino acid patterns among a set of proteins structures (independent of their sequence similarity). This new tool does not require any previous structural knowledge about ligands, and all data are organized in a high-performance graph database. The input can be a text file with the PDB access codes or a zip file of PDB coordinates regardless of the origin of the structural data: X-ray crystallographic experiments or in silico homology modeling. The results are presented as lists of sequence patterns that can be further analyzed within the web page. We tested the accuracy and suitability of 3D-PP using two sets of proteins coming from the Protein Data Bank: (a) Zinc finger containing and (b) Serotonin target proteins. We also evaluated its usefulness for the discovering of new 3D-patterns, using a set of protein structures coming from *in silico* homology modeling methodologies, all of which are overexpressed in different types of cancer. Results indicate that 3D-PP is a reliable, flexible and friendly-user tool to identify conserved structural motifs, which could be relevant to improve the knowledge about protein function or classification. The web server can be freely utilized at https://appsbio.utalca.cl/3d-pp/.

Keywords: conserved patterns; similarity; 3D-patterns

1. Introduction

Most drugs interact with more than one molecular target [1,2]. This fact is usually considered an undesired feature since it might be related to the side effects of pharmacological treatments. However, current trends in drug discovery have put hope and considerable effort into the development of multitarget compounds, due to the improved efficacy and safety profiles shown by some promiscuous drugs [3–8]. In this context, several computational approaches to predict the polypharmacological profile of either novel or known drugs have been developed, most of which are based on two main methodological strategies. In the first case, methods are based on ligand characteristics, for example, the search of compounds showing similar pharmacological/molecular activities with known drugs, those that represent the ligands as a bi-dimensional graph and look for similarities in databases using graph-based techniques and those based on the three-dimensional (3D) similarities of ligands. The second approach is centered on target(s) features and involves methods that use the known 3D structure of proteins to perform inverted docking, structure-based pharmacophore searching and the evaluation of binding sites similarities [8–10].

The usefulness of assessing structural similarities of ligand binding sites in different proteins, aimed to target clustering or drug development, is supported by the fact that the structure of proteins is several times more conserved than their sequences [11–13]. Furthermore, even in those cases where a close evolutionary relationship exists between two proteins, it might be possible that their global sequences and structures were not conserved and only share partial 3D-patterns, which define, in most cases, the function of such proteins. Indeed, the comparative analysis of important 3D protein patterns such as binding sites, catalytic sites and protein-protein interaction motifs, have been recently used to, for example, identify putative off-targets of known drugs, the design of polypharmacological compounds and drug re-purposing [14,15]. For these aims, several computational tools have been developed [16–25], which, in general, require a known query (ligands/binding sites) for their searching processes. Thus, these algorithms usually utilize (only) the orthosteric binding site in proteins, annotated motifs and/or previously known functional residues to make similarity assessments. This represents a weakness of the current tools, since some evidence indicates that a conserved 3D arrangement of amino acids might be enough to consider such a 3D-pattern as functionally relevant, even if no prior knowledge of their biological activity is available (e.g., protein cavities/pockets that may serve as allosteric sites) [20,26–31].

Thus, unveiling and comparing all local structural patterns (including those unknown or previously unobserved) into a set of protein structures could be more informative for the discovery, search and characterization of conserved 3D-patterns than exploring only previously known sites. In a recent report [32], we described a strategy for the exhaustive searching of similar 3D-patterns between two protein structures, which allowed the discovery of some conserved structural residue arrangements between proteins that differ in their function, structure and tissue localization but that share the same endogenous ligand and perform complementary physiological functions [33]. This type of finding, along with the increasing availability of structural data (more than 130,000 protein structures in the Protein Data Bank [34] and more than 3 million homology models in the SWISS-MODEL Repository [35]), represent an opportunity to use and develop structure-based methods for the classification, description and discovery of conserved 3D amino acid patterns among multiple protein structures.

Here we present 3D-PP, a new free access web server designed to discover all conserved 3D amino acid patterns among a set of protein structures. The pre-processing modules of 3D-PP were developed in Python language and all data generated are processed and organized automatically in a scalable, high-performance graph database [36]. Remarkably, this kind of database has shown better performance than relational databases, particularly when problems must be realistically modeled through, for example, the use of properties in a graph mode analysis [37–41].

2. Results

To demonstrate the applicability of 3D-PP, in the following sections we show the results of two different examples in which the existence of known and unknown 3D-patterns are assessed in a set of proteins. Also, as a benchmark analysis, we tried to replicate the same experiments with other available tools.

2.1. Known Small 3D-Patterns

We used a dataset of 46 protein structures, all of which contain the PROSITE Zinc finger C3H1-type motif (https://prosite.expasy.org/PDOC50103) (PDBids: *1m9o, 1rgo, 2cqe, 2d9m, 2d9n, 2e5s, 2fc6, 2rhk, 2rpp, 3d2n, 3d2q, 3d2s, 3jb9, 3tp2, 3u1l, 3u1m, 3u9g, 4c3b, 4c3d, 4c3e, 4cyk, 4ii1, 4yh8, 5elh, 5elk, 5gmk,*

5lj3, 5lj5, 5lqw, 5mps, 5mq0, 5mqf, 5u6h, 5u6l, 5u9b, 5wsg, 5y88, 5ylz, 5z58, 6bk8, 6dnh, 6eoj, 6exn, 6fbs, 6ff4, 6fuw; range of PDB resolutions: 1.5 to 5.9 Å). This sequence motif is composed of three cysteines and one histidine amino acids, which are located in the primary sequence as defined by the following regular expression *C-x(8)-C-x(5)-C-x(3)-H*. At a structural level, this motif represents a small 3D-pattern, is highly conserved and shows chemical coordination of the residues with one Zinc ion. Usually known as Zinc finger, this pattern is essential for the folding stabilization of this kind of protein structure [42]. After the simultaneous evaluation of these 46 protein structures, 3D-PP identified 737,793 sites corresponding to 43,305 3D-patterns organized in 47,203 clusters. As shown in Figure 1, the 3C1H was the most represented 3D-pattern with a protein coverage value of $PCv = 95.7\%$.

Search Patterns (RegExp):

Show 10 entries

Pattern	In Prot	Not In Prot	% Protein Coverage	# Total Sites	# Clusters	% Max Cluster
3C1H	44	2	95.7	152	1	100
2C1F1H	41	5	89.1	123	1	100
2C1H1Y	37	9	80.4	85	2	100
2C1G1H	35	11	76.1	74	1	100
1E1L1R1T	34	12	73.9	323	2	100
1E1K2L	31	15	67.4	625	1	100
1A1L2V	31	15	67.4	266	1	100
1C1F1H1L	31	15	67.4	106	2	100
1A1E1F1L	30	16	65.2	177	2	90
1K1L1T1V	29	17	63.0	429	1	100

Showing 1 to 10 of 299 entries   Previous 1 2 3 4 5 … 30 Next

Figure 1. Coverage of 3D-patterns identified in the Zinc finger C3H1-type protein structures. Figure 1 shows the list of all 3D-patterns detected and several criteria for filtering.

This value means that this 3D-pattern was found in the vast majority of the proteins' structures (44 of 46 proteins). Also, this pattern grouped in only one cluster (cluster coverage $CCv = 100\%$; Figure 1), which denotes that in those 44 proteins structures, there is at least one site whose 3D topological conformation does not exceed the root mean square deviation (RMSD) threshold defined by the user (4.5 Å in this example; Supplementary Data, Figure S1). This RMSD threshold is an important input parameter of our software because it allows to discriminate between 3D-patterns that contain similar components (i.e., amino acid residues) but exhibit different topological conformations (i.e., they are not in the same spatial localization/order). Thus, in 3D-PP even though several 3D-patterns might show a high level of protein coverage (PCv), they will appear grouped in different clusters with low coverage (CCv) if they show a high structural and/or topological diversity. In this example, only one cluster formed by 152 sites was detected in the 3D-Pattern 3C1H, denoting high structural conservation (Figure 2) and irregular sequence localization. As shown in Figure 3, the common 3D-pattern can appear in different locations of the sequences (blue and green boxes in Figure 3). Also, even though the 3D-pattern found corresponds to sites structurally conserved it can occur with differential sequence order in the global protein sequence (red and orange boxes in Figure 3).

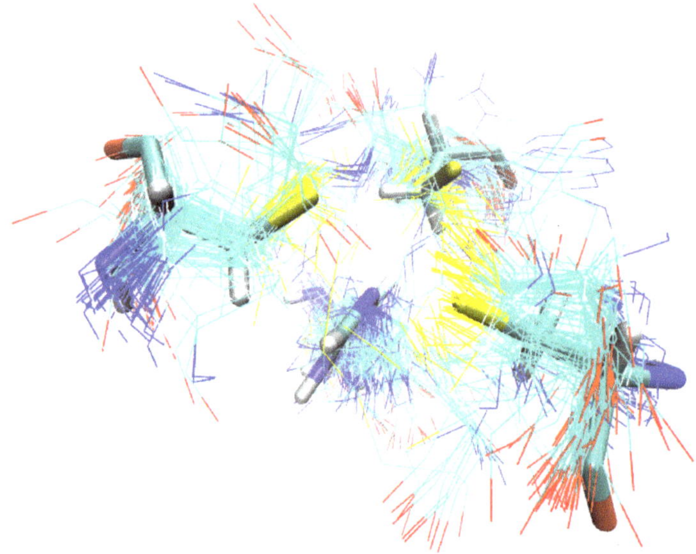

Figure 2. Structural alignment of the 152 sites detected in the cluster 3C1H-1. This result is delivered by 3D-PP using an interactive Jsmol Viewer.

Interestingly, 122 of the 152 detected sites were confirmed by the presence of the *C-x(8)-C-x(5)-C-x(3)-H* pattern in the primary sequence of the proteins analyzed and also by the appearance of the respective Zinc ion in coordination with three cysteine and one histidine amino acid in the corresponding crystal structures (confirmed using the PDBsum server [43]). The remaining sites detected by 3D-PP have similar structural features to the confirmed sites but either the protein structure does not have a co-crystallized Zinc ion or the sequence localization of the residues in the sites does not match with the corresponding PROSITE pattern (Table 1 and Supplementary Data, Table S1).

Table 1. Number of sites containing the Zinc finger C3H1-type motif at the sequence (PROSITE) and structural (3D-PP and PDBsum) levels. The last column (A & B & C) shows those sites that satisfy the sequence pattern C-x(8)-C-x(5)-C-x(3)-H (**A**), those discovered by our software that matched with the previously described sites (**B**) and those in which PDBsum shows coordination with the Zinc ion (**C**).

Item	PROSITE (A)	3D-PP (B)	PDBsum(C)	A & B	A & C	B & C	A & B & C
amount of sites	125	152	124	123	124	122	122
%	100%	–	–	94.8%	99.2%	97.6%	97.6%

In the detailed analysis of the new sites unveiled by our software, we remark the following particular cases:

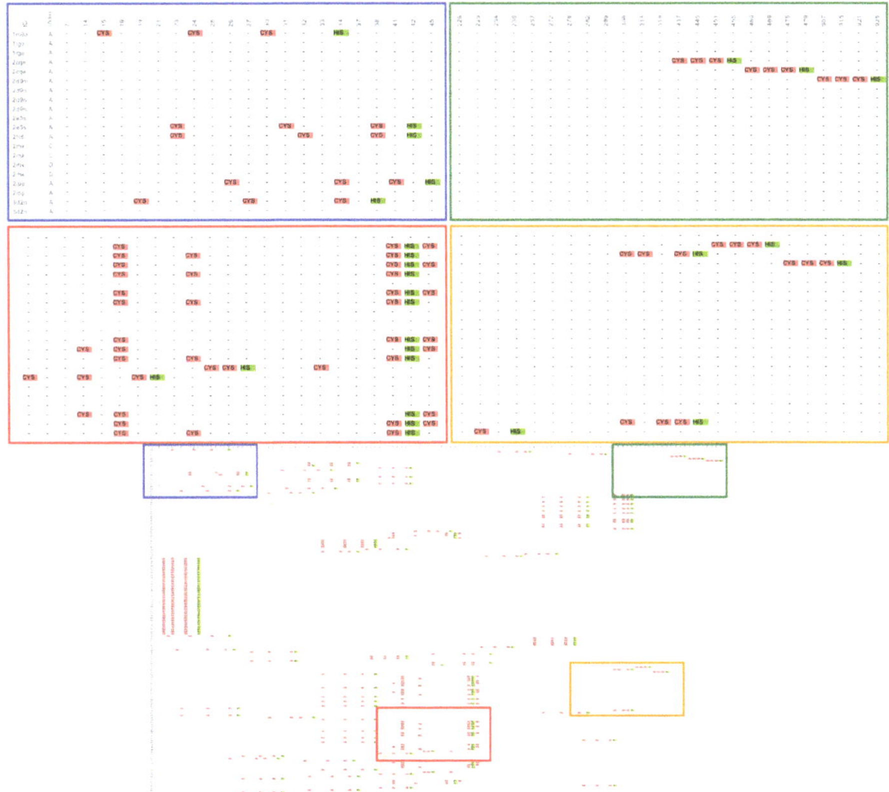

Figure 3. Sequence representation of the structural alignment of the 152 sites detected in the cluster 3C1H-1. This result is delivered by 3D-PP. The diffuse lower box shows the entire representation of the sequence alignment of the 152 sites found. The blue and green boxes show a zoom denoting the PDBids, the chain and the original PDB residue number of each site. The red and orange boxes exemplifies that some 3D-patterns found exhibit the expected sequence order (C-x(8)-C-x(5)-C-x(3)-H), whereas other sites, while having the same structural orientation, do not match with the canonical Zinc finger C3H1-type motif.

2.1.1. Putative New Zinc Ion Coordination Sites

For the protein structure with PDBid:2D9N, three sites were detected by 3D-PP. Two of them were confirmed at both sequence and structural levels and the third was only found by our software (Supplementary Data, Table S1, PDBid:2D9N). This new site, which is formed by the residues Cys68, Cys76, Cys82 and His70, shares the 3 cysteine residues with a known/confirmed site but involves a different histidine residue (His70 instead His86). As shown in Figure 4, this new identified site might keep the coordination of the Zinc ion in cases in which, for example, a punctual specific mutation of the residue His86 occurs. It should be noted that the calculated pKa of His86 (which forms the canonical Zinc coordination site) and His70 (the new putative site) was below 6, indicating that both residues are mostly deprotonated and therefore are able to establish coordination with the Zinc ion. Thus, in theory, the Zinc ion might be "moving" between both sites, since both offer a favorable environment to stabilize its binding. In the same line, the other 29 sites with similar features were discovered by our software (Supplementary Data, Table S1, Tag "New Site" in column "Comments").

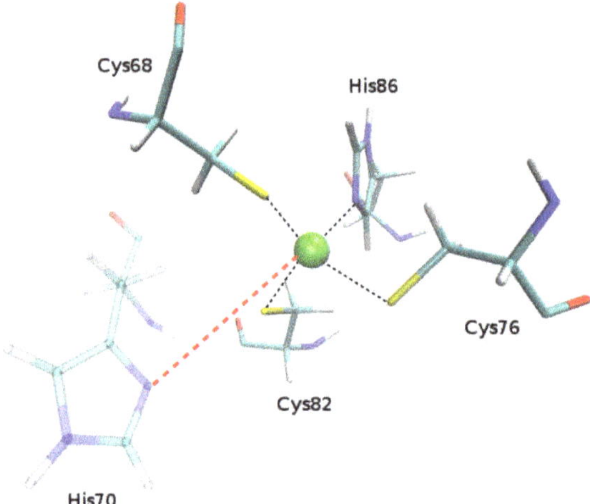

Figure 4. Putative new site discovered by 3D-PP. In the figure the residues that form the new site detected are shown. The green sphere represents the Zinc ion.

2.1.2. Promiscuous Binding Sites

Another remarkable result was the identification of two Cadmium ion binding sites that appeared in the same 3D-pattern cluster as the Zinc ion binding sites. These two sites belong to the crystal structure of the Essential Transcription Antiterminator M2-1 Protein of the human respiratory syncytial virus (PDBid:4c3d). As shown in the Supplementary Data, Figure S2, this structure effectively contains two Cadmium ions co-crystallized, which are coordinated with 3 cysteine and 1 histidine residues. These results are in agreement with previous reports that show that Zinc ions can be interchanged by Cadmium ions in some enzymes [44], indicating that this 3D-pattern can act as a promiscuous binding site. It is worth pointing out that 3D-PP does not use the information about ligand/ions co-crystallized with the protein structures and only works with the 3D-patterns found from the virtual grid of coordinates (see Materials and Methods Section).

2.1.3. Not Found Patterns

As we indicated above, in 2 of the 46 protein structures submitted it was not possible to identify 3D-patterns with the components 3C1H. These proteins, namely pre-mRNA-processing-splicing factor 8 of Human (PDBid:5MQF) and Yeast (PDBid:5LQW), were the biggest structures evaluated in this set of data. Both structures are biological assemblies obtained through cryogenic electron microscopy at a resolution of 5.9 Å and 5.8 Å , respectively. As we confirmed in our detailed analysis, low resolution—in general—limits the possibility of obtaining all the coordinates of residue side chains, some hydrogen bonds and small ligands such as metal ions. In the case of these proteins, most chains have only the atomic coordinates for the backbone and unfortunately our software cannot detect 3D-patterns without considering the side chain of protein residues.

2.2. Serotonin Target Proteins

Serotonin (5-Hydroxytryptamine; 5-HT) is a biogenic amine which is found in the gastrointestinal tract, blood platelets and the central nervous system (CNS). In the CNS, 5-HT acts as neurotransmitter and is released into the synaptic cleft where it interacts with specific 5-HT receptors (5-HTRs) to activate different signal transduction pathways [45]. After that, 5-HT is pumped back into the nerve terminals by the 5-HT transporter (SERT) and/or is metabolized by the enzyme monoamine oxidase type-A (MAO-A) [46]. Even though these three types of proteins (5-HTRs, SERT and MAO-A) have distinct functions, different sequences and diverse structural folding, they share 5-HT as the primary endogenous ligand. As observed in the matrix of amino acids' sequence identity (Supplementary Data Table S2), the range of pair-wise alignment among these proteins does not exceed 22%. In addition, their multiple sequence alignment (MSA; Supplementary Data Figure S4) only shows 15 residues conserved but with very disperse localization. Therefore, biologically relevant results cannot be obtained with these sequence alignment methods. To test our software, we submitted the crystal structures of the human SERT (PDBid:5I6X), MAO-A (PDBid:2BXR) and 5-HT$_{2A}$ receptor (PDBid:6A93) using the following input parameters: *St*: 2 Å, *Rt*: 5 Å, *RMSDt*: 4.5 Å, *Dt*: 2 Å and *Mc*: 80%. Despite protein differences, 3D-PP was able to detect several 3D-patterns with a 100% of coverage; one of them, the 3D-pattern 1*D*1*G*1*L*1*Q*, shows two clusters with 100% and 33% of *CCv* (Cluster Coverage), respectively. The first has four sites composed of one aspartate, one glycine, one leucine and one glutamine amino acids. These sites have an RMSD lower than 2.5 Å, show a similar 3D topological conformation (Figure 5A), their residues are unsorted on each primary sequence (Figure 5B) and their structural localization corresponds, for SERT and 5-HTR2A, at the extracellular side (Figure 5C,D), whereas in MAO-A, the site was detected in the protein surface (Figure 5E). The presence of aspartate residues on these sites could be significant because this type of amino acid has been shown to be critical, for example, in the inner binding site (Asp-98 [47]) and the antidepressant binding site of SERT (Asp-400 [48]), in the binding sites of the 5-HT receptors (Asp-155 [49]) and in the substrate/inhibitor cavity of MAO-A (Asp-328, Asp-132 [50]). Thus, these sites could represent a useful starting point for the design of allosteric multi-target drugs ([51]).

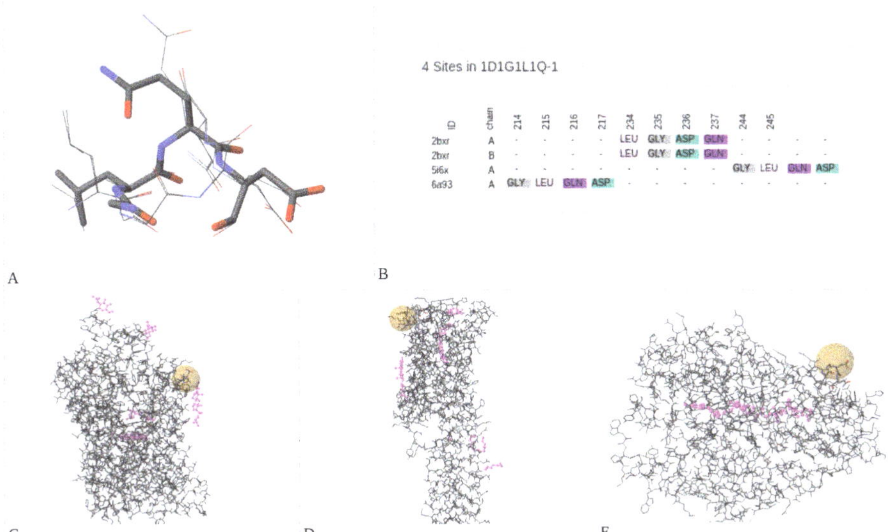

ID	chain	214	215	216	217	234	235	236	237	244	245
2bxr	A	·	·	·	·	LEU	GLY	ASP	GLN	·	·
2bxr	B	·	·	·	·	LEU	GLY	ASP	GLN	·	·
5i6x	A	·	·	·	·	·	·	·	·	GLY	LEU
6a93	A	GLY	LEU	GLN	ASP	·	·	·	·	·	·

Figure 5. Conserved 3D-pattern among the serotonin target proteins. **A** shows the structural alignment of the four sites forming the 3D-pattern 1*D*1*G*1*L*1*Q*. **B** shows a representation of a sequence alignment of the structurally aligned sites that form the 3D-pattern 1*D*1*G*1*L*1*Q*. **C**, **D** and **E** show the structural localization of the sites forming the 3D-pattern 1*D*1*G*1*L*1*Q* on the global structure of SERT, 5HTR2A and MAO-A, respectively.

2.3. Finding/Discovering Unknown 3D-Patterns on Homology Model Structures

In this case, we tried to discover conserved 3D-patterns among 10 protein structures generated through the SwissModel server (homology models). All of these proteins are over-expressed in different types of cancer (breast, prostate, lung, gastric, etc) and correspond, for example, to the insulin-like growth factor 1 receptor, the macrophage-stimulating protein receptor and the aurora kinase B, among others [52]. After the assessment with 3D-PP, our results showed the existence of several common 3D-patterns in these proteins (high PCv coverage; Max PCv = 80%) but many of them showed high structural or topological diversity (low CCv coverage). Nevertheless, the most structurally conserved 3D-pattern has a cluster with sites occurring in 8 of the 10 homology models submitted (cluster $1E1G2L - 14$, CCv = 100%; Figure 6A). The conservation of this 3D-pattern ($1E1G2L$; one glutamate, one glycin and two leucine amino acids; Figure 6B), is attractive since it might represent an event of convergent evolution which could be useful for establishing a functional annotation [53], the design of new poly-pharmacological anticancer drugs [4] and/or protein structure-based diagnosis [54]. As discussed above, this conserved 3D-pattern was detected in spite of their non-conserved sequence order (Figure 6C).

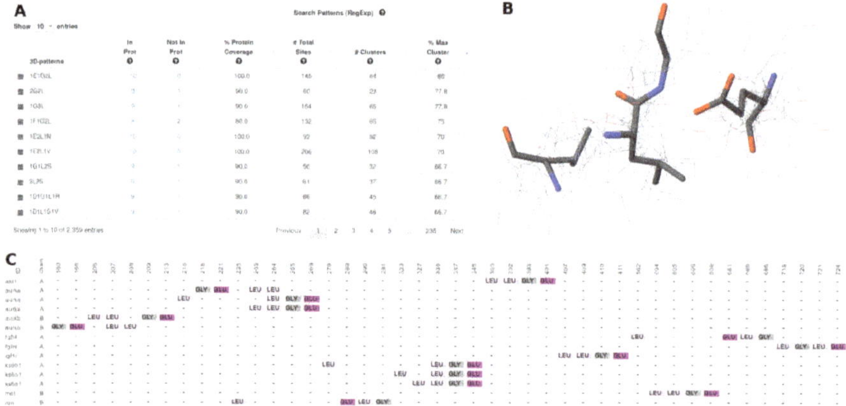

Figure 6. Coverage of 3D-patterns detected in the homology models of over expressed proteins in some cancer types. **A** shows the list of conserved 3D-patterns. **B** shows the structural alignment of the sites forming the 3D-pattern 1E1G2L. **C** shows a representation of a sequence alignment of the structurally aligned sites that form the 3D-pattern 1E1G2L.

2.4. Comparison with other Methods for the Search and Description of Amino Acid Patterns

Table 2 summarizes some features of computational tools aimed at the search of structural protein patterns, with comments regarding the results obtained when the same data set used in this work was evaluated.

In general terms, none of the software indicated in Table 2 was able to perform the same analysis as 3D-PP. Nevertheless, they were included in the benchmark, since they are the currently available algorithms with most similar performances/objectives as compared with 3D-PP. In spite of this, it seems probable that with using MMDB and VAST+ tools in combination with ProBIS (and a series of additional processing), results similar to those of 3D-PP may be obtained.

Table 2. Comparative table of similar tools.

Tool	X-ray/Homology Model	Require Known Pattern/Ligand	Comments
Catalytic site identification—a web server to identify catalytic site structural matches throughout PDB [20]	Yes/No	Yes	The most similar section to replicate our experiments with this tool is "Find CSA catalytic sites in your proteins." However, it was not possible to obtain the results because only one PDB structure can be processed at the same time. When we tried to upload a homology model the server returned a "Server Error (500)".
MMDB and VAST+ [55]	Yes/No	Yes	This tool allows the finding of 3D-patterns of amino acids in only one PDB structure at the same time. It is not possible to identify conserved 3D-patterns on a set of protein structures.
IMAAAGINE [19]	Yes/No	Yes	This tool allows searching 3D-patterns of amino acids in the entire PDB database. The user must define a structural description of query. It is not possible to identify conserved 3D-patterns on a set of protein structures.
PatternQuery [23]	Yes/No	Yes	This tool allows for the searching of 3D-patterns of amino acids in the PDB database or in a particular data set of protein structures. The user must define a detailed structural description of the query (known 3D-pattern). It is not possible to identify conserved 3D-patterns on a set of protein structures.
ProBiS [24]	Yes/Yes	No	This tool search all the 3D-patterns of amino acids associated to a ligand or functional annotations, present in a queried protein structure. Then these 3D-patterns are searched on the entire PDB database. It is not possible to identify conserved 3D-patterns on a set of protein structures.
Geomfinder [32]	Yes/Yes	No	This tool compares all the possible 3D-patterns of amino acids of one protein structure with all the possible 3D-patterns of a second protein structure. The identification of the 3D-patterns works separately on each pair of protein structures, and the results are not matched. Therefore, it is not possible to identify conserved 3D-patterns in a set of protein structures.
MultiBind [30]	Yes/Yes	No	This tool identifies similar 3D-patterns among a list of PDBids. With this tool, we could find conserved 3D-patterns, but the server did not work with our data sets. In the case of the homology models, the measures can not be assessed because our structures do not contains ligands. The server returns the following comment "No valid ligand. Can not define the query binding site".

3. Materials and Methods

3D-PP discovers conserved 3D protein patterns among an arbitrary set of structures uploaded by the user. For this, the user must define the following five threshold parameters:

- Spacing Threshold (*St*): This value is used to create the Virtual Grid of Coordinates and defines, how broad and rigorous will be the exploration of 3D-patterns. For instance, a *St* = 0.5, means that every 0.5 Å in the 3D space of each protein structure, a new virtual coordinate of reference will be created. In all cases analysed in this work (Zinc finger C3H1-type containing proteins, serotonin target proteins and structures obtained from homology models), *St* values were 0.8, 2 and 0.8 Å, respectively.

- Radius Threshold (*Rt*): This term represents the limits of the size of the 3D-patterns searched. Low *Rt* values are used to detect small binding sites (e.g., 3 Å), whereas high values allow identification of bigger sites (e.g., 7 Å). In the two cases analyzed in this work (Zinc finger C3H1-type containing proteins, serotonin target proteins and structures obtained from homology models), *Rt* values were 3, 5 and 2 Å, respectively.

- Displacement Threshold (*Dt*): This value is used to expand the size and shape for the exploration of the 3D-patterns. By default, this value is set in 0, which means that only the spherical 3D-patterns are searched. If the user changes this value; for example, *Dt* = 2, two new virtual centers will be considered for the searching of 3D-patterns. This option allows the obtaining of seven new elliptical/oval zones that will be explored to detect non spherical 3D-patterns (Supplementary Data, Figure S3).

- RMSD Threshold (*RMSDt*): This value is used for clustering the 3D-patterns detected and represents a measure of structural variability for the sites composing each 3D-pattern. As mentioned in the Results, this parameter allows the comparison of a 3D-pattern with those, containing the same components (i.e., amino acid residues), previously found by 3D-PP. Thus, if the new site exceeds the threshold values defined by the user (*RMSDt*) when comparing it with the previously found site, a new cluster of the same 3D-pattern is created. Otherwise, the new 3D-pattern is included in the same cluster as that previously found. Therefore, this parameter is crucial for 3D-PP accuracy since it allows discrimination between 3D-patterns that contain similar components but exhibit a different topological conformations (i.e., amino acid residues which are not in the same spatial localization/order).

- Minimum Coverage (*Mc*): This value allows the showing of only the 3D-patterns with a coverage value equal to or higher than *Mc*.

The sites, each one defined as a structural arrangement of residues, form different structural clusters in the same 3D-pattern. Each cluster has a central feature named coverage (*Cv*), which represents the conservation level among the evaluated proteins. For example, a *Cv* value of 100% denotes a cluster formed by sites occurring in all the assessed proteins and whose structural orientation/conformation show high similarity. Detailed information about the architecture and the essential components of 3D-PP are shown in the Supplementary Data, Figure S5.

3.1. Grid of Virtual Coordinates

One grid of virtual coordinates (*GvC*) is modeled for each protein structure submitted (Figure 7). This *GvC*, generated by the function *FIND_SITE* (Figure 8), is used for the searching of the 3D-patterns and confers to 3D-PP the ability to prescind from any previous knowledge about the ligands or binding sites in the protein structures.

Briefly, each *GvC* is constructed as follows:

- the min and max values of the X, Y, and Z axis of each structure are obtained.
- a virtual box whose size is determined by the previous values is defined (Figure 7A).

- the virtual box is filled by reference coordinates (x, y, z) distanced between them by an user-defined value (e.g., $St = 2$ Å, Figure 7B).
- only the reference coordinates that show at least four residues surrounding (at a user-defined distance, e.g., $Rt = 5$ Å) will be considered for the final grid (Figure 7C.)

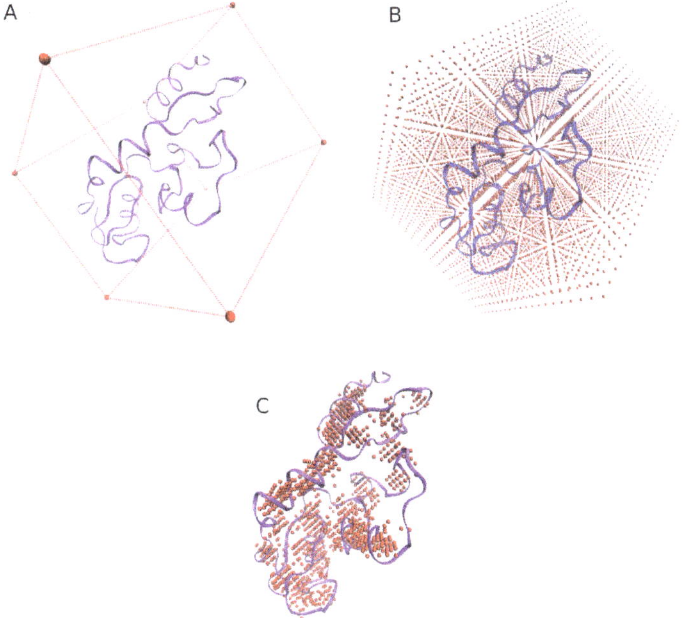

Figure 7. Grid of Virtual Coordinates (GvC). Letters **A**, **B**, and **C**, show the process of creating a Grid of Virtual Coordinates for each protein structure. The red spheres represent the reference coordinates from which the searching of 3D-patterns will be done.

3.2. Protein Preprocessing

This step represents the core of 3D-PP since it is responsible for the identification of all possible sites (arrangements of structurally related amino acids) and generates all input data for the graph database. It is worth noting that the identification of the sites is independent of the order of the amino acids sequences of proteins. This pre-processing considers all chains of each protein separately and utilizes the *GvC* previously generated. The *GvC* is used as follows:

- residues of the proteins surrounding each coordinate of the *GvC* until a user-defined distance (*Rt*) are grouped.
- groups of residues with at least four components are considered as a site.
- a vector with the list of residues is defined for each site. Then, the sites are transformed into a representation of components through a sorted alphabetical list which contains the one letter code of the amino acid and the amount of occurrences of the same amino acid (e.g., the site "H31:I32:K10:K90:L11:L12:L7:P92:S3:S8:T9" is transformed into "1H1I2K3L1P2S1T").
- If two different sites match in their representation of components (e.g., "1H1I2K3L1P2S1T"), the RMSD between these two sites is measured. If the RMSD exceeds the threshold values defined by the user (*RMSDt*), a new cluster of the same 3D-pattern is created. On the contrary, a new

site is added to the current cluster. This step is implemented to avoid two sites having the same components but different 3D conformations, being grouped in the same cluster. It should be noted that if the user set too permissive *RMSDt* values (high values), there are more possibilities for grouping sites with different structural topologies; thus, many false positives can occur.

```
 1: function FIND_SITES(args, geom_centers, atoms, chain)
 2:     min_res ← 4                                                    ▷ O(1)
 3:     temp_sites ← [ ]                                               ▷ O(1)
 4:     sites ← [ ]                                                    ▷ O(1)
 5:     /* gets the limits (min,max) x, y and z */
 6:     limit ← calculates_xyz_limits(geom_centers)                   ▷ O(1)
 7:     /* creates a kdtree (3d) for multidimensional searching */
 8:     kdtree ← create_kdtree(atoms)                                 ▷ O(1)

 9:     /* generates the virtual coordinates and find sites */
10:     for xi = limit.xmin to limit.xmax step args.step do          ▷ O(N)
11:       for yi = limit.ymin to limit.ymax step args.step do        ▷ O(M)
12:         for zi = limit.zmin to limit.zmax step args.step do      ▷ O(T)
13:           /* default searching (the virtual coordinate) */
14:           point ← [xi, yi, zi]                                    ▷ O(1)
15:           defaut_site ← kdtree.search(point, args.radius)         ▷ O(1)
16:           if length(defaul_site) > 0 then                        ▷ O(1)
17:             temp_sites.add(default_site)                          ▷ O(1)

18:           /*** extend the search from this virtual coordinate? ***/
19:           if args.extend > 0 then                                ▷ O(1)
20:             /* executes seven new searching */
21:             site   ←   calc_extends(args, min_res, kdtree, default_site, [xi +
                   (args.extend), yi, zi], [xi − (args.extend), yi, zi], [xi + (args.extend * 2), yi, zi], [xi −
                   (args.extend * 2), yi, zi])                        ▷ O(1)
22:             if site then                                          ▷ O(1)
23:               temp_sites.add(site)                                ▷ O(1)

24:             site   ←   calc_extends(args, min_res, kdtree, default_site, [xi, yi +
                   (args.extend), zi], [xi, yi − (args.extend), zi], [xi, yi + (args.extend * 2), zi], [xi, yi −
                   (args.extend * 2), zi])                            ▷ O(1)
25:             if site then                                          ▷ O(1)
26:               temp_sites.add(site)                                ▷ O(1)

27:                 ...

28:           for i = 0 to length(temp_sites) do                     ▷ O(N)
29:             if length(temp_sites[i]) >= min_res then             ▷ O(1)
30:               if temp_sites[i] not in sites then                 ▷ O(1)
31:                 obj_site ← create_site(temp_sites[i], chain)
32:                 sites.add(obj_site)                               ▷ O(1)
33:     return sites
```

Figure 8. Pseudocode of the *FIND_SITES* function of 3D-PP. Figure 8 shows the pseudocode of the function for the searching of 3D-patterns without any previous knowledge about the ligands or binding sites in the protein structures. On each line of the algorithm is indicated the computational complexity.

3.3. Creation of the Graph Databases

For each protein structure submitted, a new graph database is created simultaneously using parallel programming approaches. In these databases, the new sites identified are stored as a new node (SITE node; Supplementary Data, Figure S6A). Then, the main graph database, which is an extension of the first model, is used for the unification of data (e.g., 3D-patterns and sites of the protein 1, 3D-patterns and sites of the protein 2, etc.). For this, all the SITE node attributes are used to create or connect the corresponding PATTERN nodes, CLUSTER nodes and finally, to establish the edges SITE_IN_CLUSTER, CLUSTER_IN_PATTERN, and PATTERN_IN_PROTEIN (PATTERN_IN_PROTEIN; Supplementary Data, Figure S6B).

It is important to note that the PATTERN node with the highest amount of PATTERN_IN_PROTEIN edges represents the 3D-pattern with the highest coverage value. Moreover, if this PATTERN node has few CLUSTER_IN_PATTERN edges it is possible to estimate that the sites that are part of this 3D-pattern have a high level of structural and topological conservation. On the contrary, many CLUSTER_IN_PATTERN edges indicate a high level of structural diversity.

3.4. Result Visualization

The first level of results shows, as a graph and dynamic data tables, all 3D-patterns discovered in the set of protein structures submitted. Additionally, the user can search sub-patterns of interest through a simple regular expression query. For instance, the regular expression ^2C.*2H$, will detect all the sub-patterns that begin with 2C and finish with 2H, with any character in between, which represents a 3D-pattern containing precisely two cysteines, two histidines and any other amino acids.

Once the measures have been done, every 3D-pattern has the following ranking features available:

- *In Prot*: The number of proteins in which a specific 3D-pattern was detected.
- *Not In*: The number of proteins in which a specific 3D-pattern was not detected.
- *% Protein Coverage(PCv)*: Level of conservation of a 3D-pattern in the set of proteins evaluated. The PCv is calculated as follow:

$$PCv = In\ Prot / (amount\ of\ proteins\ submitted)$$

  A high PCv value (e.g., 80%) indicates that a pattern containing a certain type of residues is found in many proteins (e.g., 80% of the proteins analyzed). It is worth noting that the sites composing a 3D pattern found do not necessarily exhibit the same structural topology in all the proteins in which such a pattern occurs.
- *# Total Sites*: Amount of sites (arrangement of residues) which are part of a specific 3D-pattern.
- *# Clusters*: This value represents the structural variability of a 3D-pattern. Thus, a low number of clusters denotes low variability and, on the contrary, a high number of clusters is indicative of several structural conformations (with different topologies) of sites forming a 3D-pattern.
- *% Max. Cluster*: Represents the cluster with the highest coverage on each 3D-pattern.

The second level of results appears in the exploration of a particular 3D-pattern. Here, all clusters identified for the selected 3D-pattern are shown as a dynamic data table, where the following features are available:

- *#Sites*: Amount of sites (arrangement of residues) which are part of a specific cluster.
- *In Prot*: The number of proteins that contain a particular 3D-pattern.
- *% Cluster Coverage(CCv)*: Level of conservation of a cluster in the set of proteins belonging to a particular 3D-pattern. The CCv is calculated as follows:

$$CCv = In\ Prot / (amount\ of\ proteins\ into\ a\ particular\ cluster)$$

  A high CCv value (e.g., 80%) indicates that a pattern with the same structural topology is present in most of the proteins (80%) of the corresponding cluster.
- *Sequence Alignment*: This button shows a multiple sequences-based alignment of the residues of each site of a specific cluster.
- *Structural Alignment*: This button displays a jsmol viewer with multiple structural-based alignments of the residues of each site of a particular cluster.

The last level of results is displayed selecting a particular cluster. Here, all the sites grouped into a specific cluster are shown as a dynamic data table and the following features are available:

- *Site*: Information of the name and number of the residues forming the site.
- *Protein*: Name of the protein where the site was detected. This variable can be the PDBid or the name of the file, in the case of homology models.
- *Chain*: The chain where the site was detected.
- *RMSD*: Root mean square deviation of atomic positions of the particular site against the reference site.
- *SiteID*: Referential coordinate of the *GvC* from where a specific site was detected.
- *ViewSite*: This button shows a jsmol viewer loading the protein and highlighting the residues corresponding to a particular site.

4. Conclusions

In this work, we present 3D-PP, a new free access web server for discovering and recognition of all similar 3D amino acid patterns among a set of protein structures. Our software has three main features that confer competitive advantages as compared with other similar computational tools: **(a)** 3D-PP does not require previous structural knowledge about ligand(s), motif(s) or binding site(s); **(b)** 3D-PP utilizes a scalable, high-performance graph database; **(c)** 3D-PP can be used with protein structures from both experimental biophysics techniques (X-ray crystallography, NMR, etc.) and in silico homology modeling. Also, the results are shown as simple and intuitively dynamic lists of sequence/structural patterns that can be further analyzed within the web page.

We performed three representative types of uses of 3D-PP. **(I)** In the first case, using a set of protein structures containing the small 3D-pattern knows as Zinc finger, our software was able to detect almost all (98%) Zinc finger C3H1-type contained in the PROSITE database and described in crystal structures. Also, 3D-PP unveiled several new sites that have similar structural features to the known sites but which neither have a Zinc ion in the original structure nor a match between the sequence of these sites and the established sequence pattern for this type of motif. Thus, our results indicate that 3D-PP discovered new putative Zinc ion binding sites. As discussed in the Results, some of these new identified sites might serve to enhance the robustness of a crucial biological structure-derived function, by keeping the coordination of the Zinc ion in cases in which, for example, a punctual specific mutation might occur; **(II)** In the second case, we discovered some conserved 3D-patterns in the serotonin target proteins. This finding is significative considering that these proteins (5-HTRs, SERT and MAO-A) have distinct functions, different sequences and diverse structural folding; **(III)** In the third case, we found some conserved 3D-patterns in a set of protein structures coming from the in silico homology models methodologies. Considering that the X-ray structures solved until March 2019 reach a coverage of nearly 50% of the human proteome, the use of homology models substantially improves the scope of these kinds of structure-based methods. In this case for example, our criteria of selection was as ample as "Proteins overexpressed in different types of cancer'," which indicates the versatility of 3D-PP.

It is important to mention at least two limitations of 3D-PP. First, it should be noted that to identify two (or more) 3D-patterns as conserved, 3D-PP considers only sites that contain the same components (amino acid residues). It is known that, for instance, some promiscuous drugs/ligands can interact with more than one target even if the corresponding binding sites are not composed of identical amino acids but of residues with similar properties (e.g., hydrophobicity, acid or basic character, aromatic character, etc.). Therefore, 3D-patterns with "similar" structural and functional properties, but with a different composition, will not be detected by 3D-PP. The other limitation is that 3D-PP gives no information about the accessibility/drugability of the conserved 3D-patterns identified. Therefore, if a 3D-pattern is either embedded into the protein structure or in a relatively inaccessible location, it could be unproductive to try to develop compounds aimed to act at that site. Beyond these limitations, and considering as a basic idea that protein structure is more conserved than sequence, 3D-PP appears to be a flexible and user-friendly tool for identifying conserved structural motifs, which could be relevant to improve our knowledge of protein function or classification.

Supplementary Materials: Supplementary materials can be found at http://www.mdpi.com/1422-0067/20/13/3174/s1.

Author Contributions: Conceptualization, G.N.-V. and M.R.-P.; Methodology, J.-L.L.-P. and A.V.-J; Software, A.V.-J. and G.N.-V.; Validation, A.V.-J. and G.N.-V.; Resources, J.-L.L.-P. and M.R.-P.; Writing–original draft preparation, G.N.-V.; Writing–review and editing, G.N.-V. and M.R.-P.

Funding: This research was funded by the Fondo Nacional de Desarrollo Científico y Tecnológico (FONDECYT) grant number 1170662 (M.R.-P. and G.N.-V.).

Conflicts of Interest: The authors declare no conflict of interest.

References

1. Jasial, S.; Hu, Y.; Bajorath, J. Determining the degree of promiscuity of extensively assayed compounds. *PLoS ONE* **2016**, *11*, e153873. [CrossRef] [PubMed]
2. Mei, Y.; Yang, B. Rational application of drug promiscuity in medicinal chemistry. *Future Med. Chem.* **2018**, *10*, 1835–1851. [CrossRef] [PubMed]
3. Mencher, S.K.; Wang, L.G. Promiscuous drugs compared to selective drugs (promiscuity can be a virtue). *BMC Clin. Pharmacol.* **2005**, *5*, 3. [CrossRef] [PubMed]
4. Knight, Z.A.; Lin, H.; Shokat, K.M. Targeting the cancer kinome through polypharmacology. *Nat. Rev. Cancer* **2010**, *10*, 130–137. [CrossRef] [PubMed]
5. Peters, J.U. Polypharmacology—Foe or friend? *J. Med. Chem.* **2013**, *56*, 8955–8971. [CrossRef] [PubMed]
6. Kumari, S.; Mishra, C.B.; Tiwari, M. Polypharmacological Drugs in the Treatment of Epilepsy: The Comprehensive Review of Marketed and New Emerging Molecules. *Curr. Pharm. Des.* **2016**, *22*, 3212–3225. [CrossRef] [PubMed]
7. Antolin, A.A.; Workman, P.; Mestres, J.; Al-Lazikani, B. Polypharmacology in Precision Oncology: Current Applications and Future Prospects. *Curr. Pharm. Des.* **2016**, *1*, 6935–6945. [CrossRef] [PubMed]
8. Chang, M.R.; Ciesla, A.; Strutzenberg, T.S.; Novick, S.J.; He, Y.; Garcia-Ordonez, R.; Frkic, R.L.; Bruning, J.B.; Kamenecka, T.M.; Griffin, P.R. A unique polypharmacology nuclear receptor modulator blocks inflammatory signaling pathways. *ACS Chem. Biol.* **2019**, *14*, 1051–1062. [CrossRef]
9. Lavecchia, A.; Cerchia, C. In silico methods to address polypharmacology: Current status, applications and future perspectives. *Drug Discov. Today* **2016**, *21*, 288–298. [CrossRef]
10. Chaudhari, R.; Tan, Z.; Huang, B.; Zhang, S. Computational polypharmacology: A new paradigm for drug discovery. *Expert Opin. Drug Discov.* **2017**, *12*, 279–291. [CrossRef]
11. Illergård, K.; Ardell, D.H.; Elofsson, A. Structure is three to ten times more conserved than sequence—A study of structural response in protein cores. *Proteins Struct. Funct. Bioinform.* **2009**, *77*, 499–508. [CrossRef] [PubMed]
12. Ingles-Prieto, A.; Ibarra-Molero, B.; Delgado-Delgado, A.; Perez-Jimenez, R.; Fernandez, J.M.; Gaucher, E.A.; Sanchez-Ruiz, J.M.; Gavira, J.A. Conservation of protein structure over four billion years. *Structure* **2013**, *21*, 1690–1697. [CrossRef] [PubMed]
13. Pu, L.; Govindaraj, R.G.; Lemoine, J.M.; Wu, H.C.; Brylinski, M. DeepDrug3D: Classification of ligand-binding pockets in proteins with a convolutional neural network. *PLoS Comput. Biol.* **2019**, *15*, e1006718. [CrossRef] [PubMed]
14. Skolnick, J.; Gao, M.; Roy, A.; Srinivasan, B.; Zhou, H. Implications of the small number of distinct ligand binding pockets in proteins for drug discovery, evolution and biochemical function. *Bioorg. Med. Chem. Lett.* **2015**, *25*, 1163–1170. [CrossRef] [PubMed]
15. Ehrt, C.; Brinkjost, T.; Koch, O. Impact of Binding Site Comparisons on Medicinal Chemistry and Rational Molecular Design. *J. Med. Chem.* **2016**, *59*, 4121–4151. [CrossRef] [PubMed]
16. Schmitt, S.; Kuhn, D.; Klebe, G. A new method to detect related function among proteins independent of sequence and fold homology. *J. Mol. Biol.* **2002**, *323*, 387–406. [CrossRef]

17. Jambon, M.; Andrieu, O.; Combet, C.; Del??age, G.; Delfaud, F.; Geourjon, C. The SuMo server: 3D search for protein functional sites. *Bioinformatics* **2005**, *21*, 3929–3930. [CrossRef] [PubMed]

18. Nebel, J.C.; Herzyk, P.; Gilbert, D.R. Automatic generation of 3D motifs for classification of protein binding sites. *BMC Bioinform.* **2007**, *8*, 321. [CrossRef] [PubMed]

19. Nadzirin, N.; Willett, P.; Artymiuk, P.J.; Firdaus-Raih, M. IMAAAGINE: A webserver for searching hypothetical 3D amino acid side chain arrangements in the Protein Data Bank. *Nucleic Acids Res.* **2013**, *41*. [CrossRef]

20. Kirshner, D.A.; Nilmeier, J.P.; Lightstone, F.C. Catalytic site identification—A web server to identify catalytic site structural matches throughout PDB. *Nucleic Acids Res.* **2013**, *41*, 256–265. [CrossRef]

21. Ghoorah, A.W.; Devignes, M.D.; Smaïl-Tabbone, M.; Ritchie, D.W. KBDOCK 2013: A spatial classification of 3D protein domain family interactions. *Nucleic Acids Res.* **2014**, *42*. [CrossRef] [PubMed]

22. Brylinski, M. eMatchSite: Sequence Order-Independent Structure Alignments of Ligand Binding Pockets in Protein Models. *PLoS Comput. Biol.* **2014**, *10*, e1003829. [CrossRef] [PubMed]

23. Sehnal, D.; Pravda, L.; Svobodová Vařeková, R.; Ionescu, C.M.; Koča, J. PatternQuery: Web application for fast detection of biomacromolecular structural patterns in the entire Protein Data Bank. *Nucleic Acids Res.* **2015**, *43*, 383–388. [CrossRef]

24. Konc, J.; Janežič, D. ProBiS tools (algorithm, database, and web servers) for predicting and modeling of biologically interesting proteins. *Prog. Biophys. Mol. Biol.* **2017**, *128*, 24–32. 2017.02.005. [CrossRef] [PubMed]

25. Awale, M.; Reymond, J.L. Web-based tools for polypharmacology prediction. *Methods Mol. Biol.* **2019**, *1888*, 255–272._15. [CrossRef] [PubMed]

26. Russell, R.B. Detection of protein three-dimensional side-chain patterns: New examples of convergent evolution. *J. Mol. Biol.* **1998**, *279*, 1211–1227. [CrossRef] [PubMed]

27. Stark, A.; Russell, R.B. Annotation in three dimensions. PINTS: Patterns in Non-homologous Tertiary Structures. *Nucleic Acids Res.* **2003**, *31*, 3341–3344. [CrossRef] [PubMed]

28. Spriggs, R.V.; Artymiuk, P.J.; Willett, P. Searching for patterns of amino acids in 3D protein structures. *J. Chem. Inf. Comput. Sci.* **2003**, *43*, 412–421. [CrossRef]

29. Polacco, B.J.; Babbitt, P.C. Automated discovery of 3D motifs for protein function annotation. *Bioinformatics* **2006**, *22*, 723–730. [CrossRef]

30. Shulman-Peleg, A.; Shatsky, M.; Nussinov, R.; Wolfson, H.J. MultiBind and MAPPIS: Webservers for multiple alignment of protein 3D-binding sites and their interactions. *Nucleic Acids Res.* **2008**, *36*. [CrossRef]

31. Nadzirin, N.; Gardiner, E.J.; Willett, P.; Artymiuk, P.J.; Firdaus-Raih, M. SPRITE and ASSAM: Web servers for side chain 3D-motif searching in protein structures. *Nucleic Acids Res.* **2012**, *40*. [CrossRef] [PubMed]

32. Núñez-Vivanco, G.; Valdés-Jiménez, A.; Besoaín, F.; Reyes-Parada, M. Geomfinder: A multi-feature identifier of similar three-dimensional protein patterns: A ligand-independent approach. *J. Cheminform.* **2016**, *8*, 19. [CrossRef] [PubMed]

33. Núñez-Vivanco, G.; Fierro, A.; Moya, P.; Iturriaga-Vásquez, P.; Reyes-Parada, M. 3D similarities between the binding sites of monoaminergic target proteins. *PLoS ONE* **2018**, *13*, e200637. 0200637. [CrossRef] [PubMed]

34. Berman, H.M.; Battistuz, T.; Bhat, T.N.; Bluhm, W.F.; Bourne, P.E.; Burkhardt, K.; Feng, Z.; Gilliland, G.L.; Iype, L.; Jain, S.; et al. The protein data bank. *Acta Crystallogr. Sect. D Biol. Crystallogr.* **2002**, *28*, 235–242. [CrossRef] [PubMed]

35. Kiefer, F.; Arnold, K.; Künzli, M.; Bordoli, L.; Schwede, T. The SWISS-MODEL Repository and associated resources. *Nucleic Acids Res.* **2009**. [CrossRef] [PubMed]

36. Martínez-Bazan, N.; Gómez-Villamor, S.; Escalé-Claveras, F. DEX: A high-performance graph database management system. In Proceedings of the 2011 IEEE 27th International Conference on Data Engineering Workshops, Hannover, Germany, 11–16 April 2011; doi:10.1109/ICDEW.2011.5767616.

37. Have, C.T.; Jensen, L.J.; Wren, J. Are graph databases ready for bioinformatics? *Bioinformatics* **2013**, *29*, 3107–3108. [CrossRef] [PubMed]

38. Lysenko, A.; Roznovǎţ, I.A.; Saqi, M.; Mazein, A.; Rawlings, C.J.; Auffray, C.; Auffray, C.; Charron, D.; Hood, L.; Hood, L.; et al. Representing and querying disease networks using graph databases. *BioData Min.* **2016**, *9*, 23. [CrossRef]

39. Ghrab, A.; Romero, O.; Skhiri, S.; Vaisman, A.A.; Zimányi, E. GRAD: On Graph Database Modeling. *arXiv* **2016**, arXiv:1602.00503.

40. Messina, A.; Fiannaca, A.; La Paglia, L.; La Rosa, M.; Urso, A. BioGraph: A web application and a graph database for querying and analyzing bioinformatics resources. *Bmc Syst. Biol.* **2018**, *12* (Suppl. 5), 98. [CrossRef]

41. Fabregat, A.; Korninger, F.; Viteri, G.; Sidiropoulos, K.; Marin-Garcia, P.; Ping, P.; Wu, G.; Stein, L.; D'Eustachio, P.; Hermjakob, H. Reactome graph database: Efficient access to complex pathway data. *PLoS Comput. Biol.* **2018**, *14*, e1005968. [CrossRef]

42. Laity, J.H.; Lee, B.M.; Wright, P.E. Zinc finger proteins: New insights into structural and functional diversity. **2001**, *11*, 39–46. [CrossRef]

43. Laskowski, R.A.; Jabłońska, J.; Pravda, L.; Vařeková, R.S.; Thornton, J.M. PDBsum: Structural summaries of PDB entries. *Protein Sci.* **2018**, *27*, 129–134. [CrossRef] [PubMed]

44. Tang, L.; Qiu, R.; Tang, Y.; Wang, S. Cadmium-zinc exchange and their binary relationship in the structure of Zn-related proteins: A mini review. *Metallomics* **2014**, *6*, 1313–1323. [CrossRef] [PubMed]

45. Maroteaux, L.; Béchade, C.; Roumier, A. Dimers of serotonin receptors: Impact on ligand affinity and signaling. *Biochimie* **2019**, *161*, 23–33. [CrossRef] [PubMed]

46. Jonnakuty, C.; Gragnoli, C. What do we know about serotonin? *J. Cell. Physiol.* **2008**, *217*, 301–306. [CrossRef] [PubMed]

47. Felts, B.; Pramod, A.B.; Sandtner, W.; Burbach, N.; Bulling, S.; Sitte, H.H.; Henry, L.K. The two Na+ sites in the human serotonin transporter play distinct roles in the ion coupling and electrogenicity of transport. *J. Biol. Chem.* **2014**, *289*, 1825–1840. [CrossRef] [PubMed]

48. Rannversson, H.; Andersen, J.; Bang-Andersen, B.; Strømgaard, K. Mapping the Binding Site for Escitalopram and Paroxetine in the Human Serotonin Transporter Using Genetically Encoded Photo-Cross-Linkers. *ACS Chem. Biol.* **2017**, *12*, 2558–2562. [CrossRef] [PubMed]

49. Westkaemper, R.B.; Roth, B.L. Structure and Function Reveal Insights in the Pharmacology of 5-HT Receptor Subtypes. In *The Serotonin Receptors*; Humana Press: Totowa, NJ, USA, 2008; doi:10.1007/978-1-59745-080-5_2.

50. Zapata-Torres, G.; Fierro, A.; Miranda-Rojas, S.; Guajardo, C.; Saez-Briones, P.; Salgado, J.C.; Celis-Barros, C. Influence of protonation on substrate and inhibitor interactions at the active site of human monoamine oxidase-A. *J. Chem. Inf. Model.* **2012**, *52*, 1213–1221. [CrossRef]

51. Christopoulos, A. Allosteric binding sites on cell-surface receptors: Novel targets for drug discovery. *Nat. Rev. Drug Discov.* **2002**, *1*, 198–210. [CrossRef] [PubMed]

52. Zhang, J.; Yang, P.L.C.; Gray, N.S. Targeting cancer with small molecule kinase inhibitors. *Nat. Rev. Cancer* **2009**, *9*, 28–39. [CrossRef]

53. Wilkins, A.D.; Bachman, B.J.; Erdin, S.; Lichtarge, O. The use of evolutionary patterns in protein annotation. *Curr. Opin. Struct. Biol.* **2012**, *22*, 316–325. [CrossRef]

54. Bennionn, B.J.; Daggett, V. Protein conformation and diagnostic tests: The prion protein. *Clin. Chem.* **2002**, *48*, 2105–2114.

55. Madej, T.; Lanczycki, C.J.; Zhang, D.; Thiessen, P.A.; Geer, R.C.; Marchler-Bauer, A.; Bryant, S.H. MMDB and VAST+: Tracking structural similarities between macromolecular complexes. *Nucleic Acids Res.* **2014**, *42*, D297–D303. [CrossRef]

International Journal of
Molecular Sciences

Article

Accurate Representation of Protein-Ligand Structural Diversity in the Protein Data Bank (PDB)

Nicolas K. Shinada [1,2,3], Peter Schmidtke [2] and Alexandre G. de Brevern [3,*] 

[1] SBX Corp., Tokyo-to, Shinagawa-ku, Tokyo 141-0022, Japan; shinada@sbx-corp.com
[2] Discngine SAS, 75012 Paris, France; peter.schmidtke@discngine.com
[3] INSERM, UMR_S 1134, DSIMB, Univ Paris, INTS, Laboratoire d'Excellence GR-Ex, 75015 Paris, France
*   Correspondence: alexandre.debrevern@univ-paris-diderot.fr; Tel.: +33-1-44493000

Received: 28 January 2020; Accepted: 20 March 2020; Published: 24 March 2020

Abstract: The number of available protein structures in the Protein Data Bank (PDB) has considerably increased in recent years. Thanks to the growth of structures and complexes, numerous large-scale studies have been done in various research areas, e.g., protein–protein, protein–DNA, or in drug discovery. While protein redundancy was only simply managed using simple protein sequence identity threshold, the similarity of protein-ligand complexes should also be considered from a structural perspective. Hence, the protein-ligand duplicates in the PDB are widely known, but were never quantitatively assessed, as they are quite complex to analyze and compare. Here, we present a specific clustering of protein-ligand structures to avoid bias found in different studies. The methodology is based on binding site superposition, and a combination of weighted Root Mean Square Deviation (RMSD) assessment and hierarchical clustering. Repeated structures of proteins of interest are highlighted and only representative conformations were conserved for a non-biased view of protein distribution. Three types of cases are described based on the number of distinct conformations identified for each complex. Defining these categories decreases by 3.84-fold the number of complexes, and offers more refined results compared to a protein sequence-based method. Widely distinct conformations were analyzed using normalized B-factors. Furthermore, a non-redundant dataset was generated for future molecular interactions analysis or virtual screening studies.

Keywords: protein-ligand complexes; dataset; clustering; structural alignment; refinement

---

1. Introduction

Protein structures are the support of essential biological functions. They are highly dynamic macromolecules and adopt an ensemble of conformations during their lifetime. Multiple resolution techniques have been elaborated to access their three-dimensional structures. X-ray crystallography and Nuclear Magnetic Resonance spectroscopy (NMR) are the most common and efficient resolution methods. The obtained structures are stored and freely available for the scientific community in the Protein Data Bank (PDB) [1], a widely used public database since the 1970s. A significant increase in structure deposition throughout its existence is observed, e.g., going from 54,000 protein structures in 2008 to 160,000 in 2020. PDB does not exclusively contain protein structures, ligands are also displayed in PDB structures, resulting in a larger number of protein-ligand complexes. They are widely used in structure-based drug discovery [2]. Structures of ligand complexes are used for drug design purpose, e.g., they can be used to train scoring functions of protein-ligand interactions [3]. They are also critical in the understanding of the underlying principles of intermolecular interactions, e.g., the recent analyses of halogen interactions between proteins and ligands [4]. These structures are also often utilized to benchmark novel methods in the realm of molecular modeling.

Nevertheless, a major difficulty to perform a proper benchmark for a specific method using resources, such as the PDB, is to ensure an unbiased protein dataset, i.e., specific non-redundant datasets must be produced. Multiple methodologies exist to evaluate and generate such non-redundant protein datasets using underlying amino acid sequence information, e.g., PDBSelect [5] or PISCES [6]. Heuristics have also been proposed to be quick and usable for large datasets, e.g., BLASTCLUST [7] or CD-HIT [8]. Only a very limited number tried to take into account the protein structure, e.g., PAPIA [9]. Today, tools available on the PDB website allow non-redundant dataset retrieval using sequence similarity measures alone. As protein structures are, in a certain extent, subjectively created models, their recurrence can improve the confidence for a structure. Even so, these repetitions, or redundancy, can induce bias. It is widely acknowledged within the PDB by the scientific community, yet ill-considered. The only studies related to this subject are focused on conformational ensembles, such as NMR structures, corresponding to 8.5% of the PDB [10,11], which are, by definition, highly similar models.

While most of the previous methods focus only on protein sequences, proteins bound to DNA, to RNA, to small molecules, or to amino acids containing post-translational modifications (PTM) [12] are more difficult to analyze due to their diversity. For instance, in structures of protein–DNA complexes, proteins can have easily reach thousands of amino acids while a DNA structure of more than 15 bp is rare [13]. The situation is similar to protein-ligand complexes and directly affects their analyses.

Today, a few tools exist to gather proper protein-ligand complexes datasets. The Binding Mother of All Databases (MOAD) [14] includes 25,769 high-quality (resolution better than 2.5 Å and biologically relevant ligands) protein-ligand complexes taken from the September 2017 PDB. They address the question of redundancy by looking at the protein sequence and using molecular fingerprints coupled with Tanimoto coefficient regarding the ligands [15]. PDBBind [16] provides yearly releases and contains currently 17,900 biomolecular complexes in the 2017 version. They proposed a limited number of proteins defining a 'core set' to try to handle the question of redundancy curated manually [17]. The scPDB [18], an annotated database of binding sites in the PDB, contains 4782 proteins and 6326 ligands in its 2017 release. In its original publication [19], absence of redundancy is mentioned in their dataset without provided metrics. While these databases offer refined protein structures, none of them explore and assesses the structural diversity of their complexes.

Previous work by Wallach and Lilien in 2009 [20] already focused on this particular issue. To improve the quality of binding models extracted from PDB complexes, a non-redundant dataset was generated, considering sequence similarity for the protein part (BLASTp) and small molecule fingerprint similarity metrics. However, they do not consider cases where identical ligands bind to different binding sites on the same protein. Furthermore, no structural assessment was performed in their study. The last update of the dataset was performed in 2013. Drwal and coworkers have recently published a study on 2911 complexes from the PDB including 1079 fragments and 1832 small molecules highlighting fragment binding mode conservation in 74% of the dataset [21]. Small element substitutions on fragment have little to no impact to the fragment-binding mode and interaction patterns appear to be maintained.

Here, we propose a first quantitative evaluation of the structural redundancy observed in PDB focusing on protein-ligand complexes. Basic statistics on overrepresented proteins and molecules are derived. A specific clustering is performed to define the accurate number of unique complexes resulting in the generation of a refined dataset for molecular interaction studies or virtual screening protocols. Finally, we discuss and illustrate some of the surprising findings.

2. Results

2.1. Initial Dataset

2.1.1. Statistics

The initial database query retrieved 110,735 interacting complexes from 3decision™ software. Multiple filtering steps such as ligand size, single bond ratio, protein chain size, and number of residues in contact with ligand (as described in Materials and Methods) were performed. This phase contributed to a reduction of our initial dataset to 92,475 protein-ligand complexes (see Figure 1). These protein-ligand sets are spanned across 39,411 PDB entries. At this stage, one ligand can be bound to more than one protein chain; thus, each case of multimeric complex was then separated culminating in 104,777 individual protein-ligand conformations, i.e., monomeric data (see Figure 1). These units were then analyzed to define different sub-datasets.

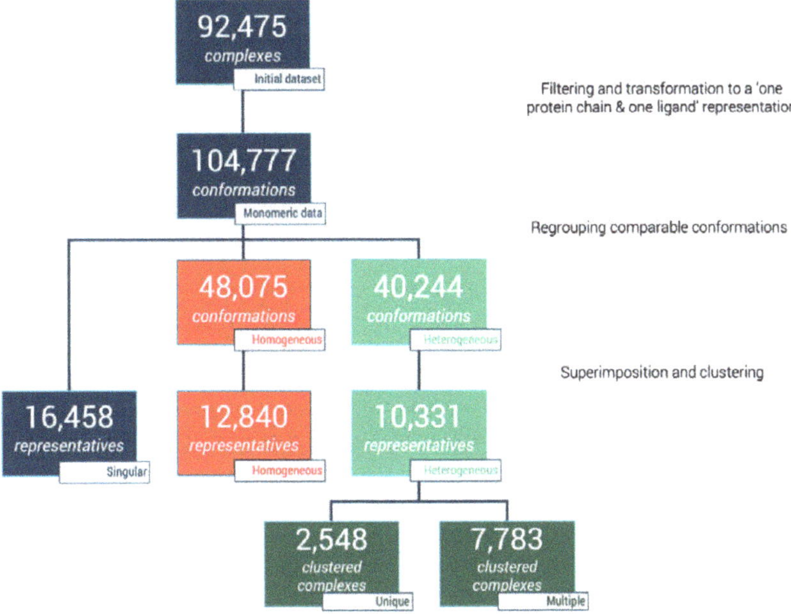

Figure 1. Flowchart of dataset clustering and characterization of protein–ligand complexes.

2.1.2. Diversity

Unsurprisingly, from the multimeric complex dataset, the most represented ligand in the PDB is the heme (PDB residue code Protoporphyrin IX Containing FE (HEM)) with 9088 occurrences, accounting for a representation of 7.9%. Expected popular ligands are present among the 10 most frequent molecules in the PDB, such as nucleobase derivatives, e.g., Adenosine-5′-Triphosphate (ATP), Nicotinamide-Adenine-Dinucleotide (NAD), Flavin Mononucleotide (FMN), S-Adenosyl-L-Homocysteine (SAH) (see Figure 2A). Interestingly, using SMILES identity, a large number of 17,135 unique small molecules are found. Consequently, this unbalanced ligand distribution is reflected among our protein representation where nucleotide coenzymes receptors and heme receptors are also frequently observed (Figure 2B). The strong occurrences of these specific ligands and receptors culminate logically in a large presence of their corresponding complexes. Heme bound to nitric oxide synthase, with respectively 9,088 and 921 occurrences, are represented by 501 distinct conformations (see Figure 2C).

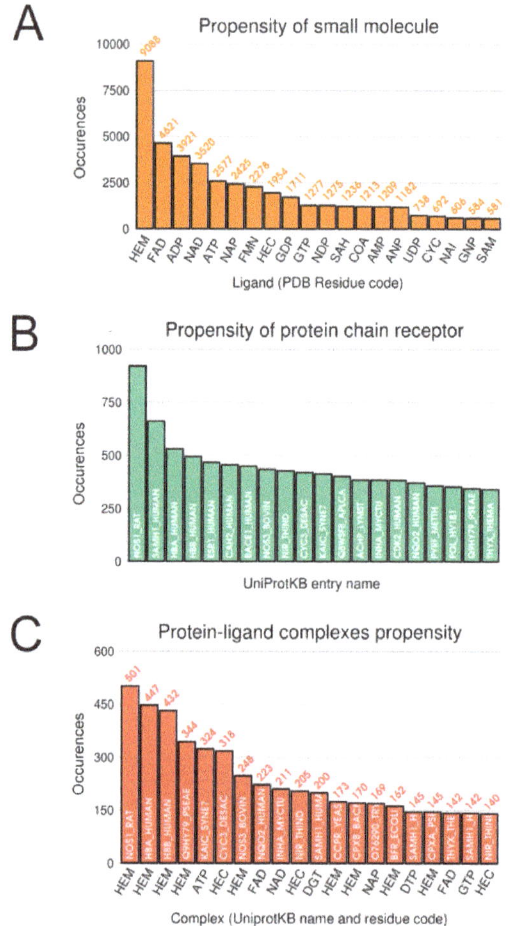

Figure 2. Top 20 distributions in monomeric dataset. (**A**) Distribution of ligands. (**B**) Distribution of protein chain (with UniProt IDs). (**C**) Distribution of protein-ligand complexes (with UniProt IDs and residue code).

2.2. Singular Protein-Ligand Complexes

The simplest type of complexes to analyze are the singular complexes as they were defined as such when no other identical ligand or identical proteins were found for these complexes, namely the *singular* dataset. Moreover, 15.7% of our dataset (16,458 out of 104,777 monomeric complexes) fit this description (see Figure 1).

Distribution of ligands among the singular dataset reveals nucleobase-like molecule remains the top representation with a combined 739 occurrences for Phosphoaminophosphonic Acid-Adenylate Ester (ANP), AMP, Adenosine-5'-Diphosphate (ADP), and ATP. Heme is slightly underrepresented compared to its overall distribution in our initial dataset with 174 cases for HEM residue code and 152 instances for Heme C (HEC) residue code. Moreover, 84.7% of these small molecules have only one representation in this unique dataset.

Those 16,458 complexes represent 5239 distinct protein chains, 2232 only present once in this subset. The remaining 3007 protein chains feature, on average, 4.73 distinct ligands with a distribution largely unbalanced. As an example, two distinct binding sites of the carbonic anhydrase enzyme are interacting with 259 distinct ligands. One of these binding sites is illustrated in Figure 3, where three different ligands are shown interacting with the same pocket. Similar redundancy is observed for prothrombin, β-secretase 1, and cyclin-dependent kinase 2 with, respectively, 204, 199, and 195 occurrences.

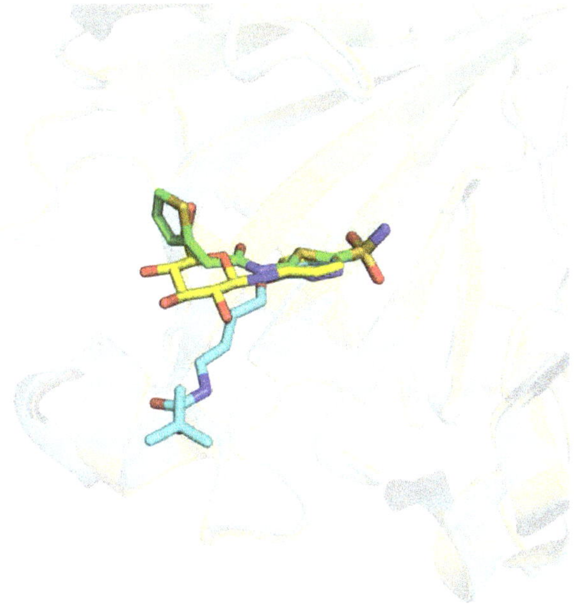

Figure 3. Three-dimensional (3D) representation of three different ligands bound to Carbonic Anhydrase 2 receptor (Protein Data Bank (PDB) IDs 4iwz in green, 2rfc in blue, 2hl4 in yellow).

These singular complexes shown here highlight general tendencies towards specific protein targets in biological research field. Recurrent proteins bound with distinct ligands underlines important binding residues and conserved molecular fragment used in ligand optimization. For instance, 95.1% of prothrombin proteins found in the PDB involve residue W215 and 94.1% A190 in the binding mechanism. Other less frequent residues involve E217 in 55.0% of complexes and F227 in 24.5% of complexes (see Figure A1).

2.3. Protein-Ligand Complexes Groups

The remaining units represent 88,319 conformations, i.e., 84.2% of our dataset. Conformations with identical ligands, protein chains, and similar binding sites were grouped and compared to each other within the group.

At first, an initial number of 18,478 groups (of complexes) were generated containing between 2 and 501 conformations. As each group features at least two units, distribution of conformations count per complex was analyzed (see Figure 4). It shows an unbalanced representation, where complexes with few conformations are predominant: the 9542 of groups (51.6%) featuring two conformations represent 21.6% of the 88,319 conformations. Still, complexes with more than 30 conformations tally for 17.8% conformations and only 1.4% of the groups; the biggest group is again for brain nitric oxide synthase, with 501 occurrences.

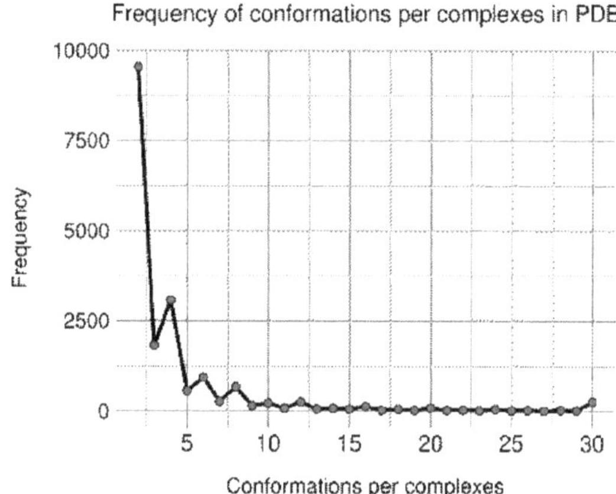

Figure 4. Distribution of observed conformations per complexes in non-singular dataset.

Then, a structural alignment and quantification was performed between each unit within one group on common binding residues, e.g., $\frac{n(n-1)}{2}$ comparisons per group with n being number of conformations. Over 1,130,000 superimpositions and weighted fragment root mean square deviation (wRMSD$_f$) computations were performed. Groups were then split between homogeneous binding modes, i.e., identical conformations, and heterogeneous binding poses, i.e., sharing some similarity.

2.3.1. Homogeneous Complexes

Homogeneous complexes are defined as groups for which each conformation is identical to each other, i.e., conserved binding modes. Each comparison among one group results with a wRMSD$_f$ value below 1.0 Å and no distance between aligned fragments was greater than 1.5 Å.

Moreover, 12,840 groups were considered as homogeneous in our dataset, equivalent to 48,075 units (45.9% of our monomeric dataset, see Figure 1). As those complexes display identical binding modes, a 3.75-fold reduction can be processed when considering only one representative per group, i.e., 12,840 unique representatives.

Multiple group sizes are represented across these complexes. A large number of redundant conformations for one complex is expected in NMR structures. However, the largest group is composed of 173 units across 158 unique PDB X-ray entries. One representative of this heme bound to mitochondrial cytochrome C peroxidase complex is available through PDB entry 1aen.

Almost every superimposition made in the homogeneous subset is associated with a good binding residues alignment. However, some interesting cases, 160 complexes (1.6%), displayed a significant number of distinct binding residues; thus, not selected for superimposition. Specific visual inspection of those cases indicates: (i) due to protein inner flexibility, these binding residues are identified in only one conformation as the detection depends on a distance threshold (e.g., highlighted by distance in Figure 5) and (ii) other residues can be unresolved in one of the structures (corresponding to residues colored in red in Figure 5). It must be noted that since ligand-binding modes remain identical in homogeneous groups, missing residues and ambiguous residue detection do not impact neither binding site superposition nor assessment.

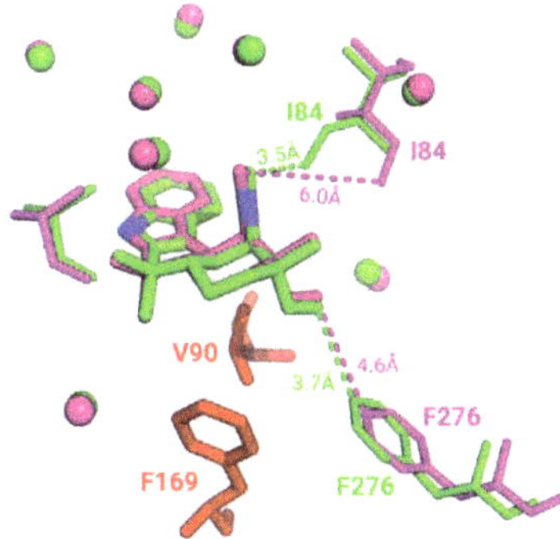

Figure 5. Three-dimensional representation of WelO5 protein in complex with 6CU ligand stick representation (PDB IDs 5iqv green and 5iqu magenta). Residues in PDB entry 5iqu that are unresolved in PDB entry 5iqv are colored in red. Closest distances observed are shown in dashes. Superposed residues are displayed in spherical representation.

This descriptor describes the similarity in pocket shape. Moreover, 97.1% of our homogeneous comparisons display a pocket RMSD less than 0.5Å, i.e., structurally conserved pocket. Overall, structure superposition result in identical pocket shape with an average pocket RMSD of 0.18Å with a standard deviation of 0.13 Å.

Only 615 comparisons out of 233,417 (0.26%) have a pocket RMSD greater than 1.0 Å. These specific cases are largely caused by flexible secondary structure, such as loop highlighted in Figure 6. Visual inspection of flavin adenine dinucleotide receptor with Flavin-Adenine Dinucleotide (FAD) ligands shows a significant number of binding residues that are conserved. However, residues such as G397 belong to a flexible loop leading to a 11.7 Å distance between the two Cαs after alignment (highlighted by orange dashes on Figure 6). E49 and V395 also displayed significant differences with 7.4 Å and 3.0 Å shifting. Interestingly, these deviations can highlight either potential multiple binding roles due to their proximity with the ligand in either cases or space filling characteristics.

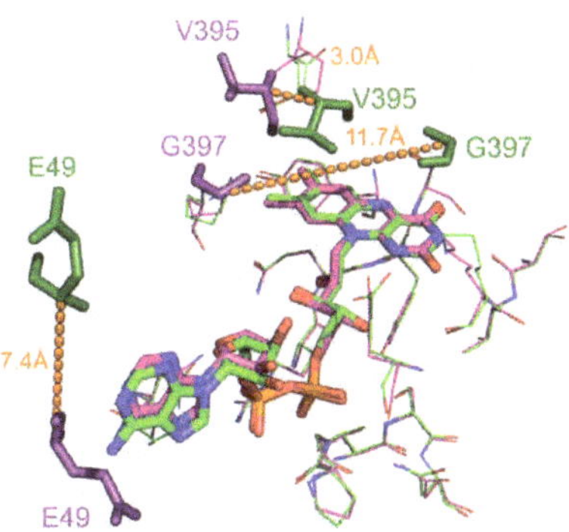

Figure 6. Three-dimensional representation of superposed Flavin-Adenine Dinucleotide (FAD) binding sites of flavin adenine dinucleotide receptor. Structurally conserved residues are displayed in 'lines' representation. Flexible residues are displayed in 'stick' representation and in darker colors with their corresponding deviation highlighted by dashes (PDB IDs 1cqx in green and 3ozv in purple).

2.3.2. Heterogeneous Complexes

Finally, the remaining 5638 complexes corresponding to 40,244 conformations (38.4% of our dataset), present at least one comparison where either wRMSD$_f$ is greater than 1.0 Å or one distance between compared fragments above 1.5 Å. To avoid unnecessary clustering, if every comparison within one group was greater than 2.0 Å, complexes were automatically considered as distinct. Clustering was performed in situations where both high and low wRMSD$_f$ values were observed to extract the representatives poses.

Filtering and clustering resulted 10,331 distinct binding modes across those 40,244 conformations, corresponding to an interesting 3.89-fold reduction. Moreover, three subgroups were distinguished from these: (i) 2548 complexes (17,602 conformations) with only one cluster, i.e., one representative, (ii) 2360 unique conformations from complexes with only distinct conformations (wRMSD$_f$ > 2.0 Å), and (iii) 5423 representative conformations originated from complexes with scattered clusters size. The latter accounts originally for 2006 complexes represented by 20,283 conformations, with a high number of complexes represented by less than five similar representatives (Figure 7).

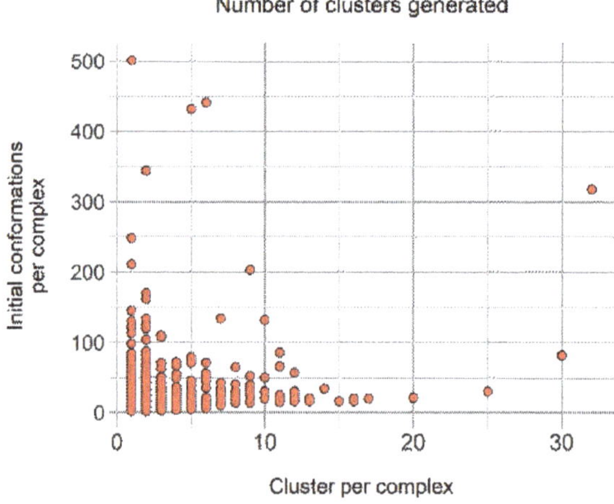

Figure 7. Distribution of number of generated clusters relative to the original binding poses count to be clustered in group B. Propensity of cluster sizes.

Figure 7 underlines the number of representative units generated from clustering relative to the number of initial conformations available. Interestingly, despite having initially 22 complexes with more than 100 initial distinct conformations, none leads to more than 35 conformations, and only four have more than 20 distinct conformations.

The most redundant complex in our dataset, 501 distinct structures of heme bound to brain nitric oxide synthase, was clustered into one representative binding modes, with an average wRMSD$_f$ of 0.37 Å (standard-deviation 0.18).

Intriguing results arise especially in the study of NMR structures. For instance, Figure 8 highlights 12 distinct binding modes identified for xylose isomerase protein (PDB id 1xlf) among 30 protein-ligand conformations. The biggest cluster being composed of six similar units. A cluster of three conformations in Figure 8A displays excellent superposition with an average wRMSD$_f$ of 0.73Å, while alignments on other representatives display significant structural deviation (three examples highlighted in Figure 8B).

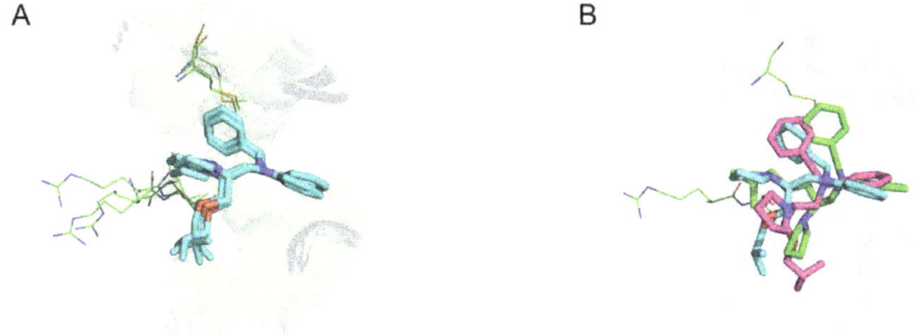

Figure 8. Three-dimensional representation of Cardiac Troponin with Bepridil (PDB ID 1lxf), ligand represented in 'stick' representation, binding residues in 'lines' representation: (**A**) Cluster of three identical ligand conformations (average weighted fragment root mean square deviation (wRMSD$_f$) 0.67 Å), (**B**) three distinct conformations extracted from same NMR structural model (average wRMSD$_f$ 2.90 Å).

Finally, structure assessment on small clusters was performed using B-factors. Clusters featuring three or more identical conformations were considered as structurally stable and therefore devoid of bias in atom positioning during resolution. For the remaining clusters, with three or less conformations, B-factors were extracted and normalized, corresponding to 7672 X-ray structures with a resolution greater than 1.5 Å.

Moreover, 1199 conformations displayed a ligand normalized B-factor value above 2.0Å^2, an arbitrary threshold but described as clearly flexible by Bornot et al. [22]. The average normalized B-factor values calculated on both backbone and side-chain residues indicate mostly rigid or intermediate environment for these ligands. Surprisingly, only four cases (highlighted in red in Figure 9) display both mobile ligand and binding sites and only 77 cases featured flexible side-chains (above 2.0 Å^2). This observation illustrates an overall binding site rigidity in opposition to ligand flexibility. Overall, 1211 conformations from our initial 7642 representatives can be categorized as cautionary due to high flexibility in either the ligand or protein counterpart. Of course, the positioning of some of these ligands could be attributed to low resolutions or poor fitting of the ligands in the electron density map.

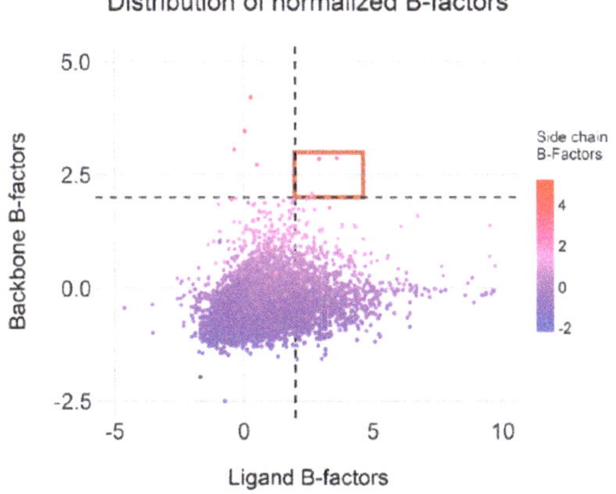

Figure 9. Normalized B-factor distributions for complexes with distinct conformations (high Root Mean Square Deviation (RMSD) values) and complexes with less than three clustered conformations. Cases with high flexibility in ligand and pocket are highlighted in red.

2.4. Non-Redundant Dataset Generation

This study focusing on binding modes diversity results in the generation of a non-redundant dataset of protein-ligand complexes based on those refined binding modes with no systematic bias. The criteria to select the representative for each cluster is an aggregate ranking between structure resolution, the maximum number of contacting residues and the averaged wRMSD$_f$ for each conformation. Following the previous described step (singular dataset and representative of the two categories of non-singular datasets), the proposed non-redundant dataset of 39,629 complexes. Hence, this supervised approach leads to a pertinent and significant 2.64-fold reduction over the initial dataset. The final list (see Appendix A) contains PDB IDs, ligand residue code, and number as defined in the PDB, and their corresponding conformation in case of NMR structures or alternate conformations.

A critical comparison must be done with the most classical approach to define a non-redundant dataset, i.e., a simple sequence identity threshold without consideration of ligand. It would have produced a dataset of 9997 complexes, i.e., a considerable loss in regard to the final proposed dataset. Associating ligand similarity to the process, 30,873 distinct complexes would have been generated, a 22.1% decrease compared to our final non-redundant dataset. This discrepancy comes from multiple circumstances, such as (i) considering multiple binding sites per protein chain with the same ligand generating specific instances (ii) differentiating conformations through structural comparison of distinct binding modes.

Our final dataset includes 31,846 complexes with only one binding mode observed in the PDB and 4366, characterized by multiple representative units. Moreover, 95.0% of those recurring complexes have less than five representative conformations. Similarly to sequence-based approach, our non-redundant dataset still retains 9997 distinct protein chains. Carbonic anhydrase 2, β-secretase 1, and cyclin-dependent kinase 2 proteins are among the recurrent complexes identified with 332, 297 and 272 occurrences respectively. These values are mostly due to the diversity of ligands crystallized with these proteins of high interest.

To note, 16,771 distinct ligands are identified, with heme still being one of the most recurrent ligands. However, compared to the 9088 instances of heme observed in our initial dataset, clustering has reduced heme representation by a factor 8.4 (see Figure 10). Similarly, flavin adenine ligand (FAD) occurrences have been reduced by 6.4-fold indicating a significant redundancy for the most frequent small molecules. Hence, it should be noted that this difference of 22% in terms of occurrence reflects the fact that the lists are very different. In the list based only on the sequences, the cases noted as redundant are taken as a single entry. Likewise, related proteins are automatically eliminated even if they can be involved in very different interactions.

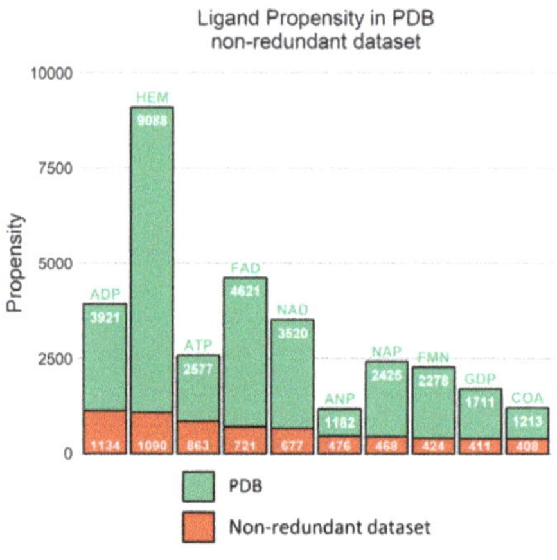

Figure 10. Distribution of 10 most frequent ligands observed in PDB and in non-redundant dataset.

3. Discussion

Throughout this study, we've highlighted the diversity and conservation of ligand binding poses in complex with identical protein. The number of distinct binding modes in the PDB can be reduced to 39,629 complexes, a significant decrease compared to the initial dataset.

While protein-ligand datasets were generated in the past such as scPDB [19], they only focus on specific criteria such as resolution, drug-like ligand, and binding residue. Generation of similar dataset using PICSES webserver can result in more than 130,000 protein chains, but mainly do not take into account the question of the ligand as it is only using protein sequence [6].

Our analysis features both structural analysis and supervised dataset generation that can be used to various degrees. Results and non-redundant dataset, available in supporting information, can be used for multiple purposes such as highlighting conserved and mobile residues in the binding mechanism for specific protein (see Figure 6). Ligand diversity for one specific binding site (see Figure 3) can also be easily explored using our results. Using the distinct representative conformations of one complex (displayed in Figure 8) to refine ensemble-docking results is another way our results can be exploited in the future. Our approach takes into account the possibility that a specific ligand interacts with different sites of the same protein. This opportunity had been rarely taken into account, but could be of great help in the search for catalytic, binding, or regulatory allosteric exosite [23,24].

The multiple levels of redundancies observed across PDB relative to protein-ligand context have also been underlined. Specific protein chains are largely overrepresented and were analyzed thoroughly in our dataset. Hence, the 501 conformations of the most recurrent complex, heme bound to brain nitric oxide synthase, are all structurally identical from a binding perspective. Consequently, the 9088 occurrences of heme complexes can be represented by 1090 unique binding modes.

Furthermore, our methodology is not biased by small discrepancies in the protein chain. Missing residues or insertion of a chimeric peptide still lead to similar binding modes that are well retrieved in our study. Structures available in poor resolutions can be validated by the co-occurrence of other redundant structures with high structural similarity for instance. We can also notice the specific interest of clusters of distinct ligand conformations. Indeed, they can reflect attractive dynamics and specifics of the binding affinity. Using different computational approaches such as scoring docking functions and molecular dynamics, they would be stimulating cases to apprehend their different binding affinities.

A comparison between structural similarity metric can also be discussed. RMSD computed with atoms is generally used in structural comparison studies. This approach requires matching predefined atom labels to compute deviation distance. We were surprised to found high RMSD for many entries that were clearly identical. Indeed, sometimes atom labels were inverted across different entries leading to high distance over perfectly superposed fragment. Figure 11 illustrates such instance where after superposition, two ligands are aligned but atom annotation in PDB are completely inverted in rings, e.g., C11 or F1 atoms, resulting in an atomic RMSD of 2.9 Å compared to a wRMSD$_f$ value of 0.8 Å. Pearson correlation coefficient computed across these two metrics in our dataset resulted in a correlation factor of 0.64, indicating a moderate correlation. Using wRMSD$_f$ in our case avoid this type of bias induced by annotations.

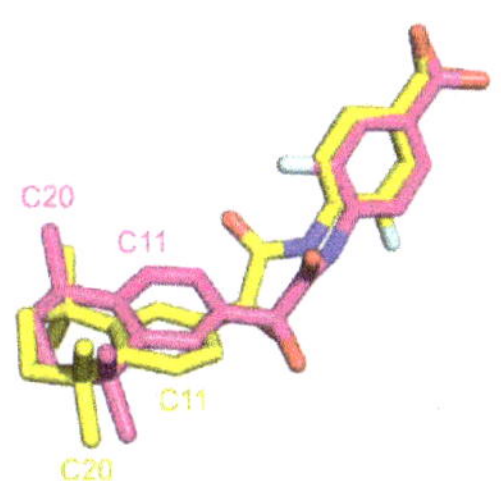

Figure 11. Atom label inverted in two PDB entries (PDB IDs 4lbd in magenta and 1exx in yellow).

Nonetheless, fragment RMSD has its own limitations, aromatic rings plane rotation for instance aren't well characterized. Two perpendicular aromatic rings superimposed on their center of mass will not be quantitatively different despite the potential change in interaction geometry and, consequently, nature.

Threshold used to define structural similarity can also be discussed. Indeed, evaluation of docking approach often uses a RMSD (atom-based) value less than 2.0 Å to consider a ligand pose similar to the native state. However, RMSD represent an average-like measure of similarity. Therefore, a value of 2.0 Å can highlight various cases: (i) a shift of the entire molecule as illustrated in Figure 12A (ii) a structurally conserved ligand region in the binding site and a significant deviation of some fragments in Figure 12B.

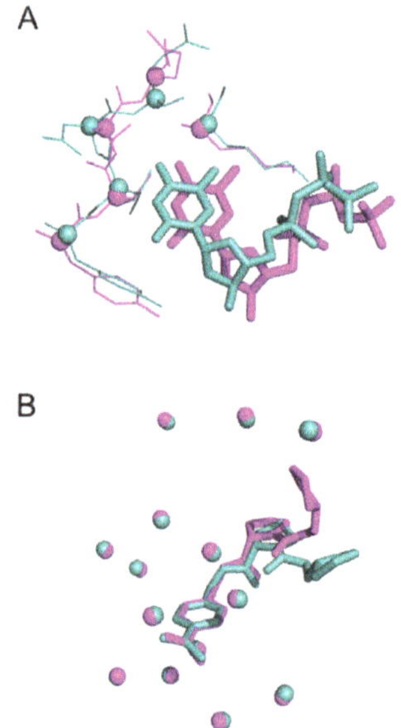

Figure 12. Three-dimensional representation of (**A**) effector TTP deviation bound to Ribonucleotide (PDB IDs 3hnd in blue and 3hnf in magenta, wRMSD$_f$ 2.08 Å), (**B**) fragment shift in trypsin S3 pocket (3ljj in blue 2zft in magenta, wRMSD$_f$ 1.97 Å).

4. Materials and Methods

The 128,843 protein structures were downloaded from the Protein Data Bank website; they were obtained by X-ray crystallography, NMR and cryogenic electron microscopy (cryo-EM) methods. They were processed and analyzed using our knowledge-based database Discngine 3decision™. Regarding NMR ensembles, each model was processed separately as a distinct protein-ligand conformation. Structures were annotated using multiple sources, such as UniProt [25], ChEMBL [26], PFAM [27], and Prosite [28]. Each structure was assigned to one (or multiple, e.g., alternative splicing events or chimeric protein) UniProt reference sequence(s) using the structure residue sequence and UniProt reference sequence.

Those annotations allowed a precise description of the specificity of one particular protein chain, macromolecules and ligands, mainly defined as heteroatoms, across multiple structure entries. No condition filter was applied regarding the ligand type, natural or designed ligand. This decision, purely subjective, was done to offer the broadest possible view of the data available to the scientific community. Similarly, the 3decision software automatically detected crystal contacts observed at the ligand level, and ligands whose positioning was suspicious were not considered for the study. For our current study, complexes were selected by narrowing to a set of specific ligands. Only molecules with at least one ring, a molecular weight between 250 Da and 850 Da, and single bond fraction below 90%, were retained in this dataset. Ligands must have had no covalent bond to be selected. To discard ambiguities, protein chains interacting with the small molecule were characterized using their UniProtKB identifier [25].

An initial dataset was constituted by taking all the protein chains binding to at least one ligand. One protein chain per ligand was considered. If one ligand was bound to multiple protein chains, i.e., multimeric complex, then each couple chain-ligand was split and considered as a distinct unit.

Two complexes were grouped and compared to each other when both protein chains and ligands were considered as identical. In our case, two protein chains were identical when their UniProtKB was equal; ligands were considered as similar using their canonicalized SMILES fingerprint [29].

Nonetheless, having two identical proteins and ligands doesn't necessarily represent the same binding sites, e.g., one hypothetical case involving two identical ligands bound to the two extremities of the same protein chain. Thus, molecular contacts were calculated between protein chain and ligand atoms to highlight interacting residues for every unit. The classical sum of van der Waals radii + 1.0 Å between ligand and protein atoms was used as distance threshold. To be sure, we are comparing the same binding pockets; at least 3 common interacting residues across two different units were required to make a comparison.

Thus, two similar complexes were superimposed through a structural alignment performed on common binding residues Cα. To perform such alignment, a minimum of 3 identical binding residues in the two different pockets was required. The superposition optimization followed the Kabsch algorithm [30] and was iteratively performed by favorably weighting spatially close residues at each step.

A critical issue here is the heterogeneous atom labeling that can be a source of error in root mean square deviation (RMSD) calculation due to inverted atom labeling, particularly in ligands. To quantify structural similarity between two units, a custom root-mean square deviation (RMSD) was calculated on both ligands. Molecular fragments were generated from ligands with their corresponding center of mass calculated. Aromatic rings and functional groups were preserved while long aliphatic chains were split into multiple fragments. The Weighted fragment Root Mean Square Deviation (wRMSD$_f$) similarity metric was then computed given by the formula:

$$\text{wRMSD}_f = \sqrt{\sum_{i=1}^{N} \frac{n_{atoms\_i}}{n_{atoms\_lig}} \delta_i^2} \tag{1}$$

With N number of fragments, n_{atoms_i} the number of atoms within the fragment i, n_{atoms_lig} the total number of atoms within the molecule and δ_i the distance between the fragment's center of masses. Molecular fragmentation was performed in Pipeline Pilot R2 2017 [31].

Euclidean distances between fragments' centers of mass were also computed. RMSD of binding residue alpha carbon was also retrieved to assess the (i) the superimposition quality, and (ii) the structural diversity in binding pockets, i.e., grasp the potential mobility of some residues in the pocket.

Using these calculated metrics, rules were defined to characterize two units as similar: (i) a wRMSD$_f$ value below a threshold of 1.0 Å and (ii) no distance greater than 1.5 Å between two fragments' center of masses. As large number of conformations can sometimes be compared between each other for highly recurrent complexes, hierarchical clustering (with complete-linkage as agglomeration method) was performed using wRMSD$_f$ measure as a distance metric using R 3.4.4 [32].

Protein-ligand conformations were finally clustered in three categories. Units that do not share similar ligand, similar chain receptor or similar binding residues, i.e., binding sites, with other complexes, were defined as singular complexes. Cases where every conformation of one complex was identified as identical, i.e., limited structural deviation, were labeled as homogeneous complexes. Groups of conformations resulted from clustering were qualified as heterogeneous.

B-factors were retrieved from PDB structure and normalized as done in [22]. In our study, only normalized B-factors of binding residues and ligand were considered. B-factors were averaged for each residue and then averaged for each binding pockets.

5. Conclusions

Throughout this study, the structural diversity and redundancy of protein-ligand binding modes was assessed in the PDB for the first time. This survey of 104,777 monomeric complexes highlights the widely acknowledged redundancy in the protein-ligand context. Clustering and filtering processes have led to the description of 39,629 specific binding modes of unique protein-ligand complexes, a 2.64-fold reduction relative to the original dataset. While this type of study has been performed in the past on a smaller dataset, 2911 complexes by Drwal et al. [21], it is the first time such analysis was performed on a large scale. This research's purpose can be used both from an analytical perspective, e.g., machine learning dataset, and applicative prospect, i.e., efficiently improving a protein-ligand complex query in a database. Such implementation will be integrated in future 3decision release.

Although the methodology offers great and innovative results compared to the approaches currently used, further improvements can be explored. For instance, RMSD of residue side chain can be computed to correlate with the observed molecular interaction. Ligand wRMSD$_f$ metric can be revised by quantifying the plane rotation shifting using basic atom distance-based matching method. Finally, a layered quantification of redundancy using protein classification such as species information or PFAM classification would allow to group complexes with highly similar binding site such as kinase for instance. We also would like to explore the possibility in future developments to look specifically at allosteric compounds implicated in the investigation of catalytic, binding, or regulatory exosite [33].

Author Contributions: Conceptualization, N.K.S., P.S., and A.G.d.B.; Formal analysis, N.K.S.; Funding acquisition, P.S. and A.G.d.B.; Investigation, N.K.S., P.S., and A.G.d.B.; Methodology, N.K.S. and A.G.d.B.; Resources, N.K.S. and Peter Schmidtke; Supervision, P.S. and A.G.d.B.; Visualization, N.K.S.; Writing–original draft, N.K.S. and A.G.d.B.; Writing–review & editing, N.K.S., P.S., and A.G.d.B. All authors have read and agree to the published version of the manuscript.

Funding: This work was funded by grants from the Ministry of Research (France), Discngine S.A.S., University Paris Diderot, Sorbonne, Paris Cite (France), University of La Reunion, Reunion Island, National Institute for Blood Transfusion (INTS, France), National Institute for Health and Medical Research (INSERM, France), and Labex GR-Ex. The Labex GREx, reference ANR-11-LABX-0051, is funded by the program "Investissements d'Avenir" of the French National Research Agency, reference ANR-11-IDEX-0005-02. A.G.d.B. acknowledges Indo-French Centre for the Promotion of Advanced Research/CEFIPRA for Collaborative Grant 5302-2.

Acknowledgments: N.K.S. acknowledges support from ANRT and Discngine S.A.S.

Abbreviations

ACHP_LYMST	Acetylcholine-binding protein (*Lymnaea stagnalis*)
ADP	Adenosine-5′-Diphosphate
AMP	Adenosine Monophosphate
ANP	Phosphoaminophosphonic Acid-Adenylate Ester
ATP	Adenosine-5′-Triphosphate
BACE1_HUMAN	Beta-secretase 1 (Human)
BFR_ECOLI	Bacterioferritin (*Escherichia coli*)
CAH2_HUMAN	Carbonic anhydrase 2 (Human)
CCPR_YEAST	Cytochrome c peroxidase, mitochondrial (*Saccharomyces cerevisiae*)
CDK2_HUMAN	Cyclin-dependent kinase 2 (Human)
COA	Coenzyme A
CPXA_PSEPU	Camphor 5-monooxygenase (*Pseudomonas putida*)
CPXB_BACMB	Bifunctional cytochrome P450/NADPH–P450 reductase (*Bacillus megaterium*)

CYC	Phycocyanobilin
CYC3_DESAC	Cytochrome c3 (*Desulfuromonas acetoxidans*)
DGT	2′-Deoxyguanosine-5′-Triphosphate
ESR1_HUMAN	Estrogen receptor (Human)
FAD	Flavin-Adenine Dinucleotide
FMN	Flavin Mononucleotide
GDP	Guanosine-5′-Diphosphate
GNP	Phosphoaminophosphonic Acid-Guanylate Ester
GTP	Guanosine-5′-Triphosphate
HBA_HUMAN	Hemoglobin subunit alpha (Human)
HBB_HUMAN	Hemoglobin subunit beta (Human)
HEC	Heme C
HEM	Protoporphyrin IX Containing FE
INHA_MYCTU	Enoyl-[acyl-carrier-protein] reductase [NADH] (*Mycobacterium tuberculosis*)
KAIC_SYNE7	Circadian clock protein kinase KaiC (*Synechococcus elongatus*)
NAD	Nicotinamide-Adenine-Dinucleotide
NAI	1,4-Dihydronicotinamide Adenine Dinucleotide
NAP	NADP Nicotinamide-Adenine-Dinucleotide Phosphate
NDP	NADPH Dihydro-Nicotinamide-Adenine-Dinucleotide Phosphate
NIR_THIND	Cytochrome c-552 (*Thioalkalivibrio nitratireducens*)
NOS1_RAT	Nitric oxide synthase, brain (Rat)
NOS3_BOVIN	Nitric oxide synthase, endothelial (*Bos taurus* (Bovine))
NQO2_HUMAN	Ribosyldihydronicotinamide dehydrogenase [quinone] (Human)
O76290_TRYBB	Pteridine reductase (*Trypanosoma brucei brucei*)
POL_HV1B1	Gag-Pol polyprotein (Human immunodeficiency virus type 1 group M subtype B)
PYRF_METH	Orotidine 5′-phosphate decarboxylase (*Methanothermobacter thermautotrophicus*)
Q8WSF8_APLCA	Soluble acetylcholine receptor (*Aplysia californica*)
Q9HY79_PSEAE	Bacterioferritin (*Pseudomonas aeruginosa*)
SAH	S-Adenosyl-L-Homocysteine
SAM	S-Adenosylmethionine
SAMH1_HUMAN	Deoxynucleoside triphosphate triphosphohydrolase SAMHD1 (Human)
THYX_THEMA	Flavin-dependent thymidylate synthase (*Thermotoga maritima*)
TPP	Thymidine 5′-triphosphate
UDP	Uridine-5′-Diphosphate

Appendix A

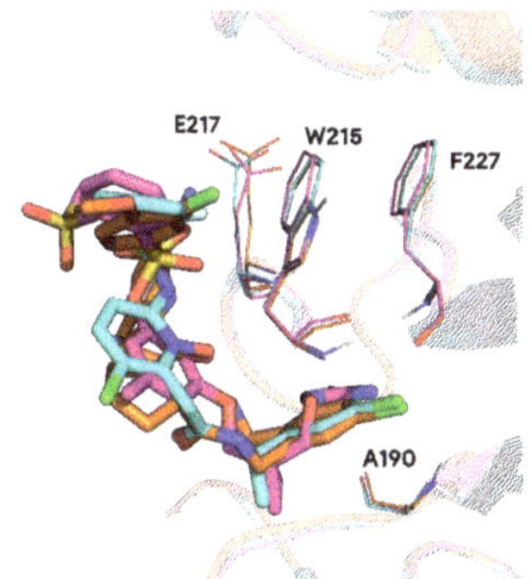

Figure A1. 3D visualization of conserved residues (lines) in the prothrombin ligand (sticks) binding process (PDB ids 1t4u in pink, 1z71 in blue and 3u9a in orange).

References

1. Berman, H.M.; Westbrook, J.; Feng, Z.; Gilliland, G.; Bhat, T.N.; Weissig, H.; Shindyalov, I.N.; Bourne, P.E. The Protein Data Bank. *Nucleic Acids Res.* **2000**, *28*, 235–242. [CrossRef] [PubMed]
2. Langer, T.; Hoffmann, R.D. *Pharmacophores and Pharmacophore Searches*; Wiley-VCH Verlag GmbH & Co. KGaA: Weinheim, Germany, 2006.
3. Sotriffer, C.A.; Gohlke, H.; Klebe, G. Docking into knowledge-based potential fields: A comparative evaluation of DrugScore. *J. Med. Chem.* **2002**, *45*, 1967–1970. [CrossRef] [PubMed]
4. Shinada, N.K.; de Brevern, A.G.; Schmidtke, P. Halogens in Protein-Ligand Binding Mechanism: A Structural Perspective. *J. Med. Chem.* **2019**, *62*, 9341–9356. [CrossRef] [PubMed]
5. Griep, S.; Hobohm, U. PDBselect 1992-2009 and PDBfilter-select. *Nucleic Acids Res.* **2010**, *38*, D318–D319. [CrossRef]
6. Wang, G.; Dunbrack, R.L., Jr. PISCES: A protein sequence culling server. *Bioinformatics* **2003**, *19*, 1589–1591. [CrossRef]
7. NCBI. Documentation of the BLASTCLUST-Algorithm. Available online: https://www.ncbi.nlm.nih.gov/Web/Newsltr/Spring04/blastlab.html (accessed on 23 November 2019).
8. Li, W.; Godzik, A. Cd-hit: A fast program for clustering and comparing large sets of protein or nucleotide sequences. *Bioinformatics* **2006**, *22*, 1658–1659. [CrossRef]
9. Akiyama, Y.; Onizuka, K.; Noguchi, T.; Ando, M. Parallel Protein Information Analysis (PAPIA) System Running on a 64-Node PC Cluster. *Genome Inform. Ser. Workshop Genome Inform.* **1998**, *9*, 131–140.
10. Sikic, K.; Carugo, O. CARON–average RMSD of NMR structure ensembles. *Bioinformation* **2009**, *4*, 132–133. [CrossRef]
11. Calvanese, L.; D'Auria, G.; Vangone, A.; Falcigno, L.; Oliva, R. Analysis of the interface variability in NMR structure ensembles of protein-protein complexes. *J. Struct. Biol.* **2016**, *194*, 317–324. [CrossRef]
12. Craveur, P.; Rebehmed, J.; de Brevern, A.G. PTM-SD: A database of structurally resolved and annotated posttranslational modifications in proteins. *Database (Oxford)* **2014**, *2014*. [CrossRef]
13. Schneider, B.; Cerny, J.; Svozil, D.; Cech, P.; Gelly, J.C.; de Brevern, A.G. Bioinformatic analysis of the protein/DNA interface. *Nucleic Acids Res.* **2014**, *42*, 3381–3394. [CrossRef] [PubMed]
14. Benson, M.L.; Smith, R.D.; Khazanov, N.A.; Dimcheff, B.; Beaver, J.; Dresslar, P.; Nerothin, J.; Carlson, H.A. Binding MOAD, a high-quality protein-ligand database. *Nucleic Acids Res.* **2008**, *36*, D674–D678. [CrossRef] [PubMed]
15. Smith, R.D.; Clark, J.J.; Ahmed, A.; Orban, Z.J.; Dunbar, J.B., Jr.; Carlson, H.A. Updates to Binding MOAD (Mother of All Databases): Polypharmacology Tools and Their Utility in Drug Repurposing. *J. Mol. Biol.* **2019**, *431*, 2423–2433. [CrossRef] [PubMed]
16. Wang, R.; Fang, X.; Lu, Y.; Yang, C.Y.; Wang, S. The PDBbind database: methodologies and updates. *J. Med. Chem.* **2005**, *48*, 4111–4119. [CrossRef]
17. Liu, Z.; Li, Y.; Han, L.; Li, J.; Liu, J.; Zhao, Z.; Nie, W.; Liu, Y.; Wang, R. PDB-wide collection of binding data: current status of the PDBbind database. *Bioinformatics* **2015**, *31*, 405–412. [CrossRef]
18. Desaphy, J.; Bret, G.; Rognan, D.; Kellenberger, E. sc-PDB: A 3D-database of ligandable binding sites–10 years on. *Nucleic Acids Res.* **2015**, *43*, D399–D404. [CrossRef]
19. Kellenberger, E.; Muller, P.; Schalon, C.; Bret, G.; Foata, N.; Rognan, D. sc-PDB: An annotated database of druggable binding sites from the Protein Data Bank. *J. Chem. Inf. Model.* **2006**, *46*, 717–727. [CrossRef]
20. Wallach, I.; Lilien, R. The protein-small-molecule database, a non-redundant structural resource for the analysis of protein-ligand binding. *Bioinformatics* **2009**, *25*, 615–620. [CrossRef]
21. Drwal, M.N.; Bret, G.; Perez, C.; Jacquemard, C.; Desaphy, J.; Kellenberger, E. Structural Insights on Fragment Binding Mode Conservation. *J. Med. Chem.* **2018**, *61*, 5963–5973. [CrossRef]
22. Bornot, A.; Etchebest, C.; de Brevern, A.G. Predicting protein flexibility through the prediction of local structures. *Proteins* **2011**, *79*, 839–852. [CrossRef]
23. Guarnera, E.; Berezovsky, I.N. Allosteric sites: Remote control in regulation of protein activity. *Curr. Opin. Struct. Biol.* **2016**, *37*, 1–8. [CrossRef] [PubMed]
24. Guarnera, E.; Berezovsky, I.N. Toward Comprehensive Allosteric Control over Protein Activity. *Structure* **2019**, *27*, 866–878. [CrossRef] [PubMed]

25. Boutet, E.; Lieberherr, D.; Tognolli, M.; Schneider, M.; Bansal, P.; Bridge, A.J.; Poux, S.; Bougueleret, L.; Xenarios, I. UniProtKB/Swiss-Prot, the Manually Annotated Section of the UniProt KnowledgeBase: How to Use the Entry View. *Methods Mol. Biol.* **2016**, *1374*, 23–54. [CrossRef] [PubMed]
26. Gaulton, A.; Hersey, A.; Nowotka, M.; Bento, A.P.; Chambers, J.; Mendez, D.; Mutowo, P.; Atkinson, F.; Bellis, L.J.; Cibrian-Uhalte, E.; et al. The ChEMBL database in 2017. *Nucleic Acids Res.* **2017**, *45*, D945–D954. [CrossRef] [PubMed]
27. Finn, R.D.; Coggill, P.; Eberhardt, R.Y.; Eddy, S.R.; Mistry, J.; Mitchell, A.L.; Potter, S.C.; Punta, M.; Qureshi, M.; Sangrador-Vegas, A.; et al. The Pfam protein families database: Towards a more sustainable future. *Nucleic Acids Res.* **2016**, *44*, D279–D285. [CrossRef]
28. Sigrist, C.J.; de Castro, E.; Cerutti, L.; Cuche, B.A.; Hulo, N.; Bridge, A.; Bougueleret, L.; Xenarios, I. New and continuing developments at PROSITE. *Nucleic Acids Res.* **2013**, *41*, D344–D347. [CrossRef]
29. Weininger, D. SMILES, a chemical language and information system. 1. Introduction to methodology and encoding rules. *J. Chem. Inf. Comput. Sci.* **1988**, *28*, 31–36. [CrossRef]
30. Kabsch, W. A solution for the best rotation to relate two sets of vectors. *Acta Crystallogr. A* **1976**, *32*, 922–923. [CrossRef]
31. Dassault Systèmes, BIOVIA, Pipeline Pilot, R2, San Diego, CA, USA. 2017.
32. R core Team. R: A Language and Environment for Statistical Computing. 2017. Available online: https://www.R-project.org/ (accessed on 23 November 2019).
33. Guarnera, E.; Berezovsky, I.N. Allosteric drugs and mutations: Chances, challenges, and necessity. *Curr. Opin. Struct. Biol.* **2020**, *62*, 149–157. [CrossRef]

MDPI

St. Alban-Anlage 66

4052 Basel

Switzerland

Tel. +41 61 683 77 34

Fax +41 61 302 89 18

www.mdpi.com

International Journal of Molecular Sciences Editorial Office

E-mail: ijms@mdpi.com

www.mdpi.com/journal/ijms